Operations Research Proceedings 2008

Bernhard Fleischmann • Karl Heinz Borgwardt
Robert Klein • Axel Tuma
Editors

Operations Research Proceedings 2008

Selected Papers of the Annual International Conference of the German Operations Research Society (GOR) University of Augsburg, September 3–5, 2008

Springer

Prof. Dr. Bernhard Fleischmann
University of Augsburg
Faculty of Economics and
Business Administration
Department of Production and Logistics
Universitätsstraße 16
86159 Augsburg
Germany
bernhard.fleischmann@wiwi.uni-augsburg.de

Prof. Dr. Karl Heinz Borgwardt
University of Augsburg
Institute of Mathematics
Universitätsstraße 14
86159 Augsburg
Germany
karl.heinz.borgwardt@math.uni-augsburg.de

Prof. Dr. Robert Klein
University of Augsburg
Faculty of Economics and
Business Administration
Department of Mathematical Methods
Universitätsstraße 16
86159 Augsburg
Germany
robert.klein@wiwi.uni-augsburg.de

Prof. Dr. Axel Tuma
University of Augsburg
Faculty of Economics and
Business Administration
Department of Production
and Environmental Management
Universitätsstraße 16
86159 Augsburg
Germany
axel.tuma@wiwi.uni-augsburg.de

ISBN 978-3-642-00141-3 e-ISBN 978-3-642-00142-0
DOI 10.1007/978-3-642-00142-0
Springer Dordrecht Heidelberg London New York

Library of Congress Control Number: 2009926697

Cover design: WMXDesign GmbH, Heidelberg, Germany

Printed on acid-free paper

Springer is part of Springer Science+Business Media (www.springer.com)

Preface

The international conference "Operations Research 2008", the annual meeting of the German Operations Research Society (GOR), was held at the University of Augsburg on September 3-5, 2008. About 580 participants from more than 30 countries presented and listened to nearly 400 talks on a broad range of Operations Research.

The general subject "Operations Research and Global Business" stresses the important role of Operations Research in improving decisions in the increasingly complex business processes in a global environment. The plenary speakers Morris A. Cohen (Wharton School) and Bernd Liepert (Executive Board of KUKA Robotics) addressed this subject. Moreover, one of the founders of Operations Research, Saul Gass (University of Maryland), gave the opening speech on the early history of Operations Research.

This volume contains 93 papers presented at the conference, selected by the program committee and the section chairs, forming a representative sample of the various subjects dealt with at Operations Research 2008. The volume follows the structure of the conference, with 12 sections, grouped into six "Fields of Applications" and six "Fields of Methods and Theory". This structure in no way means a separation of theory and application, which would be detrimental in Operations Research, but displays the large spectrum of aspects in the focus of the papers. Of course, most papers present theory, methods and applications together.

Like at the conference, the largest number of papers falls into the Section "Discrete and Combinatorial Optimization" (11 papers) and into the Logistics Sections "Supply Chain and Inventory Management" (11 papers) and "Traffic and Transportation" (18 papers), which are closely related to the global business issue. The papers of the winners of the Diploma and Dissertation Awards have already been published in the OR News No. 34 of November 2008, edited by GOR.

We would like to thank everybody who contributed to the great success of the conference, in particular the authors of the papers and all speakers of the conference, the program committee and the section chairs who have acquired the presented papers and refereed and selected the papers for this volume.

Moreover, we express our special thanks to our staff members Dipl.-Wirtsch.-Inform. Oliver Faust, Dipl.-Kfm. Christoph Pitzl, Dipl.-Wirtsch.-Inform. Claudius Steinhardt, Dipl.-Math. oec. Thomas Wörle and particularly to Marion Hauser and Dipl.-Wirt.-Inf. Ramin Sahamie for preparing the final manuscript of this volume. We are grateful for the pleasant cooperation with Barbara Feß and Dr. Werner Müller from Springer and their professional support in publishing this volume.

Augsburg, December 2008

Bernhard Fleischmann
Karl Heinz Borgwardt

Robert Klein
Axel Tuma

Commitees

Programm Commitee

Karl Heinz Borgwardt (U Augsburg)

Bernhard Fleischmann (U Augsburg)

Horst W. Hamacher (TU Kaiserslautern)

Robert Klein (U Augsburg)

Stefan Nickel (U Saarbrücken)

Erwin Pesch (U Siegen)

Axel Tuma (U Augsburg)

Brigitte Werners (U Bochum)

Local Organizing Commitee

Karl Heinz Borgwardt

Bernhard Fleischmann

Robert Klein

Axel Tuma

Scientific Sections and Section Chairs

Fields of Application

Finance and Accounting
Hermann Locarek-Junge (TU Dresden), Gunther Friedl (TU München)

Health Care and Environment
Frank Schultmann (U Siegen)

Production and Service Management
Heinrich Kuhn (KU Eichstätt-Ingolstadt)

Scheduling and Project Management
Christoph Schwindt (TU Clausthal)

Supply Chain and Inventory Management
Herbert Meyr (TU Darmstadt)

Traffic and Transportation
Herbert Kopfer (U Bremen)

Fields of Methods and Theory

Applied Probability and Stochastic Programming
Rüdiger Schultz (U Duisburg-Essen)

Business Informatics, Decision Support and Artifical Intelligence
Leena Suhl (U Paderborn)

Discrete and Combinatorial Optimization
Jörg Rambau (U Bayreuth)

Forecasting, Econometrics and Game Theory
Ulrich Küsters (KU Eichstätt-Ingolstadt)

Linear, Nonlinear and Vector Optimization
Petra Huhn (U Bonn)

Network Optimization
Anita Schöbel (U Göttingen)

Contents

Part IV Scheduling and Project Management

Part V Supply Chain and Inventory Management

Part IX Discrete and Combinatorial Optimization

Part XII Network Optimization

Part I

Finance and Accounting

Smoothing Effects of Different Ways to Cope with Actuarial Gains and Losses Under IAS 19

Matthias Amen

Universität Bielefeld, Universitätsstraße 25, DE-33615 Bielefeld, Germany.
Matthias.Amen@web.de

1 Research Question

International accounting for pension obligations is one of the most complex issues in accounting. According to the International Accounting Standard (IAS) 19 'employee benefits' the defined benefit obligation has to be measured by the *projected unit credit method* (IAS 19.64). Experience deviations from the expected values for future salary/wage increases, pension increases, fluctuation, and mortality as well as changes in the discount rate cause actuarial gains and losses, that are explicitly considered within the IAS 19 accounting system. Actuarial gains and losses are unavoidable to a certain degree and may cause fluctuations in pension costs. IAS 19 offers several ways to cope with actuarial gains and losses. The so-called *corridor approach* is explicitly designed to smooth pension costs (IAS 19 BC 39; see [1, p. 127]).
Generally, smoothing is contrary to the overall objective to provide information that is useful for economic decision. The research question of this paper is to quantify the smoothing effects of the different ways to cope with actuarial gains and losses. Furthermore, we suggest a modification that results into a moderate smoothing, but without a loss of decision usefulness, because it avoids an artificial source of actuarial gains and losses and reduces complexity.
As we are interested in pure accounting effects we focus on unfunded pension plans. Thus, the results are not influenced by the design of the entities contributions to an external fund as well as the investment policy of such a fund.

B. Fleischmann et al.(Eds.), *Operations Research Proceedings 2008*,
DOI: 10.1007/978-3-642-00142-0_1, © Springer-Verlag Berlin Heidelberg 2009

2 Different Ways to Cope with Actuarial Gains and Losses Under IAS 19

The current IAS 19 provides different ways to cope with actuarial gains and losses:

1. *Corridor approach with amortizing recognition*: cumulation and recognition of a corridor excess over the expected remaining working lives of the participating individuals (IAS 19.92)
2. *Corridor approach with immediate recognition*: cumulation and a faster recognition of a corridor excess, in particular recognition of a corridor excess in the current period (IAS 19.93, 93A)
3. *Immediate recognition in profit and loss* (IAS 19.93, 93A)
4. *Equity approach*: immediate recognition outside profit and loss in a separate statement directly within equity (IAS 19.93A)

The standard method of IAS 19 is the *corridor approach* with a amortization of a corridor excess over the expected remaining working lives of the participating individuals (no 1). For the considered unfunded pension plans, the corridor is 10 % of last years defined benefit obligation (IAS 19.92). This method defines the minimum amount of actuarial gains and losses to be recognised in the current period. IAS 19.93 allows a faster systematical recognition of cumulated actuarial gains and losses even within the corridor width of 10 %. This means that (a) we can amortize a corridor excess within a shorter period (IAS 19.93A explicitly allows to recognise a corridor excess in the current period – no 2), or (b) we can apply a corridor width lower than 10 %, or (c) a combination of both, (a) and (b). In fact, the *immediate recognition in profit and loss* (no 3) could be interpreted as the extreme case of the *corridor approach* with a corridor width of 0 % and an immediate recognition of the corridor excess.

Generally, the aim of the *corridor approach* is to avoid huge fluctuations in pension costs caused by errors in estimating the actuarial assumptions. The *corridor approach* requires additional records outside the financial statement and increases complexity of the system of accounting for pension obligations. The *equity approach* (no 4) has been adopted from the British Financial Reporting Standard (FRS) 17 'Retirement Benefits' in 2004. Because of the immediate recognition of actuarial gains and losses outside profit and loss, the pension costs are never affected by actuarial gains and losses. As a consequence, the *equity approach* reduces complexity and achieves the maximal possible smoothing of pension costs but at the price of incompleteness of costs that is contrary to 'clean accounting'.

3 Design of a Simulation Study

An analytical approach for quantifying the smoothing effects of the different approaches is not possible. Therefore, Monte Carlo simulation analysis is the method of choice. The simulation model has been characterized in [3]. We consider a *regenerating workforce*, in which an individual substitutes for an other individual who fluctuates, retires or dies. Each simulation run consists of 500 iterations.

The assumptions concerning the financial and non financial parameters are based on official German statistics (For a detailed reference see [2].). In order to generate mutually compatible financial assumptions we simulate the vector-autoregressive model presented in [3]. Contrary to [2] and [3] we focus only on smoothing. Therefore, we first have to differentiate smoothing from uncertainty which is shown in standard graphical simulation outcomes.

We are not interested in the variation of pension costs at a certain point in time during all iterations of a simulation run (uncertainty), but we are interested in the variation of pension costs during the simulation period within a single iteration of the simulation run (smoothing). Furthermore, to measure smoothing, it is not adequate to take the statistical 'variance' of the pension costs at the different points in time of a single iteration (see Fig. 1).

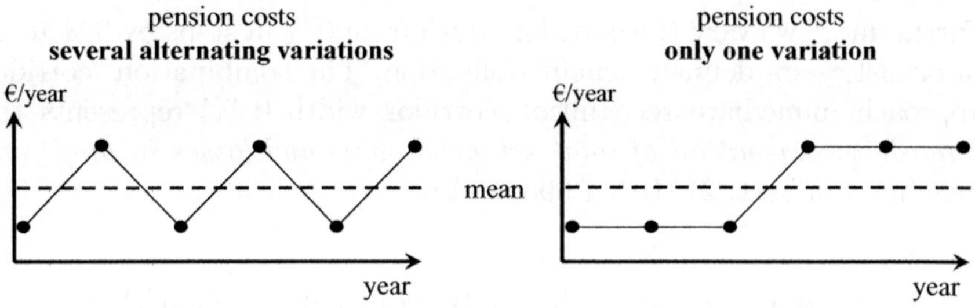

Fig. 1. Identical variance but different smoothing

For illustration, both diagrams of Fig. 1 show a time series of pension costs that have an identical statistical 'variance'. It is obvious that the diagram on the right hand side has only one variation and therefore a better smoothing than the diagram on the left hand side with several alternating variations. Therefore, we use the *annual average absolute changes of the pension costs* as a *smoothing criterion*. A low value of this criterion indicates a high smoothing.

Among the assumptions, the discount rate is a kind of artificial source of variations of the pension costs. IAS 19.78 requires to determine the discount rate from current market yields of long-term highly quality corporate bonds. The *projected unit credit method* is one of several possible cost allocation methods. In actuarial science and practice this method applies a *constant discount rate.* IAS 19 takes this long-term allocation technique from actuarial science and mixes it with the short-term 'fair value' idea. As a result, we get huge fluctuations in the pension costs that are due to fluctuations of the discount rate. We have to keep in mind, that the decision for a special discount rate does not have any cash consequences and controlls only the allocation of the total pension payments to the periods. Furthermore, the estimation of the value of the defined benefit obligation by one of several possible methods does not become more precise, just by taking a discount rate based on current market yields. Therefore, besides the *variable discount rate* required by IAS 19.78, we also consider a *constant discount rate* according to the original actuarial concept of the *projected unit credit method.*

In the simulation study we compare the following approaches using a *variable discount rate* as well as a *constant discount rate*:

- *Corridor approach with amortizing recognition* of the corridor excess
- *Corridor approach with immediate recognition* of the corridor excess
- *Equity approach*

Furthermore we vary the corridor width from 0 % in steps by 5 % to 20 % of last years defined benefit obligation. The combination 'corridor approach/immediate recognition/corridor width 0 %' represents the *immediate recognition of total actuarial gains and losses in profit and loss* (no 3 in Sect. 2) (IAS 19.93, 93A).

4 Quantitative Results for a Regenerating Workforce

The following Fig. 2 shows the quantitative results for the simulation analysis for a *regenerating workforce.*

As a remark, we have to remember, that the *equity approach* excludes actuarial gains and losses from the pension costs. Therefore, it's smoothing is an experimentally generated lower bound for the smoothing criterion of the other approaches. In Fig. 2 the big } describes the spread of smoothing, that is offered by the current IAS 19. In addition, the ◯ represents the *immediate recognition of actuarial gains and losses in profit and loss* in case of a *constant discount rate.*

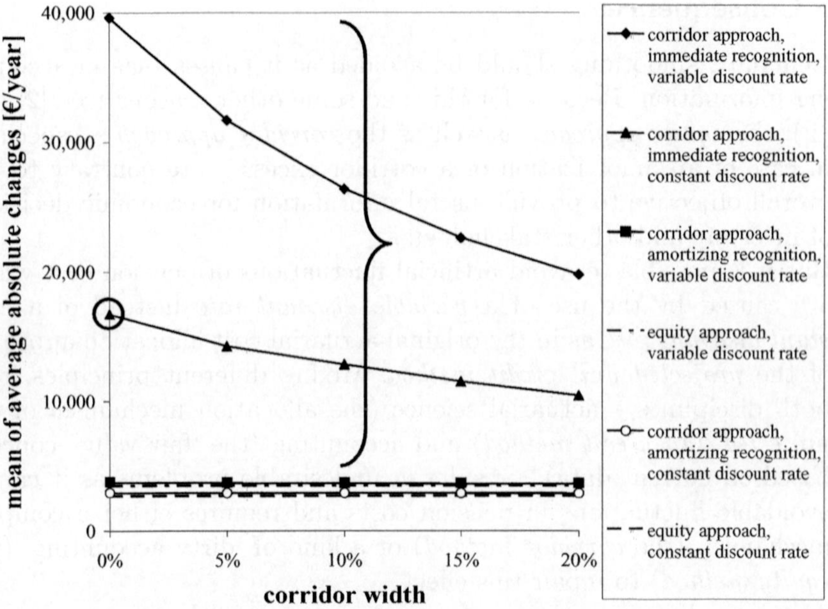

Fig. 2. Smoothing of pension costs

The quantitative results are the following:

- Regardless of the kind of discount rate or the corridor width, there is only a small difference between the *corridor approach with amortizing recognition* of the corridor excess and the *equity approach.*
- For the *corridor approach with amortizing recognition* of the corridor excess there is only little difference in the smoothing criterion if we change from a *variable discount rate* to a *constant discount rate.*
- For the *corridor approach with immediate recognition* of the corridor excess there is a considerable stronger smoothing when we apply a *constant discount rate* instead of a *variable discount rate.*
- The most important quantitative result is, that smoothing achieved by using a *constant discount rate* in case of an *immediate recognition of actuarial gains and losses in profit and loss* (illustrated by the circle ○ in Fig. 2) is within the range that is accepted by the current IAS 19 (illustrated by the big } in Fig. 2).

5 Consequences

Generally, smoothing should be avoided as it causes bias in accounting information. Because for this and some other reasons (see [2] and [3]), the *equity approach* as well as the *corridor approach* – especially in case of an amortization of a corridor excess – are contrary to the overall objective to provide useful information for economic decisions of investors and other stakeholders.

But it is possible to avoid artificial fluctuations of pension cost which are caused by the use of a *variable discount rate* instead of a *constant discount rate* as in the original actuarial cost allocation approach of the *projected unit credit method.* Mixing different principles from both disciplines – actuarial science (the allocation mechanism of the *projected unit credit method*) and accounting (the 'fair value' concept based on current data) – results in undesirable problems as it causes avoidable fluctuations in pension costs and requires either a complex mechanism (the *corridor method*) or a kind of 'dirty accounting' (the *equity method*) to repair this effect.

The *immediate recognition of actuarial gains and losses in profit and loss* while applying a *constant discount rate* for measuring the defined benefit obligation results to a moderate smoothing that is within the range accepted under the current IAS 19. Thus, this simple modification should be regarded as reasonable alternative for the amendment of IAS 19. It is just the application of a actuarial cost allocation method that avoids artificial fluctuations in pension costs at a reduced complexity.

As a final remark, the recently issued discussion paper 'Preliminary Views on Amendments to IAS 19 Employee Benefits' (March 2008) copes with three approaches that differ in the way the amount of actuarial gains and losses is recognized in profit and loss or in other comprehensive income. The *corridor approach* as well as the *equity approach* could be expected to be repealed. Unfortunately, a change toward a *constant discount rate* is not yet addressed by the International Accounting Standards Board (IASB).

References

1. Ballwieser W (2006) IFRS-Rechnungslegung. Vahlen, München
2. Amen M (2007) Simulation based comparison of existent IAS 19 accounting options. European Accounting Review 16:243–276
3. Amen M (2008) Modeling and analyzing the IAS 19 system of accounting for unfunded pensions. In: Kalcsics J, Nickel S (eds) Operations Research Proceedings 2007. Springer, Heidelberg

A Regime-Switching Relative Value Arbitrage Rule

Michael Bock and Roland Mestel

University of Graz, Institute for Banking and Finance
Universitaetsstrasse 15/F2, A-8010 Graz, Austria
{michael.bock,roland.mestel}@uni-graz.at

1 Introduction

The relative value arbitrage rule, also known as "pairs trading" or "statistical arbitrage", is a well established speculative investment strategy on financial markets, dating back to the 1980s. Today, especially hedge funds and investment banks extensively implement pairs trading as a long/short investment strategy.[1]

Based on relative mispricing between a pair of stocks, pairs trading strategies create excess returns if the spread between two normally co-moving stocks is away from its equilibrium path and is assumed to be mean reverting, i.e. deviations from the long term spread are only temporary effects. In this situation, pairs trading suggests to take a long position in the relative undervalued stock, while the relative overvalued stock should be shortened. The formation of the pairs ensues from a cointegration analysis of the historical prices. Consequently, pairs trading represents a form of statistical arbitrage where econometric time series models are applied to identify trading signals.

However, fundamental economic reasons might cause simple pairs trading signals to be wrong. Think of a situation in which a profit warning of one of the two stocks entails the persistant widening of the spread, whereas for the other no new information is circulated. Under these circumstances, betting on the spread to revert to its historical mean would imply a loss.

To overcome this problem of detecting temporary in contrast to longer lasting deviations from spread equilibrium, this paper bridges the literature on Markov regime-switching and the scientific work on statistical

[1] For an overview see [7, 3].

B. Fleischmann et al.(Eds.), *Operations Research Proceedings 2008*,
DOI: 10.1007/978-3-642-00142-0_2,

arbitrage to develop useful trading rules for "pairs trading". The next section contains a brief overview of relative value strategies. Section 3 presents a discussion of Markov regime-switching models which are applied in this study to identify pairs trading signals (section 4). Section 5 presents some preliminary empirical results for pairs of stocks being derived from DJ STOXX 600. Section 6 concludes with some remarks on potential further research.

2 Foundations of Relative Value Strategies

Empirical results, documented in the scientific literature on relative value strategies, indicate that the price ratio $Rat_t = \left(P_t^A / P_t^B\right)$ of two assets A and B can be assumed to follow a mean reverting process [3, 7]. This implies that short term deviations from the equilibrium ratio are balanced after a period of adjustment. If this assumption is met, the "simple" question in pairs trading strategies is that of discovering the instant where the spread reaches its maximum and starts to converge. The simplest way of detecting these trading points is to assume an extremum in Rat_t when the spread deviates from the long term mean by a fixed percentage. In other cases confidence intervals of the ratio's mean are used for the identification of trading signals.[2] Higher sophisticated relative value arbitrage trading rules based on a Kalman filter approach are provided in [2, 1].

Pairs trading strategies can be divided into two categories in regard to the point in time when a trade position is unwinded. According to conservative trading rules the position is closed when the spread reverts to the long term mean. However, in risky approaches the assets are held until a "new" minimum or maximum is detected by the applied trading rule.

However, one major problem in pairs trading strategy - besides the successful selection of the pairs - stems from the assumption of mean reversion of the spread. Pairs traders report that the mean of the price ratio seems to switch between different levels and traditional technical trading approaches often fail to identify profit opportunities. In order to overcome this problem of temporary vs. persistent spread deviations, we apply a Markov regime-switching model with switching mean and switching variances to detect such phases of imbalances.

[2] See [3].

3 Markov Regime-Switching Model

Many financial and macroeconomic time series are subject to sudden structural breaks [5]. Therefore, Markov regime-switching models have become very popular since the late 1980s. In his seminal paper Hamilton [4] assumes that the regime shifts are governed by a Markov chain. As a result the current regime s_t is determined by an unobservable, i.e. latent variable. Thus, the inference of the predominant regime is based only on calculated state probabilities. In the majority of cases a two-state, first-order Markov-switching process for s_t is considered with the following transition probabilities [6]:

$$prob\left[s_t = 1 | s_{t-1} = 1\right] = p = \frac{\exp\left(p_0\right)}{1 + \exp\left(p_0\right)} \tag{1}$$

$$prob\left[s_t = 2 | s_{t-1} = 2\right] = q = \frac{\exp\left(q_0\right)}{1 + \exp\left(q_0\right)}, \tag{2}$$

where p_0 and q_0 denote unconstrained parameters. We apply the following simple regime-switching model with switching mean and switching variance for our trading rule:

$$Rat_t = \mu_{s_t} + \varepsilon_t, \tag{3}$$

where $E\left[\varepsilon_t\right] = 0$ and $\sigma^2_{\varepsilon_t} = \sigma^2_{s_t}$.
To visualize the problem of switching means figure 1 plots a time series of a scaled price ratio, where the two different regimes are marked. The shaded area indicates a regime with a higher mean (μ_{s_1}) while the non-shaded area points out a low-mean regime (μ_{s_2}).
Traditional pairs trading signals around the break point BP would suggest an increase in Rat_{BP} implying a long position in Anglo American PLC and a short position in XSTRATA PLC. As can be seen in figure 1 this trading position leads to a loss, since the price of the second stock relative to the price of the first stock increases.

4 Regime-Switching Relative Value Arbitrage Rule

In this study we suggest applying Markov regime-switching models to detect profitable pairs trading rules. In a first step we estimate the Markov regime-switching model as stated in equation (3). As a byproduct of the Markov regime-switching estimation we get the smoothed probabilities $P\left(\cdot\right)$ for each state. Based on these calculated probabilities we identify the currently predominant regime. We assume a two-state process for the spread and interpret the two regimes as a *low* and

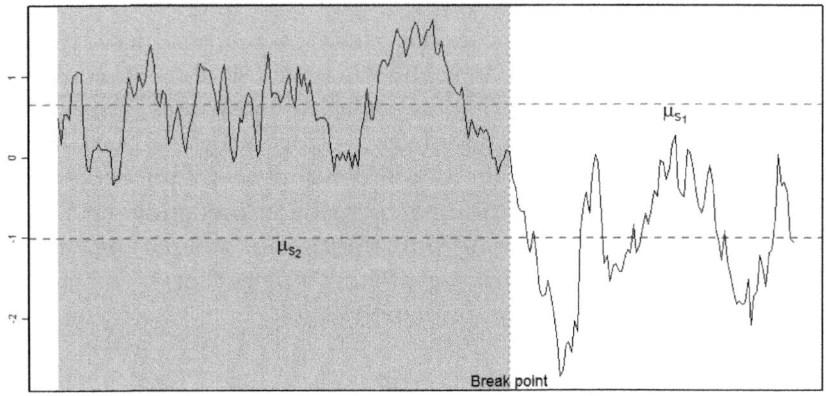

Fig. 1. Scaled ratio of the stock prices of Anglo American PLC and XSTRATA PLC from 2006-12-01 to 2007-11-15. The ratio exhibits a switching mean. The shaded area indicates the high mean regime.

a *high* mean regime. In consequence, we try to detect the instant where the spread Rat_t reaches a local extremum. As a matter of convenience, we adopt the traditional pairs trading approach that a minimum or maximum is found when the spread deviates from the mean by a certain amount. However, we extend the traditional rule by considering a low and a high mean regime, and so we create a regime dependent arbitrage rule. A trading signal z_t is created in the following way:

$$z_t = \begin{cases} -1 \quad if \quad Rat_t \geq \mu_{s_t} + \delta \cdot \sigma_{s_t} \\ +1 \quad if \quad Rat_t \leq \mu_{s_t} - \delta \cdot \sigma_{s_t}, \end{cases} \tag{4}$$

otherwise $z_t = 0$. We use δ as a standard deviation sensitivity parameter and set it equal to 1.645. As a result, a local extremum is detected, if the current value of the spread lies outside the 90% confidence interval within the prevailing regime. The interpretation of the trading signal is quite simple: if $z_t = -1$ $(+1)$ we assume that the observed price ratio Rat_t has reached a local maximum (minimum) implying a short (long) position in asset A and a long (short) position in asset B.

Probability Threshold

To evaluate the trading rule dependent on the current regime (*low* or *high* mean), we additionally implement a probability threshold ρ in our arbitrage rule. Therefore, the regime switching relative value arbitrage rule changes in the following way:

$$z_t = \begin{cases} -1 \ if \ \ Rat_t \geq \mu_{low} + \delta \cdot \sigma_{low} \wedge P(s_t = low|Rat_t) \geq \rho \\ +1 \ if \ \ Rat_t \leq \mu_{low} - \delta \cdot \sigma_{low} \end{cases} \quad (5)$$

otherwise $z_t = 0$, if s_t is in the *low mean regime*. In the *high mean regime* a trading signal is created by:

$$z_t = \begin{cases} -1 \ if \ \ Rat_t \geq \mu_{high} + \delta \cdot \sigma_{high} \\ +1 \ if \ \ Rat_t \leq \mu_{high} - \delta \cdot \sigma_{high} \wedge P(s_t = high|Rat_t) \geq \rho \end{cases} \quad (6)$$

otherwise $z_t = 0$. The probabilities $P(\cdot)$ of each regime indicate whether a structural break is likely to occur. If the probability suddenly drops from a high to a lower level, our regime switching relative value arbitrage rule prevents us from changing the trading positions the wrong way around, so that a minimum or a maximum is not detected too early. The probability threshold value is set arbitrarily. Empirical results suggest a setting for ρ ranging from 0.6 to 0.7. Therefore, the trading rule acts more cautiously in phases where the regimes are not selective.

5 Empirical Results

The developed investment strategy is applied in a first data set to the investing universe of the DJ STOXX 600. Our investigation covers the period 2006-06-12 to 2007-11-16. We use the first 250 trading days to find appropriate pairs, where we use a specification of the ADF-test for the pairs selection. The selected pairs[3] are kept constant over a period of 50, 75, 100 and 125 days. However, if a pair sustains a certain accumulated loss (10%, 15%), it will be stopped out. To estimate the parameters of the Markov regime-switching model we use a rolling estimation window of 250 observations.
For reasons of space, only one representative example will be quoted. Table 1 demonstrates the results of the regime-switching relative value arbitrage rule for the second term of 2007. In this period the best result (average profit of 10.6% p.a.) is achieved by keeping the pairs constant over 125 days and by a stop loss parameter of 15%. The setting of 50 days with a stop loss of 10% generates an average loss of -1.5% p.a. It should be noted that the trading and lending costs (for short selling) have not been considered in this stage of the study.

[3] A number of 25 was detected. One asset is only allowed to occur in 10% of all pairs because of risk management thoughts.

Table 1. Annualized descriptive statistics for the over all selected pairs averaged results of the second term of 2007. # denotes the number of pairs not leading to a stop loss.

panel	50 days		75 days		100 days		125 days	
stop loss	10%	15%	10%	15%	10%	15%	10%	15%
μ	-0.01470	0.00555	0.05022	0.07691	0.03038	0.05544	0.08196	0.10617
σ	0.17010	0.17568	0.18181	0.19042	0.19105	0.20417	0.21265	0.23059
min	-0.40951	-0.40951	-0.29616	-0.38670	-0.23157	-0.23402	-0.19000	-0.22644
1Q	-0.21938	-0.18922	-0.15507	-0.08395	-0.23157	-0.10718	-0.19000	-0.19000
2Q	0.00000	0.00823	0.00000	0.09495	-0.02199	0.05935	0.06252	0.06785
3Q	0.18277	0.18277	0.21685	0.21685	0.18718	0.18718	0.23587	0.23587
max	0.79171	0.79171	0.84898	0.84898	0.60407	0.60407	1.27242	1.27242
#	23	24	21	24	18	23	15	19

6 Conclusion

In this study we implemented a Markov regime-switching approach into a statistical arbitrage trading rule. As a result a regime-switching relative value arbitrage rule was presented in detail. Additionally, the trading rule was applied for the investing universe of the DJ STOXX 600. The empirical results, which still remain to be validated, suggest that the regime-switching rule for pairs trading generates positive returns and so it offers an interesting analytical alternative to traditional pairs trading rules.

References

1. Binh Do, Robert Faff, and Kais Hamza. A new approach to modeling and estimation for pairs trading. Monash University, Working Paper, 2006.
2. Robert J. Elliott, John van der Hoek, and William P. Malcolm. Pairs trading. Quantitative Finance, 5:271-276, 2005.
3. Evan G. Gatev, William N. Goetzmann, and K. Geert Rouwenhorst. Pairs trading: performance of a relative value arbitrage rule. Review of Financial Studies, 19(3):797-827, 2006.
4. James D. Hamilton. A new approach to the economic analysis of nonstationary time series and the business cycle. Econometrica, 57:357-384, 1989.
5. James D. Hamilton. Time Series Analysis. Princeton University Press, Princeton, 1994.
6. Chang-Jin Kim and Charles R. Nelson. State-space models with regime switching. The MIT Press, Cambridge, 1999.
7. Ganapathy Vidyamurthy. Pairs Trading: quantitative methods and analysis. Wiley, Hoboken, 2004.

Performance Measurement, Compensation Schemes, and the Allocation of Interdependence Effects

Michael Krapp, Wolfgang Schultze, and Andreas Weiler

University of Augsburg, Faculty of Business Administration and Economics, D-86135 Augsburg
{quantitative-methoden|wpc}@wiwi.uni-augsburg.de

1 Introduction

In practice, firms often exhibit divisionalized structures in which headquarters delegate decision rights to divisional managers. In this paper, we examine the problem of allocating interdependence effects stemming from interdivisional trade. For this, we analyze a model in which a divisionalized firm contracts with two managers to operate their divisions and to make relationship-specific investment decisions. Contracts can be based on both divisional profits and hence depend on the allocation of interdependence effects. In line with transfer pricing literature, we discuss a centralized as well as a decentralized setting with respect to the allocation authority.
Issues of mechanism design concerning divisional trade are extensively discussed in the literature.[1] Most related to our paper is [1] which shows that profit sharing induces managers to make first-best investment decisions in a decentralized setting. However, profit sharing imposes extra risk on the managers and therefore may not be optimal. Our paper extends the analysis of [1] by incorporating moral hazard problems with respect to investment decisions. Further, we distinguish between different organizational designs.

[1] Cf., for instance, [1], [2], [3], [4], [6], and [7]. With respect to relation-specific investments, it was shown that negotiated transfer prices result in efficient trade as long as information is symmetric between divisional managers. However, this does not imply first-best investment decisions in general. Edlin/Reichelstein [5] show that efficient investment decisions are attainable when both managers can commit to contracts prior to making their investment decisions. However, in line with [1], we assume that not all necessary parameters can be specified in advance.

B. Fleischmann et al.(Eds.), *Operations Research Proceedings 2008*,

DOI: 10.1007/978-3-642-00142-0_3,

2 The Model and Benchmark Solution

We analyze the performance measurement problem of a two-divisional firm with a risk-neutral principal, a downstreaming division (division 1), and an upstreaming division (division 2). Both divisional managers are risk-averse and effort-averse with respect to their relationship-specific investment decisions. Further, in line with [3] and [6], we consider a linear-quadratic scenario and adopt the well-known LEN assumptions. Then, division manager i ($i = 1, 2$) strives to maximize $\mathrm{E}(w_i) - \frac{\alpha_i}{2}\mathrm{Var}(w_i) - \frac{1}{2}v_i I_i^2$, where w_i denotes the compensation, α_i the coefficient of risk-aversion, and I_i the relationship-specific investment decision of manager i; v_i measures her effort-aversion. W.l.o.g. we set the reservation utilities of both managers to zero.
We assume that both divisional profits π_i depend on the allocation of the interdependence effect t: $\pi_1 = R(\vartheta, q, I_1) - t - \frac{1}{2}I_1^2 + \varepsilon_1$ and $\pi_2 = t - C(\vartheta, q, I_2) - \frac{1}{2}I_2^2 + \varepsilon_2$, where $R(\vartheta, q, I_1) = (a(\vartheta) - \frac{1}{2}bq + I_1)q$ and $C(\vartheta, q, I_2) = (c(\vartheta) - I_2)q$. Further, $\varepsilon = (\varepsilon_1, \varepsilon_2)$ denotes the vector of noise terms, where $\varepsilon_i \sim N(0, \sigma_i^2)$ and $\mathrm{Cov}(\varepsilon_1, \varepsilon_2) = \varrho\sigma_1\sigma_2$.[2] The variable q denotes the quantity transfered from division 2 to division 1. To avoid trivial solutions, we assume $a(\vartheta) > c(\vartheta)$ for all feasible values of ϑ. The state variable ϑ can be observed ex post (after contracting and making investment decisions) by the division managers only. In contrast, the distribution of ϑ is ex ante common knowledge. For convenience, we assume $\mathrm{Cov}(a(\vartheta), \varepsilon_i) = \mathrm{Cov}(c(\vartheta), \varepsilon_i) = 0$ and $\mathrm{Var}(a(\vartheta)) = \mathrm{Var}(c(\vartheta)) = \mathrm{Cov}(a(\vartheta), c(\vartheta)) = \sigma_\vartheta^2$.
In line with the LEN model, we restrict our analysis to linear compensation contracts, i.e. $w_i = \underline{w}_i + w_{i1}\pi_1 + w_{i2}\pi_2$. Note that contracts placing equivalent weights on both divisional profits ($w_{ii} = w_{ij}$) implicate profit sharing. Since we aim at studying the allocation of interdependence effects, it is appropriate to distinguish between a centralized and a decentralized setting. Figures 1 and 2 depict the event sequences for both designs. In the centralized setting, central management allocates the interdependence effect by determining t as well as q at date 1 subject to incomplete information w.r.t. ϑ.
In contrast, in the decentralized setting, central management delegates allocation authority as well as the determination of the transfer quantity to the divisions. Here we assume the divisional managers to bargain about the allocation after observing ϑ. Divisional managers are hence able to respond to the realization of the state variable ϑ. Therefore, a

[2] In contrast to [1], we allow for a possible risk interdependence between both divisions.

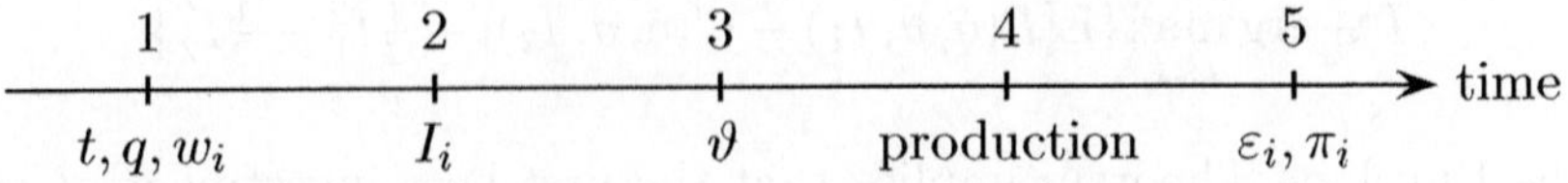

Fig. 1. Sequence of events in the centralized setting

flexibility gain is attained from the perspective of central management.[3] Since information is symmetric between the divisional managers, we re-

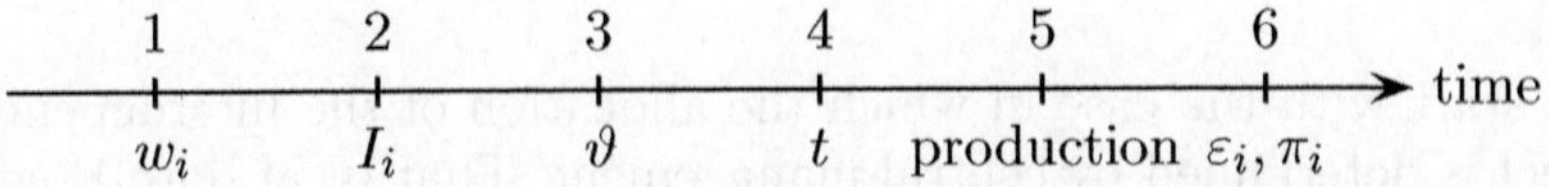

Fig. 2. Sequence of events in the decentralized setting

strict our analysis w.l.o.g. to the case in which both managers possess equal bargaining power and therefore to the Nash bargaining solution:[4]

$$t = \tfrac{1}{2}[q(a(\vartheta) + c(\vartheta) - \tfrac{1}{2}bq + I_1 - I_2) + \varepsilon_1 - \varepsilon_2] \,. \tag{1}$$

Based on the negotiated allocation, the downstreaming division 1 determines the transfer quantity. Provided this allocation mechanism, the divisional profits

$$\pi_i = \tfrac{1}{2}[q(a(\vartheta) - c(\vartheta) - \tfrac{1}{2}bq + I_1 + I_2) + \varepsilon_1 + \varepsilon_2] - \tfrac{1}{2}I_i^2 \tag{2}$$

will be realized at date 6.

Before we examine these settings in detail, we derive some benchmark results by abstracting from agency problems. Then, the efficient trade (given investments I_1 and I_2),

$$\hat{q}(I_1, I_2) \in \arg\max_{q \geq 0}\{R(q, \vartheta, I_1) - C(q, \vartheta, I_2)\} \,, \tag{3}$$

is given by

$$\hat{q}(I_1, I_2) = \frac{a(\vartheta) - c(\vartheta) + I_1 + I_2}{b} \,. \tag{4}$$

Our assumptions assure that $\hat{q}$ is unique and interior for all I_1 and I_2. The firm's investment choice can now be characterized as follows: Let $\hat{I} = (\hat{I}_1, \hat{I}_2)$ denote the vector of efficient investment decisions. Thus, $\hat{I}$ satisfies the condition

---

[3] However, this advantage is reduced by a control loss, cf. [8].

[4] A similar assumption is made in [1].

$$\hat{I} \in \arg\max_{I_1,I_2}\{\mathrm{E}[R(\hat{q},\vartheta,I_1) - C(\hat{q},\vartheta,I_2)] - \tfrac{1}{2}I_1^2 - \tfrac{1}{2}I_2^2\}\,. \tag{5}$$

The Envelope Theorem implies that the first-best investments $\hat{I}$ exist and backward induction yields $\hat{I}_i = \mathrm{E}(\hat{q})$.
In the following section, we solve the performance evaluation problem in the context of different mechanism designs.

3 Centralized vs. Decentralized Allocation Authority

We start with the case in which the allocation of the interdependence effect is determined by central management. That is, at date 1, central management fixes the underlying performance evaluation system, the allocation rule t and the transfer quantity q by solving the program

$$\max_{w_i,q,t}(1-w_{11}-w_{21})\mathrm{E}(\pi_1|q,t)+(1-w_{22}-w_{12})\mathrm{E}(\pi_2|q,t)-(\underline{w}_1+\underline{w}_2) \tag{6}$$

$$\text{s.t.}\ \ I_i \in \arg\max_{\tilde{I}}\{\mathrm{E}_{\vartheta,\varepsilon}(w_i|q,t) - \tfrac{v_i}{2}\tilde{I}_i^2 - \tfrac{\alpha_i}{2}\mathrm{Var}_{\vartheta,\varepsilon}(w_i|q,t)\}, i=1,2 \tag{7}$$

$$\mathrm{E}_{\vartheta,\varepsilon}(w_i|q,t) - \tfrac{v_i}{2}I_i^2 - \tfrac{\alpha_i}{2}\mathrm{Var}_{\vartheta,\varepsilon}(w_i|q,t) \geq 0,\ i=1,2\,, \tag{8}$$

where constraints (7) ensure that the managers' investment choices are incentive compatible and the constraints (8) guarantee the managers their reservation utility. Obviously, the participation constraints hold in equality when choosing an appropriate fixed compensation.
The following proposition summarizes our main results regarding the centralized setting.[5]

Proposition 1 (Centralized Allocation Authority).

i) First-best investment decisions can only be induced if $v_i = 0$.
ii) Investment decisions are independent of t *and* w_{ij}.
iii) Investment decisions are independent of w_{ii} *iff* $v_i = 0$ *and* $w_{ii} \neq 0$.

From the perspective of performance evaluation, there is no need to base w_i also on the profit of division $j \neq i$. Further, note that investment decisions are independent of the allocation t if central management is equipped with allocation authority. Additionally, central management can fix the optimal transfer quantity by backward induction and making use of the Envelope Theorem:

$$q = \frac{\mathrm{E}[a(\vartheta) - c(\vartheta)] + I_1 + I_2}{b + \alpha_1\sigma_\vartheta^2(w_{11} - w_{12})^2 + \alpha_2\sigma_\vartheta^2(w_{22} - w_{21})^2}\,. \tag{9}$$

[5] We omit the proofs. The authors will provide all proofs upon request.

As a consequence, even under full information, central management will only choose the first-best efficient trade if (i) both agents are risk-neutral ($\alpha_i = 0$) or (ii) by implementing full profit sharing ($w_{ii} = w_{ij}$). Both cases are equivalent. Therefore, profit sharing itself does not provide any incentives for investment decisions, however, it mitigates distortions in q caused by the trade-off between risk sharing and investment incentives.

We now turn to the analysis of the decentralized setting. In this case, central management delegates decision rights to the divisional managers and determines the performance evaluation system by solving the program

$$\max_{w_i}(1 - w_{11} - w_{21})\mathrm{E}(\pi_1) + (1 - w_{22} - w_{12})\mathrm{E}(\pi_2) - (\underline{w}_1 + \underline{w}_2) \quad (10)$$

$$\text{s.t. } q \in \arg\max_{\tilde{q}}\{\mathrm{E}_\varepsilon(w_1) - \tfrac{v_1}{2}I_1^2 - \tfrac{\alpha_1}{2}\mathrm{Var}_\varepsilon(w_1)\} \quad (11)$$

$$I_i \in \arg\max_{\tilde{I}}\{\mathrm{E}_{\vartheta,\varepsilon}(w_i) - \tfrac{v_i}{2}\tilde{I}_i^2 - \tfrac{\alpha_i}{2}\mathrm{Var}_{\vartheta,\varepsilon}(w_i)\},\ i = 1,2 \quad (12)$$

$$\mathrm{E}_{\vartheta,\varepsilon}(w_i) - \tfrac{v_i}{2}I_i^2 - \tfrac{\alpha_i}{2}\mathrm{Var}_{\vartheta,\varepsilon}(w_i) \geq 0,\ i = 1,2\,, \quad (13)$$

where (11) and (12) are the incentive compatibility constraints and (13) are the participation constraints.

Since both divisional managers bargain about the allocation under symmetric information, it is straightforward to see that this bargaining process leads to first-best efficient trade $\hat{q}(I)$ given investments I. The following proposition states our results for the decentralized setting.

Proposition 2 (Decentralized Allocation Authority).

i) Divisional managers will always make efficient trade decisions if the allocation is based on a bargaining process under symmetric information.

ii) Investment decisions depend on the expected allocation process and are first-best iff $v_i = 0$ and a full profit sharing policy is applied.

These results are in line with [1] if we abstract from moral hazard issues. However, from an optimal contracting perspective, a firm-wide performance evaluation system imposes extra risk on the managers. In contrast to the centralized setting in which central management is able to trade-off risk sharing and to control investment decisions by fixing trade quantities, central management looses degrees of freedom to solve this problem within a decentralized setting. On the other hand, divisional managers are able to directly respond to the realization of the state variable. Hence, a flexibility gain is attained.

4 Concluding Remarks

We have shown that the allocation process for interdependence effects between divisions usually cannot be substituted by performance evaluation systems. In centralized settings, however, allocation processes and fixed payments interact. Then, central management can allocate the interdependence effect in order to influence decision making and adjust the divisional managers' compensations accordingly. Furthermore, we have shown that the design of optimal performance evaluation systems essentially depend on the underlying allocation authority.
Our results suggest that different allocation procedures (given optimal performance evaluation systems) dominate each other depending on certain conditions. In particular, these are the disutilities of effort, the variance of the state parameter, the parameters of risk aversion of the managers, and the risk interdependence between both divisions. Further analyses concerning these issues are on our research agenda.

References

1. Anctil MA, Dutta S (1999) Negotiated transfer pricing and divisional vs. firm-wide performance evaluation. Accounting Review 74:87–104
2. Baldenius T (2000) Intrafirm Trade, Bargaining Power, and Specific Investments. Review of Accounting Studies 5:27–56
3. Baldenius T, Reichelstein SJ (1998) Alternative Verfahren zur Bestimung innerbetrieblicher Verrechnungspreise. Zeitschrift für betriebswirtschaftliche Forschung 50:236–259
4. Baldenius T, Reichelstein SJ, Sahay SA (1999) Negotiated versus cost-based transfer pricing. Review of Accounting Studies 4:67–91
5. Edlin AS, Reichelstein SJ (1995) Specific investment under negotiated transfer pricing: An efficiency result. Accounting Review 70:275–291
6. Pfeiffer T (2003) Corporate-Governance-Strukturen interner Märkte. Deutscher Universitäts-Verlag, Wiesbaden
7. Pfeiffer T, Wagner J (2007) Die Rekonstruktion interner Märkte, das Dilemma der pretialen Lenkung und spezifische Investitionsprobleme. Zeitschrift für betriebswirtschaftliche Forschung 59:958–981
8. Williamson OE (1985) The economic institutions of capitalism. Free Press, New York

Coordination of Decentralized Departments with Existing Sales Interdependencies

Christian Lohmann

Institute of Production Management and Managerial Accounting,
Ludwig-Maximilians-University Munich, 80539 Germany
lohmann@lmu.de

1 Introduction

This study regards a company which consists of two decentralized business units forming a value network. One business unit produces and sells a main product and the other business unit produces and sells a complementary by-product. The company pursues a firm-wide differentiation strategy. The following model assumes that the business unit being responsible for the by-product can implement this firm-wide differentiation strategy by improving, for example, the quality or the functionality of the by-product. Because of the complementary relationship of the products, the by-product and the main product obtain a unique selling proposition through a specific investment in the by-product. The specific investment is totally defrayed by that business unit acting on its own authority, but it increases the revenue of both business units. Therefore, the allocation of the profit induced by the specific investment is not made fairly. An underinvestment problem arises which endangers the objective of firm-wide profit maximization. Coordination instruments are used to improve the reconciliation between separated business units. This article compares a contribution margin based, a revenue based and a quantity based investment contribution mechanism as coordination instruments for achieving goal congruence between the considered business units inducing efficient production and investment decisions as well as overall profit maximization. The recent literature mostly focuses on profit sharing (e.g. [3]), revenue sharing (e.g. [1] and [2]) or transfer pricing (for an overview see [4]) to coordinate a two-stage value chain of a decentralized company with existing production interdependencies. In contrast to that, this study

B. Fleischmann et al.(Eds.), *Operations Research Proceedings 2008*,
DOI: 10.1007/978-3-642-00142-0_4,

regards production and investment decisions with existing sales interdependencies in a value network.
The remainder of this paper is organized as follows: Section 2 presents the framework and the solution of an equilibrium model using a contribution margin based, a revenue based and a quantity based investment contribution mechanism. In section 3, the performance of these coordination mechanisms is compared on the basis of the expected overall firm profit. Circumstances are identified under witch each investment contribution mechanism dominates the others.

2 Model

2.1 Assumptions

The considered company consists of two decentralized business units. Business unit A produces the main product at constant unit cost c_A and sells it on an anonymous market; business unit B produces a complementary by-product at constant unit cost c_B and sells it also on an anonymous market. A is organized as a profit center and states the production and sales quantity of the main product. Because of the complementary relationship between the two products, the quantity decision of A determines also the production and sales quantity of the by-product. B adapts the sales volume generating a non-negative expected profit. This model assumes a one-to-one relationship. Furthermore, B is organized as an investment center and decides about its specific investment that can increase the revenue of both business units. The risk-neutral decision makers of the business units are compensated according to their reported success after profit allocation by using a contribution margin based, a revenue based or a quantity based investment contribution mechanism given by the headquarter of the company. Therefore, they maximize the expected profit of their own business unit.
The assumed multiplicative demand function with constant price elasticity $\epsilon = 2$ is given by

$$p_A(x, I, \eta) = \sqrt{\frac{a\sqrt{I}}{x}} + \eta \quad \text{with the attributes} \quad \frac{\partial p_A(x, I, \eta)}{\partial x} < 0 \quad ,$$

$$\frac{\partial p_A(x, I, \eta)}{\partial I} > 0 \quad \text{and} \quad \frac{\partial^2 p_A(x, I, \eta)}{\partial^2 I} < 0.$$

The demand function of the main product depends on the quantity x, the specific investment I, the parameter a reflecting the market

size as well as the random variable η. The random variable η shows the uncertainty of the market conditions at the selling moment, the realized sales price is uncertain. The random variable η with mean $\mu = 0$ implies that the expected revenue of the main product is given by $R_A(x, I) = \sqrt{ax\sqrt{I}}$. The expected revenue of the by-product follows similarly with $R_B(x, I) = \sqrt{bx\sqrt{I}}$.
Fig 1. shows the decision making and time sequence of the model.

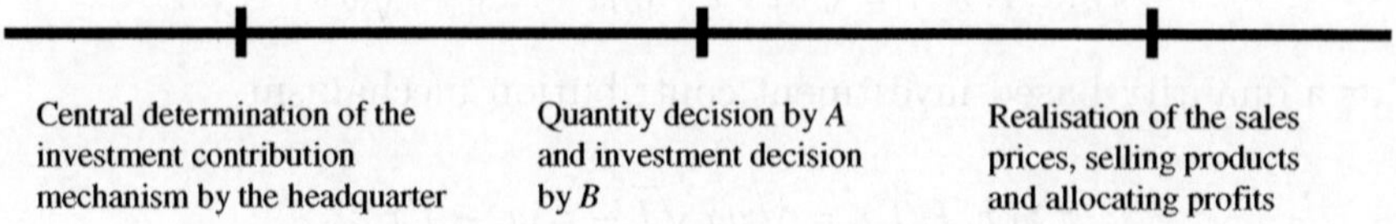

Fig. 1. Decision making and time sequence

At the outset, the headquarter of the company chooses an investment contribution mechanism to coordinate the quantity and investment decisions maximizing the expected overall firm profit. It determines a transfer payment by fixing a contribution margin based, a revenue based or a quantity based investment contribution. Then, the investment center B makes its specific investment. Without additional information, profit center A states the sales volume x, which is also produced and sold by B. Finally, the market price is realized and the profits are allocated.
The specific investment is not observable and can not ascertain ex post because of the uncertainty of the market conditions. But each decision maker knows the quantity respectively the investment decision problem of the other manager. Therefore, they are able to calculate optimal decisions in their view considering a given investment contribution mechanism. The decision makers are planning their decisions on the basis of their expected allocated profits. The following analysis uses an equilibrium model solving the several decision problems by backward induction (see [5]).

2.2 Solution

The starting points of the analysis are the expected profit functions of A and B using a contribution margin based investment contribution mechanism

$$\pi_A^{CM}(x, I, \tau^{CM}) = (1 - \tau^{CM}) \left(\sqrt{ax\sqrt{I}} - c_A x\right)$$
$$\pi_B^{CM}(x, I, \tau^{CM}) = \sqrt{bx\sqrt{I}} - c_B x - I + \tau^{CM} \left(\sqrt{ax\sqrt{I}} - c_A x\right),$$

a revenue based investment contribution mechanism

$$\pi_A^R(x, I, \tau^R) = (1 - \tau^R)\sqrt{ax\sqrt{I}} - c_A x$$
$$\pi_B^R(x, I, \tau^R) = \sqrt{bx\sqrt{I}} - c_B x - I + \tau^R \sqrt{ax\sqrt{I}},$$

or a quantity based investment contribution mechanism

$$\pi_A^Q(x, I, T) = \sqrt{ax\sqrt{I}} - c_A x - Tx$$
$$\pi_B^Q(x, I, T) = \sqrt{bx\sqrt{I}} - c_B x - I + Tx.$$

A specifies the optimal production and sales quantity depending on the investment level I and the allocation parameter of the used investment contribution mechanism. Anticipating the quantity decision, B determines its specific investment depending on the allocation parameter of the used investment contribution mechanism. Finally, the headquarter of the company sets the allocation parameter maximizing the expected overall firm profit. The allocation parameters are calculated for a contribution margin based, a revenue based and a quantity based investment contribution mechanism with

$$\tau^{CM} = 1,$$
$$\tau^R = \frac{2ac_B^2 + (a - 3\sqrt{ab})c_A c_B - 3(a + 3\sqrt{ab})c_A^2}{2(2ac_A c_B + ac_B^2)} + \frac{\sqrt{a(a + b + 2\sqrt{ab})c_A^2(9c_A^2 + 2c_A c_B + c_B^2)}}{2(2ac_A c_B + ac_B^2)} \quad \text{and}$$
$$T = \frac{3(a + \sqrt{ab})c_B - (4b + 5\sqrt{ab})c_A + \sqrt{a(3a + b + 2\sqrt{ab})}(c_A + c_B)}{2a + 4b + 8\sqrt{ab}}.$$

The decision interdependencies can be resolved and the investment level, the production and sales quantity, the expected profits of the business units as well as the expected overall firm profit can be calculated.

3 Performance Evaluation

The expected overall firm profit reflects the ability of the investment contribution mechanisms to induce efficient investment and quantity decisions. Therefore, the investment contribution mechanisms are compared on the basis of the overall firm profit. Fig. 2 shows the overall firm profit depending on relative unit costs (c_A and c_B) and relative market sizes (a and b) using a contribution margin based, a revenue based and a quantity based investment contribution mechanism determined by the headquarter of the company. Depending on relative unit costs and relative market sizes the implementation of every single investment contribution mechanism can be expedient. The expected overall firm profit without using an investment contribution mechanism (π^{no}) can always be increased by one of these investment contribution mechanisms. As can be seen from Fig. 2, there are three areas dominated by a single investment contribution mechanism.

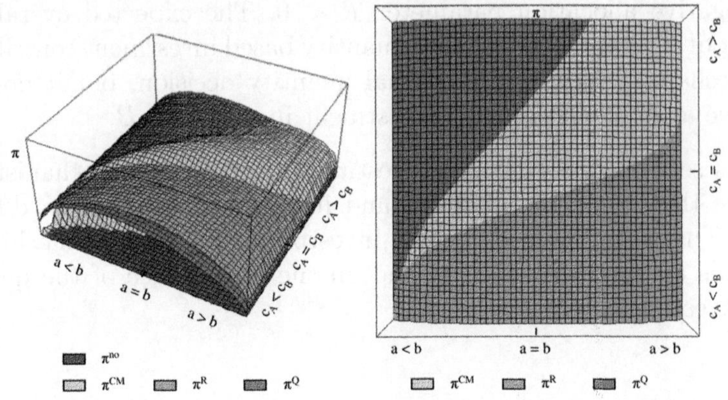

Fig. 2. Expected overall firm profit depending on relative unit costs and relative market sizes, with $c_A \in \,]0, 2[\,$, $c_B \in \,]0, 2[\,$, $c_A + c_B = 2$, $a \in \,]0, 20[\,$, $b \in \,]0, 20[$ and $a + b = 20$

1. $c_A > c_B \cap a > b$: The relation of the unit costs $c_A > c_B$ and the relation of the market sizes $a > b$ indicate the circumstances in which the contribution margin based investment contribution mechanism performs best. In this situation, A makes already a sufficient optimal quantity decision. A contribution margin based investment contribution mechanism does not influence the quantity decision, but it causes a higher investment level for any given $\tau^{CM} > 0$. To induce an efficient investment decision by B and to increase the

expected overall firm profit, the headquarter shifts the whole contribution margin of A to B by using a contribution margin based investment contribution mechanism with $\tau^{CM} = 1$.

2. $c_A < c_B$: If the unit costs c_A are smaller than the unit costs c_B, a revenue based investment contribution mechanism should be used by the headquarter. Without any investment contribution mechanism, the production and sales quantity exceeds the optimal quantity level. A revenue based investment contribution mechanism affects the quantity as well as the investment decisions. In this situation, a positive allocation parameter $\tau^R = 1$ reduces the production and sales quantity decided by A and increases the investment level of B. As result, a revenue based investment contribution mechanism increases the expected overall firm profit.
3. $c_A > c_B \cap a < b$: In this situation, a quantity based investment contribution mechanism is used to influence the quantity decision of A without any negative effect on the investment decision of B. A determines a larger production and sales quantity for a given negative allocation parameter $T < 0$. The expected overall firm profit increases, because the quantity based investment contribution mechanism induces an optimal quantity decision, but it does not give a counterproductive investment incentive to B.

In this setting, the analyzed investment contribution mechanisms influence the considered quantity and investment decisions in different ways. Therefore, the use of these investment contribution mechanisms depends on their influence to and on the importance of the quantity and investment decisions.

References

1. Chwolka A, Simons D (2003) Impacts of Revenue Sharing, Profit Sharing and Transfer Pricing on Quality-Improving Investments. European Accounting Review 12: 47-76
2. Martini JT (2007) Verrechnungspreis zur Koordination und Erfolgsermittlung. Deutscher Universitäts-Verlag, Wiesbaden
3. Milgrom P, Roberts J (1992) Economics, Organization and Management. Prentice Hall International, New Jersey
4. Pfaff D, Pfeiffer T (2004) Verrechnungspreise und ihre formal-theo-re-tische Analyse. Zum State of the Art. Die Betriebswirtschaft 64: 296-319
5. Selten R (1975) Reexamination of the Perfect Concept for Equilibrium Points in Extensive Games. International Journal of Game Theory 4: 25-55

Empirical Examination of Fundamental Indexation in the German Market

Max Mihm

Dresden University of Technology, Department of Finance
mihm@finance.wiwi.tu-dresden.de

Summary. Fundamental Indexation is the name of an index methodology that selects and weights index constituents by means of fundamental criteria like total assets, book value or number of employees. This paper examines the performance of fundamental indices in the German equity market during the last 20 years. Furthermore the index returns are analysed under the assumption of an efficient as well as an inefficient market. Index returns in efficient markets are explained by applying the three factor model for stock returns of [2]. The results show that the outperformance of fundamental indices is partly due to a higher risk exposure, particularly to companies with a low price to book ratio. By relaxing the assumption of market efficiency, a return drag of capitalisation weighted indices can be deduced. The index methodology implies an investment strategy that benefits from well known market anomalies like the value effect without relying on active portfolio management. Furthermore under the assumption of an inefficient market there is an added value of fundamental indices.

1 Introduction

Traditional stock market indices weight companies by means of market capitalisation, mostly corrected by a free float factor. A low index turnover and good investability qualify these indices as suitable underlyings for index replicating funds or derivatives. Furthermore cap weighted indices are essential benchmarks since they reflect the average return of a certain stock market.

However, a new index approach named Fundamental Indexation triggered a lively and controversal debate among the fund and indexing industry. The constituents of fundamental indices are selected and weighted according to fundamental factors like revenues or total assets

B. Fleischmann et al.(Eds.), *Operations Research Proceedings 2008*,
DOI: 10.1007/978-3-642-00142-0_5, © Springer-Verlag Berlin Heidelberg 2009

just to mention a few. [1] shows that these indices outperform traditional cap weighted indices by 1.66 to 2.56 percentage points annually over a 43 years period. While critics of fundamental indices say the return advantage is delivered through higher systematic risks, advocates explain superior performance by inefficient markets and a return drag of traditional index approaches.
In this paper five different fundamental indices that differ in terms of their selection and weighting criteria are calculated and compared against a cap weighted and an equal weighted index in the German market over the last 20 years. The empirical results are presented in Section 3. The cap weighted index represents the traditional approach to construct a stock index, while the equal weighted index shows the differences of fundamental indices to a naïve investment strategy. Furthermore, the empirical data is used to explain performance differences. In Section 4 the analysis is conducted under the assumption of an efficient as well as an inefficient market.

2 Data and Index Methodology

The data set consists of all German companies and covers the period 1 January 1988 to 23 July 2007. In case of more than one share classes, the one with the biggest capitalisation is included in the index universe. Index rebalancing is performed annually, while only those companies with a track record of at least one year are taken into account.
All indices encompass 100 companies in total and are constructed as total return indices. In addition to the traditional market cap weighted index (MK100) there is one equal weighted index (GG100) as well as five fundamental indices based on the criteria revenues (UM100), number of employees (MA100), total assets (GK100), book value of equity (EK100) and dividend payment (DV100). According to these criteria the index portfolio is rebalanced annually. All indices except the cap weighted approach are weighting their index constituents market independently, that is there is no direct link between price and weight.

3 Results

The index return statistics summarised in table 1 show that the cap weighted index approach exhibits the lowest historical returns. Fundamental indices realise geometric mean returns that exceed the market return (MK100) by 1.87 to 3.79 percentage points annually after rebalancing costs.

With returns that exceed the market cap weighted index returns, while having similar standard deviations, the Modern Portfolio Theory suggests a dominance of fundamental indices. All performance measurements, namely the Sharpe Ratio Jensen Alpha and Treynor Ratio, indicate that fundamental indices exhibit a superior performance over the last 20 years. The same applies for the equal weighted index, based on its low standard deviation.

Table 1. Index Performance

Index	Index value 23/07/2007	Geom. mean return	Geom. mean return after costs	Standard deviation	SHARPE ratio	JENSEN alpha	TREYNOR ratio
MK100	732.01	10.08%	9.84%	17.45%	0.415	-	0.0724
GG100	930.05	11.61%	10.89%	12.06%	0.658	3.06	0.1239
UM100	1310.88	13.30%	12.82%	17.11%	0.602	3.37	0.1089
MA100	1319.72	13.30%	12.79%	17.53%	0.587	3.36	0.1070
GK100	1505.98	14.13%	13.63%	17.32%	0.632	4.16	0.1161
EK100	1134.56	12.58%	12.09%	17.83%	0.543	2.29	0.0962
DV100	1116.62	12.52%	11.71%	16.14%	0.567	2.71	0.1019

Further analysis show that the return differences are statistically significant for the fundamental indices. Interestingly for longer holding periods the probability of superior returns for the fundamental indices increases.[1] This effect does not apply for the equal weighted GG100 index.

The different characteristics which can be seen from the past performances are certainly caused by differing index compositions. Fundamental indices are characterised by a tilt towards companies with low valuation levels, while the equal weighted index puts emphasis on companies that are small in terms of market capitalisation.

The calculated fundamental indices feature superior performance characteristics in the German stock market of the last 20 years. Thus the empirical results confirm findings of former studies on fundamental indexation.[2] Past performances do not allow any conclusions on future

[1] This time horizon effect is not illustrated here. For detailed information see [5]

[2] See [4], [1] or [7] for the US market. For studies on other markets compare [3]

developments. However, by explaining returns through a reasonable model, under the assumptions of the model predictions can be made.

4 Analysis

The index returns are analysed in two ways. Firstly, an efficient market is presumed. Under this assumption the three factor model based on [2] is used to explain returns based on the risk exposure of each index. The second approach looses the assumption of an efficient market and implies a mean reverting price process. Such a price process can explain return discrepancies of the calculated index methodologies. The author considers both perspectives since none of them can be negated. Every test of market efficiency includes a joint hypothesis which does not allow unambiguous conclusions.

4.1 Efficient Market

A multiple regression based on the three factor model shows the risk exposures of the index portfolios. The model takes three systematic risks into account. Besides the market excess return $(r_t^m - r_t^f)$ that is also incorporated in the CAPM, there is a SMB_t (small minus big) and a HML_t (high minus low) factor. They refer to the higher risk of small firms and companies with low valuation levels respectively. Equation 1 illustrates the regression equation that is used to calculate the risk exposures for each index i.

$$r_t^i - r_t^f = \hat{\alpha}^i + \hat{\beta}_m^i * (r_t^m - r_t^f) + \hat{\beta}_{SMB}^i * SMB_t + \hat{\beta}_{HML}^i * HML_t + \epsilon_t^i \quad (1)$$

The regression parameters are summarised in table 2. The equal weighted index has a high risk exposure to the SMB factor due to its bias towards small cap companies. Fundamental indices exhibit relative high risk exposures to companies with low valuation levels, expressed by the HML factor beta.

However, a significant part of the superior returns of fundamental indices are not explained by systematic risk factors. There is a high regression alpha which contributes to the outperformance of fundamental indices without being based on any kind of systematic risk. Importantly within the model assumptions this proportion of the return can not be considered as persistent since in an efficient market there is no sustainable return without systematic risk. This conclusion is based on the presumptions that the market is efficient **and** the model incorporates all risk factors and hence calculates fair values.

Table 2. Regression Parameters

i	$\hat{\alpha}^i$	$\hat{\beta}^i_m$	$\hat{\beta}^i_{SMB}$	$\hat{\beta}^i_{HML}$	R^2
MK100	0.0056%	0.980***	-0.046***	-0.008	0.985
GG100	0.0248%	0.773***	0.227***	0.099***	0.878
UM100	0.0299%	0.924***	-0.002	0.251***	0.940
MA100	0.0268%	0.988***	0.069**	0.225***	0.925
GK100	0.0497%**	0.954***	0.044	0.187***	0.900
EK100	0.0236%*	0.937***	-0.089***	0.206***	0.971
DV100	0.0295%*	0.853***	-0.050**	0.195***	0.944

*, ** or *** indicate that the null hypothesis of an arithmetic mean of zero is rejected with a 10%, 5% or 1% level of significance respectively.

4.2 Inefficient Market

In inefficient markets prices differ from fair values. The likelihood that market participants identify irrational prices increases with the extent of mispricing. Consequently, prices revert to their fair value in inefficient markets, which lead to a mean reverting price process.

Given the fact that prices are mean reverting, [6] shows that a return drag of market cap weighted indices can be deduced. Certainly there is no way to definitely identify over- and undervalued companies. However, there is a systematic failure in the capitalisation weighted index methodology. Since the weighting is dependent on market valuation, a positive mispricing will automatically lead to a higher weight in the index. Consequently, compared to market independently weighted indices, traditional market cap weighted indices relatively overweight overvalued companies and relatively underweight undervalued companies. Regardless of differences in exposure to systematic risks, whenever prices are mean reverting, the market cap weighted index is expected to exhibit inferior returns in comparison to fundamental or equal weighted indices.

Under the assumption of an inefficient market with mean reverting prices, the return advantage of fundamental indexes that is not based on systematic risk factors can be considered as persistent. This conclusion differs fundamentally from the implications in an efficient market where the regression alpha was considered as a non-sustainable part of the index returns.

5 Conclusion

The empirical results as well as the conclusions of the analysis indicate that fundamental indices are contributing to the universe of passive investment products. Without being actively managed, fundamental indices capture a value premium through placing emphasis on companies with low valuation levels. The regression shows that this significantly accounts for the superior performance of fundamental indices.
Apart from the risk exposure to value companies, there is a return advantage for fundamental indices when prices are mean reverting. Given this assumption fundamental indices deliver a real added value since investors do not have to bear any additional risk to justify the extra return.

References

1. R. D. Arnott, J. Hsu, and P. Moore. Fundamental indexation. Financial Analysts Journal, 61(2):p83-99, 2005. ISSN 0015198X.
2. E. F. Fama and K. R. French. Common risk factors in the returns on stocks and bonds. Journal of Financial Economics, 33(1):3-56, Feb. 1993.
3. J. Hemminki and V. Puttonen. Fundamental indexation in Europe. Journal of Asset Management, 8(6):p401-405, 2008. ISSN 14708272.
4. V. Lowry. Fundamentally better. Journal of Indexes, March/April, 2007.
5. M. Mihm. Überprüfung der "fundamental indexation" am deutschen Markt, Diploma Thesis, Dresden University of Technology, 2008.
6. J. Treynor. Why Market-Valuation-Indifferent indexing works. Financial Analysts Journal, 61(5):p65-69, 2005. ISSN 0015198X.
7. P. Wood and R. Evans. Fundamental profit based equity indexation. Journal of Indexes, 2nd quarter, 2003.

Trading in Financial Markets with Online Algorithms

Esther Mohr[1] and Günter Schmidt[1+2*]

[1] Saarland University, P.O. Box 151150, D-66041 Saarbrücken, Germany, em@itm.uni-sb.de

[2] Hochschule Liechtenstein, Fürst-Franz-Josef-Strasse, FL-9490 Vaduz, Liechtenstein,
gs@itm.uni-sb.de

Summary. If we trade in financial markets we are interested in buying at low and selling at high prices. We suggest an active reservation price based trading algorithm which tries to solve this type of problem. The effectiveness of the algorithm is analyzed from a worst case point of view. We want to give an answer to the question if the suggested algorithm shows a superior behaviour to buy-and-hold policies using simulation on historical data.

1 Introduction

Many major stock markets are electronic market places where trading is carried out automatically. Trading policies which have the potential to operate without human interaction are often based on data from technical analysis [5, 3, 4]. Many researchers studied trading policies from the perspective of artificial intelligence, software agents or neural networks [1, 6]. In order to carry out trading policies automatically they have to be converted into trading algorithms. Before a trading algorithm is applied one might be interested in its performance. The performance of trading algorithms can basically be analyzed by three different approaches. One is Bayesian analysis, another is assuming uncertainty about asset prices and analyzing the trading algorithm under worst case outcomes. This approach is called competitive analysis [2]. The third is a heuristic approach where trading algorithms are analyzed by simulation runs based on historical data. We apply the second and the third approach in combination.

* corresponding author

B. Fleischmann et al.(Eds.), *Operations Research Proceedings 2008*,
DOI: 10.1007/978-3-642-00142-0_6,

The reminder paper is organized as follows. In the next section different trading policies for a multiple trade problem are introduced. Section 3 presents detailed experimental findings from our simulation runs. In the last section we finish with some conclusions.

2 Multiple Trade Problem

In a multiple trade problem we have to choose points of time for selling current assets and buying new assets over a known time horizon. The horizon consists of several trading periods i of different types p with a constant number of h days. We differ between $p = 1, 2, \ldots, 6$ types of periods numbered with $i = 1, \ldots, n(p)$ and length h from $\{7, 14, 28, 91, 182, 364\}$ days, e.g. period type $p = 6$ has length $h = 364$ days. There is a fixed length h for each period type p, e.g. period length $h = 7$ corresponds to period type $p = 1$, period length $h = 14$ corresponds to period type $p = 2$, etc.

We differ between three trading policies. Two elementary ones are Buy-and-Hold ($B + H$), a passive policy, and Market Timing (MT), an active policy. The third one is a Random ($Rand$) policy. To evaluate the policies' performance empirically we use an optimal algorithm called Market (MA) as a benchmark. We assume that for each period i there is an estimate of the maximum price $M(i)$ and the minimum price $m(i)$. Within each period $i = 1, \ldots, n(p)$ we have to buy and sell an asset at least once. The annualized return rate $R(x)$, with x from $\{MT, Rand, B + H, MA\}$ is the performance measure used. At any point of time a policy either holds an asset or overnight deposit.

In order to describe the different policies we define a holding period with respect to MT. A holding period is the number of days h between the purchase of asset j and the purchase of another asset j' ($j' \neq j$) by MT. Holding periods are determined either by reservation prices $RP_j(t)$ which give a trading signal or by the last day T of a period.

MARKET TIMING (MT). Calculates $RP_j(t)$ for each day t for each asset j based on $M(i)$ and $m(i)$. The asset j^* MT buys within a period is called MTasset. An asset j^* is chosen by MT if $RP_{j^*}(t) - p_{j^*}(t) = \max\{RP_j(t) - p_j(t) | j = 1, \ldots, m\}$ and $p_{j^*}(t) < RP_{j^*}(t)$. Considering $RP_{j^*}(t)$ MT must decide each day t whether to sell MTasset j^* or to hold it another day: the first offered asset price $p_{j^*}(t)$ with $p_{j^*}(t) \geq RP_{j^*}(t)$ is accepted by MT and asset j^* is sold. If there was no signal by $RP_{j^*}(t)$ within a period trading must be executed at the last day T of the period, e.g. MT must sell asset j^* and invest asset j' ($j' \neq j^*$).

RANDOM (*Rand*). Buys and sells at randomly chosen prices $p_{j^*}(t)$ within the holding period.

BUY AND HOLD ($B+H$). Buys j^* at the first day t and sells at the last day T of each period.

MARKET (MA). Knows all prices $p_{j^*}(t)$ of a period in advance. Each holding period MA will buy the MTasset at the minimum possible price $p_{min} \geq m(i)$ and sell at the maximum possible price $p_{max} \leq M(i)$.

The performance of the investment policies is evaluated empirically.

3 Experimental Results

Simulations of the trading policies discussed in Section 2 are run for all six period types with number $n(p)$ from $\{52, 26, 13, 4, 2, 1\}$ and length h. Clearly the benchmark policy MA cannot be beaten. Simulations are run on Xetra DAX data for the interval 2007/01/01 to 2007/12/31 in oder to find out

(1) if MT shows a superior behaviour to buy-and-hold policies
(2) the influence of m and M on the performance of MT

Two types of $B+H$ are simulated. (MT_{B+H}) holds the MTasset within each period and (Index$_{B+H}$) the index over the whole time horizon. MT_{B+H} is synchronized with MT, i.e. buys the MTasset on the first day and sells it on the last day of each period. Index$_{B+H}$ is a common policy and often used as a benchmark. In addition the random policy *Rand* buys and sells the MTasset on randomly chosen days within a holding period.

We first concentrate on question (1) if MT shows a superior behaviour to MT_{B+H} and Index$_{B+H}$. Simulation runs with two different reservation prices are carried out, called A and R. For calculating both reservation prices estimates from the past are used, i.e. in case of a period length of h days m and M are taken from these h days which are preceding the actual day t^* of the reservation price calculation, i.e. $m = \min\{p(t)|t = t^* - 1, t^* - 2, \ldots, t^* - h\}$ and $M = \max\{p(t)|t = t^* - 1, t^* - 2, \ldots, t^* - h\}$. Table 1 displays trading results under transaction costs. For MA, MT and *Rand*) transaction costs are the same; all follow the holding period of MT. The MT policy for both reservation prices, R and A, dominates MT_{B+H} and Index$_{B+H}$ in two cases (1 and 4 weeks). MT_{B+H} dominates MT and Index$_{B+H}$ in two cases (6 and 12 months). Index$_{B+H}$ dominates MT and MT_{B+H}

Table 1. Annualized return rates for different period lengths

Historic R	Annualized Returns Including Transaction Costs					
Policy	1 Week $n(7) = 52$	2 Weeks $n(14) = 26$	4 Weeks $n(28) = 13$	3 Months $n(91) = 4$	6 Months $n(182) = 2$	12 Months $n(364) = 1$
MA	418.18%	138.40%	201.61%	47.93%	72.95%	61.95%
MT	41.08%	1.37%	54.86%	6.08%	32.39%	31.35%
MT_{B+H}	9.70%	0.50%	17.18%	15.80%	45.30%	35.29%
Index$_{B+H}$	20.78%	20.78%	20.78%	20.78%	20.78%	20.78%
$Rand$	-23.59%	-21.23%	17.18%	-18.23%	6.20%	15.42%
Historic A	Annualized Returns Including Transaction Costs					
Policy	1 Week $n(7) = 52$	2 Weeks $n(14) = 26$	4 Weeks $n(28) = 13$	3 Months $n(91) = 4$	6 Months $n(182) = 2$	12 Months $n(364) = 1$
MA	437.14%	164.44%	201.61%	50.27%	75.27%	61.94%
MT	31.52%	13.37%	57.02%	2.09%	45.28%	34.50%
MT_{B+H}	7.45%	11.53%	17.18%	15.80%	45.29%	35.28%
Index$_{B+H}$	20.78%	20.78%	20.78%	20.78%	20.78%	20.78%
$Rand$	-1.49%	-12.97%	5.36%	-20.80%	24.37%	12.64%

in two cases (2 weeks and 3 months). MT generates the best overall annual return rate when applied to 4 weeks. In case R MT_{B+H} generates the worst overall annual return rate when applied to 2 weeks, in case A when applied to 1 week. MT_{B+H} improves its performance in comparison to Index$_{B+H}$ and MT proportional to period length h. The longer the period the better the relative performance of MT_{B+H}. MT outperforms Index$_{B+H}$ in two-thirds and MT_{B+H} in one-thirds of the cases. If period length $h \leq 4$ MT outperforms MT_{B+H} in all cases and if $h > 4$ MT_{B+H} outperforms MT in all cases. Index$_{B+H}$ outperforms MT_{B+H} in half the cases. If we consider the average performance we have 27.86% for MT, 20.78% for Index$_{B+H}$, and 20.63% for MT_{B+H} in case R and 30.63% for MT, 20.78% for Index$_{B+H}$, and 22.09% for MT_{B+H} in case A. MT is best on average. On average MT shows a superior behaviour to $B+H$ policies under the assumption that m and M are based on historical data.

In general we assume that the better the estimates of m and M the better the performance of MT. Results in Table 1 show that the longer the periods the worse the relative performance of MT. This might be due to the fact that for longer periods historical m and M are worse estimates in comparison to those for shorter periods. To analyze the influence of estimates of m and M simulations are run with the observed m and M of the actual periods, i.e. we have optimal estimates. Results shown in Table 2 have to be considered in comparison to the results for historic estimates in Table 1. Now we can answer question (2) discussing the influence of m and M on the performance of MT. In all cases the returns of policy MT improve significantly when estimates

Table 2. Annualized returns for optimal historic estimates

Clairvoryant R	Annualized Returns Including Transaction Costs					
Policy	1 Week $n(7) = 52$	2 Weeks $n(14) = 26$	4 Weeks $n(28) = 13$	3 Months $n(91) = 4$	6 Months $n(182) = 2$	12 Months $n(364) = 1$
MA	418.18%	315.81%	280.94%	183.43%	86.07%	70.94%
MT	102.60%	87.90%	76.10%	81.38%	55.11%	54.75%
MT_{B+H}	9.70%	-4.40%	22.31%	19.79%	45.30%	35.29%
Index$_{B+H}$	20.78%	20.78%	20.78%	20.78%	20.78%	20.78%
Rand	-23.59%	-101.3%	-10.67%	47.37%	46.08%	15.42%

Clairvoryant A	Annualized Returns Including Transaction Costs					
Policy	1 Week $n(7) = 52$	2 Weeks $n(14) = 26$	4 Weeks $n(28) = 13$	3 Months $n(91) = 4$	6 Months $n(182) = 2$	12 Months $n(364) = 1$
MA	437.14%	317.87%	271.57%	153.68%	66.33%	76.14%
MT	119.77%	98.11%	85.65%	63.61%	46.55%	62.65%
MT_{B+H}	6.21%	-4.40%	27.16%	19.79%	45.30%	35.29%
Index$_{B+H}$	20.78%	20.78%	20.78%	20.78%	20.78%	20.78%
Rand	-34.04%	-24.39%	-19.67%	52.93%	26.01%	37.18%

of m and M are improved. For all period lengths MT is always better than MT_{B+H} and Index$_{B+H}$. The estimates of m and M are obviously of major importance for the performance of MT.

4 Conclusions

To answer the questions from section 3 24 simulation runs were performed. In the clairvoyant test set MT outperforms $B+H$ in all cases even under transaction costs. Tests on historical estimates of m and M show that MT outperforms $B+H$ in one-thirds of the cases and also on average. We conclude that if the period length is small enough MT outperforms $B+H$. It is obvious that the better the estimates of m and M the better the performance of MT. Results show that the shorter the periods, the better the estimates by historical data. As a result, the performance of MT gets worse the longer the periods become. It turned out that the shorter the periods the less achieves MT in comparison to MA. A MT trading policy which is applied to short periods leads to small intervals for estimating historical m and M. In these cases there is a tendency to buy too late (early) in increasing (decreasing) markets and to sell too late (early) in decreasing (increasing) markets due to unknown overall trend directions, e.g. weekly volatility leads to wrong selling decisions during an upward trend.

The paper leaves some open questions for future research. One is that of better forecasts of future upper and lower bounds of asset prices to improve the performance of MT. The suitable period length for estimating m and M is an important factor to provide a good trading signal. Simulations with other period lengths for estimating m and M could be of interest. Moreover, the data set of one year is very small. Future research should consider intervals of 5, 10, and 15 years.

References

1. Chavarnakul, T., Enke, D.: Intelligent technical analysis based equivolume charting for stock trading using neural networks. Expert Systems and Applications 34, 1004–1017 (2008)
2. El-Yaniv, R., Fiat, A., Karp, R., Turpin, G.: Competitive analysis of financial games. IEEE Symposium on Foundations of Computer Science, 327–333 (1992)
3. Ratner, M., Leal, R.P.C.: Tests of technical trading strategies in the emerging equity markets of Latin America and Asia. Journal of Banking and Finance 23, 1887–1905 (1999)
4. Ronggang, Y., Stone P.: Performance Analysis of a Counter-intuitive Automated Stock Trading Strategy. In: Proceedings of the 5th International Conference on Electronic Commerce, pp. 40–46. ACM International Conference Proceeding Series, 50 (2003)
5. Shen, P.: Market-Timing Strategies that Worked. Working Paper RWP 02-01, Federal Reserve Bank of Kansas City, Research Division (May 2002)
6. Silaghi, G.C., Robu, V.: An Agent Policy for Automated Stock Market Trading Combining Price and Order Book Information. In: ICSC Congress on Computational Intelligence Methods and Applications, pp. 4-7. (2005)

Part II

Health Care and Environment

A New Model and Tabu Search Approach for Planning the Emergency Service Stations

Ayfer Başar, Bülent Çatay, and Tonguç Ünlüyurt

Sabanci University, Orhanlı, Tuzla, 34956, Istanbul
ayferbasar@su.sabanciuniv.edu, {catay,tonguc}@sabanciuniv.edu

Summary. The location planning of emergency service stations is crucial, especially in the populated cities with heavy traffic conditions such as Istanbul. In this paper, we propose a Backup Double Covering Model (BDCM), a variant of the well-known Maximal Covering Location Problem, that requires two types of services to plan the emergency service stations. The objective of the model is to maximize the total population serviced using two distinct emergency service stations in different time limits where the total number of stations is limited. We propose a Tabu Search (TS) approach to solve the problem. We conduct an extensive experimental study on randomly generated data set with different parameters to demonstrate the effectiveness of the proposed algorithm. Finally, we apply our TS approach for planning the emergency service stations in Istanbul.

1 Introduction

The location planning of emergency medical service (EMS) stations is crucial, since an effective planning of these stations directly affects human life protection. In the last 30 years, a lot of research effort has been spent in the literature to plan the locations of both fire brigade and EMS stations. [1] and [7] provide a good review of these studies. In this paper, we propose a Backup Double Covering Model (BDCM), a variant of the well-known Maximal Covering Location Problem, that requires two types of services. The proposed Backup Double Covering Model (BDCM) is conceptually similar to Maximal Covering Location Model in [3], Double Coverage Model in [5], and Backup Coverage Model in [8]. Metaheuristic approaches have been successfully employed for solving such models, e.g. [5] proposed a Tabu Search (TS) algorithm to plan the EMS stations in Montreal and [4]compared the performance of Ant Colony Optimization to that of TS in Austria. In this paper,

B. Fleischmann et al.(Eds.), *Operations Research Proceedings 2008*,

DOI: 10.1007/978-3-642-00142-0_7, © Springer-Verlag Berlin Heidelberg 2009

we propose a TS approach and test its performance on both randomly generated data and data gathered for Istanbul.

2 Backup Double Coverage Model

For location planning of EMS stations, we propose BDCM where two types of service requests are fulfilled. Our aim in having a double covering model is to provide a backup station in case no ambulance is available in the closer station. In the proposed model, the objective is to maximize the total population serviced within t_1 and t_2 minutes ($t_1 < t_2$) using two distinct emergency service stations where the total number of stations is limited. If a region is covered by any emergency service stations, we assume that the whole population in this region is covered. BDCM originally proposed by [2] is as follows: M: set of demand regions, N: set of location sites, K: Maximum number of EMS stations to be opened and P_j: Population of region j.

$$a_{ij} = \begin{cases} 1, \text{ if station in location } i \text{ can reach region } j \text{ in } t_1 \text{ time units} \\ 0, \text{ otherwise} \end{cases}$$

$$b_{ij} = \begin{cases} 1, \text{ if station in location } i \text{ can reach region } j \text{ in } t_2 \text{ time units} \\ 0, \text{ otherwise} \end{cases}$$

Decision variables:

$$x_i = \begin{cases} 1, \text{ if a station is opened in location } i \\ 0, \text{ otherwise} \end{cases}$$

$$y_j = \begin{cases} 1, \text{ region } j \text{ is double covered} \\ 0, \text{ otherwise} \end{cases}$$

$$\max \sum_{j \in M} P_j y_j \tag{1}$$

subject to

$$\sum_{i \in N} x_i \leq K, \tag{2}$$

$$\sum_{i \in N} a_{ij} x_i - y_j \geq 0, \ \forall j \in M \tag{3}$$

$$\sum_{i \in N} b_{ij} x_i - 2y_j \geq 0, \ \forall j \in M \tag{4}$$

$$x_i \in \{0,1\}, \ \forall i \in N, y_j \in \{0,1\}, \ \forall j \in M \tag{5}$$

The objective of the model is to maximize the population which is double covered with a backup station. Constraint 2 imposes the total number of stations that can be opened. Constraints 3 ensure that any demand point must be covered in t_1 minutes in order to be covered multiple times. Constraints 4 ensure that y_j takes the value 1 if location j is double covered by two distinct stations. Constraints 5 show that all the decision variables are binary.

3 Tabu Search Approach

TS is a local search technique that was originally developed by [6]. Using an initial feasible solution TS investigates the neighbors of the existing solution in each iteration in an attempt to improve the best solution obtained so far by trying to escape local optima. Thus, new candidate solutions are generated by using different neighborhood search methods. In order to avoid the repetition of the same solutions, TS forbids a given number of moves by keeping these moves in a tabu list. The moves in the tabu list are not accepted unless they provide solutions better than a pre-determined aspiration level.

In our TS approach, three initialization methods are utilized for comparison. A random method, where we randomly select K stations among potential locations; a steepest-ascent method, where essentially pairs of stations are opened that gives the maximum additional double coverage per station; and a Linear Programming (LP) relaxation method, where the relaxation of the model is solved and integer x_i's in addition to maximum fractional x_i are fixed at 1 and the resulting model is solved until the maximum number of stations are opened.

The outline of the TS algorithm is as follows. First an initial solution is obtained using one of the methods described above. Then we find the station pair whose closing and opening provides the largest objective function value. We decided to use two separate tabu lists, one of which for the station opened and the other for the closed one. If they are not in the tabu list, we do the exchange and update objective function value if necessary. If at least one move is in tabu list, the moves are executed if the aspiration criteria is satisfied. Otherwise, we repeat the above steps. To avoid cycling, we replace the station to be closed with the station resulting in the least decrease in the current objective function if the current objective function value remains same during the last k_1 iterations. If the best-so-far objective function value does not improve during the last k_2 iterations, we perform random diversification by randomly closing and opening a station. The diversification mechanism improves the solution quality significantly. This procedure is repeated for k_3 iterations.

4 Experimental Study

After making experiments on randomly generated data, we decided to use $k_1 = 5$, $k_2 = 15$ and $k_3 = 5000$. The tabu list size is chosen as 7 and aspiration level of 100% of the best solution is used.

A set of problem instances with different number of potential stations and demand points are generated to test the efficiency of the proposed TS. The algorithms are coded in C++ and executed on 1.7 GHz Intel Celeron with 512 MB RAM. Our data set includes problems with different number of potential stations and demand points are generated 200, 300, 400, and 500 demand regions. The demand regions are distributed uniformly within a square area. The total number of potential sites is set equal to 100%, 75%, and 50% of the number of demand points. For each demand point-location site configuration we have generated 5 problem instances. Thus a total of 60 problem instances were generated. The average speed of the ambulances is assumed to be 40 km/h and Euclidean metric is assumed as the distance measure. Using these data a_{ij} and b_{ij} values are obtained. The populations of the demand regions were generated from an exponential distribution with mean 1000. The values of t_1 and t_2 are set equal to 5 minutes and 8 minutes, respectively, as determined by the Directorate of Instant Relief and Rescue (DIRR).

The results are compared with respect to different initialization mechanisms as well as against solutions obtained by OPL Studio 5.5 with CPLEX 11.0 (will be referred as CPLEX). First, we investigate the performance of the initialization heuristics benchmarked against the solution obtained using CPLEX. While the random heuristic gives a gap of 54.89% on the average, steepest-ascent and LP-relaxation heuristics' performances are similar: 6.95% and 7.18%, respectively. The gap is calculated as (CPLEX solution/Initialization heuristic solution)-1.

Next we investigate the performance of TS approach. In Table 1, we report the average results of all 60 problem instances. In these experiments, CPLEX time limit is set to 600 seconds for problems with less than 300 potential locations and 1200 seconds for others. TS1, TS2, and TS3, respectively, refer to the TS with the random, steepest-ascent, and LP-relaxation initialization approaches, respectively. As seen in Table 1, all three TS approaches provide good results in comparison with

Table 1. Results for random instances

		CPLEX	TS1		TS2		TS3	
Regions	Potential locations	Time (s)	% Gap	Time (s)	% Gap	Time (s)	% Gap	Time (s)
200	200	31	0.35	90	0.26	93	0.20	141
200	150	9	0.10	71	0.00	75	0.04	123
200	100	6	0.05	45	0.00	49	0.05	84
300	300	558	0.20	298	0.17	299	0.03	392
300	225	18	0.04	232	0.23	227	0.21	308
300	150	11	0.69	157	0.82	152	0.51	223
400	400	1200	0.33	646	0.14	584	0.33	874
400	300	257	0.35	509	0.66	469	0.12	710
400	200	54	0.35	306	0.25	325	0.34	499
500	500	1200	0.00	1182	0.09	1254	0.12	1618
500	375	872	0.65	1136	0.50	1001	0.44	1416
500	250	162	0.69	686	0.47	597	0.45	960
Average		365	0.32	447	0.30	427	0.24	612

the solutions found by CPLEX whose average computation time is 365 seconds.

5 Planning The Locations of EMS Stations in Istanbul

Since Istanbul is a large and populated city, we agreed on a quarter-wise analysis with the DIRR. This corresponds to a total of 710 quarters, 243 in the Asian side and 467 in European side. We forecasted the population for each quarter based on the data provided by Turkish Statistical Institute (TÜIK). Reachability data (a_{ij}, b_{ij}) for the quarters were collected by the help of the experienced ambulance drivers of the DIRR. We assume that each quarter is a potential station site. Furthermore, the response across European and Asian sides is not allowed. The number of stations is determined as 35 by the DIRR. CPLEX solved this problem in 50 seconds. This rather short solution time is possibly due to the fact that Istanbul data have certain characteristics different than the random data. The computational results for Istanbul are shown in Table 2.

6 Conclusion

In this study, we present a mathematical model to plan the locations of EMS stations. Since this problem is intractable for large-scale cases, we propose a TS solution approach. We test the performance of the

Table 2. Results for Istanbul

	CPLEX	TS1	TS2	TS 3
% Coverage	74.75	73.77	74.63	74.60
Time (s)	50	182	166	193
% Gap	-	1.30	0.16	0.20

TS with different initialization methods on randomly generated data as well as the data we collected for Istanbul. The results show that our TS approach with either initialization method provide good results compared to the solutions obtained using CPLEX. Further research on this topic may focus on the multi-objective modelling of the problem by considering the investment and operating costs of the stations and ambulances. Another interesting extension would be the multi-period version of the problem, where there is a maximum number of additional stations that can be opened at every period.

References

1. L. Brotcorne, G. Laporte, and F. Sement. Ambulance location and relocation models. European Journal of Operational Research, 147:451-463, 2003.
2. B. Çatay and A. Başar and T. Ünlüyurt. İstanbul'da Acil Yardım İstasyonları ve Araçlarının Planlanması. İBB Projem İstanbul Projesi Sonuç Raporu, 2007. (in Turkish).
3. R.L. Church and C.S. ReVelle. The maximal covering location problem. Papers of the Regional Science Association, 32:101-118, 1974.
4. K.F. Doerner, W.J. Gutjahr, R.F. Hartl, M. Karall, and M. Reimann. Heuristic solution of an extended double-coverage ambulance location problem for austria. Central European Journal of Operational Research, 13:325-340, 2005.
5. M. Gendreau, G. Laporte, and F. Semet. Solving an ambulance location model by tabu search. Location Science, 5:75-88, 1997.
6. F. Glover. Heuristic for integer programming using surrogate constraints. Decision Sciences, 8:156-166, 1977.
7. J.B. Goldberg. Operations research models for the deployment of emergency services vehicles. EMS Management Journal, 1:20-39, 2004.
8. K. Hogan and C.S. ReVelle. Concepts and applications of backup coverage. Management Science, 34:1434-1444, 1986.

Integration of Carbon Efficiency into Corporate Decision-Making

Britta Engel, Grit Walther, and Thomas S. Spengler

Technische Universität Braunschweig, Institut für Produktion und Logistik, Katharinenstr. 3, 38106 Braunschweig, Germany
{b.engel,g.walther,t.spengler}@tu-bs.de

Summary. Concerning increasing carbon emissions and resulting climate change discussions, methods to assess corporate carbon emissions gain importance in research and practice. A transparent and integrated method for decision support to control ratios like eco-efficiency does not exist. Against this background, a concept for the implementation of carbon emission accounting indicators into decision-making processes has been developed, which consists of different aggregation levels and their connections. Within this contribution, a model for the level of internal planning is presented and applied to assess carbon reduction provisions of a dyeing company.

1 Introduction

In research and practice, instruments to assess a company's performance in terms of economic and ecological aspects are gaining importance. These instruments could be classified in external performance measurement and internal planning models. On the level of external performance measurement, ratios for emission accounting (und thus reduction) gain importance. However, these ratios are highly aggregated and do not deliver any information necessary for improvement measures. In order to improve the ratio over time, it is not sufficient that the ratio is calculated and published once a year. Instead, the ratio has to be implemented into decision making processes. On the level of decision making processes, indicators are treated isolated (e.g Life Cycle Assessment - LCA), which leads to a lack of acceptance. Decision makers don't know how to consider ecological aspects. Since 2005, companies of some branches trade with emission certificates. But studies have shown, that decision relevant emission reduction costs are widely unknown [1].

B. Fleischmann et al.(Eds.), *Operations Research Proceedings 2008*,

DOI: 10.1007/978-3-642-00142-0_8,

The two groups of instruments show clearly, that there is a loophole between external performance measurement and internal planning models. A hierarchical model has been developed to overcome this loophole as presented within [2]. The assessment of companies, either conducted by non-governmental organizations or the company itself is based on external and historical data (level of external performance measurement). On the medium level (level of internal performance measurement), organizational units like sites are assessed in terms of economic and ecological aspects. Since the ratio is disaggregated in the different dimensions, e.g. value added and carbon emissions, improvement potentials can be analyzed. So far, external as well as internal evaluations and benchmarking are based on ex-post data of sites and companies. Thus, the performance is only known subsequently, and improvements are no longer possible. Therefore, it is the aim at the bottom level (level of internal planning models) to plan and implement efficient alternatives ex-ante. Therefore, it is the aim of this paper to shape the level of internal planning models. A planning model, which considers both economic and ecological aspects, is presented. Finally, the model is applied to the case study of a textile supply chain.

2 Internal Planning Model

The task on the level of internal planning models is to identify relevant emission reduction options and to assess these options in terms of economic and ecological aspects. If an integrated consideration of ecological aspects is carried out in literature, ecological aspects are as of now integrated in four ways: One group of methods expresses ecological aspects in monetary terms by the current or expected price for emission certificates. An examples can be found in [3]. However, emission prices are very volatile, and many companies are taking part in emission trading. The second group applies weights to every objective (for an example see [4]), aiming at a balance between conflicting objectives. However, defining these weights is a very subjective procedure. In a third group, ecological aspects are regarded by constraints e.g. limits for maximal emissions [5]. This form of integration cannot be acceptable within the meaning of sustainable development, since (reduction) targets are not given in form of legal regulations for all relevant emissions and limit values solely set by companies are as subjective as weights. Finally, a new perception can be found in literature recently. Following the question of how much we have to spend in order to improve the ecological quality, trade-offs and efficiency frontiers are calculated [6].

To develop a more specific model, details about the emission reduction alternatives need to be known. Emissions can be reduced directly inside the company or indirectly along the supply chain or within company networks. In the following, we are focusing on direct emission reduction alternatives only. Direct carbon emissions reductions can be classified in input-orientated measures, technology-orientated measures, application of end-of-pipe technologies, reduction of the production volume or qualitative modification of the product portfolio [7]. Since energy consumption represents roughly 85 percent of total greenhouse-gas-emissions, we are focusing in the following on these input-orientated measures.

To assess carbon emission reduction options in the field of energy consumption, variable costs of the fuels have to be considered as well as investments, if new installations need to be done. The integrated assessment results in two objective functions: Minimization of emissions E and total costs C. Each fuel source has a specific emission factor (e_i), heat value (h_i) and specific costs (c_i). The total emissions are calculated by multiplication of the emissions factors and the amount used of each fuel (x_i). Costs resulting from an investment are depreciations (D_i), interest (I_i) and operating costs (O_i). However, cheaper fuels usually have a higher emission rate than the more expensive fuels. In the mathematical formulation of the emission-reduction-assessment (ERA)-model, which is strongly simplified, we impose the assumptions that a company can choose between different fuels i and that the fuels can be combined.

$$\min \sum_{i=1}^{n} x_i * c_i + (D_i + I_i + O_i) * y_i \qquad \min \sum_{i=1}^{n} e_i * x_i \tag{1}$$

s.t.

$$\sum_{i=1}^{n} h_i * x_i \geq T_i \tag{2}$$

$$x_i \leq M * y_i \qquad \forall i \tag{3}$$

$$y_i \leq x_i \qquad \forall i \tag{4}$$

$$x_i \geq 0, \qquad y_i \in \{0,1\} \qquad \forall i \tag{5}$$

The total energy demand (T_i) needs to be fulfilled. Constraint (3) guarantees, that a fuel can only be used, if the corresponding investment is done (y_i is a binary decision variabe, M a very big number). If emissions are minimized and costs are disregarded, it needs to be guaranteed, like constituted within constraint (4), that there is only an investment if the corresponding fuel is used.

The solution of the model presented above can be conducted with one of the approaches described above. Since subjective weighting, monetization and diverse treatment of ecological and economic aspects should be avoided here, the idea of visual exploration of the efficient frontier and trade-offs is selected. As a result of this calculation, the decision-maker knows all efficient solutions, i.e. solutions for which there are no improvements in one objective possible without associated deterioration in the other objective. Thus, the decision-maker can now freely decide based on his/her specific trade-offs.

3 Case Study

A dyeing-company in India is chosen as a case study since a huge amount of thermal energy is needed to dye textiles. The chemicals need to be heated to high temperatures in order to penetrate into the textiles.
The company has an heat energy demand of 1.339.200 kWh/a. Currently, only diesel is used for heating. Doing so, 329 t/a of carbon emissions are produced. The annual costs for heating amount to 68.200 EUR/a. Diesel can be substituted by natural gas. After technical retooling, it is possible to substitute diesel by LPG (other tanks are needed). Furthermore, a combined heat and power unit (CHPU) can be installed, which produces heat and electricity. The CHPU can be operated with natural gas or diesel. The investment for a CHPU is much higher than for the combustion alternatives, but the demand of electricity can be fulfilled as well. To enable a comparison between electricity producing CHPUs and only heat producing constructions, credits are given for every kWh electricity produced for both carbon emissions and costs [8]. Emission credits, for also producing electric energy, are given to the CHPU by multiplying its emissions with the proportion of heating energy in the usable energy production (heating and electricity). Credits for costs are considered by the market price per kWh electricity, which are subtracted from the specific costs of the fuel. The energy demand can be obtained by one of these fuels or linear combination of the alternatives. Parameters of the fuels are shown in table 1.
To calculate the efficient frontier, the two objective functions are optimized separately. The efficient points between these two optima are calculated by taking the values of emission as constraint for the cost function. In figure 1, the current situation and the linear approximation of the efficient frontier for the energy consumption of the dyer is shown. As can be seen, to reach the efficient frontier, the lowest total

Table 1. Parameters for calculation (Values for gas is given per m^3, credits for CHPU are already included)

Fuel	D_i [EUR/a]	I_i [EUR/a]	O_i [EUR/a]	c_i [EUR/l]	e_i [kg CO_2/l]	h_i [kWh/l]
Diesel	0	0	2.304	0,55	2,650	10,80
LPG	6.117	4.588	6.422	0,19	1,631	12,70
Natutral gas	2.778	2.084	1.878	0,23	2,041	11,06
CHPU (diesel)	17.907	13.430	18.802	0,50	1,292	10,80
CHPU (natural gas)	11.587	8.690	12.166	0,18	1,204	11,06

costs of 34.590 EUR/a can be achieved with the highest emission rate of 247 t CO_2/a. In that case, natural gas is used for heating. Contrary to that, the lowest emission level of 146 t CO_2/a goes along with 54.238 EUR/a. To reach that point, a CHPU is installed and natural gas is used. Between these two extreme points another efficient point exist, which can be achieved by usage of LPG. However, there is a trade-off between emissions and costs.

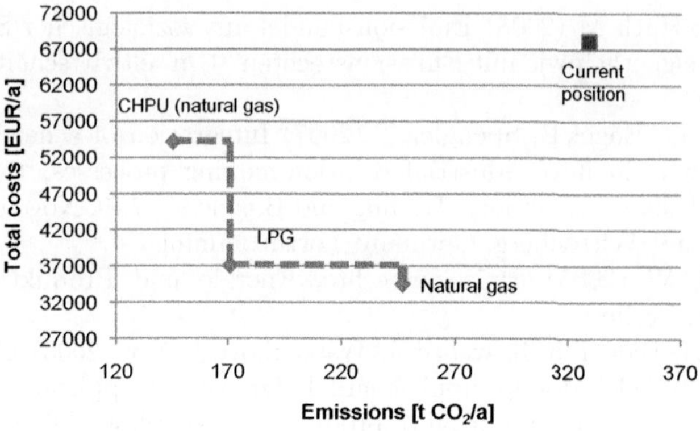

Fig. 1. Trade-offs between carbon emissions and annual costs

The calculated efficient frontier presented in figure 1 consists of points which are efficient in terms of Pareto efficiency. Within the efficient frontier the dashed line is not acceptable considering ecological aspects, because carbon emissions can be freely disposed. To reach the efficient frontier, the decision maker can now choose between the presented trade-offs. If he prefers lower emissions, he would choose a position on the upper left side of the frontier and use a CHPU with natural gas for the heating processes. If he prefers to reduce annual total costs, he would choose a point on the lower right side of the frontier and therefore use natural gas. A compromising solution would be the usage of LPG. The position, or definite point on the frontier chosen, depends on the actual situation respectively the performance measurement results of the site assessment.

4 Conclusions

Economic and ecological aspects need to be considered on different decision-making levels. On the lower disaggregated level, ex-ante planning with simultaneous consideration of economic and ecological aspects was done applying a bi-objective model. Thereby, no weighting or monetization of the objectives was carried out, but the efficient frontier was calculated allowing the decision-maker to opt on trade-offs between economic and ecological results.

References

1. Knoll L, Huth M (2008) Emissionshandel aus soziologischer Sicht: Wer handelt eigentlich wie mit Emissionsrechten? Umweltwirtschaftsforum 2, 16:81–88
2. Walther G, Engel B, Spengler T (2007) Integration of a new emission-efficiency ratio into industrial decision-making processes, Proceedings des Workshops "Emissions Trading and Business", 7.-9. November 2007, Lutherstadt Wittenberg, Germany, Forthcomming
3. Fichtner W (2005) Emissionsrechte, Energie und Produktion. Erich Schmidt, Berlin
4. Krikke H, Bloemhof-Ruwaard J, Wassenhove L N v (2003) Concurrent product and closed-loop supply chain design with an application to refrigerators, International journal of production research 41, 16: 3689–3719
5. Mirzaesmaeeli H (2007) A Multi-Period Optimization Model for Energy Planning with CO_2 Emission Consideration, University of Waterloo Ontario, Canada
6. Quariguasi J F N, Walther G, Bloemhof-Ruwaard J M, Numen v J A E E, Spengler T (2007) A methodology for assessing eco-efficiency in logistics networks, European Journal of Operational Research. Forthcoming
7. Betz R, Rogge K, Schleich J (2005) Flexible Instrumente zum Klimaschutz, Emissionsrechtehandel, Umweltministerium Baden-Württemberg
8. Fritsche U, Matthes F, Rausch L, Witt J, Stahl H, Jenseit W, Hochfeld C (1996) GEMIS 4.01 Gesamt-Emissions-Modell integrierter Systeme, Darmstadt

Production Planning Under Economic and Ecologic Objectives - A Comparison in Case Studies from Metals Production

Magnus Fröhling, Frank Schwaderer, and Otto Rentz

Institute for Industrial Production (IIP), Universität Karlsruhe (TH), Hertzstraße 16, D-76187 Karlsruhe,
{magnus.froehling,frank.schwaderer,otto.rentz}@wiwi.uni-karlsruhe.de

1 Introduction

Competition forces industrial companies to continuously improve resource efficiency. In the field of recycling of by-products from metal industries coke and coal are the most important primary resources. Hence, improvements can be achieved either by reducing the necessary coke and coal input or by maximizing the production output. As the use of coke and coal is also responsible for the major part of their CO_2 emissions, the enhancement of the resource efficiency also leads to a reduction of these. It is the aim of this contribution to further analyze the relationship between economic and environmental objectives such as the maximization of the contribution margin or the minimization of CO_2 emissions in operational production planning. [1] The analysis is carried out for two case studies. We use an approach based on a problem adequate modeling of the production processes enabling to model the input-output relationships between used resources and outputs (products and by-products). In the following section the general methodological approach is explained before the case studies are described. Finally conclusions are drawn.

[1] In this context reduction of emissions is possible by adapting the operation of the regarded industrial process. Potentials by changing the process itself or adding end-of-pipe techniques are not considered.

B. Fleischmann et al.(Eds.), *Operations Research Proceedings 2008*,

DOI: 10.1007/978-3-642-00142-0_9,

2 Methodological Approach

In the process industry in general and metallurgical production processes in particular a problem adequate process modeling is necessary for production planning approaches. It is important to sufficiently describe the relationship between the masses and compositions of the input materials and the masses and composition of the resulting (by-) products. For this purpose thermodynamic simulation models of the relevant production processes are developed with the flowsheet simulation system Aspen Plus (cf. e.g. [1]). The modeling bases on collected process data. This data is used to calibrate and validate the models. By sensitivity analyses systematic and independent data series for the output flows are calculated. Multiple linear regression analyses (cf. e.g. [3]) are carried out to determine production functions which describe the dependencies between the feed and the output flows. These form the core of the planning models. Thus, specific features of the underlying processes are regarded within economic planning approaches.

3 Analyses for the Production of Pig Iron from Ferrous Wastes

In the iron and steel industry large amounts of ferrous wastes, e.g. dusts and sludges, accrue which, according to European law, have to be utilized if economically feasible. The first case study[2] focuses on the utilization of ferrous wastes with low zinc contents in a blast furnace process. The plant produces pig iron and a by-product which is characterized by high zinc concetrations. This zinc contentrate is sold to the zinc industry. Further revenues are generated by the use of the blast furnace flue gas for electricity generation. Inputs of the process are especially coke and fluxes for which costs arise. Further costs arise from adsorbents which are used in the gas cleaning facilities and the treatment of accruing slag and dusts. The input and output flows depend on the blending of the ferrous wastes (cf. figure 1).
The economic objective function of the blending problem maximizes the contribution margin which is calculated from the stated cost and revenues (cf. [4]). For the calculation of the CO_2 emissions the following assumptions are made: CO_2 emissions originate from the sinter strand where the carbon content in the feed is nearly totally converted to CO_2. With the exception of an average amount of carbon left in the pig iron (3.75%) the coke used in the blast furnace is also converted to CO_2.

[2] DK Recycling und Roheisen GmbH, Duisburg

The objective in this case is the minimization of the resulting emissions under the condition that a certain amount of residues is processed. The model calculates the composition of the input feed i.e. the amount of each raw material. Thereby residues with different zinc and iron contents can be used. Restrictions comprise the input-output functions, technical requirements, minimal or maximal amounts of residues and the number of usable raw materials.

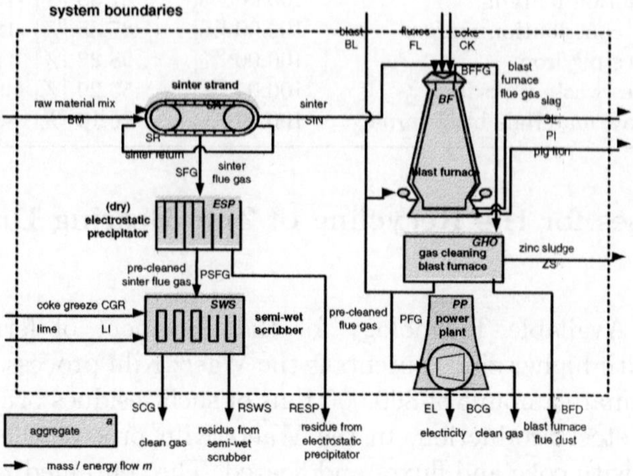

Fig. 1. Relevant aggregates and mass and energy flows in case study 1

The CO_2 emission reduction objective leads to a reduction potential of approx. 3% in comparison to the economic optimisation. Due to the changes in the feed approx. 1.8% less of pig iron is produced. Thus, the specific CO_2 emissions per ton of pig iron are solely reduced by 1.5% (cf. table 1). Nevertheless, this has significant influence on the cost and revenues. The contribution margin is decreased by 12.6% due to lower revenues for utilized wastes which are not compensated by reduced costs for coke. Though having a significant negative impact on the economic results, the possible improvement concerning the reduction of CO_2 emissions is limited. Hence, due to the dominating influence of the costs of coke the CO_2 emissions of the economic optimized process operation are comparable to the minimization of CO_2 emissions.

Table 1. Selected Results

	Contribution margin maximisation	CO_2 emission minimisation	Deviation
Raw materials	1350 [t]	1350 [t]	0.00 [%]
Pig iron	646.53 [t]	635.05 [t]	-1.78 [%]
CO_2	1615.85 [t]	1563.03 [t]	-3.27 [%]
CO_2 specific	2.5 [t/t_iron]	2.46 [t/t_iron]	-1.52 [%]
Contribution margin	100.00 [%]	87.35 [%]	-12.65 [%]
Specific contribution margin	100.00 [%]	87.35 [%]	-12.65 [%]
Revenues pig iron	100.00 [%]	98.22 [%]	-1.78 [%]
Revenues waste utilisation	100.00 [%]	50.29 [%]	-49.71 [%]
Costs raw materials blast furnace	100.00 [%]	96.19 [%]	-3.81 [%]

4 Analyses for the Recycling of Zinc Bearing Dusts and Sludges

The Best Available Technology for the treatment of ferrous metal residues with higher zinc contents is the Waelz Kiln process. In Europe an approximate amount of 860,000 tons of such residues originate from secondary steel production. In the Waelz Kiln process these residues are mixed with coke and fluxes and heated. The contained zinc oxide is reduced, vaporised and re-oxidised in the kiln atmosphere. It leaves the kiln with the process gas and is separated by bag filters as Waelz oxide which then can be further processed by leaching. As a by-product a so called Waelz slag accrues.
The reference company of our second case study[3] operates four such recycling plants. The company has to allocate the residues from different sources to the plants. The determination of this allocation has to regard transport and production planning aspects in an integrated way (cf. figure 2). Costs accrue on the one hand for the transportation of the residues to the Waelz plants. On the other hand, depending on the allocation of the residues and therefore on the chemical composition of the kiln feed, specific amounts of coke and fluxes are required which are both connected with costs. The quantity and quality of the produced Waelz oxide and the costs for leaching, the revenues for the selling of the Waelz oxide as well as the mass of slag, causing utilization fees, depend also on the feed. On base of the approach described in section 2 an optimization model for the described problem is developed.

[3] BEFESA Steel Services GmbH, Duisburg

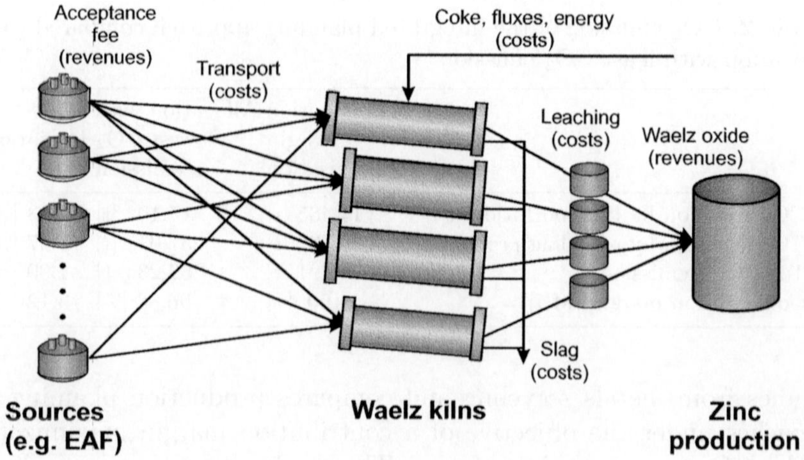

Fig. 2. Illustration of the planning problem

In the following we compare the results of the economic optimization (contribution margin) to a minimization of CO_2 emissions. The latter case regards the reduction potential which can be achieved by reducing transports and saving coke within the Waelz process. It is assumed that the total amount of residues has to be recycled. The model considers capacity limits as well as technical constraints.
To calculate the CO_2 emissions it is assumed that 90% of the used coke is transformed to CO_2 as some coke remains unreacted in the slag and the Waelz oxide. We assume transportation by truck and estimate the transport CO_2 emissions on base of the average CO_2 emissions per 40 t truck in Germany of the year 2000 (cf. [2]).
Table 2 shows the CO_2 emissions and contribution margin of the economic optimization results compared to the results with the lowest CO_2 emissions. In the latter case the additional emissions caused by transport are overcompensated by the reduction of used coke. The CO_2 reduction potential beyond the economic approach is therefore only about 1%. Thus, aiming at the most economic solution leads to an ecological (related to CO_2) only little improvable allocation and operation of the processes.

5 Conclusions

In the metals recycling industries the resource consumption especially of coke and coal is a dominating cost driver and responsible for the major part of the CO_2 emissions. This contribution analyses two case

Table 2. CO_2 emission of the integrated planning approach compared to the allocation with least CO_2 emission

	Contribution margin maximisation	Allocation with lowest CO_2 emission	Deviation
CO_2 emission by transportation [t]	14,485 [t]	15,199 [t]	4.93 [%]
CO_2 emission by recycling process [t]	240,327 [t]	237,035 [t]	-1.37 [%]
Total CO_2 emission	254,791 [t]	252,234 [t]	-1.00 [%]
Contribution margin [MU]	100 [%]	96,88 [%]	-3.12 [%]

studies from metals recycling and compares production planning approaches under the objective of a contribution margin maximization and a CO_2 emission minimization. The results show that an exclusive consideration of the emission reduction target leads to only small emission reduction potentials compared to the economic optimization. At the same time economic results can be siginificantly worse. Additional measures like emission trading certificates will therefore have very little impact on operational production planning.

References

1. Fröhling M (2006) Zur taktisch-operativen Planung stoffstrombasierter Produktionssysteme - dargestellt an Beispielen aus der stoffumwandelnden Industrie. Deutscher Universitäts-Verlag: Wiesbaden.
2. German Federal Environmental Agency (2008) ProBas data base, URL: http://www.probas.umweltbundesamt.de, Dessau July 2008.
3. Hartung J, Elpelt B, Klösener KH (1998) Statistik: Lehr- und Handbuch der angewandten Statistik. Oldenbourg: Munich;
4. Rentz O, Fröhling M, Nebel F, Schultmann F, Engels B (2006): Integrierter Umweltschutz in der Metallerzeugung: Simulationsgestützte operative Produktionsplanung zur Optimierung metallurgischer Abfallverwertungsprozesse. Universitätsverlag Karlsruhe: Karlsruhe

Production Planning with Common Parts in Closed-Loop Supply Chains

Jenny Steinborn, Grit Walther, and Thomas S. Spengler

Institut für Produktion und Logistik, Technische Universität Braunschweig, Katharinenstr. 3, 38106 Braunschweig
{j.steinborn,g.walther,t.spengler}@tu-bs.de

Summary. Common parts in different product variants can lead to decreased procurement costs if recovery strategies are applied. Thereby new parts are substituted by recovered ones. Since long- or mid-term supplier contracts for these new parts exist demand as well as return forecasts are essentially to negotiate contracts. In this article forecasting issues are addressed and a planning procedure to calculate the need of new parts is given.

1 Introduction

The costs of material are a considerable part of the costs for the production of a product. Hence, the procurement of material has a determining impact on the efficiency of production and sales processes in the supply chain (SC). By implementing product and component recovery management in terms of enhancing the SC to a closed-loop supply chain (CLSC) exactly these costs can be decreased. Existing attempts show advantages of such strategies for the spare part management. Moreover, the substitution of new components by recovered ones gains importance, particularly for products with short life cycles and many product variants including common parts [6].
However, in order to capture arising economic potentials by sale of recovered products or by substitution of new components, the recovery strategies have to be implemented into forward production planning processes at the OEM as well as in procurement processes along the SC. Thereby, not only demand but also return of common parts has to be considered ahead of negotiations of frame contracts with suppliers. Otherwise, a surplus of common parts exists and the potential of returned parts cannot be exploited.

B. Fleischmann et al.(Eds.), *Operations Research Proceedings 2008*,

DOI: 10.1007/978-3-642-00142-0_10,

Accordingly, forecasts are needed not only for demand but also for return of common parts. Subsequently, a planning procedure calculating the need of new parts is to be implemented comparing the forecasted demand and return. Since the quality of a return determines the effort and cost of the recovery, not all returned products and parts can be recovered under economic viewpoint. Hence, the quality of returns need to be considered as well while determining the need for new parts. Thus, uncertainties have to be taken into account with regard to demand and return quantities, but also with regard to the quality of the returned products and parts. This becomes even more complicated if product recovery takes place, i.e. if different states of product demand have also be taken into account. Thus, complexity of planning procedures increases tremendously.
Against this background, a planning procedure to determine the need of new parts is developed in the following. Product variant forecasts are addressed first to provide a basis for common part forecasts. Thereupon different possibilities to include the quality of returns into the planning procedure are regarded.

2 Forecasts and Component Substitution

At tactical level demand and return forecasts are inevitably. Existing forecast approaches in CLSC mainly concern the amount and time of product returns. Here, the availability of data which leads to forecasts with entire or incomplete information (e.g. [2], [4]) or forecasts with general and product-individual data (e.g.[5]) is discussed. Since returned amounts depend on sales of the past periods life cycle data like time of sale and time of utilisation are included in forecasts (e.g. [4]). An enhancement of such product forecasts to determine the need of new parts occurs rarely (e.g. [1], [3]).
Nevertheless, existing approaches do not consider the quality of products or components in return as well as demand. Therefore, the integration of quality aspects in demand and return forecasts is discussed in the following.
A demand forecast for the different product variants and its components is already done in SC when setting up supplier contracts. At this, through bill explosion the demand of common parts is determined. However, common parts can not only be used in different variants of new products, but also in recovered products. Thereby recovered products can differ in quality (e.g. as-new, used, repaired) depending on the market segment where they are sold. Thus, demand forecasts for

common parts must be enhanced taking different product variants of new products into account as well as different qualities of recovered products.
In the following we assume that an exponential demand function $F_i^D(t)$ for every new and recovered product variant i with parameters α_i^D, β_i^D and γ_i^D can be prepared.

$$F_i^D(t) = \gamma_i^D e^{\left(-\frac{t-\alpha_i^D}{\beta_i^D}\right)^2} \tag{1}$$

A comparable function can be developed for product return. However, while the parameters of the demand functions are determined through time series analysis of past variants, the parameters α_i^R, β_i^R and γ_i^R of return functions depend on former sales of the same variant and thus on the demand function. Hence $F_i^R(t)$ can be derived from $F_i^D(t)$ due to changes in parameters.
Parameter α describes the movement on the x-axis. Since returns arrive after the demand was satisfied $\alpha_i^D < \alpha_i^R$ must be fulfilled. The difference in α_i^D and α_i^R determines the expected value of time of product usage. A stretching or compression along the x-axis caused by parameter β can be neglected if an average time of usage is assumed ($\beta_i^D = \beta_i^R$). Parameter γ forms the stretching or compression along the y-axis. Since not all sold products return, a compression of the curve occurs ($\gamma_i^D \geq \gamma_i^R$).
An aggregated confrontation of the demand and the return gives a first clue for the new part need and the potential of substitution. Functions of demand $F_i^D(t)$ and return $F_i^R(t)$ over time t for the different variants i are determined first. If the amount ν_{ip} of common part p is included in product variant i, the demand function $F_p^D(t)$ for common part p can be determined as follows:

$$F_p^D(t) = \sum_i \nu_{ip} F_i^D(t) \tag{2}$$

Analogously the return of the common part $F_p^{R-max}(t)$ arises by the aggregation of all returns of product variants:

$$F_p^{R-max}(t) = \sum_i \nu_{ip} F_i^R(t) \tag{3}$$

The minimal amount of a part $purchase_p^{min}$ to be constituted in supply contracts in order to cover a time span (a, b) is thus given by the following integral:

$$purchase_p^{min} = \int_a^b F_p^D(t) - F_p^{R-max}(t)\,dt \tag{4}$$

The maximum amount to be constituted $purchase_p^{max}$ is automatically given by the demand function without consideration of returns.

$$purchase_p^{max} = \int_a^b F_p^D(t)\,dt \tag{5}$$

The difference between $purchase_p^{min}$ and $purchase_p^{max}$ thus constitutes the potential of substitution of new common parts by recovered ones. The time of utilisation determines the quality of a component to a large extent. Utilisation time of components depends on the utilisation time of product variant each having individual average times of usage. However, components can be used in various new product variants as well as in recovered product variants. Hence, for the determination of the need of new parts an estimation of the number of returning components with their remaining time of utilisation is needed for every common part. Such a forecast is very costly and, besides, it assumes the storage or possibility of the inquiry of data of utilisation for each single common part. Therefore, data on the fraction of reusable common parts in returned products is estimated out of past data on returns. Thus an average value r_p is determined. Thereby a return function $F_p^{R-prog1}(t)$ is derived and the amount of new parts to be considered in supply contracts ($purchase_p^{prog1}$) is calculated depending on the average fraction of reusable parts.

$$F_p^{R-prog1}(t) = r_p \sum_i \nu_{ip} F_i^R(t) \tag{6}$$

$$purchase_p^{prog1} = \int_a^b F_p^D(t) - F_p^{R-prog1}(t)\,dt \tag{7}$$

Nevertheless, the fraction of reusable common parts may not be steady over time. If a common part is introduced, i.e. procured for the first time, products contain only new components of this common part. Thus, the condition of returned components of this common part will be good for nearly every returning component. Consequently all components can be used for recovery. After a certain time, some components have been used (recovered) several times but there are also newly produced components used. Hence, a lower quality level will result taking into account the recovery, demand and the life span of the component. However, by the end of the production of a certain common part the average quality of the return will decrease, because new parts are no longer produced. Instead of the average value r_p a function $r_p(t)$ can thus be used to consider quality aspects. Thereby, a return function $F_p^{R-prog2}(t)$ results for the determination of the amount ($purchase_p^{prog2}$) to be negotiated in supply contracts:

$$F_p^{R-prog2}(t) = r_p(t) \sum_i \nu_{ip} F_i^R(t) \tag{8}$$

$$purchase_p^{prog2} = \int_a^b F_p^D(t) - F_p^{R-prog2}(t)\, dt \tag{9}$$

Due to the quality considerations presented above the potential described formerly through $purchase_p^{min}$ and $purchase_p^{max}$ decreases. The coherences explained above are shown in figure 1.

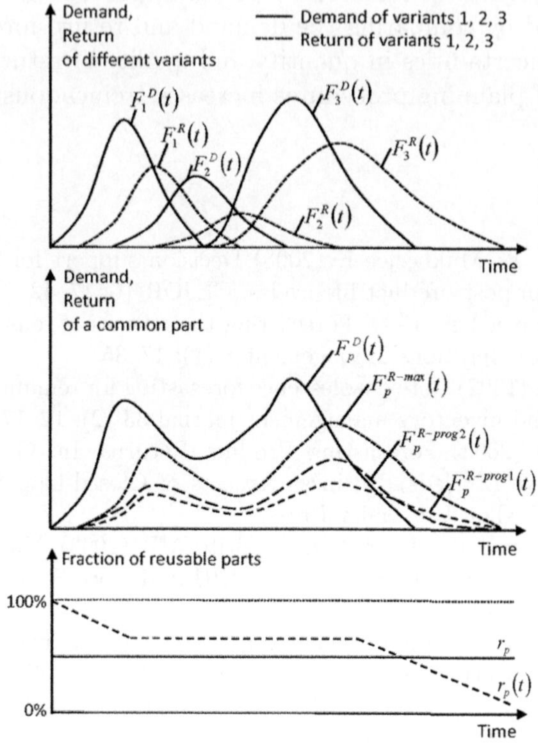

Fig. 1. Influence of return quality on component substitution

The need of new parts to constitute in contracts can be determined through the procedure explained above. A variance of this constituted amount can result from uncertainties in amount, time and quality. These variances can lead to a surplus in new parts (storage and or material recycling of reusable common parts) or to shortfalls. Shortfalls result in a non-satisfaction of the demand and are to be avoided, if possible, by acquisition from the spot market. Nevertheless, high costs emerge in such a case. Thus, decisions have to be taken regarding the trade-off between economic potential and substitution of new parts by recovered ones and economic risks resulting from uncertainties in quality and quantity of return.

3 Conclusion

In this article questions concerning forecast to determine future demand and return are addressed. A planning procedure calculating the new part need by comparing the demand and return forecast is given. Because of uncertainties in quantity and quality of returned products, complexity of planning procedures increases tremendously.

References

1. Inderfurth K, Mukherjee K (2008) Decision support for spare parts acquisition in post product life cycles. CEJOR 16: 17–42
2. Kelle P, Silver EA (1989) Forecasting the returns of reusable containers. Journal of Operations Management 8 (1): 17–35
3. Krupp JA (1992) Core obsolescence forecasting for remanufacturing. Production and inventory management journal 33 (2): 12–17
4. Toktay B (2003) Forecasting Product Returns. In: Guide VDR, Van Wassenhove LN (eds) Business Aspects of Closed-loop Supply Chains. Carnegie Mellon University Press
5. Toktay B, van der Laan EA, de Brito MP (2003) Managing Product Returns: The Role of Forecasting. ERIM Report Series Reference No. ERS-2003-023-LIS
6. Walther G, Steinborn J, Spengler T (2008) Variantenvielfalt und Lebenszyklusbetrachtungen im Remanufacturing. To be published in: Tagungsband des Workshops der GOR-Arbeitsgruppen Entscheidungstheorie und -praxis und OR im Umweltschutz, Shaker

Part III

Production and Service Management

Production Planning in Continuous Process Industries: Theoretical and Optimization Issues

Krystsina Bakhrankova

Molde University College, Postboks 2110, 6402 Molde, Norway
krystsina.bakhrankova@himolde.no

1 Introduction and Research Purpose

A solid theoretical basis related to production planning for different process systems has been established in the literature. However, continuous flow process industries with non-discrete items remain least researched with respect to specific theoretical and optimization issues. The purpose of this work is to locate continuous non-discrete production within the theoretical frameworks and apply planning techniques on a concrete industrial example. Subsequently, two mathematical optimization models are developed and analyzed in conjunction with their practical implications and computational results.[1] The first formulation allows for a better utilization of the production capacity with respect to energy costs, whereas the second one maximizes the production output.

1.1 Theoretical Frameworks and Issues

As previously cited, continuous flow production systems are least researched with respect to specific theoretical and optimization issues [1]. This development is explained by a common contrasting of discrete and non-discrete industries with less consideration to the variety of the latter (e.g., oil refineries, chemicals, food, roofing, glass and fiberglass). This disbalance is a consequence of a univocal acceptance of the classic product-process matrix [2], which contributes to unawareness of its shortcomings and forecloses promising research directions. This is further exacerbated by the mounting body of works that hinder the industry type distinctions and concentrate on discrete manufacturing [3].

[1] None of the absolute values can be presented for confidentiality reasons

B. Fleischmann et al.(Eds.), *Operations Research Proceedings 2008*,

DOI: 10.1007/978-3-642-00142-0_11,

Concurrently, due to process complexities observed within this production type, the focus has rested with process control and automation. Moreover, planning solutions are hard to obtain and their feasibility set is limited. Therefore, lesser emphasis has been put on production planning in such environments.

2 Description of Production System

This paper considers a processing industry company in Europe. Using a set of specified materials, the company manufactures a range of commodity products that are distributed to its customers via one or two distribution levels. The researched manufacturing facility is a non-discrete flow shop, where every product follows a fixed process-specific routing via a number of machines and intermediary tanks while product transfer is conducted via a network of pipes (Figure 1).

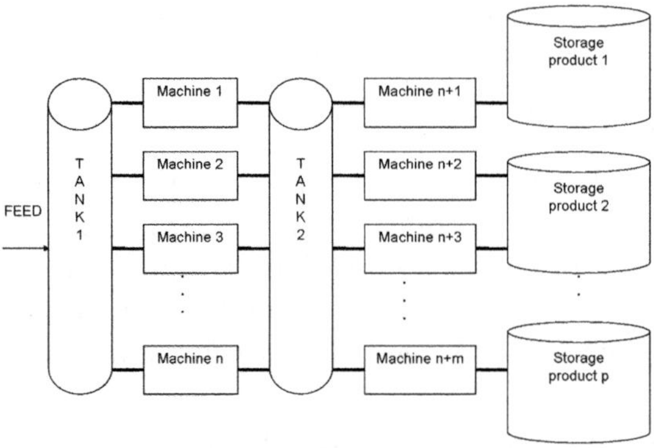

Fig. 1. Generalized scheme of the researched production system

All the products can be split in two main groups, depending on processing technology and quality. Yet, production machinery utilized at the plant can be divided in three groups: machines configured to make a subset of the first product group, machines tuned for a subset of the second product group and hybrid machines making a subset combination of both groups. The characteristics of the machinery contribute to uniqueness of the incumbent production system. For each of the machines and each of the products it is configured to manufacture, no stoppage, set-up or change-over is required in order to switch between

any products. This feature makes the production system completely continuous. Moreover, none of the produces are wasted as they can be reprocessed or mixed to meet the final product specifications.
As previously observed within the body of production planning literature and supported by methodical analysis of continuous process industry applications, the described production problem remains unexplored [1]. The classic production planning problems that come close to the researched environment are mainly employed for discrete manufacturing and concentrate on line balancing, line configuration and jobs sequencing [4]. Even within the continuous production type the focus rests with a myriad of lot sizing problems, where the majority of production settings require set-ups with related downtimes or constitute a batch process (e.g., [5] and [6], for details cf. [1]).

3 Problem Characteristics and Optimization Issues

Provided the system's characteristics, a single-stage multi-product multi-period mixed integer linear programming model was formulated. The model is focused on better capacity utilization at the final processing stage, minimizing the total energy costs (essential component of the total production costs) and synchronizing production and distribution planning [7]. In the following, a selection of constraints illustrating the system's peculiarities is presented and discussed.[2]

3.1 Discrete Machine Capacities

In the researched system, machine capacities are discrete - it is rarely possible to match a given planned production quantity with a specific machine production rate or a combination of machines' rates. Relative to the deterministic planned production quantities q_{pt} of product p in period t, over- P^+_{pt} or underproduction P^-_{pt} will occur most of the time (here, r_{mp} defines individual machine capacities for each product, while Z_{mpt} is a binary variable indicating machine-product assignment in period t).

$$\sum_{m \in M} r_{mp} Z_{mpt} - q_{pt} = P^+_{pt} - P^-_{pt}, \forall p \in P, \forall t \in T \qquad (1)$$

Thus, a distinct production rate is attached to the binary variable, where $\sum_{m \in M} r_{mp} Z_{mpt}$ is not continuous and, thus, cannot take any value as assumed in the classic production planning situations [8] - the

[2] Here M, P and T are respective sets of machines, products, and time periods

value area is a finite set of discrete points, not an infinite area of continuous values. Therefore, two additional constraints (2) and (3) and a binary indicator variable Y_{pt} are introduced to guarantee that over- and underproduction do not occur simultaneously. Here, μ and λ are empirically obtained control parameters, bounding respectively overproduction ($Y_{pt} = 1$) and underproduction ($Y_{pt} = 0$) and cutting the solution space to solve problems of reasonably large size very quickly:

$$P_{pt}^{+} \leq \mu Y_{pt}, \forall p \in P, \forall t \in T \tag{2}$$

$$P_{pt}^{-} \leq \lambda(1 - Y_{pt}), \forall p \in P, \forall t \in T \tag{3}$$

Within the horizon, the total underproduction relative to the planned production quantities is limited to κ_p values obtained by utilizing both empiric figures and post-optimization analysis (neither back-order nor a stock-out situation is acceptable as the customer service level requirement is 100%).

$$\sum_{t \in T} P_{pt}^{-} \leq \kappa_p \sum_{t \in T} q_{pt}, \forall p \in P \tag{4}$$

3.2 Quality Constraints

The final product quality is of the foremost importance in the researched production system. It must be ensured due to variability of raw materials' characteristics, consequent changes in product yields and potential off-grade production. For each product, the quality condition is satisfied by requesting the minimum production length of three consecutive periods on the same machine:

$$Z_{mp,t+1} + Z_{mp,t+2} \geq 2Z_{mpt} - Z_{mp,t-1} - Z_{mp,t-2}, \forall m \in M, \forall p \in P, \forall t \in T \tag{5}$$

3.3 Summary of Testing Results

The model formulation has been tested with the use of CPLEX 9.0.0 solver on the real data provided by the plant utilizing Dell OptiPlex GX260 computer with Intel Pentium 4 processor, where the number of analyzed products is within 20 and the number of machines is within 40. As previously discussed [7], the model reflects the essence of the production system and provides energy cost savings for the horizons of 2 and 3 weeks relative to the budgeted levels,[3] while rendering losses for the 1-week time span (Table 1).

[3] The absolute values are significant, though the relative figures are rather small

Table 1. Energy cost savings (% to the budget), minimization model

Horizon / Instance	1	2	3	4
1 week	**(1.06)**	(1.13)	(1.41)	(1.53)
2 weeks	2.64	2.65	**3.43**	2.45
3 weeks	**4.15**	1.72	3.49	2.27

As to the CPU times, for the 1 week horizon the range is 0.37-1.92 seconds, for 2 weeks - 6.70-232.42 seconds, and for 3 weeks - 231.00-578.40 seconds or 3.85-9.64 minutes (the best obtainable given a limit of 10 minutes). Thus, a specialized algorithm is needed to solve problems spanning longer time periods.

4 Capacity Maximization

The following model of the researched system maximizes production output regardless of the costs involved:

$$\max \sum_{m\in M}\sum_{p\in P}\sum_{t\in T} r_{mp}Z_{mpt} \tag{6}$$

subject to assignment, inventory balance, quality (5) and machine availability constraints as defined in [7]. The formulation is warranted by a practical need to estimate the total available production capacity due to machine maintenance and product demand variations. The same four data instances for 1 week horizon were tested (Table 2).

Table 2. Summary of the test results, maximization model

Measure	Instance 1	Instance 2	**Instance 3**	Instance 4
Cost savings (% vs. budget)	(1.79)	(4.12)	**0.74**	(3.06)
Total CPU time (seconds)	47.03	42.86	**22.17**	81.66

Here, Instance 3 does not only provide marginal energy cost savings, it also yields the second largest total output among the four alternatives short of only 0.49% vs. Instance 2, where the largest loss is incurred. The CPU times were reasonable within this horizon, yet they exceeded the span of 24 hours, when tested on the 2-week instances. Though the maximization formulation has fewer binary variables compared to the minimization model, it has a less tightly constrained solution space with no cuts. Therefore, to allow the solution of this problem for the planning horizons spanning more than one week, it is necessary to design cuts and, ultimately, construct a specialized algorithm for this problem.

5 Conclusions and Further Research

This paper located continuous non-discrete production within the theoretical frameworks and applied planning techniques on a concrete industrial example. In particular, it addressed specific theoretical and optimization issues pertinent to the researched production environment. Subsequently, a gist of two models - cost minimization and capacity maximization - was presented, illustrating the system's peculiarities and the models' practical use. The future research on the subject will focus on formulating mathematical models for multi-stage production systems and developing solution algorithms for the defined problems.

References

1. Bakhrankova K, Gribkovskaia I, Haugen KK (2007) Production planning in continuous process industries. In: Halldorsson A, Stefansson G (eds) Proceedings of the 19th Annual Conference for Nordic Researchers in Logistics. NOFOMA 2007, Reykjavik, Iceland, CD-ROM: 69–84
2. Hayes RH, Wheelwright SC (1979) Link manufacturing process and product life cycles. Harvard Business Review 57:133–140
3. Silver EA, Pyke DF, Peterson R (1998) Inventory management and production planning and scheduling. John Wiley & Sons, New York
4. Hax AC, Candea D (1984) Production and inventory management. Prentice-Hall, New Jersey
5. Schmenner RW (1993) Production and operations management. Macmillan Publishing Company, New York
6. Cooke DL, Rohleder TR (2006) Inventory evaluation and product slate management in large-scale continuos process industries. Journal of Operations Management 24:235–249
7. Bakhrankova K (2008) Planning, productivity and quality in continuous non-discrete production. In: Kujala J, Iskanius P (eds) Proceedings of the 13th International Conference on Productivity and Quality Research. ICPQR 2008, Oulu, Finland: 109–123
8. Nahmias S (2005) Production and operations analysis. McGraw-Hill Irwin, Boston

Cost-Oriented Models of Network Industries Price Regulation

Eleonora Fendekova[1] and Michal Fendek[2]

[1] Department of Business Economics, University of Economics Bratislava, Dolnozemská 1, 852 35 Bratislava, Slovakia, nfendek@dec.euba.sk

[2] Department of Operations Research and Econometrics, University of Economics Bratislava, Dolnozemská 1, 852 35 Bratislava, Slovakia, fendek@dec.euba.sk

Summary. The existence of pure monopoly in network industries increases the róle of regulation mechanisms in connection with objectification and increase in their social effectiveness. The objective of regulation mechanisms is to find an appropriate proportion between price and product supply of network industry under assumption of the existence competitive market. With regard to analysis of equilibrium in network industries models it is important to point out that except for competition policy protection the state fulfils another specific task - regulation of network industries.
The aim of the paper is to examine the equilibrium conditions in the market of network industries. The state influences proportional relations between price and supply of network industry production.
The conditions for equilibrium of network industries and methods of their regulations will be examined in the paper. The stress will be laid on the regulation on the base of cost - return over costs regulation

1 Introduction

This paper will deal with price regulation scheme on the basis of return over costs regulation. Return over costs is scheme of natural monopoly regulation, which is in principle different form the model of regulation on the basis of rate of return. It derives the barrier for the not exceedement of reasonable profit only from the part of the regulated entity's input activities, namely from the volume of investment. This undesirably motivated monopoly to disproportionate increase of capital investment, which was of course contra productive.
Return over costs regulation sets the maximum profit margin for regulated firm on the basis of its overall costs. We can see that there is

B. Fleischmann et al.(Eds.), *Operations Research Proceedings 2008*,
DOI: 10.1007/978-3-642-00142-0_12,

certain analogy between this form of regulation and regulation on the basis of the rate of return. However the difference is that return over costs regulation does not prefer particular cost group, but uses the overall costs.
The idea, that firm's profit is in this case some kind of function of its costs is of course mystification. This scheme, however, effectively hinders natural monopoly from asserting such combination of its supply and monopolistic market price, which would allow it to make inappropriate profit in comparison with exerted costs.

2 Cost-oriented Models of Network Industries Price Regulation

In short, the keystone of return over costs regulation is that regulator as the base for regulated entity's reasonable profit definition sets its overall costs and defines reasonable profit as a certain allowed percentage *RoC* of its costs. Analytically we can express this condition as

$$RoC \times n(q) \geq \pi(q)$$

or

$$RoC \times (w \times L + r \times K) \geq \pi(q)$$

where
q – production
$n(q)$ – function of the total costs of the firm, $n : R \to R$
$\pi(q)$ – profit function $\pi : R \to R$
L – labor (production factor)
K – capital (production factor)
w – labor price
r – capital price
RoC – reasonable profit margin set by the regulator corresponding to the unit costs.
Regulated output and regulated price in the return over costs environment are calculated by solving the following mathematical programming task

$$\pi(q) = \pi\left(f(K,L)\right) = p\left(f(K,L)\right) \times f(K,L) - w \times L - r \times K \to \max, \quad (1)$$

subject to

$$p\left(f(K,L)\right) \times f(K,L) - w \times L - r \times K - RoC \times (w \times L + r \times K) \leq 0, \quad (2)$$

$$K, L \in R_{\geq 0}. \quad (3)$$

Solution of the optimization task (1)–(3) is optimal consumption volume of production factors labor L^* and capital K^*, on the basis of which, with the help of the production function, the regulated optimal volume of output q^*_{RoC} is quantified

$$q^*_{RoC} = f(K^*, L^*)$$

And regulated optimal price p^*_{RoC} with the help of the price-demand and production function on the basis of the relation

$$p^*_{RoC} = p(q^*_{RoC}) = p(f(K^*, L^*))$$

While also in this regulation approach the rate of return on revenues defined by the parameter RoC is respected, i.e. exogenous control parameter set by the regulator.
Let us now transform the optimization task (1)–(3) with two variables L, K into a task with one variable q in a following way

$$\pi(q) = p(q) \times q - n(q) \to \max, \tag{4}$$

Subject to

$$p(q) \times q - n(q) - RoC \times n(q) \leq 0, \tag{5}$$

$$q \in R_{\geq 0}, \tag{6}$$

where
$t(q) = p \times q$ – function of revenues of the firm, $t : R \to R$
$n(q) = nv(q) + n_F$ – function of the total costs of the firm, $n : R \to R$
$nv(q)$ – function of the variable costs of the firm, $nv : R \to R$
n_F – fixed costs of the firm, $n_F \in R$
RoC – reasonable profit margin set by the regulator corresponding to the unit costs.
On the basis of the substitution we can reformulate the optimization task (4)–(6) to

$$\pi(q) = t(q) - n(q) \to \max \tag{7}$$

subject to

$$t(q) - n(q) - RoC \times n(q) \leq 0, \tag{8}$$

$$q \in R_{\geq 0}. \tag{9}$$

Task (7)–(9), or task (4)–(6) analytically describes the situation that is geometrically interpreted on the Fig. 1. Graphs of the basic functions describing production and cost attributes of the regulated firm and the attributes of the relevant market as well are taken from the return on output regulation model.

Revenues function in the situation on the Fig. 1. Reaches its maximum in the point q^*, which represents supply of the firm in the elastic demand zone. Firm is selling its goods for a relative high price p^* in the elastic demand zone, i.e. in the zone for positive values of the marginal revenues function $t(q)$.
Regulated firm has the tendency to set its decision making parameters in a way the limit set by the regulator would allow it to reach maximum profit. As we can seen on the Fig. 1, margin of the regulated profit is represented by the curve, which is a transformation of the total cost curve. We get it by multiplying the function values by regulation parameter RoC, i.e. by the regulatory allowed portion of the profit from the costs.
Let us note that RoC parameter is not in percentage but in relative expression. So it has the value $RoC \in (0,1)$. In the situation presented on the Fig. 1 the firm will have volume of the output q_{RoC} and with greater volume of the output by the lower price p_{RoC} it will reach lower profit.
If the regulator would decide to use more strict regulation conditions and enforce lower allowed rate of profit RoC_N , $(RoC > RoC_N)$ upon the regulated entity, the regulated entity would henceforth increase its output on the volume q_{RN} by decreasing production price p_{RN} and by gradual decrease of the profit. However price decreased and volume of output increase at the same time encourages the growth of the social welfare.

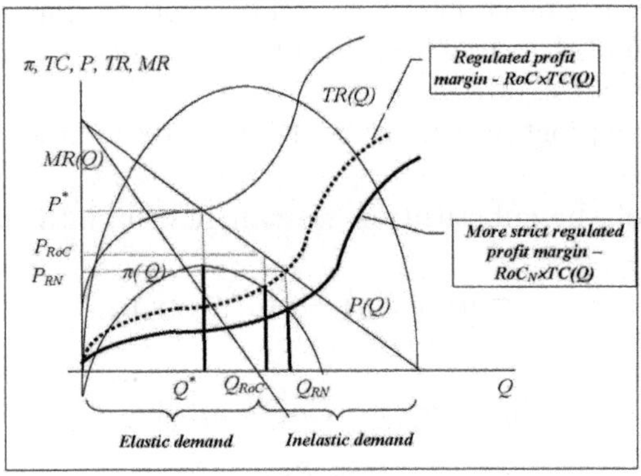

Fig. 1. Regulation on the return over costs basis – *RoC*

The comparison of the products market price and production costs of the firm also leads to interesting conclusion while using this type of the regulation. Let us explore the reasonable profit margin in return over costs regulation from this aspect again.

$$RoC \times TC(Q) \geq \pi(Q). \tag{10}$$

In this condition we express the total cost function of the firm analytically. We get reformulated condition expression (10):

$$RoC \times (nv(q) + n_F) \geq p(q) \times q - (nv(q) + n_F).$$

And after further modification

$$p(q) \times q \leq RoC \times (nv(q) + n_F) + (nv(q) + n_F),$$

$$p(q) \times q \leq (1 + RoC) \times (nv(q) + n_F). \tag{11}$$

After division of the equation by the supply volume $q > 0$ we get

$$p(q) \leq \frac{(1 + RoC) \times (nv(q) + n_F)}{q},$$

$$p(q) \leq (1 + RoC) \times \frac{(nv(q) + n_F)}{q},$$

$$p(q) \leq (1 + RoC) \times np(q), \tag{12}$$

where

$$np(q) = \frac{nv(q) + n_F}{q}, \qquad q > 0, \qquad \text{are total average costs of the firm.}$$

Relation (12) represents substantial feature of the return over costs regulation, which also explains already mentioned mystification about direct relation of the regulated reasonable profit margin and total costs of the natural monopoly.

3 Conclusion

On the basis of the relation (12) and from the geometrical interpretation of the optimization task solution (1)–(3) we can formulate the following important conclusions about firm behavior in the condition of the return over costs regulation:

1. In general, return over costs regulation constructs reasonable profit margin for the regulated entity on the basis of the proportional portion of its total expended costs. This proportional portion is defined by the *RoC* parameter. So primary it encourages the producer to produce greater volume of supply by the lower price, which is increasing social welfare.

2. The particular optimal position of the regulated firm is determined by the characteristics of the cost function, which is directly related to the character of the profit function on one side, because

$$\pi(q) = t(q) - n(q).$$

And on the other side by the characteristics of the price-demand function $p(q)$, which specifies elastic and inelastic demand zones.
3. From the relation (12) we can see that regulated firm can set its production parameters, production prices and consumption of production factors only in a manner for its production market price to be up to *RoC* percent greater than the average unit costs of production. We can see that ineffective cost increase of the firm, in accordance with regulatory relation of this method, would be albeit creating room for reasonable profit increase however the validity of the relation (12) needs to be ensured and such combination of supply production price found, that would ensure its consumption.

It is obvious that in elastic demand zone, i.e. by positive marginal revenues, regulated firm produces greater volume of output compared to nonregulated firm and tries not to waste the production factors.

References

1. Bien F (2008) Systemwechsel im Europäischen Kartellrecht. In: Der Betrieb N' 46/2000. Düsseldorf: Verlagsgruppe Handelsblatt
2. Fendek M, Fendeková E (2008) Mikroekonomická analźa. Bratislava: IURA Edition
3. Fendeková E (2006) Oligopoly a regulované monopoly. Bratislava: IURA Edition
4. (1993) Mehr Wettbewerb auf allen Märkten. Zehntes Hauptgutachten der Monopolkommission 1992/1993. Baden-Baden: Nomos Verlaggeselschaft
5. O'Sullivan A, Sheffrin S, Perez P (2006) Microeconomics: Principles, Applications, and Tools. New York: Prentice Hall
6. Pepall L, Richards DJ, Norman D (2004) Industrial Organization: Contemporary Theory and Practice (with Economic Applications). New York: South-Western College Publishing
7. Shy Oz (2001) The Economics of Network Industries. Cambridge: Cambridge University Press
8. Tirole J (1997) The Theory of Industrial Organisation. Massachusetts, Cambridge: The MIT Press
9. Train KE (1995) London: Optimal Regulation. The MIT Press

Production Chain Planning in the Automotive Industry

Claas Hemig and Jürgen Zimmermann

Clausthal University of Technology, Institute of Management and Economics, Operations Research Group, Julius-Albert-Str. 2, D-38678 Clausthal-Zellerfeld,
{claas.hemig,juergen.zimmermann}@tu-clausthal.de

Summary. We allocate a given production workload between L production lines producing P products with a subsequent buffer of limited capacity by modeling the problem as a classical transportation problem. The model is embedded in a Dynamic Programming approach that finds a cost-optimal solution exploiting the given flexibility of an automotive plant with respect to production capacity, production volume and staff. Preliminary computational results show that our approach solves real-world instances within some hours.

1 Introduction

In the last two decades the automotive market dramatically changed from a sellers to a buyers market and the automotive is no longer a mass product but a symbol of individuality. Hence, automotive Original Equipment Manufacturers (OEMs) installed flexible manufacturing systems to be able to produce and deliver individually ordered cars more flexibly and more quickly. Currently, OEMs are in a position to adjust production time and speed and to use flexible working hours, hirings and dismissals. Moreover, it is possible to shift production workload and workers between production facilities, typically production lines inside a plant. This variety of flexibility is limited to technical restrictions, labor legislation and in-plant agreements, concluded e.g. with the staff association or a labor union. An automotive plant typically consists of a body shop, a paint shop and a final assembly. In the body shop raw component parts are welded in a highly automated process to create a car body. After an elaborate painting in the paint shop, the production process is finished in the final assembly by installing the engine, gearbox, and interior decoration etc. Product specific buffers of limited capacity are located between subsequent shops to decouple

B. Fleischmann et al.(Eds.), *Operations Research Proceedings 2008*,
DOI: 10.1007/978-3-642-00142-0_13,

the corresponding production processes [4, 1]. The planning horizon of three to five years is divided into T periods, typically weeks or months. Our topic is to find a cost-optimal solution by determining the production time and speed, the number of workers to hire and dismiss, and the production volume for each line and period. In the following, we restrict ourselves to the paint shop and its dependency to the final assembly shop and the corresponding buffer.

2 The Basic Approach

For each period and line we determine a production time, e.g. eight hours per day, and a production speed as well as the production volume for each product incorporating the subsequent buffer capacities. Additionally, we determine the amount of workers to be hired or dismissed, distinguishing between permanent and temporary workers, and we decide whether to displace permanent staff from one production line to another. Associating the periods with the stages of a Dynamic Programming (DP) approach [2], we are able to find a cost-optimal solution for the outlined planning problem with L production lines and P different products. Later on, we extend this approach with a combined buffer for all products, i.e. with a buffer not specific to the products. The production time is a result of the selected so-called shift model which determines the number of shifts per day and their lengths. Additionally, each shift model is associated with some cost values, e.g. the premiums for overtime and night shift. The production speed is given by the platform configuration of the line which indicates the average percentage of platforms carrying a car body. Being aware of the production time and speed we are able to quantify the production capacity of the considered production line. Taking into account all lines, the overall capacity must be sufficient to produce the required demand for products given by the subsequent shop, the final assembly. The permanent workers are required for an output quality on a constantly high level, they are well-protected against dismissal, and earn higher wages than their temporary colleagues. Worker demand peaks can be met, first, by displacements of permanent workers from one line to another and, second, by additional temporary workers. As stated in agreements with the labor unions, the fraction of the temporary workers in the overall staff may not exceed a given value. Incorporating the contracted hours, the shift model at a line determines the amount of over- or undertime the workers have to work in each period. Accumulating these values over all periods elapsed, we get the running total of the so-called working time account which is bounded up- and downwards due to in-plant agreements. In order to avoid oversized data the working time account of a line is stored as the average of all workers currently assigned to that line. Summarizing, a state in our DP approach consists of the selected shift model and platform configuration, the current buffer stock for each product in the subsequent buffer, the number of temporary and permanent workers, and their average working time account. A decision in our DP approach contains the selected shift model and platform configuration, the number of hirings, dismissals and displacement of workers as well as the production volume for each line and product. Especially the buffer stock, the number of workers, and the working time account are responsible for the "curse of dimensionality" invoking a huge number of states and decisions during the DP approach [5]. Hence, we discretize the state variables mentioned above to reduce the number of decisions and states. In addition, we cease to enumerate all feasible decisions and restrict ourselves to the meaningful ones. For this, we focus on the subproblem of distributing the production workload between the lines in the paint shop in a certain period t assuming that the demand for each product is given from the production schedule of the final assembly. Incorporating all production lines, we enumerate over all combinations of shift models and platform configurations with enough production capacity. For each of those combinations we determine meaningful values for the production volume for each line, product, and shift in the considered period. The corresponding subproblem can be modeled as a classical Transportation Problem (TP). We use the solution of the TP to identify the decision concerning the production volume during the DP approach. The decision is completed with the help of a heuristic considering the hirings, dismissals and displacements of staff between the production lines. This approach provides a "cost-optimal" solution keeping in mind the disaggregation into subproblems and the heuristic for the staff decision. In the upcoming section we describe the subproblem of allocating the production workload in detail.

3 Allocating the Production Workload

The workload in the paint shop is determined by the production schedule in the final assembly and the buffer capacity located between the two shops. To provide a buffer-feasible production schedule we have to ensure feasibility for every single shift of the considered period because typically the buffer has a capacity of about one shift. We initially model the problem as a classical transshipment problem where lines supply and products ask for production capacity. The buffers are represented by transshipment nodes. In particular, we introduce a supply node L_{ls}

for each line $l \in \{1, \ldots, L\}$ in each shift $s \in \{1, \ldots, S\}$ with a supply equal to the corresponding production capacity available. Analogously, we introduce a demand node P_{ps} for each product $p \in \{1, \ldots, P\}$ in each shift s with a demand equal to the corresponding product demand arising from the production schedule of the final assembly. To model product specific buffers we introduce a transshipment node B_{ps} for each product p and shift s. The buffer capacities – reduced by the minimum safety stock – act as the maximal transshipment capacity of nodes B_{ps}, while the opening stock is modeled as an additional supply node B^-_{p0} for each product. To balance the transshipment problem, we introduce a dummy demand node C representing, roughly spoken, the excess capacity. More precisely, it represents the sum of, first, the difference between supply of production capacity and demand for products and, second, the difference between the buffer stocks previous to the first and after the last shift of the considered period. The corresponding graph is rather sparse because transportation is restricted to only a few connections as follows. Each supply node L_{ls} is adjacent to P_{ps} and B_{ps} if and only if product p can be produced on line l. Transportation over these arcs represents the production to fulfill the demand for products and filling the buffer, respectively. Additionally, L_{ls} is connected to C for all l and s to model unused capacity in the particular shift. The buffer stock $B_{p,s-1}$ can be, first, used to meet the demand for products in the next shift and, second, stored for future shifts. These two utilizations are represented by arcs from $B_{p,s-1}$ to P_{ps} and B_{ps}, respectively. Note that B_{pS} is only adjacent to C for all products. The amount of transportation over these arcs equals the closing stock of the appropriate product. All arcs are valued with appropriate production or storage cost values. Using the reduction in [3, p. 170] we can transform the given transshipment problem into a TP. For this we introduce a supply node B^-_{ps} and a demand node B^+_{ps} for each transshipment node B_{ps}. The supply and demand of the new nodes equals the maximal transshipment capacity of B_{ps}. Incoming arcs in B_{ps} are transformed into incoming arcs in B^+_{ps}, the analogy holds for outgoing arcs and B^-_{ps}. B^-_{ps} represents the buffer stock of product p at the end of shift s, B^+_{ps} at the beginning of shift s. Finally, we introduce an arc from B^-_{ps} to B^+_{ps} for each product p and shift s which models unused buffer capacity in the particular shift. Next, we extend the problem with a combined buffer for all products with a maximal capacity of $\overline{B}$. To clarify this transformation, we illustrate the resulting graph for two lines, two products, and two shifts in Figure 1. The additional restriction regarding the combined buffer means that for each shift s the inequality

$$\sum_{(i,j)\in X_s} x_{ij} \leq \overline{B} \qquad (1)$$

must hold, where X_s contains all arcs incident with B^+_{ps} except the arcs from B^-_{ps} for all products p.

By adding this restriction the problem looses the property of being a TP, but by adding $2S$ nodes appropriately we reobtain an equivalent TP. Hence, we extend X_s with the arcs incident with $B^+_{p\sigma}$ for all p and $1 \leq \sigma \leq s-1$ as well as $P_{p\sigma}$ for all p and $1 \leq \sigma \leq s$. The resulting set of arcs X'_s is represented by the solid ones in Figure 1(a) for $s=1$ and in Figure 1(b) for $s=2$.

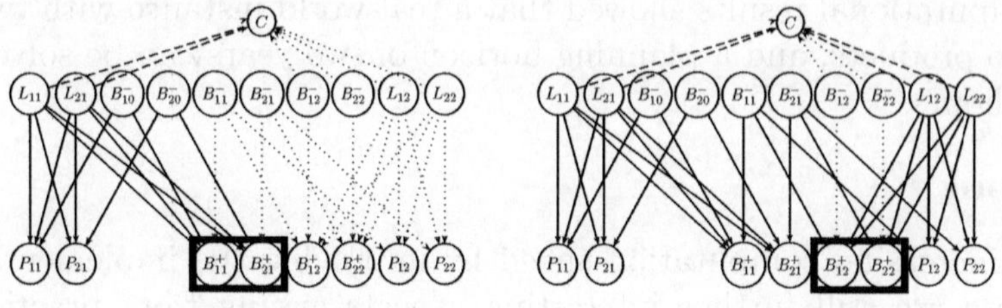

(a) $s=1$ (b) $s=2$

Fig. 1. The combined buffer for $L = P = S = 2$

At the same time we enlarge the right hand side of (1) by the demand of the nodes just mentioned, so that (1) is equivalent to

$$\sum_{(i,j)\in X'_s} x_{ij} \leq \overline{B} + \sum_{p=1}^{P} \left(\sum_{\sigma=1}^{s-1} (B^+_{p\sigma} + P_{p\sigma}) \right) + P_{ps}. \qquad (2)$$

It is easy to verify that the upper bound implied by (2) is equivalent to a lower bound on the sum of transportation units on the dashed arcs in Figures 1(a) and 1(b). This means that up to each shift some fraction of the line capacities may not be used, neither for fulfilling demand for products nor for increasing the buffer stock. The transformation described allows us to consider a set of arcs all incident with one node, namely C. Therefore we gain some upper bounds on partial sums. Additionally, the set of arcs used for shift s is a subset of the set used for shift $s+1$, so we can use the idea of Wagner [6]: We introduce an additional supply node C^-_s as well as a demand node C^+_s for each shift and and reobtain a TP. The demand and supply of the nodes added equal the maximal amount of the line capacities that can be left unused or that can be used to increase the buffer stock in all following shifts $s+1, \ldots, S$. The resulting graph is illustrated in Figure 2.

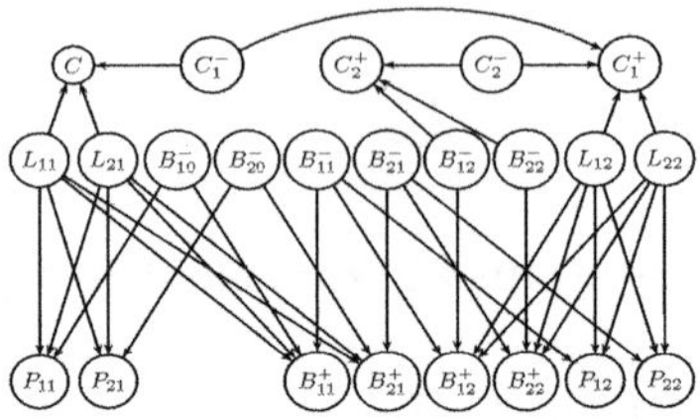

Fig. 2. The resulting graph for $L = P = S = 2$

Summarizing, we showed that we are able to model the arising subproblem and its extension as a classical TP which can be solved efficiently. The solution is integrated into our DP approach as described. Preliminary computational results showed that a real-world instance with two lines, two products, and a planning horizon of two years can be solved in some hours.

4 Outlook

Next to increasing computation speed by improving the implementation, there are still further interesting aspects arising from practice whose integration into our approach would provide an additional benefit: First, we would like to integrate several buffers where different products can be stored, e.g. one combined buffer for product 1 and 2, and another one for product 2 and 3. Second, it is possible to save an enormous amount of money by switching off a line l for a shift s which is equivalent to setting the amount of transportation from L_{ls} to C_s^+ equal to the supply of L_{ls}.

References

1. Askar, G.: Optimierte Flexibilitätsnutzung in Automobilwerken, Clausthal University of Technology, Dissertation, 2008
2. Bellman, R. E.: Dynamic Programming. New Jersey : Princeton, University Press, 1957
3. Lawler, E. L.: Combinatorial Optimization: Networks and Matroids. New York : Holt, Rinehart, and Winston, 1976
4. Meyr, H.: Supply chain planning in the German automotive industry. In: OR Spectrum 26 (2004), Nr. 4, S. 447–470
5. Powell, W. B.: Approximate Dynamic Programming: Solving the Curses of Dimensionality. John Wiley & Sons, 2007 (Wiley Series in Probability and Statistics)
6. Wagner, H. M.: On a Class of Capacitated Transportation Problems. In: Management Science 5 (1959), Nr. 3, S. 304–318

Two-Dimensional Guillotineable-Layout Cutting Problems with a Single Defect - An AND/OR-Graph Approach

Vera Neidlein[1], Andréa C.G. Vianna[2], Marcos N. Arenales[3], and Gerhard Wäscher[1]

[1] Dept. of Management Science, Faculty of Economics and Management, Otto-von-Guericke-University Magdeburg, Postbox 4120, 39106 Magdeburg, Germany,
{vera.neidlein,gerhard.waescher}@ww.uni-magdeburg.de

[2] Departamento de Computação, Faculdade de Ciências, Unesp, Av. Eng. Luiz Edmundo Cariijo Coube, 14-01, 17033-360 - Bauru - SP, Brazil,
vianna@fc.unesp.br

[3] Department of Applied Mathematics and Statistics, University of Sao Paulo, Campus Sao Carlos, Av. do trabalhador saocarlense, 400, 13560-970 - Sao Carlos - SP, Brazil,
arenales@icmc.usp.br

1 Introduction

In this paper, a specific cutting problem will be considered that has only received very limited attention in the literature so far, namely one in which the plate to be cut down contains a (rectangular) defective region. For the corresponding cutting problem without defects, Morabito et al. (cf. [3]) have introduced a solution method which is based on the AND/OR-graph approach. We suggest several modifications of this method which allow for dealing with a plate that contains a single defect, a problem type introduced by Carnieri et al. (cf. [1]).

2 Problem Definition

The problem to be discussed here is a variant of the so-called (unconstrained, guillotineable-layout) Two-Dimensional Rectangular Single Large Object Placement Problem (2D_UG_SLOPP; cf. [6]): Let a (weakly heterogeneous) set of small items be given which are grouped into relatively few classes (types) in which the items are of identical

B. Fleischmann et al.(Eds.), *Operations Research Proceedings 2008*,
DOI: 10.1007/978-3-642-00142-0_14,

size (l_i, w_i) and value v_i, $i = 1, \ldots, m$. The small items have to be laid out orthogonally on a single large object (stock plate) of given dimensions (L, W) such that the small items do not overlap and lie entirely within the large object. Any such layout is called a cutting pattern. The number of times each item type appears in a cutting pattern is not limited and can even be 0. The item types have a fixed orientation and the pattern must be guillotineable. A cutting pattern is to be provided which maximizes the total value of the small items in the pattern. This problem is known to be NP-hard as it is a generalization of the classic Single Knapsack Problem (cf. [2]). Unlike in the standard 2D_UG_SLOPP, we now assume that the large object contains a defective region which is represented by a rectangle (l_d, w_d) whose edges run in parallel to the edges of the large object and whose position on the large object is known.

3 An AND/OR-Graph Approach to the 2D_UG_SLOPP

3.1 Fundamentals

The solution approach suggested here is an extension of the AND/OR-graph approach introduced by Morabito et al. for the (standard) 2D_UG_SLOPP without defects (cf. [3]). In short, this approach consists of a representation of the solutions of the cutting problem by means of a specific graph (AND/OR-graph) and a Branch & Bound search. An AND/OR-graph is a directed graph $G = (V, E)$ with a set of nodes V and a set of directed arcs $E = \{e_1, \ldots, e_s\}$, where each arc e_u, $u = 1, \ldots, s$, has been assigned a set S of end nodes: $e_u = (j, S), j \in V, S \subseteq V$. In our case, S always consists of a pair of nodes. When following a path through the graph, one can choose between several arcs that emerge from a node (OR part), but one has to follow both branches of the chosen arc (AND part). This type of graph provides an appropriate tool for the representation of a cutting process, each node standing for a (stock or intermediate) plate, each arc $e_u = (j, \{p, q\})$ for a guillotine cut that separates a plate j into a pair $\{p, q\}$ of new plates. Throughout the process of realizing a cutting pattern by means of (a series of) guillotine cuts, "intermediate" rectangular plates will occur, which may have to be considered for being cut down further. The Branch & Bound search requires the determination of upper and lower bounds for the objective function value which can be generated by cutting down such plates. Given a plate $N = (L_N, W_N)$, a straightforward upper bound $UB(L_N, W_N)$ is provided as follows:

$$UB(L_N, W_N) = L_N \cdot W_N \cdot \max\left\{\frac{v_i}{l_i w_i} : i = 1, \dots, n\right\}.$$

The determination of a lower bound $LB(L_N, W_N)$ can be based on homogeneous cutting patterns which contain only small items of a single type:

$$LB(L_N, W_N) = \max\left\{v_i \cdot \left\lfloor\frac{L_N}{l_i}\right\rfloor \cdot \left\lfloor\frac{W_N}{w_i}\right\rfloor : i = 1, \dots, n\right\}$$

For a detailed description of the approach we refer to the original publication ([3]).

3.2 Modifications for a Defective Plate

The computation of the upper bound for a plate N containing a defect is based on the usable area of the plate, which is given by the size of plate N less the size of the defect. In order to compute a lower bound, N is partitioned (using guillotine cuts) into so-called (maximal) non-defective (rectangular) regions which fill the entire non-defective area of N. Fig. 1 depicts all such partitions which can be generated by guillotine cuts. It also demonstrates that all the partitions are made up of (at most) 16 different non-defective regions only. For each of the

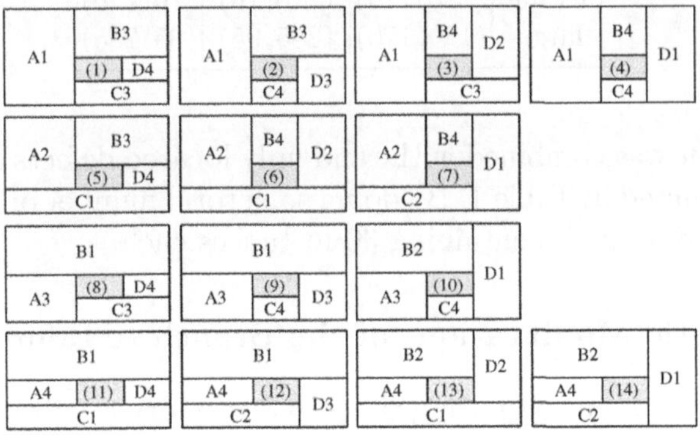

Fig. 1. Partitions of a defective plate into non-defective regions

non-defective regions, the best homogeneous cutting pattern is determined. Then, for each partition, the lower bounds for the respective non-defective regions are added up, resulting in a lower bound for each partition. Finally, the lower bound for N is computed as the maximum of the lower bounds of all 14 partitions. A similar, yet less elaborate approach has been presented by Vianna and Arenales (cf. [5]).

4 Implementation and Design of Numerical Experiments

4.1 Problem Classes and Generation of Problem Instances

In order to demonstrate the solution quality of the proposed method, extensive numerical experiments on randomly generated problem instances have been performed. Initially, 18 main problem classes were defined by combination of the following problem characteristics:

- dimensions of the large objects (in length units): $(600, 600)$, $(900, 400)$, $(1200, 300)$
- number of item types: 5, 10
- size of item types (in area units): $1 - 3600$ (small), $10800 - 18000$ (large), $1 - 18000$ (mixed).

By means of a problem generator specifically designed for the problem type under discussion (for details see [4]), for each main problem class, 30 realizations of item types were generated randomly which can be considered as the "core" data set of each main problem class. Then

Table 1. Dimensions of the defects

	quadratic	horizontal	vertical
small	$(60, 60)$	$(105, 35)$	$(35, 105)$
medium	$(120, 120)$	$(210, 70)$	$(70, 210)$
large	$(170, 170)$	$(285, 95)$	$(95, 285)$

each data set was combined with randomly located defects of 9 different sizes, as depicted in Table 1. By doing so, a total number of 162 problem classes was obtained, containing 30 instances each.

4.2 Heuristic Modifications of the Branch & Bound Search

In order to keep the computing times and memory requirements within reasonable limits, the Branch & Bound search has been modified heuristically in two ways (for details see Morabito et al. [3]): Only promising cuts are used, i.e. a significant improvement can be expected when looking at the bounds, and a heuristic search strategy is applied which combines a complete depth-first tree search down to a depth bound of 3 with a greedy heuristic that chooses the best branch computed so far.

4.3 Computer Hardware and Software

The algorithm has been coded using the Borland Pascal Compiler Version 7. The tests were performed on a microcomputer with a Pentium 4 processor with 1.7 GHz and 512 MB RAM.

5 Test Results

Across all 4860 problem instances, the average waste amounted to 5.0333% of the useable area of the large object (note that the optimal solutions are not known); the average computing time per instance was 10.25 seconds. Table 2 gives an overview of the results for the main problem classes, each consisting of 270 instances. These results show

Table 2. Waste and computation times for the 18 main problem classes

class no.	large object	no. item types	item type size	waste [%]	time [sec]
1	(600, 600)	5	small	1.472	9.63
2	(600, 600)	5	mixed	5.042	7.48
3	(600, 600)	5	large	9.117	3.88
4	(600, 600)	10	small	0.713	18.93
5	(600, 600)	10	mixed	3.661	17.92
6	(600, 600)	10	large	7.752	10.59
7	(900, 400)	5	small	1.600	9.49
8	(900, 400)	5	mixed	4.967	6.65
9	(900, 400)	5	large	10.588	3.84
10	(900, 400)	10	small	0.974	19.85
11	(900, 400)	10	mixed	4.045	14.33
12	(900, 400)	10	large	7.732	9.58
13	(1200, 300)	5	small	1.811	8.17
14	(1200, 300)	5	mixed	7.636	4.74
15	(1200, 300)	5	large	10.839	2.82
16	(1200, 300)	10	small	0.991	18.17
17	(1200, 300)	10	mixed	4.219	11.61
18	(1200, 300)	10	large	7.442	6.84

that a growing number of item types leads to an increase in waste as well as a decrease in computing time, due to the higher number of possible cutting patterns which have to be investigated. For the same reason, waste increases and computing time decreases for larger item types. The shape of the large object has only little influence on the percentage of waste, which increases slightly with a more rectangular shape. Looking at the detailed results for the different types of defects shows that a larger defect yields more waste and slightly less computing time, due to the smaller number of possible cutting patterns. The shape of the defect has almost no influence on computing times, but a strong one on the percentage of waste. A horizontal defect yields the largest percentage of waste, while a vertical defect yields the smallest (this is because a vertical defect divides a rectangular large object in two, leaving very little space above and below). It can be said that the size of the item types has the by far strongest influence on the solution speed and quality.

6 Outlook

To enable the algorithm to deal with larger problem instances, stricter bounds for the tree search and new heuristics are to be developed. As a next step, the algorithm can be extended to the staged or constrained 2D_UG_SLOPP with a defect, which requires new upper and lower bounds as well as an adaptation of the Branch & Bound procedure.

References

1. Carnieri C, Mendoza GA, Luppold WG (1993) Optimal Cutting of Dimension Parts from Lumber with a Defect: a Heuristic Solution Procedure. Forest Products Journal 43(9):66–72
2. Karp RM (1972) Reducibility Among Combinatorial Problems. In: Miller RE, Thatcher JW (eds) Complexity of Computer Computations. Plenum, New York:85-103
3. Morabito RN, Arenales MN, Arcaro VF (1992) An And-Or-Graph Approach for Two-Dimensional Cutting Problems. European Journal of Operational Research 58:263–271
4. Neidlein V, Wäscher G (2008) SLOPPGEN: A Problem Generator for the Two-Dimensional Rectangular Single Large Object Placement Problem With a Single Defect. Working Paper No. 15/2008, Faculty of Economics and Management, Otto-von-Guericke-University Magdeburg
5. Vianna ACG, Arenales MN (2006) O Problema de Corte de Placas Defeituosas. Pesquisa Operacional 26:185–202
6. Wäscher G, Haußner H, Schumann H (2007) An Improved Typology of Cutting and Packing Problems. European Journal of Operational Research 183:1109–1130

The Single Item Dynamic Lot Sizing Problem with Minimum Lot Size Restriction

Irena Okhrin[1] and Knut Richter[2]

[1] Juniorprofessur in Information and Operations Management, European University Viadrina, Grosse Scharrnstrasse 59, 15230 Frankfurt (Oder), Germany
irena.okhrin@euv-frankfurt-o.de

[2] Department of Industrial Management, European University Viadrina, Grosse Scharrnstrasse 59, 15230 Frankfurt (Oder), Germany
richter@euv-frankfurt-o.de

1 Introduction

In practise, production managers prefer to determine an optimal production plan by using minimum lot size restrictions instead of setup cost [3, 5]. Anderson and Cheah [1] also noticed, that in "lot sizing practice out-of-pocket setup cost are commonly accounted for by specifying a minimum batch size parameter". Therefore, the objective is to minimize the total inventory cost only with respect to the lot size restrictions, and not the sum of setup cost and inventory cost, as in mainstream models. In the paper we formulate the single item dynamic lot sizing problem with minimum lot size restriction and elaborate a dynamic programming algorithm for its solution. The preliminary computational results show that the algorithm is highly efficient and determines optimal solutions in negligible time.

2 Problem Formulation

Let us consider an uncapacitated single item dynamic lot sizing problem [2]. Instead of common set up cost in the original Wagner/Whitin paper, we introduce into the model a minimum lot size (MLS) restriction. Furthermore, we assume that production and inventory holding cost are constant and thus can be omitted in the objective function. The problem under consideration has the following view [6]:

B. Fleischmann et al.(Eds.), *Operations Research Proceedings 2008*,
DOI: 10.1007/978-3-642-00142-0_15, © Springer-Verlag Berlin Heidelberg 2009

$$\min \sum_{j=1}^{T} I_j \tag{1}$$

$$I_j = I_{j-1} + X_j - d_j, \tag{2}$$

$$Y_j L \leq X_j \leq Y_j d_{jT}, \tag{3}$$

$$Y_j \in \{0, 1\}, \tag{4}$$

$$I_j \geq 0, \quad I_0 = I_T = 0, \quad j = 1, \ldots, T. \tag{5}$$

In model (1)–(5) known parameters T, L and d_j denote the length of the planning horizon, the minimum lot size and the demand values in periods $j = 1, \ldots, T$, respectively. We assume that they are integers. The decision variables X_j, I_j and Y_j denote the production quantity in period j, the inventory level at the end of period j and the Boolean variable, which equals unity if production occurs in period j, and zero otherwise. The cumulative inventory, which represents inventory holding cost, is minimized by the objective function (1). The inventory balance equations are provided by (2). Restriction (3) models the fact, that the produced quantity in period j is either zero or at least L, where d_{jT} denotes the cumulative demand in periods from j to T and $d_{j+1,j} = 0$ for all j. Restriction (4) is obvious, and restriction (5) states that no negative inventories are allowed. Without loss of generality we assume that $I_0 = 0$. Moreover, in this paper we explore only the so-called *limited* version of the problem with $I_T = 0$. Nevertheless, all results derived for the limited problem can also be extended to the *unlimited* problem with $I_T > 0$. Because of space limitations, however, we restrict ourselves here to the examination of the limited problems only. The generalized zero-inventory property for the similar to (1)–(5) problem with minimum batch restriction was proven in [1]:

Theorem 1 (Anderson and Cheah, 1993). There exists an optimal solution in which

a) $I_{t-1}X_t(X_j - L) = 0$ for each j and t satisfying $1 \leq j < t \leq T$, where $X_j \geq L$ and $X_i = 0$ for each $j < i < t$.
b) If $X_j > L$, then $X_j = d_{jt} - I_{j-1}$ for some $j \leq t \leq T$.
c) $I_T < L$.

Statement a) of the theorem gives a generalized zero-inventory property according to which only second one of the two subsequent production values with positive inventories between them can be greater than the lower bound L. Similarly to the classical case, statement b) says that the sum of a production value, that is greater than the lower bound, and the inventory before that period will cover the cumulative demand for

some next periods. Finally, statement c) states that the final inventory is always lower than L and therefore is relevant only for unlimited problems.
We present an efficient dynamic algorithm for problem (1)–(5) which, in contrast to the solution given by Anderson and Cheah [1], does not regard inventory values I_j as states but uses a network presentation of the problem solution procedure. Main efforts are put on eliminating as many arcs as possible. To our knowledge, algorithms published so far for various LSP do not draw much attention to the structure of demand inputs. The current paper presents an attempt to reduce the complexity of the solution algorithm by considering special characteristics of demand values, such as the relation of cumulative demand to the lower bound and jumps in the demand in various periods.

3 Minimal Sub-Problems

A sub-problem SP_{it} is a part of the problem (1)–(5) on periods $i, \ldots, t$ with $I_{i-1} = I_t = 0$, where $1 \leq i$ and $t \leq T$. A sub-problem is *solvable* if $d_{it} \geq L$, while the production quantity should be at least L and there should be no final inventory. Formally, the parameter $t_i^- = \min\{j \geq i \mid d_{ij} \geq L\}$ provides the lower bound for the horizon of the sub-problem SP_{it}, because in case when $t < t_i^-$ the problem SP_{it} is unsolvable.
Definition. Sub-problem SP_{it} is *minimal* if there is no such period k, $i \leq k < t$ that $\hat{I}_{it} = \hat{I}_{ik} + \hat{I}_{k+1,t}$ and $\hat{I}_k = 0$, where $\hat{I}_{it}$ denotes the optimal cumulative inventory for periods from i to t.
In other words, whatever optimal solution of a minimal sub-problem is found, the inventories for all periods, except the last one, are positive. A sub-problem is not minimal if such a period k exists.
Corollary of Theorem 1. For a minimal sub-problem (a) at most one production value is greater than L; (b) this is the last production period for this sub-problem; (c) all other production quantities equal either 0 or L.
The concept of minimal sub-problems is very important for the development of a solution procedure, as for dynamic programming only such problems should be taken into consideration. Hence, leaving aside all non-minimal sub-problems, we reduce the complexity of the solution algorithm. To determine the upper bound for the horizon of a minimal sub-problem, we investigate further properties and introduce two critical periods. First, let us consider a period in which the cumulative demand equals a multiple of L. This period is the last one for the minimal sub-problem as its end-period inventory equals zero.

Definition. First period j^I, $i \leq j^I \leq t$ of a sub-problem SP_{it} is called the *critical period of the first type* if $d_{i,\,j^I} = m \cdot L$, where $m \in \mathbb{N}$.
Next, we consider the structure of the cumulative demand and pay special attention to big jumps. The integers $k_j = \left\lfloor \frac{d_{ij}}{L} \right\rfloor$, $i \leq j \leq t$, where $k_{i-1} = 0$, allow to determine the smallest number of minimal lots which suffice to satisfy the cumulative demand d_{ij}. The number k_t is the minimal number of lots which satisfy the total demand of the sub-problem, where all lots except the last one are of size L. If demand is satisfied in every period, then J denotes the last production period for SP_{it} and $d_{i,\,J-1} \leq (k_t - 1)L < d_{i,\,J}$ holds.
Definition. First period j^{II}, $i \leq j^{II} \leq t$ of a sub-problem SP_{it} is called the *critical period of the second type* if $k_{j^{II}} - k_{j^{II}-1} > 1$ holds. This period is critical while it is the first period when production value must be greater than L, since the cumulative demand $d_{i,\,j^{II}}$ cannot be satisfied by producing only minimal lots. According to Corollary, this should be the last production period of the minimal sub-problem. Next theorem provides the relationships between critical periods and the horizon of a minimal sub-problem.
Theorem 2 (critical periods). Let SP_{it} be a minimal sub-problem. Then

a) if j^I exists, then $j^I \geq J$ and $d_{j^I+1,\,t} < L$ holds;
b) if j^{II} exists, then $j^{II} = J$ and $d_{j^{II}+1,\,t} < L$ holds.

Based on two critical periods we can determine the upper bound t_i^+ for the horizon of the minimal sub-problem SP_{it}:

$$t_i^+ = \min_{j \in \{j^I,\, j^{II}\}} \max \left\{ r \mid d_{j+1,r} < L \right\} \tag{6}$$

In other words, in the solution algorithm there is no need to regard sub-problems with the horizon larger than given in (6) as they do not belong to any optimal solution.

4 Solution Algorithm

To solve the problem (1)–(5) we first construct a solution of a minimal sub-problem and then prove that it is optimal. Next, we provide the algorithm that splits the problem with the horizon T into series of minimal sub-problems. Finally, we prove that the solution of the problem (1)–(5) assembled from optimal solutions of its minimal sub-problems is also optimal.
So, let us construct the following solution for a minimal sub-problem SP_{it} with the last production period J.

$$
\begin{aligned}
\tilde{I}_{i-1} &:= 0, \\
\tilde{I}_j &:= (k_j+1)L - d_{ij}, & j &= i,\ldots,J-1, \\
\tilde{X}_j &:= \tilde{I}_j + d_j - \tilde{I}_{j-1}, & j &= i,\ldots,J-1, \\
\tilde{X}_J &:= d_{Jt} - \tilde{I}_{J-1}, \\
\tilde{X}_j &:= 0, & j &= J+1,\ldots,t, \\
\tilde{I}_{j-1} &:= d_{jt}, & j &= J+1,\ldots,t, \\
\tilde{I}_t &:= 0.
\end{aligned}
\tag{7}
$$

We will call solution (7) the *critical solution.*

Theorem 3 (critical solution). The critical solution of a minimal sub-problem is optimal.

Now we are ready to present the dynamic programming algorithm for solving the problem (1)–(5) by splitting it into series of minimal sub-problems. The algorithm is based on the approach proposed by Florian et al. [4] for solving a capacitated lot size problem. It rests upon the fact that an optimal solution between two nearest regeneration periods, i.e. between two periods with zero inventories, has special properties, which make the solution procedure more efficient. We reduce the complexity of our algorithm by considering only minimal sub-problems instead of revising the cumulative inventories for every value of i and t. The established bounds for the sub-problem's horizon are used to limit effectively the number of sub-problems that come into consideration. The solution algorithm for the limited problem (1)–(5) is presented below.

Step 1: $i := 1,\ t := t_1^-;\ \ F_0 := 0,\ F_j := +\infty,\ j = 1,\ldots,T$

Step 2: If $F_{i-1} + \tilde{I}_{it} \le F_t$ then $F_t := F_{i-1} + \tilde{I}_{it}$ and $i(t) := i-1$

Step 3: If $t < \min\{t_{max} - 1,\ t_i^+\}$ then $t := t+1$

return to Step 2 (8)

Step 4: If $t < T \le t_i^+$ then $t := T$

return to Step 2

Step 5: If $i = T$ or $t_i^- > T$ then *Stop,*

else $i := \max\{i,\ t_1^-\} + 1,\ t := t_i^-$ and *return to Step 2*

In algorithm (8) values F_t denote the minimal cumulative inventory in period t; $i(t)$ represent the regeneration periods for disseminating the problem into minimal sub-problems, and t_{max} provides the upper bound for the beginning of the last sub-problem.

Theorem 4 (optimal solution). Algorithm (8) generates an optimal solution of problem (1)–(5) as a series of optimal solutions of minimal sub-problems.

To scan the efficiency of the developed algorithm, we conducted a preliminary empirical study. An extensive study, as well as the comparison of our algorithm with the procedure developed in [1], is planned for the near future. In the performed tests we created 12 types of problems and for every type randomly generated three instances. The achieved results are very promising, especially for the case when the minimum lot size is not considerably greater than the average demand.

5 Conclusions

The paper continues the analysis of a special uncapacitated single item lot sizing problem, where a minimum lot size restriction, instead of the setup cost, guarantees a certain level of the production lots. The detailed analysis of the model and investigation of particularities of the cumulative demand structure allowed us to develop a solution algorithm based on the concept of minimal sub-problems. Furthermore, we presented an optimal solution of a minimal sub-problem in the explicit form and proved that it serves as a building block for the optimal solution of an initial problem.

References

1. Anderson EJ, Cheah BS (1993) Capacitated lot-sizing with minimum batch sizes and setup times. International Journal of Production Economics 30-31:137–152
2. Brahimi N, Dauzere-Peres S, Najid NM, Nordli A (2006) Single item lot sizing problems. European Journal of Operational Research 168:1–16
3. Dempe S, Richter K (1981) Polynomial algorithm for a linear discontinuous knapsack problem. Wissenschaftliche Informationen der Sektion Mathematik, TH Karl-Marx-Stadt No. 25
4. Florian M, Klein M (1971) Deterministic production planning with concave cost and capacity constraints. Management Science 18:12–20
5. Richter K, Bachmann P, Dempe S (1988) Diskrete Optimierungsmodelle. Effektive Algorithmen und Näherungslösungen, VEB Verlag Technik, Berlin
6. Richter K, Gobsch B (2005) Kreislauf-Logistik mit Losgrößenrestriktionen. Zeitschrift für Betriebswirtschaft 4:57–79

On the Relation Between Industrial Product-Service Systems and Business Models

Alexander Richter and Marion Steven

Chair of Production Management, Ruhr-University Bochum,
Universitätsstraße 150, 44780 Bochum
marion.steven@rub.de[†]

1 Introduction

Companies especially in B-to-B markets increasingly focus on the value generated for customers through innovative business models instead of merely selling products. Following such customer-oriented strategies dissolves the boundary of products and services. Most companies' offerings can at best be characterized as bundles of products and services, which we refer to as Industrial Product-Service Systems (IPSS) in this paper. IPSS are problem solutions for B-to-B markets, which consist of customized configurations of product and service parts tailor-made to meet individual customer needs. These product and service parts of IPSS exert a mutual influence on each other, owing to an integrated development and operation. The possibility of adjusting an original configuration along the IPSS-life-cycle by partially substituting product and service parts (IPSS-flexibility) is of special importance with regard to IPSS.

Integrating services into the core product, however, triggers a transition from a transaction- to a relationship-based (long-term) business connection [4], which is encompassed by challenges for the business parties involved. As a consequence the contractual and implicit relations need to be re-designed, striving at reallocating risks and incentives. In this context, business models which are based on the dynamic bundles describe the design of the customer-supplier-relationship in the form of performance schemes and responsibilities. With business models concentrating on use orientation, the utilizability of the manufacturing assets is ensured. By doing so, the supplier executes business processes

[†] This research is financially supported by the German Science Foundation (DFG) through SFB/TR29.

DOI: 10.1007/978-3-642-00142-0_16,

of the customer for the first time. Facing a result-oriented business model, the supplier is fully responsible for the output value. Thus, for both business models it is distinctive that the supplier participates in the customer's risks, which is due to connection of the supplier's compensation with the output value of the manufacturing asset.
This paper aims at examining the relation between IPSS and these business models in order to determine which business model is best suited for which IPSS. In section 2 we describe a contractual problem adequate for IPSS and conduct a comparative analysis of the use-oriented and result-oriented business models in section 3.

2 An Incomplete Contract Approach to IPSS

We consider a two-period business relationship between a risk-neutral IPSS supplier and a risk-neutral IPSS customer. As illustrated in figure 1, the supplier plans and develops the product (e.g. aero engine, machine tool...) and service (e.g. maintenance, training, operation...) parts in period 1. In period 2, the IPSS is operated and used. Its configuration varies depending on the chosen business model and the realization of environmental conditions. Offering IPSS therefore takes place in a complex, multitasking environment.

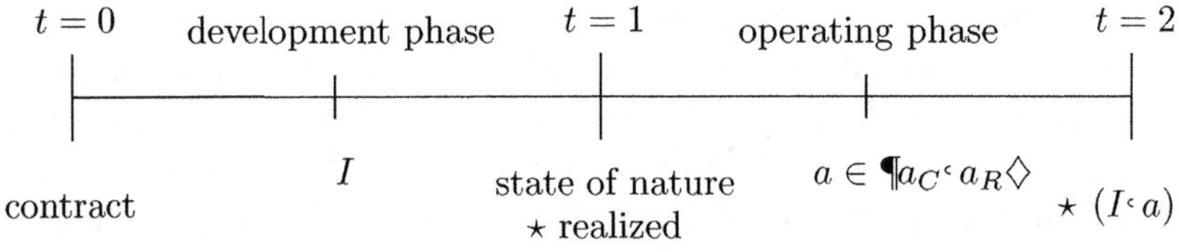

Fig. 1. Timeline

According to Williamson [7], complexity and bounded rationality make it impossible to cover for all future contingencies of a business relationship. IPSS contracts are therefore generally incomplete and may require ex post renegotiations, which generate transaction costs. As the property-rights theory states, the problem of opportunistic behavior resulting from this incompleteness can at least partially be solved by an appropriate reallocation of property rights [2]. According to this line of argumentation, the property of the manufacturing asset should stay with the IPSS supplier to generate incentives for investments into the

asset's design. This argumentation is consistent with innovative business models in which the customer "merely" acquires the value of use generated by an asset, which raises the question as to why the asset should become property of the customer at all.
However, with regard to the question of make-or-buy, property-rights theory is limited exclusively to the hold-up-problem. Contracts are being designed with the goal of setting suitable ex ante incentives.[3] Unanticipated ex post events cannot influence the sunk actions. Considerations regarding complexity and the degree of incompleteness are therefore not decisive for determining the optimal contract.
The flexibility of reacting to unanticipated events during the use and operating phase, however, is an essential characteristic of IPSS. Furthermore, considering property-rights is ill-suited to delineate the business models, as company boundaries are fixed. Organizing business relationships to efficiently deal with unanticipated events is the motivation behind the adaptation theory beginning with Simon [5]. In contrast to property-rights theory, the operative decision in period 2 can no longer be contractually regulated ex post, which is why an ex ante contractual allocation of decision rights across company boundaries is of importance.[4] Thus, the allocation of decision rights influences the decision to be made ex post, as this decision is made exclusively with regard to the interest of the party in control. Generally, this decision does not have to be efficient. Consequently, the decision rights should be transferred to the party which maximizes the use of the overall business relationship with its decision.
It can be concluded that adaptation theory is suitable to account for aspects of flexibility and complexity. This, however, at the expense of integrating ex-ante actions in the form of specific investments. These actions are again central to property rights theory, which is why we claim the need for an integrated approach of both theories for IPSS.[5]

3 Which Business Models for Which IPSS?

To illustrate the contractual problem of IPSS we consider a simple model, in which both ex ante and ex post actions are neither observable nor verifiable. Therefore, contracts can only specify the distribution (α) of the value generated for the customer (Π) and the allocation of the decision rights regarding the ex post action (a). We need to answer,

[3] We define *ex ante* and *ex post* relative to realization of the state of nature.
[4] See for more details on a framework of "contracting for control" Gibbons [1].
[5] See for an integrative framework of different theories of the firm Gibbons [1].

which party should be given authority over decision a and therewith decide about reconfiguring the system. As assumed by our model suppliers have the decision rights in the result-oriented business models, while customers have the decision rights in use-oriented business models. This is based on the assumption that in result-oriented business models the production responsibility lies solely with the supplier and that customers are not involved in the production process. Choosing an allocation of decision rights is therefore equivalent to choosing a business model.

After the contract is signed in $t = 0$, the supplier plans and develops the IPSS which triggers investments I for coordinating the development processes. Costs of these investments are simply given by $\kappa(I) = I$.

In $t = 1$ two events are possible, a "good" and a "bad" one, $\theta \in \{\theta_G, \theta_B\}$. The ex ante probability of a good event (θ_G) is given by $Pr(\theta = \theta_G) = p \in [0,1]$ and can be interpreted as the completeness of the initial problem solution. In this case the value proposition for the customer is assumed to be constant and denoted by $\bar{V}$. With the converse probability $(1 - p)$ of a bad event (θ_B) unanticipated contingencies occur which require for a reconfiguration of the system. For this reason it is supposed that the value proposition in the bad event, denoted by $\underline{V}$, is less than the value proposition in the good event $\bar{V}$. The probability $(1 - p)$ can therefore be regarded as the complexity of the system, which is given exogenously [6]. System complexity results from the development task, which is determined by the number of components, the interaction between components and the system's degree of innovation. More complex systems bear even major challenges regarding the coordination for development processes, involving an increase in transaction cost.[6] This is considered by the fact that with inclining complexity a given value proposition can only met with higher investments. Although the investment costs are realized within the development phase, its utility will only be exploited during the operating phase.

After the realization of state of nature θ in case of a bad event the party in control can choose between continuation (C) and reconfiguration (R), $a \in \{a_C, a_R\}$. Let the utility of the investments made during the development phase be greater in the case of continuation than in the case of reconfiguration, so that

$$\frac{\partial \underline{V}(I, a_C)}{\partial I} > \frac{\partial \underline{V}(I, a_R)}{\partial I} > 0 \tag{1}$$

[6] See for examples regarding system complexity and its inherent transaction cost in the automotive industry Novak and Eppinger [3].

For reasons of simplicity, in the following we will only differentiate between a high (h) and a low (l) investment, $I \in \{I_l, I_h\}$. Then suppose that for a given investment level reconfiguration is better than continuation in the bad event. Consequently, the value proposition in the bad event can be put as

$$\underline{V}(I_i, a_R) > \underline{V}(I_i, a_C) > 0,\ i = \{l, h\} \tag{2}$$

Possible reconfigurations, though, are encompassed by a follow up investment $K(a_R) > 0$ for the supplier, so that conflicts of interest concerning the ex post decision between the contractual parties may arise. We therefore concentrate on the situation where $K(a_R) > \alpha[\underline{V}(I_i, a_R) - \underline{V}(I_i, a_C)]$.[7] Assuming supplier control (and therewith a result-oriented business model) we find continuation to be the dominant strategy, whereas under customer control (and therewith a use-oriented business model) reconfiguration is efficient. By applying backward induction we then can describe the total expected utility functions under supplier ($\hat{\Pi}^s$) and customer ($\hat{\Pi}^c$) control as

$$\hat{\Pi}^s(I,p) = p\bar{V} + (1-p)\underline{V}(I, a_C) - I \tag{3}$$
$$\hat{\Pi}^c(I,p) = p\bar{V} + (1-p)[\underline{V}(I, a_R) - K(a_R)] - I \tag{4}$$

Since the contractual parties share the total realized utility according to the allocation parameter α, both are interested in maximizing the expected utility $\hat{\Pi}$. Considering (3) and (4) can easily derive, that expected utility is maximum if $\underline{V}(I, a)$ is highest. Referring to (1) and (2), we discover the following trade-off: While customer control indeed leads to an efficient decision ex post but at the same time to an underinvestment, supplier control entails a loss in the ex post efficiency but provides better investment incentives. The question to be answered therefore is, when ex ante incentives are relatively more important than efficient ex post decisions.

It is easy to verify that the total expected utility function exhibits decreasing differences in I and p, i.e.

$$\frac{\partial^2 \hat{\Pi}(I,p)}{\partial I \partial p} < 0 \tag{5}$$

According to (5), (the importance of) investments are increasing with rising complexity, so that we can conclude (taking the previous assumptions into account): More complex IPSS are more likely to be developed and operated within the result-oriented business models, while

[7] The parameter α is not subject of renegotiation.

more simple IPSS are more likely to be developed and operated within use-oriented business models. This result is in line with an empirical study by Novak and Eppinger who argue that product complexity and vertical integration are complements [3].

4 Concluding Remarks

This contribution analyzed the relation between IPSS and innovative business models. We focused on the question, as to how complexity of an IPSS in the development has an effect on operation. We found, that with high complexity result-oriented business models are rather applied, with lower complexity there was a tendency toward use-orientation, in order to maximize the total utility of the business partnership.

References

1. Gibbons R. (2005) Four Formal(izable) Theories of the Firm? Journal of Economic Behavior & Organization 58: 200–245
2. Grossman S., Hart O. (1986) The Costs and Benefits of Ownership: A theory of Vertical and Lateral Integration. Journal of Political Economy 94: 691–719
3. Novak S., Eppinger S.D. (2001) Sourcing By Design: Product Complexity and the Supply Chain. Management Science 47: 189–204
4. Oliva R., Kallenberg R. (2003) Managing the Transition from Products to Services. International Journal of Service Industry Management 14: 160–172
5. Simon H.A. (1951) A Formal Theory of Employment Relationship. Econometrica 19: 293–305
6. Tadelis S. (2002) Complexity, Flexibility, and the Make-or-Buy Decision. American Economic Review 92: 433-437
7. Williamson O. (1975) Markets and Hierarchies: Analysis and Antitrust Implications. Free Press, New York, NY

Dynamic Bid-Price Policies for Make-to-Order Revenue Management

Thomas Spengler[1], Thomas Volling[1], Kai Wittek[1], and Derya E. Akyol[2]

[1] Institute for Production and Logistics, TU Braunschweig
{t.spengler,t.volling,k.wittek}@tu-bs.de

[2] Department of Industrial Engineering, Dokuz Eylul University
derya.eren@deu.edu.tr

Summary. A challenge of make-to-order (MTO) manufacturing lies in maximizing the total contribution margin by compiling the optimal order portfolio, given a fixed capacity. Particularly, bid-price based heuristics are successfully applied. The objective of this paper is to provide an extension to traditional bid-price based revenue management in MTO. Based on an analysis of the pros and cons of anticipative and reactive approaches, a hybrid approach combining both elements is proposed.

1 Introduction

Make-to-order (MTO) industries require decision support in acceptance or rejection of orders which differ from each other by specific characteristics [4]. Through evaluating customer requests with respect to the utilization of bottleneck capacity, revenue management provides such a support.

One of the methods widely used in revenue management is the selection of orders based on their contribution margin. While the potentials of such approaches have been shown for various industry settings, there is a conceptual trade-off to be solved. Using static selection criteria allows for a stable coordination of sales, which is in particular important for the MTO-industry where transactions are less standardized than those in other industries. Most MTO-companies therefore operate large and often globally distributed sales organizations, which provide technical consultancy and negotiate with the customer. If static selection criteria are used, acceptance decisions can be readily reconstructed ex-post. This facilitates the implementation of bonus payment

B. Fleischmann et al.(Eds.), *Operations Research Proceedings 2008*,
DOI: 10.1007/978-3-642-00142-0_17,

schemes, increases the transparency and therefore contributes towards coordination [2]. If real demand does not match the anticipation, however, significant losses in contribution margin are to be expected. In such a setting, dynamic approaches that update selection criteria with respect to recent information seem to be more promising, yet subject to the organization's ability to adapt to the changes. To this end, revenue management approaches for the MTO-industry need to balance requirements for recency and stability.
The objective of this paper is to introduce a revenue management approach, which allows for decision support in dynamic MTO settings and addresses the trade-off stated above. Building blocks are the classification of the demand situation and the adjustment of the capacity control at a defined point in time. The general setting is given in [3]. For the sake of simplicity, only one resource is considered.

2 Dynamic Aspects of Bid-Price Based Revenue Management

Bid-price approaches are amongst the most popular instruments to control capacity in revenue management [2]. A threshold price, referred to as bid-price, is set for each resource. This price reflects the opportunity cost of consuming one unit of bottleneck capacity. Orders are accepted, if the associated revenue exceeds the opportunity cost of their resource requirements.
The determination of opportunity costs requires knowledge on the optimal use of resources. Bid-price methods are therefore typically based on forecasts. However, forecasts are intrinsically incorrect, such that approaches are necessary to avoid misleading decision support. Two fundamental approaches can be distinguished. Anticipative approaches, like randomized linear programming (RLP), seek to incorporate possible realizations of the future demand when evaluating revenue management policies. To support order selection, the average bid-prices are computed and used throughout the booking period. As a consequence, however, in a high number of demand realizations the bid-prices are too unrestrictive. Consequently, orders are accepted that should have been declined as to optimize total contribution margin. On the contrary, if real demand falls below the forecasted demand, the bid-prices are too restrictive.
To this end, reactive approaches recompute the bid-prices, as time evolves. A reactive extension of RLP would be the re-calculation of bid-prices within a so called resolving procedure. While this results in

a better utilization of the information available, two operating conditions exist. Firstly, reactive approaches require the organization to cope with varying bid-prices. In MTO settings, this is all but trivial, as highlighted above. Secondly, reactive approaches require meaningful state information. Only if new dependable information is available, revisions of the revenue management policy should be done. This fact is illustrated in Fig. 1. Depicted are the ex-post optimal bid-prices of 1,000 demand realizations plotted versus the cumulated requested capacity up to a specified point in the booking period. We used empirical data from a high performance alloy manufacturer with a single bottleneck resource. The length of the booking period starting in $t = 0$ was set to T and we used identical assumptions regarding the normally distributed lead times across the analysis (constraint to the interval $[0, T]$). The number of arriving orders in each demand realization is sampled from the same Poisson distribution, such that capacity exceeds demand by on average 10%.

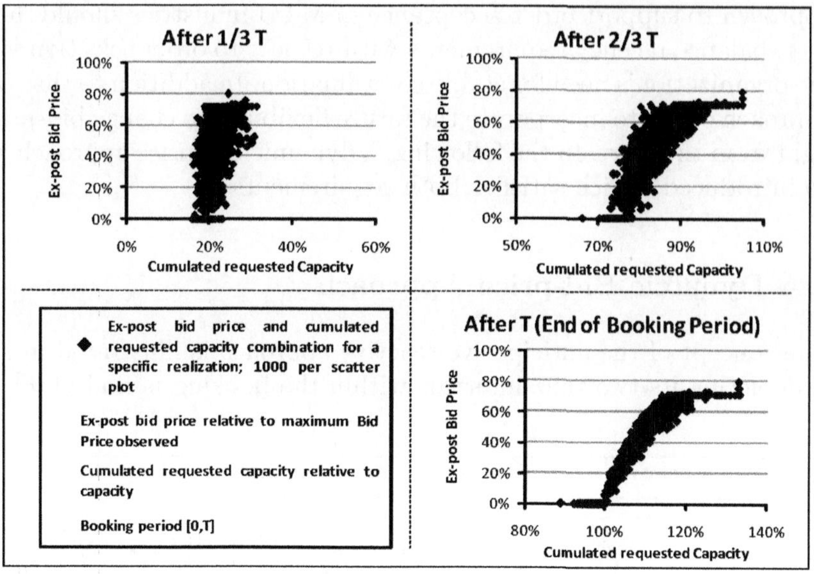

Fig. 1. Interrelation between ex-post bid-prices and demand realization

As can be seen from the analysis, there is an interrelation between the cumulated capacity requested by the orders and the ex-post bid-price. In particular, at the end of the booking period, two intervals can be distinguished. For realizations with demand falling below capacity, the ex-post bid-prices compute to zero. For all other realizations, there are

positive bid-prices. The remaining fuzziness is induced by the empirical data. A similar pattern can be obtained when the same analysis is conducted at two thirds of the booking period. Considering the situation after one third of the booking period, however, there hardly is any interrelation.
With respect to the implementation of reactive approaches, these findings are crucial. From a conceptual point of view, there are no major improvements of the planning performance to be expected from changing the bid-price early in the booking period. The underlying reason is that there is not sufficient information available. Even worse, changes to the bid-price might have to be revised, when better information is available. As a consequence, the planning nervousness is expected to increase. The situation is different at two thirds of the booking period. At this point in time, all characteristics inherent in the final shape can be identified. Adjusting the bid-price thus seems to be promising.
A recent application to extend anticipative approaches by reactive elements is given for instance in [1] for the airline industry. A suitable approach to support order acceptance in MTO-industries should, however, balance recency requirements with respect to order selection with the organization's need for stable co-ordination. In addition to that, the approach ought to incorporate the future flexibility to change bid-prices right from the start. In the following, a dynamic bid-price approach will be introduced, which satisfies both requirements.

3 A Dynamic Bid-price Approach

The concept of the anticipative-reactive approach is illustrated in Fig. 2, depicting its two time intervals within the booking period $[0, T]$.

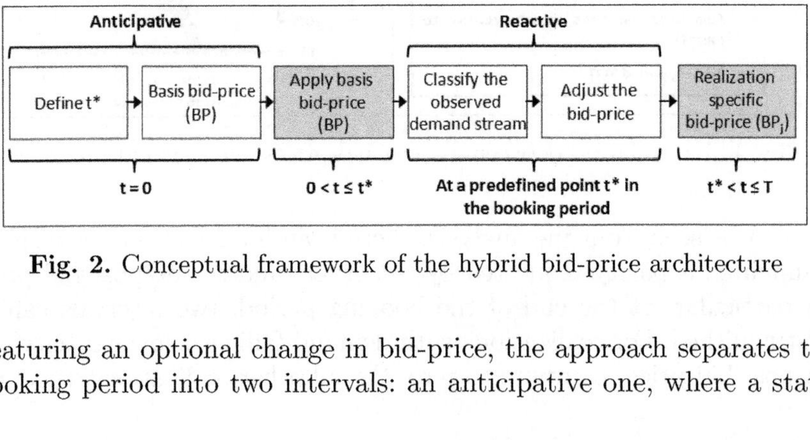

Fig. 2. Conceptual framework of the hybrid bid-price architecture

Featuring an optional change in bid-price, the approach separates the booking period into two intervals: an anticipative one, where a static bid-price referred to as basis bid-price (BP) is applied and a second reactive interval with a bid-price according to the observed demand realization in the first segment, labelled as realization specific bid-price (BP_j). The point in time, where the change in bid-price occurs, is denoted by t^*. Setting a low value for t^* is favorable, since it increases the number of orders, which are evaluated based on the specific bid-price. However, choosing t* to be early in the booking period reduces the ability to accurately predict the total demand and the associated specific bid-price (BP_j). The adjustment of bid-prices for different demand realizations is depicted in Fig. 3. If demand realized until t^* is greater than the expectation derived from the forecast, the bid-price is increased. If it is less, the bid-price is reduced.

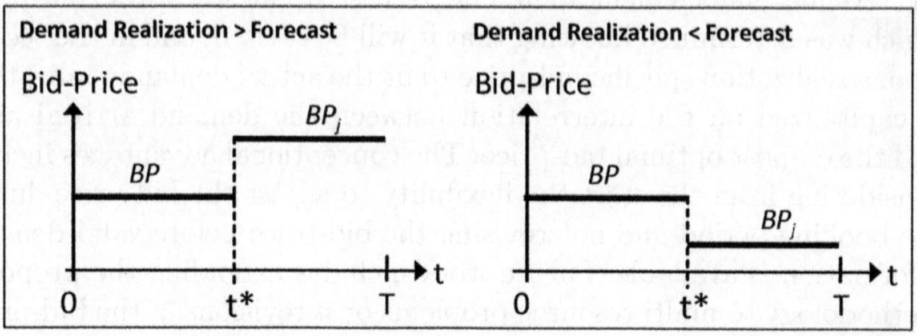

Fig. 3. Bid-price adjustment for different demand realizations

In order to apply the hybrid approach, four functionalities have to be fulfilled. These include the determination of BP and t^*, both being offline functionalities. The two online functionalities are the classification of demand and the adjustment of the bid-price. The switch point t^* can be considered as exogenous as it depends on the availability of information and can be obtained from a data analysis as presented in section 2. Using a predefined range for BP values, t*, forecast data and the assumptions stated in section 2, a simulation can be applied for an enumeration procedure to derive the optimal values for BP and BP_j simultaneously. Assuming perfect hindsight, the BP is identified, which on average yields the highest total contribution margin. To this end, realization specific bid-prices (BP_j) need to be determined for each basis bid price (BP). These can be derived from a deterministic linear programmig (DLP) procedure, given the remaining capacity at t^* and the orders to come.

With the data obtained from the simulation, models fulfilling the online functionalities of the reactive interval in the hybrid approach can be built. Two models are required, one to classify the actual demand realization as below ($BP_j = 0$) or above capacity ($BP_j > 0$). The second one to determine an appropriate value for the realization specific bid-price BP_j when the demand is classified as demand exceeding capacity ($BP_j > 0$). These models can be based on artificial neural networks (ANNs), since they have been successfully employed in a variety of classification and forecasting problems.

4 Conclusions and Outlook

In this paper, the concept of a two step dynamic bid-price method combining both anticipative as well as reactive elements is introduced for revenue management in MTO. It first applies a static bid-price, which was determined knowing that it will be revised, and in the second step, a realization specific bid-price to fit the actual demand realization. It capitalized on the interrelation between the demand arrived at t^* and the ex-post optimal bid-price. The conceptional advantages include considering from the start the flexibility to adjust the bid-price during the booking period and not revising the bid-price before valid demand information is available. Future work includes extending the proposed methodology to multi resource problems or n revisions of the bid-price.

Acknowledgement. The research of Derya Eren Akyol was supported by TUBITAK-DFG.

References

1. Klein R (2007) Network capacity control using self-adjusting bid-prices. OR Spectrum 29:pp.39–60
2. Spengler T, Rehkopf S (2005) Revenue Management Konzepte zur Entscheidungsunterstützung bei der Annahme von Kundenaufträgen. ZFP 16:pp.123–146
3. Spengler T, Rehkopf S, Volling T (2007) Revenue management in make-to-order manufacturing: an application to the iron and steel industry. OR Spectrum 29:pp.157–171
4. Spengler T, Volling T, Hintsches A (2008) Integration von Revenue Management Konzepten in die Auftragsannahme. ZFB Special Issue 4/2008: pp. 125–151

Part IV

Scheduling and Project Management

A GA Based Approach for Solving Multi Criteria Project Scheduling Problems

Felix Bomsdorf, Ulrich Derigs, and Elisabeth von Jagwitz

Department of Information Systems and Operations Research (WINFORS), University of Cologne, Pohligstr.1, 50969 Cologne, Germany
{bomsdorf,ulrich.derigs}@uni-koeln.de

Summary. In this paper we present a Genetic Algorithm (GA) for generating efficient solutions for multi criteria resource constrained project scheduling problems where conflicting regular as well as non regular performance measures have to be respected (cf. [4]). The GA-scheme is similar to the one developed by [1], i.e. the genes of the genotype do not only represent activities of the phenotype but also information on the decoding scheme and modus. Since a large number of (mostly similar) solutions is usually not of great help for a planner not all efficient solutions are maintained in order to reduce complexity. Thus, we use a "relaxed Pareto operator" which does accept efficient solutions only which differ (substantially) with respect to solution quality (and structure). We have applied this procedure to solve the Movie Shoot Scheduling Problem (MSSP) and we report first results.

1 Introduction

The main and usually also the only objective in project scheduling models is the minimization of the project makespan. Yet, in practice there exist project types where several different criteria are relevant and have to be respected. These criteria might be regular as well as non-regular, and, obviously, these criteria are also often conflicting. Applying a standard multi criteria search technique can lead to a very large number of Pareto efficient solutions, which clearly does not support a planner on deciding which project schedule to use. Thus, we have developed a multi population GA with a relaxed Pareto criterion by which the number of Pareto efficient solutions presented to the planner is reduced and which can automatically adapt its search strategy depending on the characteristics of the problem/instance.

B. Fleischmann et al.(Eds.), *Operations Research Proceedings 2008*,

DOI: 10.1007/978-3-642-00142-0_18,

2 Multi Criteria GA

In general, our multi criteria GA works similar to "standard" GAs. The specific characteristics are outlined in Algorithm 1 and are described in the following subsections.

Algorithm 1 Outline of the Multi Criteria GA

(1) Generate initial population(s): $P = \{m, p_1 ... p_n\}$ //$m :=$ multi criteria population, $p_i :=$ single criteria population
(2) Evaluation of the individuals of the initial population(s)
while iterations < maximum number of iterations **do**
    for all Population $p \in P$ **do**
        (3a) Selection and Recombination of individuals to generate new individuals
        (3b) Mutation of individuals with a certain probability
        (4) Evaluation of the new individuals
        (5) Selection of individuals for the new population
    end for
    (6) Apply local search to current "best" solution in order to further improve this solution
    if every x iterations, $x \in \mathbb{N}$ **then**
        (7a) Replace population m with "mix" of all populations
        (7b) Initialize each single criterion populations with its solution
    end if
end while

2.1 Extended Genotype, Crossover and Mutation

The genotype used by our multi criteria GA is based on the extended activity list representation proposed in [1]. The activity list is extended by two more genes, which indicate which scheduling mode and which scheduling scheme shall be used during the decoding of the activity list into a project schedule.
The gene which determines the scheduling mode indicates which of the standard modes of project scheduling, i.e. forward or backward scheduling, shall be used, while the gene determining the scheduling scheme indicates whether the serial schedule generation scheme or the parallel generation scheme shall be used.
By adding these two additional genes the GA becomes self adaptive.
The crossover of two individuals is performed as suggested by [1], i.e. a "regular" crossover where the additional genes of the son (daughter)

take the values of these genes of the father (mother). During optimization the activity list as well as the genes determining the scheduling mode and scheme can be mutated with a certain probability.

2.2 Populations

The GA makes use of two or more populations. The so-called multi criteria population m is used to maintain the relaxed Pareto efficient solutions, the other population(s) is (are) used to generate solutions which are then evaluated either using a single criterion or using a (single) aggregated objective function which has been composed from several or all of the relevant criteria.
After a certain number of iterations the different populations are mixed. All individuals from the single criterion populations are introduced to the multi criteria population m, if they fulfill the relaxed Pareto efficiency criterion.
The other populations, which use a single objective function, are initialised using individuals which are generated randomly and the so far best individual regarding this single objective function in the specific population.

2.3 Relaxed Pareto Operator

Usually there is a large number of Pareto efficient solutions, which might be very similar to each other regarding solution quality and/or solution structure. To overcome this problem of redundancy for the planner one should only accept Pareto efficient solutions if they also fulfill additional conditions regarding their solution quality (and/or their solution structure).
Under the (relaxed) Pareto criterion a solution x dominates a solution y if for at least one criterion x has a value that is better than the value for y and if for all other criteria the values of x are not (significantly) worse than the values of y. In the case of a minimization problem this can be formulated as follows:
Let $F = \{f_1 ... f_f\}$ be the set of criteria, then

$$\begin{aligned} f_i(x) &< f_i(y) \quad \text{, for at least one criterion } f_i \in F \\ r * f_j(x) &\leq f_j(y) \text{ , for all } f_j \in F \setminus f_i \\ &\qquad \text{with } 0 < r \leq 1 \end{aligned}$$

We then use domination as proposed by [3] in order to select the individuals. Using the relaxed Pareto operator it could happen that a

solution x dominates a solution y and vice versa. There are different approaches to deal with such a case, for the sake of simplicity and diversification we always maintain the newer solution in the population. As an addition to the previously described relaxation to reduce the number of accepted Pareto efficient solutions the acceptance of individuals could additionally be based on the structure of the solution, either evaluating only the position of genes in the genotype or evaluating the phenotype when comparing the structure of two individuals (for the latter cf. [7]).
Both approaches help to increase diversity of the (multi criteria) population. The adapted Pareto operator helps to reduce the number of multi criteria solutions and at the same time keeps a relative high diversification among the different criteria without needing to make any assumption about the preferences of the planner among the different criteria. Other approaches for reducing the size of populations during multi criteria optimization usually either make some kind of assumptions or require interaction with the decision maker.

3 Application to the MSSP

The multi criteria GA using an adapted Pareto operator was applied to test instances of the MSSP. Here, a set A of activities requiring a set R of resources has to be scheduled within a fixed time horizon T. For each $j \in A$ there might exist daily time windows as well as precedence relations of general type or of the type that only a break but no other activity might be performed in between two activities connected with this special type of precedence relation. Let $R_j \subseteq R, j \in A$ be the set of resources required for the entire duration p_j of activity j. Each resource $r \in R$ may only be available a certain amount of consecutive hours and require a certain amount of consecutive hours of idleness to refresh to full capacity. Also there might be daily time windows in which a resource $r \in R$ may only be available e.g. from 8 a.m. to 10 p.m., as well as periods in T in which r is unavailable. The objective of the MSSP is to create a feasible schedule, which minimizes the number of changes of locations (a specific resource) and certain other criteria as for instance the number of working days (CR) etc. These criteria contain regular as well as non-regular criteria. For a detailed model of the MSSP we refer to [2].
The problem size of an instance of the MSSP is typically relative large in comparison to the project sizes used in the PSPLib (cf. [5]), i.e. the test instance used for the evaluation contains 114 activities, 23

resource types, 200 actual resources, 1536 periods compared to the largest instances used in the PSPLib with 120 activities, 4 resource types with a total capacity of 156 and a project horizon of 700 periods. The implementation of the solution procedure for the (multi-criteria) MSSP is straightforward.

From the generic description of the GA in section 2 the number of populations and the criteria evaluated in the additional populations are parameters which can be set for the specific implementation. Also an encoding and decoding procedure has to be developed.

For the MSSP we use several populations. One which contains the relaxed Pareto efficient individuals and other populations which are each evaluated only for a single criterion: the number of changes of locations (LC), the length of gaps (LG) and the number of gaps (NG) on all resources, the number of capacity renewals (CR), time continuity (TC) and emotional continuity (EC). Reducing the number of changes of locations is generally identified as the main objective by the planner.

The decoding mechanism for scheduling activities is similar to the standard project scheduling schemes. One difference is that the calendarization, i.e. the consideration of weekdays, holidays etc., is integrated into the scheduling scheme and not done in a separate step, because the separate calendarization leads to problems (cf. [6])

Table 1 summarizes the results of the optimization with the multi criteria GA in comparison to results of the greedy indirect search technique (GIST) used in [2] and the actual schedule which was created manually. The best value for each criterion is underlined. The greedy decoder used in the GIST approach of [2] is optimized for the minimization of changes of locations. From the relaxed Pareto efficient individuals (RP) the "best" individual was chosen by evaluating a hierarchical objective function which first evaluates LC, then CR, LG, NG, TC and EC.

Table 1. Computational Analysis

	Actual schedule	GIST	RP	LC	LG	NG	CR	TC	EC
Number of changes of locations	<u>35</u>	<u>35</u>	45	48	56	72	55	59	53
Length of gaps	201	184	204	349	<u>87</u>	129	260	136	284
Number of gaps	48	49	53	68	27	<u>21</u>	52	42	61
Number of capacity renewals	<u>199</u>	247	259	267	282	320	237	273	244
Time continuity	30	40	33	38	38	36	32	<u>27</u>	34
Emotional continuity	50	61	57	58	64	56	57	49	<u>41</u>

4 Conclusions

The approach using the multi criteria GA with a relaxed Pareto operator and several populations is able to significantly reduce the number of schedules (solutions) which are presented to the planner by a decision support system while producing a high degree of diversity with respect to the different criteria within the set of presented solutions (cf. Table 1). The extended genotype which leads to an adaptive GA facilitates the application of the multi criteria GA to solving a wide range of different project scheduling problems with diverse criteria.
For future research other scheduling modes, e.g. the one used by the GIST approach, could be added, which might find better schedules in case of calendarized problems when forward or backward scheduling can lead to inferior solutions. Also the approach could be applied to other project scheduling problems and finally be generalized to a framework which can be applied to other areas of multi criteria optimization than project scheduling.

References

1. Alcaraz, J.; Maroto, C. (2006): A Hybrid Genetic Algorithm Based on Intelligent Encoding for Project Scheduling. In: Józefowska, J.; Weglarz, J. (Eds.): Perspectives in Modern Project Scheduling. Springer US, 2006, p. 249–274.
2. Bomsdorf, F.; Derigs, U. (2007): A model, heuristic procedure and decision support system for solving the movie shoot scheduling problem. In: OR Spectrum.
3. Horn, J.; Nafpliotis, N.; Goldberg, D. (1994): A Niched Pareto Genetic Algorithm for Multiobjective Optimization. In: Proceedings of the ICEC, p. 82–87.
4. v. Jagwitz, E. (2008): Mehrzieloptimierung für ein kapazitiertes Scheduling Problem mit Zeitfenstern – Entwicklung, Implementierung und Evaluation einer (Meta-)Heuristik für das Movie Shoot Scheduling Problem. Diplomarbeit, University of Cologne.
5. Kolisch, R. ; Hartmann, S.: Experimental investigation of heuristics for resource-constrained project scheduling: An update. In: European Journal of Operational Research 174 (1), Elsevier, 2006.
6. Schirmer, A.; Drexl, A.: Allocation of partially renewable resources - Concept, Capabilities, and Applications. 1997/2000.
7. Sörensen, K.; Sevaux, M. (2006): MA|PM: memetic algorithms with population management. In: Computers & Operations Research 33, p. 1214–1225.

Solving the Batch Scheduling Problem with Family Setup Times

Udo Buscher and Liji Shen

Industrial Management, Dresden University of Technology
buscher@rcs.urz.tu-dresden.de, liji.shen@tu-dresden.de

Summary. In this paper we address the machine scheduling problem involving family setup times and batching decisions. The m-machine flow shop system is considered with the objective of minimizing the makespan. We first present a mathematical formulation which is able to solve small instances. Subsequently, a tabu search algorithm is proposed with diverse neighbourhoods.

1 Introduction

Recently, considerable attention has been diverted to scheduling families of jobs on common facilities, where jobs belonging to the same family have certain similarities. In many realistic situations, negligible setup times occur among jobs of the same family. However, setups are inevitable at the start of a schedule and on each occasion when a machine switches from processing one family to another.[3] As a result, *family setup times* are usually defined as the change-over times/costs for the processing of jobs from different families.[1] In addition, the length of setup time may depend on the family of the current job as well as the family of the previous job, which is also called *sequence dependent* family setup time. *Batching*, on the other hand, refers to the decision of whether or not to schedule similar jobs contiguously.[4] Therefore, a batch is a maximal subset of jobs which share a single setup and must be processed jointly. As a result of integrating batching into the family scheduling model, advantages can be achieved due to the reduced number of setups and higher machine utilization. However, a vast majority of existing literature in this area focuses on the single machine problem. To the best of our knowledge, only special cases of the

B. Fleischmann et al.(Eds.), *Operations Research Proceedings 2008*,

DOI: 10.1007/978-3-642-00142-0_19, © Springer-Verlag Berlin Heidelberg 2009

multi-machine problem have been investigated in the previous studies. The general shop problems still remain as a challenge.
In this paper we address the m-machine flow shop scheduling problem subject to job availability and serial batching, where sequence dependent family setup times are involved. The objective is to find both batch compositions and job sequences so that the makespan is minimized. In the subsequent section we first present a mathematical formulation which is able to solve small instances. A tabu search algorithm is then proposed in section 3 with diverse neighbourhoods. Brief results are summarized in section 4.

2 A MIP Formulation

Assume that n jobs ($i = 1, \dots, n$), m machines ($k = 1, \dots, m$) and F families ($f = 1, \dots, F$) are given. Each job contains m operations and follows the same technological order. let s_{fgk} denote the sequence dependent family setup time when a job of family f immediately precedes a job of a different family g on machine k. Further, s_{0fk} represents the initial setup time in the case without preceding jobs. In addition, t_{ik} and p_{ik} are respectively the start time and processing time of job i on machine k. It is of advantage to formulate the problem under study as a standard scheduling problem first. Special emphasis is then placed on modelling family setup times as follows:

$$\min C_{max} \tag{1}$$

Subject to:

$$\sum_{f=1}^{F} \beta_{if} = 1 \quad \forall i \tag{2}$$

$$\gamma_{iik} = 0 \quad \forall i, k \tag{3}$$

$$t_{ik} \geq 0 \quad \forall i, k \tag{4}$$

$$t_{ik} \geq \left(\sum_{f=1}^{F} s_{0fk} \cdot \beta_{if} \right) \cdot \left(\sum_{j=1}^{n} \gamma_{ijk} - n + 2 \right) \quad \forall i, k \tag{5}$$

$$t_{ik} \geq t_{jk} + p_{jk} + \sum_{f=1}^{F} \sum_{g=1}^{F} s_{gfk} \cdot \beta_{if} \cdot \beta_{jg} - \gamma_{ijk} H \quad \forall i \neq j, k \tag{6}$$

$$t_{jk} \geq t_{ik} + p_{ik} + \sum_{f=1}^{F}\sum_{g=1}^{F} s_{fgk} \cdot \beta_{if} \cdot \beta_{jg} - (1 - \gamma_{ijk})\, H \quad \forall i \neq j, k \tag{7}$$

$$t_{ik} \leq t_{jk} + p_{jk} + \left[\left| \sum_{i'=1}^{n} (\gamma_{ii'k} - \gamma_{ji'k}) \right| - \beta_{if}\beta_{jf} \right] H \quad \forall i \neq j, k, f \tag{8}$$

$$t_{i(k+1)} \geq t_{ik} + p_{ik} \quad \forall i, k < m \tag{9}$$

$$C_{max} \geq t_{im} + p_{im} \quad \forall i. \tag{10}$$

Note that the parameter β_{if} equals 1 if job i belongs to family f and 0 otherwise. Therefore, constraints (2) ensure that each job is assigned to one and only one family. Constraints (3) are included for simplicity, in which γ_{ijk} is the binary variable indicating job sequences. It takes the value 1 if job i is scheduled before j on machine k and 0 otherwise. Following this definition, constraints (3) are used to predefine the value of γ_{iik} without loss of generality. Constraints (4) represent the conventional non-negativity conditions. Next, we incorporate constraints (5) to specify the start time of the first operation scheduled on machine k. The term $\sum_{f=1}^{F} s_{0fk} \cdot \beta_{if}$ ensures assigning the associated initial setup time. Since the value of $\sum_{j=1}^{n} \gamma_{ijk}$ is not greater than $n-1$, constraints (5) indicate that the first job on each machine can only start its processing after the initial setup being completed. Job sequences on machines are then determined by employing constraints (6) and (7). First, the expression $\sum_{f=1}^{F}\sum_{g=1}^{F} s_{gfk} \cdot \beta_{if} \cdot \beta_{jg}$ identifies the sequence dependent setup time corresponding to i and j. If job i is scheduled prior to j (that is, $\gamma_{ijk} = 1$), only constraints (7) are relevant, which require that job j begins upon the completion of job i and of the corresponding setup (when necessary). On the other hand, constraints (6) operate similarly if job i is processed after j. Constraints (8) integrate batching decision into the formulation. Recall that batching requires contiguous processing of jobs in the same family. If jobs i and j are scheduled successively and belong to the same family, the expression $\left| \sum_{i'=1}^{n} \left(\gamma_{ii'k} - \gamma_{ji'k}\right) \right| - \beta_{if}\beta_{jf}$ is equal to 0. Combining constraints (6) and (8), it follows $t_{ik} = t_{jk} + p_{jk}$. Jobs i and j are thus processed jointly. Constraints (9) indicate that the processing of a job on machine $k+1$ can not start until it is completed on the previous machine k. According to constraints (10), makespan is then defined as the largest completion time of all jobs on the last machine m.

3 Tabu Search Algorithm

Initial Solution

Initially, an entire family is viewed as a single batch. The problem under study thus reduces to scheduling jobs within each family and sequencing different families. Whereas the first sub-problem is solved as a conventional flow-shop problem, various constructive heuristics can be modified and applied to the latter case by defining the artificial processing time (AP) as:

$$AP_{fk} = \sum_{i \in f} p_{ik} + \frac{\sum_{g=1}^{F} s_{gfk}}{F} \quad \forall f, k. \tag{11}$$

Obviously, processing time is calculated considering the aggregate processing time of all jobs included in a family and the effective/average family setup time.

Neighbourhood Structure

In order to provide a non-permutation schedule, we adopt the concepts *block* and *internal operations* used in the job-shop system. A critical path can be divided into blocks which contain adjacent operations processed on the same machine. Except for the first and the last operations of each block, all the other operations on this path are internal.[2] Regarding the batching requirement, a block can be further refined with the emphasis on batches since operations in the same batch share a single setup. In this context, it should be noted that a block may contain complete or partial batches which can be referred to as *sub-blocks*. Formally, they are defined as follows.

Definition 1 (Sub-block). *A sub-block is a subset of the corresponding block and contains a maximal sequence of adjacent operations that belong to the same batch.*

In this neighbourhood structure, moves focus on non-internal operations and are differentiated with respect to sub-blocks. Since moves of non-internal operations have the potential of immediately improving the makespan, these moves are based on insertion techniques instead of the simple swap of two adjacent operations. More specifically, *intra-sub-block moves* indicate that operations involved in a move exclusively belong to the same sub-block. On the other hand, *across-sub-block moves* refer to insertions of operations into different sub-blocks. Finally, *inter-sub-block moves* describe interactions among individual sub-blocks. For example, 6 operations, 4 of which form a block, are grouped into three different batches in figure 1. The block thus consists of three sub-blocks: {2,3}, {4} and {5}, in which the sub-block with operation 4 is internal. Moreover, neighbours are obtained by performing various moves, where thick boxes highlight the resulting batch variation (It is assumed that operations 1, 2, 3, 5 and 6 belong to the same family).

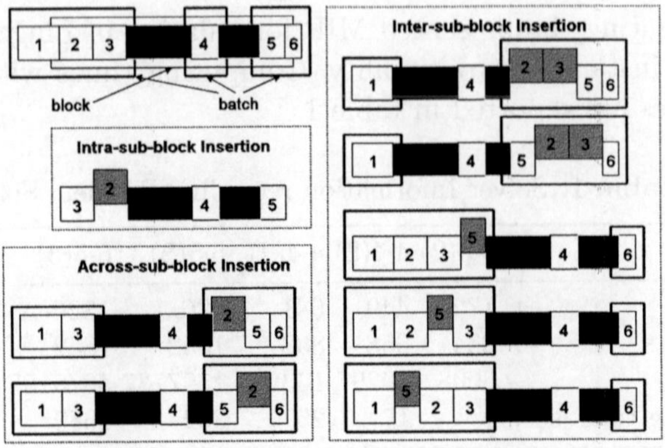

Fig. 1. Illustration of Neighbourhoods

A major advantage of across-sub-block moves is that they assist in splitting batches. Especially in the starting phase when the entire family is scheduled as a single batch, performing these moves effectively divides families into batches. On the other hand, if an operation is inserted in a non-adjacent sub-block that belongs to the same family, batch formations of this particular family can be altered. Benefits of inter-sub-block moves first consist in re-sequencing batches. Another important aspect again lies in their capability of changing batch compositions.

Conventionally, scheduling models with batching considerations are solved as two-stage problems. The importance of performing across- and inter-sub-block moves is apparent since they succeed in integrating batching into the scheduling decision.

Further Components

Tabu List. If the insertion of operation u between v and w is selected in an iteration, moves involving (v, u, w) are maintained in the tabu list for a prescribed number of iterations, which is determined by the corresponding problem size.

Aspiration Level. The tabu status of a particular move can be overwritten if the resulting makespan is smaller than the best value found so far.

Termination Criterion. The search procedure terminates after executing a given number of iterations, of which the concrete value depends on the problem size.

4 Implementation and Results

By implementing the presented MIP formulation in Lingo 9.0, we can solve small instances to optimality. Computing times with respect to problem sizes are reported in table 1.

Table 1. Solver Information regarding Problem Sizes

$n{\cdot}m(f)$	3·3(2)	4·4(2)	4·4(3)	5·5(2)	5·5(3)	6·6(2)
Variables	172	449	641	926	1326	1657
Constraints	254	658	898	1352	1852	2414
Iterations	6008	90297	143207	1197737	4875153	282688739
Computing time [sec.]	1	17	25	302	1680	63842

The proposed tabu search algorithm is coded in C++ and runs on an AMD Athlon64 2450 MHz personal computer with 4GB memory. Similar to the MIP formulation, small instances constantly reached their optima. On the other hand, large instances can also be solved within a reasonable amount of computing time.

References

1. Cheng T, Lin B, Toker A (2000) Makespan minimization in the two-machine flowshop batch scheduling problem. Naval Research Logistics 47:128–144
2. Nowicki E, Smutnicki C (1996) A fast taboo search algorithm for the job shop problem. Management Science, 42:797–813
3. Potts C, Kovalyov M (2000) Scheduling with batching: A review. European Journal of Operational Research 120:228–249
4. Potts C, Wassenhove LV (1992) Integrating scheduling with batching and lot-sizing: A review of algorithms and complexity. The Journal of the Operational Research Society 43:395–406

On Single Machine Scheduling and Due Date Assignment with Positionally Dependent Processing Times

Valery S. Gordon[1] and Vitaly A. Strusevich[2]

[1] United Institute of Informatics Problems, National Academy of Sciences of Belarus, Surganova 6, Minsk, 220012 Belarus
gordon@newman.bas-net.by

[2] School of Computing and Mathematical Sciences, University of Greenwich, Old Royal Naval College, Park Row, Greenwich, London, SE10 9LS, U.K.
V.Strusevich@greenwich.ac.uk

1 Problem Formulation

This paper addresses single machine scheduling problems in which the decision-maker controls two parameters: the due dates of the jobs and the processing times. In the problems under consideration, the jobs have to be assigned the due dates and the objective includes the cost of such an assignment, the total cost of discarded jobs and, possibly, the holding cost of the early jobs represented in the form of total earliness. The processing times of the jobs are not constant but depend on the position of a job in a schedule. We mainly focus on scheduling models with a *deterioration* effect. Informally, under deterioration the processing time is not a constant but changes according to some rule, so that the later a job starts, the longer it takes to process. An alternative type of scheduling models with non-constant processing times are models with a *learning* effect, in which the later a job starts, the shorter its processing time is. The two types of models are close but not entirely symmetric.

The jobs of set $N = \{1, 2, \dots, n\}$ have to be processed without preemption on a single machine. The jobs are simultaneously available at time zero. The machine can handle only one job at a time and is permanently available from time zero. For each job j, where $j \in N$, the value of its 'standard' or 'normal' processing time p_j is known. If the jobs are processed in accordance with a certain permutation $\pi = (\pi(1), \pi(2), \dots, \pi(n))$, then the processing time of job $j = \pi(r)$,

DOI: 10.1007/978-3-642-00142-0_20,

i.e., the job sequenced in the r–th position is given by $p_j^{[r]} = p_j g(r)$, where $g(r)$ is a function that specifies a positional deterioration effect and p_j is the *normal* or *standard* processing time. For positional deterioration, we assume that $g(1) = 1$ and $g(r) \leq g(r+1)$ for each $r, 1 \leq r \leq n-1$. Under positional *polynomial* deterioration, the actual processing time of a job j that is sequenced in position r is given by $p_j^{[r]} = p_j r^A$, where A is a given positive constant that is common for all jobs. Under positional *exponential* deterioration, the actual processing time of a job j that is sequenced in position r is given by $p_j^{[r]} = p_j \gamma^{r-1}$, where $\gamma > 1$ is a given constant representing a rate of deterioration, which is common for all jobs. For the results on scheduling problems with positionally dependent processing times defined by polynomial functions, see [1], [4, 5] and [3], and by exponential functions, see [6], [7] and [3].

The models studied in this paper belong to the area of *due date assignment* (DDA); see the survey by [2] for a recent comprehensive review of this area. Here, each job $j \in N$ has to be assigned a due date d_j, by which it is desirable to complete that job.

In all problems that we consider in this paper, the jobs of set N have to be split into two subsets denoted by N_E and N_T. The jobs of subset N_T are essentially discarded, and a penalty α_j is paid for a discarded job $j \in N_T$. In a feasible schedule S only the jobs of set N_E are sequenced, and each of these jobs is completed no later than its due date. Given a schedule, we refer to the jobs of set N_E as *early* jobs, while the jobs of set N_T are called *discarded.* The processing times of the jobs of set N_E are subject to positional deterioration. The purpose is to select the due dates for the jobs and the sequence of the early jobs in such a way that a certain penalty function is minimized. We focus on two objective functions. One of them includes the cost of changing the due dates and the total penalty for discarding jobs, i.e.,

$$F_1(d, \pi) = \varphi(d) + \sum_{j \in N_T} \alpha_j, \tag{1}$$

where π is the sequence of the early jobs, d is the vector of the assigned due dates, $\varphi(d)$ denotes the cost of assigning the due dates that depends on a chosen rule of due date assignment. Another objective function additionally includes the total earliness of the scheduled jobs, i.e.,

$$F_2(d, \pi) = \sum_{j \in N_E} E_j + \varphi(d) + \sum_{j \in N_T} \alpha_j, \tag{2}$$

2 Two Auxiliary Problems

In this section we study two auxiliary single machine sequencing problems that are closely related to the problems of our primal concern. We start with the following problem (Problem PdE). The jobs of set $N = \{1, 2, \ldots, n\}$ are processed on a single machine, and for a job $j \in N$ the normal processing time is equal to p_j. If the jobs are processed in accordance with a certain permutation $\pi = (\pi(1), \pi(2), \ldots, \pi(n))$, then the processing time of job $j = \pi(r)$ is given by $p_j^{[r]} = p_{\pi(r)} g(r)$, where $g(r)$ is a function that specifies a positional deterioration or learning. The objective is to find a permutation that minimizes the function

$$f(\pi) = \beta d(\pi) + \sum_{k=1}^{n} E_{\pi(k)}, \tag{3}$$

where β is a given positive constant, $d(\pi)$ is the completion time of the last job $\pi(n)$ and is treated as a due date common to all jobs, while

$$E_{\pi(k)} = d(\pi) - C_{\pi(k)} = \sum_{j=1}^{n} p_{\pi(j)}^{[j]} - \sum_{j=1}^{k} p_{\pi(j)}^{[j]} = \sum_{j=k+1}^{n} p_{\pi(j)}^{[j]} \tag{4}$$

is the earliness of job $\pi(k)$ with respect to that due date. Notice that by definition, $E_{\pi(n)} = 0$.
A permutation $\pi = (\pi(1), \pi(2), \ldots, \pi(n))$, such that $p_{\pi(1)} \geq p_{\pi(2)} \geq \ldots \geq p_{\pi(n)}$ is said to follow the Longest Processing Time (LPT) rule; if $p_{\pi(1)} \leq p_{\pi(2)} \leq \ldots \leq p_{\pi(n)}$ it is said to follow the Shortest Processing Time (SPT) rule.
The following theorem is valid.

Theorem 1. *For Problem PdE, if the inequality $\frac{g(u+1)}{g(u)} \geq \frac{\beta+u-1}{\beta+u}$ holds for each u, $1 \leq u \leq n-1$, then an optimal permutation can be found by the LPT rule. If $\frac{g(u+1)}{g(u)} \leq \frac{\beta+u-1}{\beta+u}$ holds for each u, $1 \leq u \leq n-1$, then an optimal permutation can be found by the SPT rule.*

The theorem immediately implies the following statement.

Corollary 1. *For Problem PdE with positional deterioration, an optimal permutation can be found by the LPT rule.*

To complete this section, we introduce another auxiliary problem, that is essentially a version of Problem PdE in which the earliness penalties are ignored, so that only the function $\beta d(\pi)$, or rather $d(\pi)$ is to be minimized. We refer to this problem as Problem Pd. The following theorem holds for this problem.

Theorem 2. *For Problem Pd with positional deterioration, the LPT rule is optimal.*

The auxiliary Problems Pd and PdE are closely related to the DDA problems to minimize the functions F_1 and F_2 defined by (1) and (2), respectively. We use the sequencing rules established for Problems Pd and PdE in designing solution procedures for our DDA problems.

3 CON and SLK Due Date Assignments

First, we consider the due date assignment problems to minimize the functions F_1 and F_2, provided that the CON due date assignment rule is employed, i.e., all jobs to be scheduled are given a common due date d. For the CON model, all due dates are equal, and we select βd as the cost function $\varphi(d)$, where β is a positive constant. Thus, we study the problems of minimizing the objective functions

$$F_1(d, \pi) = \beta d(\pi) + \sum_{j \in N_T} \alpha_j, \tag{5}$$

and

$$F_2(d, \pi) = \sum_{j \in N_E} E_j + \beta d(\pi) + \sum_{j \in N_T} \alpha_j. \tag{6}$$

For the DDA problems of minimizing the functions (5) or (6), no matter which model of positional dependence of the processing times is chosen, we may search for an optimal schedule only among those schedules in which one of the jobs is on-time, i.e., completes at its due date (otherwise, the common due date can be reduced, thereby decreasing the objective function without creating any late jobs). Thus, in any feasible schedule, the jobs of set N_E are processed consecutively starting from time zero with no intermediate idle time, and the completion time of the last of these jobs is accepted as the due date d, common to all jobs in N_E.
For a given set N_E, we need to find a sequence of the jobs in this set that minimizes the contribution of these jobs to the objective function. This can be done by solving either Problem Pd (for function (5)) or Problem PdE (for function (6)). For function (5) such a contribution is given by $\beta d(\pi)$, where $d(\pi) = \sum_{j \in N_E} p_j^{[r]}$, and in the case of positional deterioration the required permutation π of the early jobs can be found by the LPT rule in accordance with Theorem 2. For function

(6), the contribution of the early jobs to the objective function is given by $\beta d(\pi) + \sum_{j \in N_E} E_j$, which corresponds to (3), i.e., in the case of positional deterioration the required permutation π of the early jobs can be found by the LPT rule in accordance with Corollary 1. In order to deliver the overall minimum to the function (5) or (6), we also need to take into account the contribution to the objective function in the form of the total cost of the discarded jobs. This is done by designing dynamic programming algorithms with running time $O(n^2)$.

Now we consider the due date assignment problems to minimize the functions F_1 and F_2, provided that the SLK due date assignment rule is employed, i.e., for each job its due date is computed by increasing its actual processing time by a *slack* q, common to all jobs. In the SLK model, the main issue is that of choosing an appropriate value of the slack so that a certain objective function is minimized.

Formally, suppose that there are h jobs in set N_E and they are ordered in accordance with a permutation $\pi = (\pi(1), \pi(2), \ldots, \pi(h))$. Then the due date of job $j \in N_E$ scheduled in position r is defined as $d_j = p_j^{[r]} + q$. It is clear that due to positional dependence of the processing times, to guarantee that all jobs in set N_E are completed by their due dates the slack q must depend on the sequence of these jobs, i.e., $q = q(\pi)$. Since for the SLK model the due dates are assigned essentially by selecting the slack value q, in our problems of minimizing the functions (1) or (2) we select βq as the cost function $\varphi(d)$, where β is a positive constant. Thus, we study the problems of minimizing the objective functions

$$F_1(q, \pi) = \beta q(\pi) + \sum_{j \in N_T} \alpha_j, \tag{7}$$

$$F_2(q, \pi) = \sum_{j \in N_E} E_j + \beta q(\pi) + \sum_{j \in N_T} \alpha_j. \tag{8}$$

We start with establishing some structural properties of schedules for the problems under consideration. In what follows, we restrict our search for an optimal schedule only to those schedules in which at least one job is on-time, i.e., completes at its due date; otherwise, the slack can be reduced, thereby decreasing the objective function without creating any late jobs.

Theorem 3. *Let S be a schedule for the problem of minimizing one of the functions (7) or (8) in which the jobs of set N_E are processed in accordance with a permutation $\pi = (\pi(1), \pi(2), \ldots, \pi(h))$, where $h \leq n$. Then only the last job $\pi(h)$ completes on time, i.e., $C_{\pi(h)} = d_{\pi(h)}$,* *and the slack $q(\pi)$ is equal to the total processing time of all jobs that precede the on-time job $\pi(h)$, i.e., $q(\pi) = \sum_{j=1}^{h-1} p_{\pi(j)}^{[j]}$.*

Theorem 3 implies that the value of the slack depends on the sequence of jobs that precede the on-time job, but not on the on-time job itself. Thus, we should try to place each job as an on-time job and find the best schedule for the remaining jobs. The overall optimal schedule is either the best schedule for those found in each class with a fixed on-time job, or the empty schedule in which all jobs are discarded. Suppose that some on-time job $h \in N$ is fixed and set N_E is known. We need to find a sequence of the jobs in set $N_E \backslash \{h\}$ that minimizes the contribution of these jobs to the objective function. This can be done by solving either Problem Pd (for function (7)) or Problem PdE (for function (8)). Additionally, we need to take into account the total cost of the discarded jobs. Minimizing the general cost functions can be done by dynamic programming algorithms, in which the jobs are scanned in an appropriate sequence provided by Theorems 1 and 2. The running time of both algorithms is $O(n^3)$.

We also discuss how the obtained results can be extended to the models with a positional learning effect.

References

1. Biskup D. (1999) Single-machine scheduling with learning considerations. European Journal of Operational Research 115: 173–178
2. Gordon VS, Proth J-M, Strusevich VA (2004) Scheduling with due date assignment. In: Leung J Y-T (Ed) Handbook of Scheduling: Algorithms, Models and Performance Analysis. CRC Press, Boca Raton
3. Gordon VS, Potts CN, Strusevich VA, Whitehead JD (2008) Single machine scheduling models with deterioration and learning: handling precedence constraints via priority generation. Journal of Scheduling, In Press
4. Mosheiov G (2001) Scheduling problems with a learning effect. European Journal of Operational Research 132: 687–693
5. Mosheiov G (2005) A note on scheduling deteriorating jobs. Mathematical and Computer Modelling 41: 883–886
6. Wang J-B (2005) Flow shop scheduling jobs with position-dependent processing times. Journal of Applied Mathematics and Computing 18: 383-391
7. Wang J-B, Xia Z-Q (2005) Flow shop scheduling with a learning effect. Journal of the Operational Research Society 56: 1325-1330

Scheduling Component Placement Operations for Collect-and-Place Type PCB Assembly Machines

Özgür Kulak[1], Serol Bulkan[1], and Hans-Otto Günther[2]

[1] Department of Industrial Engineering, Marmara University, Göztepe Campus, 34722 Istanbul, Turkey,
ozgur.kulak@gmail.com

[2] Department of Production Management, Technical University of Berlin, Wilmersdorfer Str. 148, 10585 Berlin,
hans-otto.guenther@tu-berlin.de

Summary. Component placement is one of the most time-consuming operations in printed circuit board (PCB) assembly and dominates the cycle time of any PCB assembly line. In our study, we focus on collect-and-place machines which first collect a number of electronic components from the component magazine and then place them one-by-one onto the PCB. With this type of machinery two problems arise, i.e. generating placement tours and determining the placement sequence within each tour. To solve these problems, an efficient clustering algorithm to create placement tours and two modified nearest neighbor algorithms (MNNHs) to determine the placement sequence are proposed. The objective is to minimise the assembly cycle time per board. Numerical experiments show that high-quality solutions are obtained within very short CPU time.

1 Introduction

Most of the subproblems in PCB assembly are NP-hard and can only be approximately solved by heuristic procedures (cf. [3]). The high complexity of the PCB assembly problems also suggests its decomposition into more manageable subproblems (cf. [3], [7]). The majority of production planning software systems utilize, in some way or another, hierarchical decomposition techniques too (cf. [6]).
A type of assembly machine becoming increasingly popular in industry is the collect-and-place machine which first collects a number of electronic components from the component magazine of the machine and

B. Fleischmann et al.(Eds.), *Operations Research Proceedings 2008*,
DOI: 10.1007/978-3-642-00142-0_21,

then places them one-by-one onto the PCB. For a detailed description of the working principle of collect-and-place machines, cf. [4].

In the literature there are only few papers dealing with collect-and-place machines. [4] proposed heuristic solution procedures for scheduling the machine operations by solving the assignment problem of component feeders to slots in the component magazine first. Contrary to the solution approach of [4], our solution procedure determines the placement sequence first based on the proximity of the component locations to each other on the board. Since component locations on the board are fixed, determining the placement sequence first gives important advantages to minimize the cycle time for this machine type. In this way, the feeder assignment can also be made with certainty by taking the predetermined placement sequence as input.

[5] presented different genetic algorithm approaches (GAs) and utilized the density search construction method of [1] to generate clusters for the placement tours. We use a clustering algorithm to constitute placement tours too. However, our proposed clustering algorithm is specifically tailored for the scheduling problem of collect-and-place machines. Thus, it gives us more opportunities to minimize the cycle time. Another unique feature of our solution approach is that it has no limitation in terms of the number of placement points and their distribution on the board. Contrary to GAs, the CPU time performance of our proposed solution approach does not worsen noticeably by even a considerable increase in the number of placement points.

2 Clustering Algorithm

The clustering algorithm generates equal-size tours, the size of which is determined by the number of nozzles (typically 6 or 12) on the machine head (head capacity). This quality enables minimizing the number of tours for each PCB and decreases the cycle time considerably. The algorithm subdivides the PCB placement area by using horizontal dividing lines and constitutes clusters within these subareas. The proper number of dividing lines to obtain the best cycle time for each PCB is determined by the algorithm. This approach creates homogeneous and compact clusters and enables the machine head to complete the placement tours within less time by more benefiting from the rotational cycle time. Another advantage of this approach compared to distance or density-based algorithms in the literature is to prevent the deterioration that occurs towards the last clusters produced by those algorithms.

The proposed clustering and the first modified nearest neighbour algorithms (MNNH1) aim to construct reverse U-shaped placement routes within each cluster. Thus, they benefit from the rotational cycle time of the machine head to minimize the cycle time and to decrease the y-distances to travel between the placement points and the component magazine. The rotational cycle time is the minimum time between two successive placement operations due to the stepwise rotational movement of the revolver-type placement head.

The clustering algorithm assesses different numbers of dividing lines for obtaining the best clusters to minimize the assembly cycle time of the PCB. The necessary number of dividing lines to assess constitutes an upper limit to be determined by the algorithm in the following way. The maximum distance between two placement points to visit within the rotational cycle time of the machine head is calculated first. This value is multiplied by half of the head capacity to find the maximum length of the reverse U-shaped placement routes. The length of the PCB is divided by this maximum length and the result is rounded to the nearest integer. This value is multiplied by a parameter to find the upper limit. The parameter value of two has been found in our numerical tests as appropriate. The best cycle times have been obtained by the number of dividing lines which are less or equal to this upper limit found by using the value of two for this parameter. However, we recommend to try higher numbers, such as three or four, when the algorithm runs first time with a new data set because the best cycle times may be obtained with a higher number of dividing lines depending on the specific PCB design. Since CPU times are very short, this preliminary test can be done quite easily.

The remaining part of the algorithm is as follows: Partition the PCB placement area into equal subareas by using horizontal dividing lines. Calculate the bounds of the subareas. List placement points within each subarea. Find new y-values of the placement points by subtracting the y-level of the lower dividing line at each subarea from the original y-values of the placement points. Apply the following part of the ordered constructive heuristic of [2] for the assignment of placement points in each subarea to the various tours of the machine head: Sort the PCB points starting with the minimum of maximum (x,y) coordinate and assign the sorted PCB points consecutively starting with the top in the list to a placement tour. [1]

If the number of placement points remained for the last tour in each subarea is less than the head capacity, assign these placement points

[1] The subtour generation method of [2] for multi head placement machine.

temporarily to a matrix. After all clusters are generated in every subarea in the predescribed way, sort the placement points in this matrix in descending order of their y-values. Assign the sorted PCB points consecutively starting with the top in the list to a placement tour by taking the head capacity into account. Call MNNH, determine the cycle time and keep the result. After the algorithm assesses different numbers of dividing lines from 1 to the calculated upper limit, the best cluster combination among them having the minimum cycle time is obtained.

3 Modified Nearest Neighbour Algorithms

MNNH1 takes the clusters produced by the clustering algorithm as input. For each cluster on the PCB, determine the placement point having the minimum y-value within the points in the cluster as start point of the tour. Repeat this to determine the end point of the tour within the remaining points in the cluster. The algorithm then applies the well-known nearest neighbour procedure from the literature (NNH) to determine the placement sequence by using the predetermined start and end points as the first and last placement operation in each placement tour. The second modified nearest neighbor algorithm (MNNH2) determines only the start point as described in the MNNH1, and then applies the NNH to determine the placement sequence within the tour.

4 Computational Results

The data set used for the numerical experiments consists of 8 industrial PCBs assembled for automation control equipments. The number of components on each PCB ranges from 152 to 537. The proposed algorithms have been programmed using the MATLAB software on a PC with 1.8 GHz processor and 2038 MB RAM.
Assembly cycle times (ACT) for each PCB obtained by the proposed clustering algorithm with MNNH1 are presented in the second column of the Table 1. Percentage improvements of the proposed algorithms against the applied part of the ordered constructive heuristic are given in the last column. The average improvement is 6.28% and the average CPU time to produce a solution for a PCB is 0.12 seconds. The largest improvement in cycle time of 10.15% is achieved for the PCB having the maximum number of components in the data set.

Table 1. Assembly cycle time performance

No. of components	ACT (sec)	ACT improvement (%)
152	19.35	4.07
416	52.69	7.08
420	52.36	6.35
424	54.62	3.98
448	56.85	4.04
460	58.90	7.81
501	66.97	6.73
537	71.19	10.15

In Table 2, the percentage improvement in assembly cycle time achieved by the proposed MNNHs is presented. The average improvement of MNNH1 against NNH is 2.80%. MNNH1 performs 2.14% better than MNNH2 on average. MNNH1 gives the best perfomance with the proposed clustering algorithm and proves the effectiveness of the reverse U-shaped placement routes for the investigated machine type.

Table 2. Performance comparison of the MNNHs

No. of components	MNNH1 against NNH (%)	MNNH1 against MNNH2 (%)	MNNH2 against NNH (%)
152	3.59	2.10	1.46
416	2.16	1.78	0.38
420	3.64	2.80	0.82
424	2.89	2.32	0.56
448	2.66	2.23	0.42
460	3.36	2.46	0.89
501	1.81	1.61	0.19
537	2.26	1.82	0.43

5 Conclusions

In this paper, a novel solution approach for scheduling component placement operations of collect-and-place type PCB assembly machines is proposed. Clustering and MNNH algorithms have been developed by considering specific working principles of this machine type. Thus, the proposed solution approach is quite effective and particularly useful when the number of electronic components on a PCB is large. MNNH1 performs best with the proposed clustering algorithm and the reverse

U-shaped placement routes create an important advantage for this machine type. Contrary to other algorithms in the literature, the CPU time performance of the clustering algorithm does not worsen noticeably with the number of placement points on a PCB and their distribution. Our proposed clustering algorithm has no limitation in terms of the number of placement points and their distribution, either. These features make the algorithm superior to others and a proper candidate for industrial use. Integrating the proposed algorithms with a feeder assignment algorithm is considered as a topic for future research. The proposed clustering algorithm will enable us also to use it as a workload balancing algorithm for the dual-gantry collect-and-place machine. Different solution approaches in the literature do not provide the same opportunity to researchers.

Acknowledgement. The first author has been supported for this research by the German Academic Exchange Service (DAAD). We also would like to thank Dr. Ihsan Onur Yilmaz for providing the data.

References

1. Ahmadi S, Osman IH (2005) Greedy random adaptive memory programming search for the capacitated clustering problem. European Journal of Operational Research 162: 30-44
2. Ayob M, Kendall G (2003) An Investigation of An Adaptive Scheduling Approach for Multi Head Placement Machines, in: Proceedings of MISTA 2003, Nottingham, UK, pp. 363-380
3. Crama Y, Kolen AWJ, Oerlemans AG, Spieksma FCR (1990) Throughput rate optimization in the automated assembly of printed circuit boards. Annals of Operations Research 26: 455-480
4. Grunow M, Günther HO, Schleusener M, Yilmaz IO (2004) Operations planning for collect-and-place machines in PCB assembly. Computers & Industrial Engineering 47: 409-429
5. Kulak O, Yilmaz IO, Günther HO (2007) PCB assembly scheduling for collect-and-place machines using genetic algorithms. International Journal of Production Research, 45: 3949-3969
6. Smed J, Johnsson M, Nevalainen O (2000) A hierarchical classification scheme for electronics assembly problems, in: Proceedings of TOOLMET2000 Symposium, Oulu, Finland, pp.116-119
7. Yilmaz IO (2008) Development and Evaluation of Setup Strategies in Printed Circuit Board Assembly. Gabler, Wiesbaden

Scheduling of Multiple R&D–Projects in a Dynamic and Stochastic Environment

Philipp Melchiors and Rainer Kolisch

Lehrstuhl für Technische Dienstleistungen und Operations Management, Technische Universtität München, Arcisstr. 21, 80333 München
{philipp.melchiors,rainer.kolisch}@wi.tum.de

Summary. In R&D–departments typically multiple projects are processed simultaneously and compete for scarce resources. The situation is dynamic since new projects arrive continuously and stochastic in terms of interarrival times of projects as well as of the work content of the projects. The problem of scheduling projects in this context such that the weighted tardiness is minimized is particularly difficult and has not been covered much in the literature. The contribution of this paper is an assessment of priority rules originally proposed for the static and deterministic context in the dynamic and stochastic context.

1 Introduction

R&D–departments are under pressure to deliver results such as product specifications, prototypes etc. in a timely manner. Typically, different projects which are competing for scarce resource such as employees, computers etc. are processed simultaneously. The situation is dynamic in the sense that new projects arrive continuously and stochastic in terms of the interarrival times and of the work content (measured in units of time). An important problem which has not been covered much in the literature is the scheduling of projects in this context such that the weighted tardiness is minimized.

Multiple projects in a dynamic and stochastic context have been considered by different authors but without much consideration of scheduling decisions. Adler et al. [1] present a queueing network based approach in order to describe the processing of R&D–projects. They propose a simulation model to investigate the determinants of development time. Anavi-Isakow and Golany [2] propose two control mechanisms to reduce

B. Fleischmann et al.(Eds.), *Operations Research Proceedings 2008,*
DOI: 10.1007/978-3-642-00142-0_22, © Springer-Verlag Berlin Heidelberg 2009

the makespan of projects by maintaining either a constant number of projects (CONPIP) or a constant work content (CONTIP) in the system. Cohen et al. [3] consider the critical chain approach for determining priorities which are then used for scheduling. They provide a performance analysis for the minimum slack priority rule (MINSLK), the first come, first served priority rule (FCFS), the input control mechanisms CONPIP and CONTIP as well as queue size control (QSC). However, no comparison with other priority rules is done and the analysis does not consider a weighted tardiness objective.

Scheduling multiple projects in a static and deterministic context using priority rules has been covered by a number of authors. Kurtulus and Narula [4] compare the performance of several priority rules. Lawrence and Morton [6] propose a new priority rule which shows good results for problems with a weighted tardiness objective function.

Priority rules have also been analysed in the context of the dynamic and (in terms of the activity durations known before scheduling) deterministic job shop scheduling problem with weighted tardiness objective. This problem is a special case of the dynamic and stochastic multi-project scheduling problem. Vespalainen and Morton [7] propose two new rules and compare them with existing ones in a simulation study. Kutanoglu and Sabuncuoglu [5] analyze the performance of different priority rules with an emphasis on recently proposed priority rules.

In this paper we follow the approach of Adler et al. [1]. We make the following assumptions. First, there are project types such that projects of type p have the same precedence network and the same distribution and expected value of the interarrival times. Furthermore, the duration of a specific activity of a project type has a given distribution and expected value. Secondly, there are multiple resources where resource r has c_r servers. Each activity is processed by a single server of a specific resource r in a non-preemptive manner. Due to these assumptions, the problem can be modelled by a queueing network. The objective is to minimize the expected weighted tardiness.

In the following, different priority rules for selecting activities in the queues are compared with respect to their performance. The rules are described in Section 2 before the computational experiment is presented in Section 3. In Section 4 we will briefly present and discuss the obtained results.

2 Priority Rules

It is assumed that a single rule is used for all queues. The selection of the rules has been made with respect to their performance for related problems and the information employed. The rules presented in this paper are taken from Kurtulus and Narula [4] and Lawrence and Morton [6]. Note that the duration of an activity is equivalent to its work content since the activity is processed by a single server of a resource. Since each project j is always of a type p, the expected duration of activity i is known. The following parameters are used: w_j denotes the weight of project j, $\overline{p}_{ij}$ denotes the expected duration of activity i of project j, $\overline{p}$ denotes the expected duration of all activities of all project types, l_{ij} denotes the estimation for the resource unconstrained latest start time of activity i of project j, k denotes a look ahead parameter, t denotes the current time, U_i is a set of the unfinished activities of project i at time t, $r(i,j)$ denotes the resource used by activity i, and t_{ij} is the arrival time of activity i of project j. We can now define the priority rules as follows:

- First come, first served (FCFS): t_{ij}
- Maximum penalty (MAXPEN): w_j
- Minimum slack (MINSLK): $l_{ij} - t$
- Weighted shortest processing time (WSPT): $\frac{w_j}{\overline{p}_{ij}}$
- Rule of Lawrence and Morton [6] (LM):[1] $\frac{w_j}{\pi_{ij}} \cdot \exp\left(-\frac{(l_{ij}-t)^+}{k\overline{p}}\right)$ with $\pi_{ij} = \sum_{k \in U_i} \frac{\overline{p}_{kj}}{c_{r(i,j)}}$

3 Experimental Design

The generation of problem instances is done in two steps: First, the project types are generated. Secondly, the problem instances composed of a combination of project types and resources as well as different parameters are built. The exponential distribution is used for interarrival times and activity durations. Hence, only the expected values need to be specified. We use two project types p with total work contents W_p of 100 and 50 units. In the following, the generation of a project type with W_p=100 is described and the information on the generation of a project type with $W_p = 50$ is given in parentheses. The total number of activities is 20 (10). In the experiment $R = 3$ resources are assumed.

[1] We use the variant with uniform resource pricing and global activity costing.

W_p is distributed among the resources by prespecified percentages of 60%, 30%, and 10%. By this, we depict the fact that there are resources of different work load. The assignment of the activities of a project type to the resources is proportional to the work content. The mean durations of the activities at resource r are randomly drawn and normalized such that they add up to the work content assigned to r. Finally, the activities of a project type are randomly assigned to the nodes of three different project networks with 20 (10) nodes and an order strength of 0.5. The project networks have been generated with PROGEN/max (cf. Schwindt [8]).

In the second step the problem instances are generated with $P = 2$ project types. The following combinations of the total work content W_p (W_1, W_2) are considered: (100, 100) and (100, 50). For each (W_1, W_2)–tuple three different combinations of project types are randomly selected such that in each combination the project networks are different from each other. The weights of the projects of type p are randomly drawn from the uniformly distributed interval $[\overline{w}_p \cdot 0.75, \overline{w}_p \cdot 1.25]$ where the expected weight $\overline{w}_p$ of project type p is proportionally set to W_p such that $\overline{w}_p = 2$ (1) for $W_p = 100$ (50). To determine the due date for each project, a time span which is randomly drawn from the uniformly distributed interval $[0.9 \cdot \tau \cdot \overline{D}^u_p,\ 1.1 \cdot \tau \cdot \overline{D}^u_p]$ is added to the project arrival time. $\overline{D}^u$ represents the expected duration of a project of type p. $\overline{D}^u_p$ is calculated by using Monte Carlo simulation where resource constraints are not taken into account. Hence, for $\tau = 0$ the tightness of the due date is maximum and the objective is the minimization of the weighted expected makespan. We use tightness factors of $\tau = 0, 2$, and 4. The probability that an arrived project is of type p is a_p. We have set the (a_1, a_2)–tuples to (0.2, 0.8), (0.5, 0.5), and (0.8, 0.2). The number of servers of the three resources is $c_r = 6$, 3, and 1. c_r has been set such that the expected utilization of the resources is equal. The total arrival rate λ is controlled by the utilization per server u. u has been set to 0.6, 0.75, and 9. The number of instances has been obtained as follows: Level of (W_1, W_2) $\times$ level of τ $\times$ level of u $\times$ level of combinations of project networks $\times$ level of (a_1, a_2) $= 2 \cdot 3 \cdot 3 \cdot 3 \cdot 3 = 162$. According to preliminary tests we have set the look ahead parameter of the LM–rule to $k = 1$. The replication length has been set to 500,000 and the warm up phase to 100,000 time units. Four replication have been carried out for each instance.

4 Results

We analyse the results w.r.t. the levels of the parameters τ and u because these two factors have shown a significant impact on the results in studies of the dynamic and deterministic job shop scheduling problem and the static and deterministic multi–project scheduling problem, respectively. In the context of a dynamic and deterministic job shop with weighted tardiness objective (cf. Kutanoglu and Sabuncuoglu [5]) WSPT achieved good results when due dates were difficult to meet (high utilization or tightness) while MINSLK achieved good results when due dates were easy to meet (low utilization or tightness). A special case of the LM–rule (cf. Vepsalainen and Morton [7]) showed a better performance than MINSLK and WSPT in many cases. The LM–rule (cf. Lawrence and Morton [6]) showed better results than a number of other rules, including MINSLK, for the static and deterministic multi-project scheduling problem with weighted tardiness objective.

Table 1 reports the average weighted tardiness as well as the average rank of the priority rules for the combinations of tightness τ and utilization u. There are two main observations: First, MINSLK is the best rule in case of low utilization ($u = 0.6$), irrespectively of the tightness of the due dates, and in case of tight due dates ($\tau = 0$), irrespectively of the utilization. The LM–rule is the best rule in the case of loose due dates ($\tau = 4$) and a medium or high utilization ($u \geq 0.75$). Secondly, the performance gap between MINSLK and LM is not very large. Hence, the simpler rule MINSLK can be used if ease of implementation is important. We have the following explanations for these results. First, in the case of a stochastic problem the value of the information on the activity duration is decreased because only expected values are known. Furthermore, the priority rules WSPT and LM cannot distinguish between identical activities (having identical expected values) which belong to different projects of the same type. Secondly, if the slack of an activity is negative, the LM–rule reduces to w_{ij}/π_{ij} which makes it static while MINSLK uses the dynamic information of negative slack.

Table 1

$\tau = 0$				$\tau = 4$			
u	Rule	Weighted tardiness	Rank	u	Rule	Weighted tardiness	Rank
0.60	MINSLK	109.27	1.00	0.6	MINSLK	0.01	1.22
	LM	112.16	2.06		LM	0.01	1.44
	MAXPEN	116.88	3.28		FCFS	0.09	3.00
	WSPT	118.29	3.72		MAXPEN	1.97	4.44
	FCFS	121.89	4.94		WSPT	1.92	4.56
0.75	MINSLK	137.62	1.17	0.75	LM	0.31	1.39
	LM	141.20	1.89		MINSLK	0.31	1.61
	MAXPEN	151.22	3.11		FCFS	3.04	3.00
	WSPT	161.22	4.06		MAXPEN	14.92	4.28
	FCFS	166.83	4.78		WSPT	17.09	4.72
0.90	MINSLK	258.93	1.67	0.9	LM	30.37	1.00
	LM	260.83	1.78		MINSLK	32.56	2.00
	MAXPEN	280.56	2.61		FCFS	91.56	3.00
	WSPT	345.75	4.17		MAXPEN	115.68	4.06
	FCFS	354.20	4.78		WSPT	158.52	4.94

References

1. Adler P. S., Nguyen V., Schwerer E.. From Project to Process Management: An Empirically-based Framework for Analyzing Product Development Time. Management Science, 41/3:458–484, 1995.
2. Anavi-Isakow S., Golany B. Managing multi-project environments through constant work-in-process. International Journal of Project Management, 21: 9–18, 2003.
3. Cohen I., Mandelbaum A., Shtub A.. Multi-Project Scheduling And Control: A Process-Based Comparative Study Of The Critical Chain Methodology And Some Alternatives. Project Management Journal, 35/2:39–50, 2004.
4. Kurtulus I.S., Narula S.C.. Multi-project scheduling: Analysis of project performance. IIE Transactions, 17/2:58–66, 1985.
5. Kutanoglu E., Sabuncuoglu I.. An analysis of heuristics in a dynamic job shop with weighted tardiness objectives. International Journal of Production Research, 37/1:165–187, 1999.
6. Lawrence S. R., Morton Th. E.. Resource-constrained multi-project scheduling with tardy costs: Comparing myopic, bottleneck, and resource pricing heuristics. European Journal of Operational Research, 64:168–187, 1993.
7. Vepsalainen A. P. J., Morton Th. E.. Priority rules for job shops with weighted tardiness costs. Management Science, 33/8:1035–1047, 1987.
8. Schwindt C.. Generation of Resource-Constrained Project Scheduling Problems Subject to Temporal Constraints. Technical Report, WIOR-543, Universitaet Karlsruhe, 1998.

An Evolutionary Algorithm for Sub-Daily/Sub-Shift Staff Scheduling

Volker Nissen[1] and René Birnstiel[2]

[1] Technical University of Ilmenau, Chair of Information Systems in Service, PF 10 05 65, D-98684, Germany, volker.nissen@tu-ilmenau.de
[2] rene.birnstiel@wirksam-beraten.de

1 Introduction

Staff scheduling involves the assignment of a qualified employee to the appropriate workstation at the right time while considering various constraints. According to current research employees spend 27 to 36% of their working time unproductively, depending on the branch [10]. Major reasons include a lack of planning and controlling. Most often staff scheduling takes place based on prior experience or with the aid of spreadsheets [1]. Even with popular staff planning software employees are regularly scheduled for one workstation per day. However, in many branches, such as trade and logistics, the one-employee-one-station concept does not correspond to the actual requirements and sacrifices potential resources. Therefore, *sub-daily (including sub-shift) planning* should be an integral component of demand-oriented staff scheduling.

2 Description of the Problem

The present problem originates from a German logistics service provider which operates in a spatially limited area 7 days a week almost 24 hours a day. The employees are quite flexible in terms of working hours. There are strict regulations e.g. with regard to qualifications because the assignment of unqualified employees can lead to material damage and personal injury. Many employees are qualified to work at different workstations. Currently, monthly staff scheduling is carried out with MS EXCEL, in which employees are assigned a working-time model and a set workstation on a full-day basis. Several considerations are included, such as attendance and absence, timesheet balances, qualifications and resting times etc. The personnel demand for the workstations is subject to large variations during the day. However, employees

B. Fleischmann et al.(Eds.), *Operations Research Proceedings 2008*,
DOI: 10.1007/978-3-642-00142-0_23, © Springer-Verlag Berlin Heidelberg 2009

are generally scheduled to work at the same workstation all day, causing large phases of over- and understaffing. This lowers the quality of service and the motivation of employees and leads to unnecessary personnel costs as well as downtime. Usually, floor managers intervene on-site by relocating employees on demand. Nine different workstations need to be filled. Personnel demand is given in 15-minute intervals. A total of 45 employees are on duty, each having different start and end times from their work-time models. A staff schedule is only valid if any one employee is only assigned to one workstation at a time and if absent employees are not part of the plan. There are hard and soft constraints which are penalised with error points. The determination of error points is in practice an iterative process and not covered here. The objective is a minimisation of error points in the final solution.

3 Related Work

In [6] Ernst et al. offer a classification of papers related to the issue of staff scheduling between the years 1954 and 2004. This category *flexible demand* is characterised by little available information on schedules and upcoming orders (e.g. in inbound call centres). A demand per time interval is given as well as a required qualification. Thus, the logistics service provider problem can be classified in the group flexible demand schemes. However, it also has characteristics from the category *task assignment*. In [8] Schaerf and Meisels provide a universal definition of an employee timetabling problem. Both the concepts of shifts and of tasks are included, whereby a shift may include several tasks. Employees are assigned to the shifts and assume task for which they are qualified. Since the task is valid for the duration of a complete shift, no sub-daily changes of tasks are made. Blöchlinger [4] introduces, timetabling blocks (TTBs) with individual lengths. In this model employees may be assigned to several sequential TTBs, by which sub-daily time intervals could be represented within a shift. Blöchlinger's work also considers tasks; however, a task is always fixed to a TTB. Essentially, our research problem represents a combination of [8] (assignment of staff to tasks) and [4] (sub-daily time intervals). As a work quite related to our own research Vanden Berghe [12] presents an interesting planning approach for sub-daily time periods (flexible demand), which allows the decoupling of staff demand from fixed shifts resulting in fewer idle times. However, scheduling is not performed at the detailed level of individual workstations as in our re-search. Apparently, there exists no off-the-shelf solution approach to the kind of detailed sub-daily staff planning problem considered here, although *local search* and *constructive heuristics* were successful in the 2007 ROADEF-challenge for a similar scheduling problem from France Telecom [11]. An evolutionary algorithm (EA) for the logistics service provider problem is outlined below. We assume the reader is familiar with basics of EA [2].

4 An Evolutionary Solution Approach

Evolutionary Algorithms are broadly applicable metaheuristics, based on an abstraction of natural evolution [2]. The EA suggested here combines the efficient selection scheme from evolution strategies (ES) [3] with search operators similar to genetic algorithms (GA). ES and GA belong to the same class of metaheuristics, so merging them is relatively straightforward and can improve performance. It is important, though, that coding, search operators and selection pressure work together. For a practical application, this can today only be determined through experimental testing. To apply solution methods, the scheduling problem needs to be conveniently represented. A two-dimensional matrix is applied (table 1). The rows of the matrix signify employees and the columns signify time periods. To mark times in which an employee is not present due to his work-time model, a dummy workstation is introduced (in table 1: workstation 0). Assignment changes can only be made on non-dummy workstations, so that no absent employee is included. Within the planned period, time is viewed as discrete. An event point (a new time interval begins) occurs when the allocation requirement for one or more workstations or employee availability change. With this method the periods are not equally long, so their lengths need to be stored. Thus, the matrix of each solution consists of 45 rows and 63 columns, yielding 2,835 dimensions, but workstation assignments can only be made in 1,072 dimensions due to employee absence.

Table 1. Assignment of workstations in a two-dimensional matrix.

employee	period						
	0	1	2	3	4	5	6
1	1	1	1	1	1	1	1
2	0	0	2	2	2	2	2
3	0	0	1	1	2	2	2
4	0	0	6	6	6	6	2
5	3	3	3	2	2	0	0

The EA-population is initialized with valid solutions where workers have been assigned to workstations for an entire shift. The fitness evaluation of individuals is based on penalties for violating the constraints of the problem, such as under- or overstaffing of workstations or insufficient qualification of employees. Excessive job rotations are also punished. A (μ, λ)-selection is employed, meaning that in every generation λ offspring are generated from μ parent solutions. It is a deterministic, non-elitist selection method that prevents parents to survive to the next generation cycle. However, the best solution found during an experimental run is always stored and updated in a "golden cage". It represents the final solution of the run. Following suggestions in the literature [2] [3], the ratio μ/λ is set to 1/5 - 1/7 during the practical experiments.

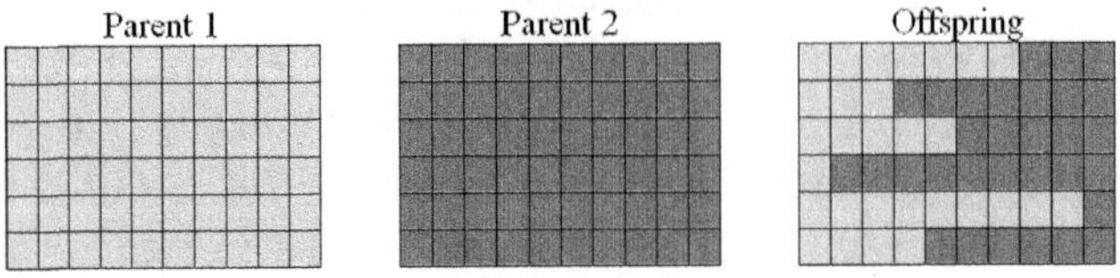

Fig. 1. Recombination operator employed.

The recombination of parents to create an offspring solution works as follows: A crossover point is determined independently and at random for each employee (row) of a solution and the associated parts of the parents are exchanged (similar to n-point crossover in GA, fig. 2). Mutation of an offspring is carried out by picking an employee at random and changing the workstation assignment for a time interval chosen at random. It must be ensured, though, that valid assignments are made w.r.t. the problem constraints. Since mutation is a rather disruptive operator it is applied with probability 0.8 to only one employee per offspring. This value is purely heuristic and based on tests, as no recommendation for this type of application is available from the literature. The EA terminates when 400.000 solutions have been inspected or the population has converged. Alg. 1 presents an overview of the approach.

Algorithm 1 Outline of evolutionary approach.

```
procedure EVOLUTIONARY ALGORITHM
    initialize the population
    calculate fitness of initial population
    repeat
        draw and recombine parent solutions
        mutate offspring
        calculate fitness for offspring
        select the new population
    until termination criterion holds
    output best solution from current run
end procedure
```

EA-results are compared to *local search* that performed well in the 2007 ROADEF-challenge [11]. It starts from the manual plan. Then, the workstation allocations are systematically and successively altered through all dimensions. The objective function value is calculated after each change. When error points are not increased, the workstation change is accepted. In a multi-start version, local search is re-initialized until roughly the same amount of solutions has been inspected as in the EA. Furthermore, the EA is compared to an own implementation of *Particle Swarm Optimization* (PSO), not detailed here for reasons of space. In PSO swarm members are assumed to be massless, collision-free particles that communicate and search for optima with the aid of a fitness function within a solution space [7] [9]. In this process each single particle together with its position embodies a solution to the problem. PSO performed well in a similar timetabling problem [5].

5 Results and Discussion

EA-experiments with different selection schemes were conducted. Table 2 presents the results for (30,200)- and (10,50)-selection. The average number of fitness evaluations (solutions inspected) as a hardware-independent measure is used to compare the computational requirements between all different solutions approaches. Thirty independent runs were performed for each of the experiments. All tests were conducted on a standard PC.

Table 2. Comparison of results for the logistic service provider problem, based on 30 independent runs for each heuristic. The best solutions (EA) are underlined.

	average error	minimal error	number of jobchanges	wrong qualification in minutes	understaffing in minutes	overstaffing in minutes (demand > 0)	overstaffing in minutes (demand = 0)	number of fitness evaluations
manual plan	11,850.00	11,850.00	0.00	0.00	2,400.00	1,950.00	3,750.00	-
local search	5,640.00	5,640.00	180.00	0.00	825.00	3,615.00	510.00	38,592
multistart local search (random workplaces per period)	5,623.75	5,552.00	242.00	0.00	750.00	3,540.00	510.00	424,512
multistart local search (random workplaces per day)	5,666.73	5,505.00	195.00	0.00	750.00	3,540.00	510.00	424,512
PSO	5,618.53	5,521.00	211.00	0.00	750.00	3,540.00	510.00	400,000
EA (10,50)-selection	5,503.33	5,487.00	177.00	0.00	750.00	3,540.00	510.00	400,010
EA (30,200)-selection	5,525.33	5,503.00	193.00	0.00	750.00	3,540.00	510.00	400,030

The manually generated full-day staff schedule results in 11,850 error points as an upper bound. The absolute optimum (minimum) of the test problem is unknown. An indication of a very good fitness value was generated by an extremely time-consuming PSO-run that investigated a total of 93,397,072 solutions and resulted in 5,482 error points. The simple local search that starts from the manual full day schedule as initial solution has the least computational requirements but is dependent on the start solution. In terms of effectiveness, local search appears inferior to the other methods. In its multi-start form, however, the quality of results is much better but at the cost of higher computational requirements. PSO generally performs at the level of multi-start local search. It is apparently difficult for PSO to reduce the number of workstation rotations and to maintain all hard constraints simultaneously while fine-tuning the solution. The evolutionary approach produces the best results with (10,50)-selection slightly superior to (30,200)-selection. Both require not more computational effort in terms of inspected solutions than PSO or multi-start local search. Thus, the EA-approach is the most effective heuristic for this application. The superior performance must be attributed to the operators of the EA since the coding is identical for all heuristics discussed here. In future research, these promising results are to be expanded by increasing the current planning horizon and creating further test problems with the aid of cooperating companies. Moreover, other heuristics from roughly comparable problems in the literature are currently being adapted to our domain and tested to further validate the results.

References

1. ATOSS Software AG, FH Heidelberg (2006) (ed.) Standort Deutschland 2006. Zukunftssicherung durch intelligentes Personalmanagement. München
2. Bäck T (2002) (ed.) Handbook of Evolutionary Computation. Institute of Physics Publishing, Bristol
3. Beyer H-G, Schwefel H-P (2002) Evolution strategies: a comprehensive introduction. Natural Computing 1: 3-52
4. Blöchlinger I (2004) Modeling Staff Scheduling Problems. EJOR 158: 533-542
5. Chu S C., Chen Y T, Ho J H (2006) Timetable Scheduling Using Particle Swarm Optimization. In: Proceedings of the International Conference on Innovative Computing, Information and Control (ICICIC Beijing 2006) Vol. 3: 324-327
6. Ernst A T, Jiang H Krishnamoorthy M, Owens B, Sier D (2002) An Annotated Bibliography of Personnel Scheduling and Rostering. Annals of OR 127: 21-144
7. Kennedy J, Eberhart R C, Shi Y (2001) Swarm Intelligence. Kaufmann, San Francisco
8. Meisels A, Schaerf A (2003) Modelling and solving employee timetabling. Annals of Mathematics and Artificial Intelligence 39: 41-59
9. Poli R (2007) An Analysis of Publications on Particle Swarm Optimization. Report CSM-469, Dep. of Computer Science, University of Essex, England
10. Proudfoot Consulting (2007) Produktivitätsbericht 2007. Company Report
11. ROADEF Challenge (2007) Technicians and Interventions Scheduling for Telecommunications. `http://www.g-scop.inpg.fr/ChallengeROADEF2007(2008-06-22)`
12. Vanden Berghe G (2002) An Advanced Model and Novel Meta-heuristic Solution Methods to Personnel Scheduling in Healthcare. Thesis, University of Gent, Belgium

Scheduling of Modular Production Lines in Mobile Terminal Manufacturing Using MILP

Frank Pettersson[1] and Janne Roslöf[2]

[1] Faculty of Technology, Åbo Akademi University,
Biskopsgatan 8, FIN-20500 Åbo, Finland
frank.pettersson@abo.fi

[2] Department of Information Technology, Turku University of Applied Sciences, Joukahaisenkatu 3 C, FIN-20520 Turku, Finland
janne.roslof@turkuamk.fi

Summary. The aim towards optimal utilization of business opportunities and manufacturing resources increases the requirements on both production planning and scheduling. One of the main goals is to maintain a high level of flexibility to permit a fast response to changed situations in the market. In this paper, a scheduling problem encountered in the flexible manufacturing of mobile terminals is discussed.

1 Introduction

In the mobile phone business the competition is focused on the ability to respond to changing market conditions promptly and flexibly. Market shares are retained or increased by introducing new designs and concepts at an increasing pace. This leads to a growing product portfolio and together with the shortened product life cycles they present a considerable challenge for efficient steering of the manufacturing processes [1].

In this paper, a real-life challenge in the production of mobile terminals is described and solved applying mathematical optimization tools. An mixed integer linear programming (MILP) -based multi-objective optimization model able to consider raw material availability and customer order specific constraints as well as staff balancing aspects is proposed.

B. Fleischmann et al.(Eds.), *Operations Research Proceedings 2008*,
DOI: 10.1007/978-3-642-00142-0_24,

2 Problem Description

The production of the core hardware components of mobile terminals is typically organized as modular production lines (MPL). These lines consist of a set of modules responsible for the different steps in the assembly. Each line is tailored to be able to produce a set of different intermediate variants with minor set-up modifications.
The short-term scheduling of MPL utilization is performed based on the following information: A) Demand for different products, B) The number of required operators for each product/line, and C) Knowledge on the alternative lines where the different intermediate product versions can be assembled. Besides minimizing tardiness of the orders, also maximal utilization of equipment and human resources, and minimization of the mean flow time of the orders are considered.
A as illustration of the problem setting, a task adopted from a real-life scheduling configuration is considered. Let us assume that a mobile terminal company has 10 production lines of three different types; 4 of type I, 4 of type II and 2 of type III. These types can be seen as different generations of production lines and they are thus the result of gradual process evolution and development. The most recent lines are the most efficient ones.
The production is planned for a period of two weeks and the operational environment is flexible due to the large extent of operators that are contracted just for the period in question. The manufacturing is run with two twelve hour shifts a day, 7 days a week. During the current planning period 14 different product variants have to be assembled and delivered. The planning is performed on shift-basis so that a line only participates in the manufacturing of one product each shift. Also due-dates are defined based on the shifts.
The configuration matrix and the order data are given in Table 1. In the upper part of the table, each row corresponds to a production line type and the columns correspond to the different orders. The numbers in the matrix indicate the required amount of operators to run a production line in case of the respective product. Accordingly, the absence of a number in the matrix denotes that the product cannot be assembled using the line type. In the lower part, the customer demand and due-dates of each product batch are given so that the numbers refer to the specific work shifts. The production relies on a subcontractor network and therefore also release dates have to be considered. The set-up times are not significant.

Table 1. Above: MLP configuration matrix for the example with 14 orders. Below: Demand, relase- and due-date for each order.

Line type	Order 1	2	3	4	5	6	7	8	9	10	11	12	13	14
1	4	4	5	4	5	3	3	6	3	3	5	3	5	4
2	4	5	6	5	6	3	5	-	3	4	5	-	5	4
3	-	5	6	-	6	3	5	6	-	4	-	-	5	4

	Order 1	2	3	4	5	6	7	8	9	10	11	12	13	14
Demand	7	12	21	14	10	10	13	10	15	12	18	21	22	25
Release date	1	1	1	3	3	9	5	11	13	13	19	17	21	21
Due date	2	4	5	7	9	12	15	16	19	20	24	25	28	30

3 Deterministic Optimization Approach

The problem can be formulated as an MILP problem as follows: A set of product orders I are to be assembled during the time horizon with discretized time periods, shifts, belonging to the set J. The set of production line types is denoted by K. In this case, the constraints are, for example, that all ordered amounts d_i of products i have to be assembled during the planning time horizon

$$\sum_{j \in J} \sum_{k \in K} e_{ik} z_{ijk} \geq d_i \qquad \forall\, i \in I \tag{1}$$

where the integer variables z_{ijk} denote the number of production lines of type k manufacturing product i at time period j. e_{ik} is an efficiency factor describing the production capacity when manufacturing product i with one MPL of line type k. Because combinations of differently efficient assembly lines may result in difficulties to exactly match the demand, a limited slack in production will be allowed. The slack s describes thus the fraction of time a line may be idle during the production of an order.

$$\sum_{j \in J} \sum_{k \in K} e_{ik} z_{ijk} \leq s \cdot d_i \qquad \forall\, i \in I \tag{2}$$

The value of the slack variable has here been set to 1.1 which indicates that, for a 12 hour shift, the production can be discontinued for maximally about one hour to ensure correct amount of products. Furthermore, the maximum number of active line types may not exceed the available number of lines n_k at any time period.

$$\sum_{i \in I} z_{ijk} \leq n_k \qquad \forall\, j \in J, k \in K \tag{3}$$

To be able to encounter the tardiness of the orders, binary variables y_{ij} are introduced that take the value one if product i is manufactured in period j.

$$\sum_{k} z_{ijk} \leq M y_{ij} \qquad \forall\, i \in I, j \in J \tag{4}$$

where M is a sufficiently big number, e.g. the total number of different lines available. The tardiness T_i of an order is defined as the number of time periods, shifts, from the due-date until the time when the last unit of the order can be delivered.

$$T_i \geq (j - dd_i) y_{ij} \qquad \forall\, i \in I, j \in J \tag{5}$$

where dd_i is the due-date of product i. The number of required operators $oper_j$ at each time period is obtained by

$$\sum_{i \in I} \sum_{k \in K} o_{ik} z_{ijk} = oper_j \qquad \forall\, j \in J \tag{6}$$

where o_{ik} contains the information of required operators (Table 1). The maximal difference between the minimum and maximum number of operators Δo required during the planning horizon is given by

$$oper_j - oper^{alloc} \leq \Delta o \qquad \forall\, j \in J \tag{7}$$

$$-oper_j + oper^{alloc} \leq \Delta o \qquad \forall\, j \in J \tag{8}$$

where $oper^{alloc}$ is the number of operators allocated for the production during the two-week period.
The main objectives are to obtain a production schedule with a minimal amount of tardy orders with an as small as possible number of operators and, furthermore, with an as even operator load as possible:

$$\min\ \Delta o \tag{9}$$

$$\min \sum_{i \in I} T_i \tag{10}$$

$$\min\ oper^{alloc} \tag{11}$$

4 Optimization Results

The solution of a multi-objective optimization problem is often obtained as a set of non-dominated solutions. None of these solutions can be said to outperform the others with respect to all objectives and the user has to select one based on the importance of the objectives. Many algorithms for solving such problems using classical optimization methods have been suggested during the last four decades [3]. In this work no a priori information of the different objectives' importance has been used, although it may be available. Instead, the used approach resembles the ε-approach by Haimes et al. [4]. The problem is reformulated into a single objective optimization task:

$$\min 100 \sum_{i \in I} T_i + \Delta o \tag{12}$$

$$\text{s.t.} \quad oper^{alloc} = p \tag{13}$$

$$\Delta o \leq \varepsilon \tag{14}$$

where p is a fixed number of operators allocated for the period (*eqs.*(1–8) shall be included in the formulation). The weights in the objective function are selected so that the influence of Δo on the objective function will never exceed that of one order being delayed one time period because the magnitude of Δo is below 100. Still, Δo will be minimized as a secondary important objective and when a solution have been obtained, ε , which initially was given a large number, is set to, $\varepsilon = \Delta o - 1$, and the problem is resolved a number of times until no feasible solution can be found. When no more non-dominated solutions can be obtained, the selected p can be changed and the procedure reiterated. The number of operators p is changed over a span of values that are expected to be sufficient.

The non-dominated solutions obtained are shown in Figure 1. The markers refer to solutions with different values on the number of operators p to hire. It can, for example, be seen that the orders can be met by a schedule requiring 28 operators with maximally one operator idle during the two-week period and with a total tardiness value of 4.

The example resulted in an MILP model with about 640 integer and 240 binary variables. The set J, with time periods, was defined so that every order can be delayed for at most 4 time periods. The model was solved on a 2000 Mhz PC with the CPLEX MILP solver [5] using the mip-emphasis set for integer feasibility and with a time limit for

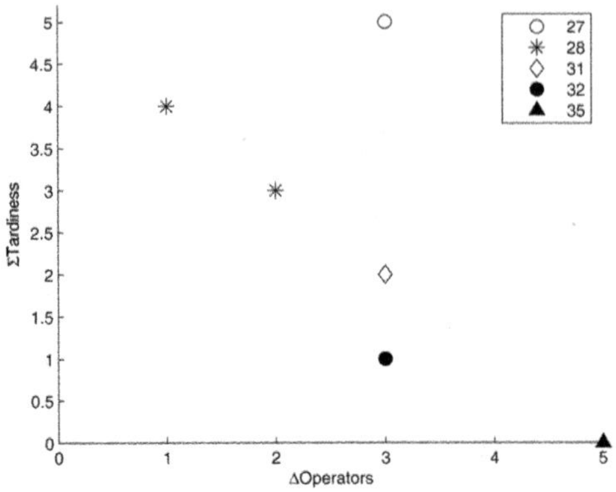

Fig. 1. Non-dominated solutions for the scheduling task. The legends correspond to the total number of operators needed.

each MILP problem set to 400 seconds. Thus, global optimality is not guaranteed.

5 Conclusions

In this paper, the production planning and scheduling challenges in mobile terminal manufacturing were discussed. A case study derived from a typical industrial problem setting was presented and solved using an MILP formulation to represent the multi-criteria optimization problem. Although the example was a simplified illustration of the production arrangement in mobile terminal manufacturing, the importance of this kind of considerations is obvious.

References

1. Partanen J, Haapasalo H (2004) Fast Production for Order Fulfillment: Implementing Mass Customization in Electronics Industry. International Journal of Production Economics, 90:213-222
2. Shah N (1998) Single- and Multisite Planning and Scheduling: Current Status and Future Challenges. Proceedings of Foundations of Computer Aided Process Operations Conference, Snowbird, USA
3. Miettinen K (1999) Nonlinear Multiobjective Optimization. Kluwer, Boston, USA
4. Haimes YY, Lasdon LS, Wismer DA. (1971) On a bicriterion formulation of the problems of integrated system identification and system optimization. IEEE Transactions on Systems, Man, and Cybernetics, 1(3):296-297
5. ILOG CPLEX 9.0 user's manual (2003), ILOG

Revenue Maximization on Parallel Machines

Malgorzata Sterna[1], Jacek Juraszek[1], and Erwin Pesch[2]

[1] Institute of Computing Science, Poznan University of Technology, Poland, 60-965 Poznan, Piotrowo 2,
{malgorzata.sterna,jacek.juraszek}@cs.put.poznan.pl

[2] Institute of Information Systems, University of Siegen, Germany, Hoeldelinstrasse 3, 57068 Siegen,
erwin.pesch@uni-siegen.de

1 Introduction

Revenue management is essentially the process of allocating resources to the right customer at the right time and the right price (cf. [9]). A slightly different approach to revenue maximization can be met in "classical" scheduling theory (cf. [3]), where the goal is to maximize the criterion value, i.e. the profit, for some given values of the problem parameters (cf. [8]). Such a model finds many practical applications. For example, a set of jobs can represent a set of customer orders which may give certain profit to a producer. Due to limited resources, modeled by a machine or a set of machines, the producer has to decide whether to accept or reject a particular order and how to schedule accepted orders in the system. Delays in the completions of orders cause penalties, which decrease the total revenue obtained from the realized orders. For this reason, maximizing revenue is strictly related to due date involving criteria (cf. [3]) such as minimizing tardiness or late work (cf. [11]). The maximum revenue objective function has been studied mostly for the single machine environment (cf. [2], [5], [8], [10]).
In our research, we investigate the problem of selecting and executing jobs on identical parallel machines in order to maximize the total revenue (profit) with the weighted tardiness penalty.

2 Problem Definition

In the work, we analyzed the problem of selecting and executing a set of n jobs $\mathcal{J} = \{J_1, J_2, ..., J_n\}$ on a set of m parallel identical machines

B. Fleischmann et al.(Eds.), *Operations Research Proceedings 2008*,

DOI: 10.1007/978-3-642-00142-0_25,

$\mathcal{M} = \{M_1, M_2, ..., M_m\}$. Each job J_j is described by the release time r_j, at which it becomes available, and the processing time p_j, for which it has to be performed without any preemption. All jobs selected for execution in the system have to be completed before their deadlines D_j. Moreover, a due date d_j is defined for each job, which determines its latest finishing time resulting in the full revenue q_j. If the completion time of a job exceeds its due date, the revenue is decreased by the weighted tardiness (with the weight w_j), which represents the fine paid by a system owner to a customer for feasible delay. The goal is to select a subset of jobs and to schedule them between release times and deadlines on machines in order to maximize the total revenue Q.

Introducing binary positional variables x_{jki}, which represent the fact of assigning job J_j ($j = 1, ..., n$) to machine M_k ($k = 1, ..., (m+1)$) on position i ($i = 1, ..., n$), we can formulate the scheduling case stated above as the mixed integer programming problem (MIP). We assume that machine M_{m+1} is a dummy machine collecting all rejected jobs giving zero revenue.

$$Maximize \qquad Q = \sum_{k=1}^{m} \sum_{i=1}^{n} Q_{ki} \tag{1}$$

$$Subject\ to: \qquad \sum_{i=1}^{n} \sum_{k=1}^{m+1} x_{jki} \leq 1 \qquad j = 1, ..., n \tag{2}$$

$$\sum_{j=1}^{n} x_{jki} \leq 1 \qquad k = 1, ..., (m+1);\ i = 1, ..., n \tag{3}$$

$$\sum_{j=1}^{n} x_{jk(i+1)} \leq \sum_{j=1}^{n} x_{jki} \qquad k = 1, ..., (m+1);\ i = 1, ..., (n-1) \tag{4}$$

$$x_{jki} \in \{0, 1\} \qquad j = 1, ..., n;\ k = 1..., (m+1);\ i = 1, ..., n \tag{5}$$

$$t_{ki} \geq 0 \qquad k = 1, ..., m;\ i = 1, ..., n \tag{6}$$

$$t_{ki} \geq \sum_{j=1}^{n} x_{jki}\, r_j \qquad k = 1, ..., m;\ i = 1, ..., n \tag{7}$$

$$t_{ki} \leq \sum_{j=1}^{n} x_{jki}\, (D_j - p_j) \qquad k = 1, ..., m;\ i = 1, ..., n \tag{8}$$

$$t_{ki} + \sum_{j=1}^{n} x_{jki}\, p_j \leq t_{k(i+1)} \qquad k = 1, ..., m;\ i = 1, ..., (n-1) \tag{9}$$

$$Q_{ki} \leq q_j - w_j\, max\{\, 0,\, t_{ki} + p_j - d_j \,\} + (1 - x_{jki})\, M \qquad j = 1, ..., n;\; k = 1, ..., m;\; i = 1, ..., n \tag{10}$$

$$Q_{ki} \leq \sum_{j=1}^{n} x_{jki}\, M \qquad k = 1, ..., m;\; i = 1, ..., n \tag{11}$$

Constraints (2) ensure that each job is assigned to exactly one position on a machine, while constraints (3) guarantee that each position is occupied by at most one job. Due to constraints (4), there is a job on a certain position in a sequence, only if the previous position is also occupied. Constraints (5) guarantee correct values of binary decision variables. The starting times for jobs are further variables in our model. They have to take non-negative values (6), exceeding the release time of a job (7), but not violating its deadline (8). Constraints (9) ensure that two consecutive jobs do not overlap. Constraints (6)-(9) are formulated only for real machines ($k \leq m$), since only jobs executed on real machines provide revenue (10), which increases the criterion value (1). Constraints (11) ensure that the positions on machines not occupied by any job give zero revenue (M denotes a big integer value).
The considered problem is NP-hard, since the problem of minimizing the weighted tardiness for a subset of accepted jobs, which is necessary to maximize the total revenue, is already intractable [7]. Similarly, the problem of selecting jobs for execution with the tardiness penalty is also NP-hard [10].

3 Solution Methods

Any method solving the problem under consideration has to decide about three issues: which jobs should be accepted, how to assign accepted jobs to machines and how to schedule jobs on particular machines. In the presented research, we proposed three strategies: branch and bound, list scheduling and simulated annealing.

List Scheduling. A list scheduling algorithm (LA) is a simple heuristic which was implemented in order to obtain an initial solution for simulated annealing as well as to determine an initial lower bound for the exact approach. At each iteration, it assigns an available job to a free machine, i.e. the job whose release time is exceeded and which still can be completed before its deadline giving some revenue. Jobs which cannot be feasibly processed or provide no revenue due to large delays are rejected. The method selects one job from a set of available jobs according to a priority dispatching rule. We implemented 7 static rules (determining priorities based on the problem parameters, e.g. the maximum revenue per a processing time unit) and 3 dynamic rules (calculating priorities based on the problem parameters and the description of a partial schedule, e.g. the maximum possible revenue at a certain time moment).

Branch and Bound. Branch and Bound (B&B) determines an optimal solution by analyzing all possible partitions of a set of jobs to machines. Since a set of machines contains also a dummy machine collecting rejected jobs, the method checks simultaneously all possible decisions on accepting/rejecting jobs. Then, for a particular partition of a job set, B&B checks all feasible permutations of jobs on machines. The partial schedules whose upper bound (determined as the value of current revenue increased with the total revenue which might be obtained from non-assigned jobs) exceeds the lower bound are suppressed. As it was mentioned, an initial lower bound is calculated by the list scheduling heuristic using all proposed priority dispatching rules.

Simulated Annealing. We propose a simulated annealing algorithm (SA) based on the classical framework of this method (cf. [4], [6]). It starts exploring the solution space from an initial schedule determined by the list scheduling heuristic with an initial temperature determined in a tuning process. The SA solution is represented as an assignment of jobs to $m+1$ machines and it is transformed to a schedule by an exact approach determining the optimal sequence of jobs on machines. At each iteration SA picks at random a neighbor solution from the neighborhood of a current schedule. A new solution is improved with a local search procedure, whose role plays a simple and fast simulated annealing method. We proposed two different move definitions for generating new solutions based on shifting or interchanging jobs between two machines. If the new schedule improves the criterion value, it is accepted, otherwise it replaces a current schedule with the probability depending on the criterion value deterioration and the current temperature. After a certain number of iterations, the temperature is decreased geometrically (an arithmetic cooling scheme appeared to be not efficient in preliminary experiments). Moreover, after a given number of iterations without improvement in the total revenue, we apply diversification, which randomly fluctuates a current solution and reheats the system allowing the search in a new area of the problem space. The method terminates after a given number of diversifications and iterations without improvement.

4 Computational Experiments

The proposed algorithms were implemented in Java JDK 1.6 MAC OS X and tested on an Intel Core 2 Duo 2.4 GHz computer. Computational experiments were performed for a set of randomly generated input instances of 4 various characteristics in terms of the distribution of job parameters over time. There were instances with "tight" release times (TR), in which most jobs were available for processing around time zero, and "loose" release times (LR), which were spread in time. Similarly, instances with "tight" due dates and deadlines (TD) contained jobs with narrow time windows in which they can be processed without delay, while in instances with "loose" dues date and deadlines (LD) the lengths of these intervals were very close to the job processing times. Test results disclosed the strong influence of the instance characteristic on the efficiency of the heuristic methods. Due to time requirements of the branch and bound algorithm, the comparison with optimal solutions was possible only for small instances ($m = 2, 3, 4$ and $\frac{n}{m} = 3, 4, 5$). It showed the high efficiency of both heuristics, especially of simulated annealing, which generated mostly optimal solutions. SA found solutions with 98.73%, and LA with 89.47% of the optimal revenue on average. In the computational experiments performed for large instances ($m = 5, 10, 15, 20$ and $\frac{n}{m} = 2, 5, 10, 15$) SA was still able to improve the quality of initial solutions generated by LA by 9.96% on average. The branch and bound algorithm was obviously a very time consuming method. It required nearly 16 hours for computational experiments with 72 instances of the small size, while SA and LA consumed negligibly short time in this case. For 320 large instances, the running time of simulated annealing was still acceptable: it varied from a few seconds

Table 1. The efficiency of the list scheduling and simulated annealing methods

	Percentage of optimum revenue in [%] ($m = 2, 3, 4$)						Improvement of initial revenue in [%] ($m = 5, 10, 15, 20$)			
	LA vs. B&B			SA vs. B&B			SA vs. LA			
$\frac{n}{m}$	3	4	5	3	4	5	2	5	10	15
TR-TD	88.43	83.34	74.62	100.00	97.66	97.62	5.66	12.62	17.46	22.10
LR-TD	92.34	93.09	90.48	97.70	99.10	98.88	0.21	5.47	10.85	8.75
TR-LD	89.89	93.36	84.73	99.39	98.67	96.59	3.87	11.95	18.14	17.23
LR-LD	100.00	91.68	91.64	100.00	98.93	98.20	0.04	4.04	9.18	11.85

to 50 minutes depending on the number of jobs and the difficulty of an instance.

5 Conclusions

We investigated the problem of simultaneous selecting and scheduling a set of jobs on a set of identical parallel machines in order to maximize the total revenue. We proposed the MIP formulation for this scheduling case and designed exact and heuristic approaches, which were tested in the extensive computational experiments. Within the future research, we will analyze a similar problem of orders acceptance and scheduling, which arises in a real world environment [1].

References

1. Asbach L, Dorndorf U, Pesch E (2007) Analysis, modeling and solution of the concrete delivery problem. European Journal of Operational Research in press: doi:10.1016/j.ejor.2007.11.011
2. Bilginturk Z, Oguz C, Salman S (2007) Order acceptance and scheduling decisions in make-to-order systems. In: Baptiste P, Kendall G, Munier-Kordon A, Sourd F (eds) Proceedings of the 3rd Multidisciplinary International Conference on Scheduling: Theory and Application. Paris
3. Blazewicz J, Ecker K, Pesch E, Schmidt G, Weglarz J (2007) Handbook on scheduling: from theory to applications. Springer, Berlin Heidelberg New York
4. Crama Y, Kolen A, Pesch E (1995) Local search in combinatorial optimisation. Lecture Notes in Computer Science 931:157–174
5. Ghosh JB (1997) Job selection in a heavily loaded shop. Computers and Operations Research 24(2):141–145
6. Kirkpatrick S, Gelatt CD, Vecchi MP (1983) Optimization by simulated annealing. Science 220(4598):671–680
7. Lawler EL (1977) A pseudopolynomial algorithm for sequencing jobs to minimize total tardiness. Annals of Discrete Mathematics 1:331–342
8. Lewis HF, Slotnick SA (2002) Multi-period job selection: planning work loads to maximize profit. Computers and Operations Research 29:1081–1098
9. McGill JI, van Ryzin GJ (1999) Revenue management. Research overview and prospects. Transportation Science 33:233–256
10. Slotnick SA, Morton TE (2007) Order acceptance with weighted tardiness. Computers and Operations Research 34:3029–3042
11. Sterna M (2006) Late work scheduling in shop systems. Dissertations 405, Publishing House of Poznan University of Technology, Poznan

A New Bottleneck-Based Heuristic for Reentrant Job Shops: A Case Study in a Textile Factory

Seyda Topaloglu and Gamze Kilincli

Department of Industrial Engineering, Dokuz Eylul University, Izmir, TURKEY
seyda.topaloglu@deu.edu.tr
gamzekilincli@yahoo.com

1 Introduction

The classical job shop assumes that each job visits a machine only once. In practice, this assumption is often violated. Recently, the reentrant job shop has become prominent in which a certain job may visit a specific machine or a set of machines more than once during the process flow. Reentrant job shops can be found in many production systems, particularly in high-tech industries such as semiconductor manufacturing. Another example is the manufacturing of printed circuit boards that require both surface-mounted devices and conventional pin-through-hole devices. It is also employed in parts that go through the painting and baking divisions alternately for different coats of paint in a painting shop. The problem of minimizing makespan in a reentrant job shop is theoretically challenging. In fact, it is NP-hard in the strong sense even for the two-machine case [1]. For the solution of job shop scheduling problems (JSSPs), exact methods such as integer programming formulations [2] and branch-and-bound algorithms [3] have been developed to produce optimal solutions. However, their worst-case computational burden increases exponentially with the size of the problem instance. As noted in Aytug et al. [4], for industrial problems the computational time of any algorithm must be short enough that the resulting schedule can be used. Hence, a variety of heuristic procedures such as dispatching rules, decomposition methods, and metaheuristic search techniques have been proposed for finding "good" rather than optimal solutions in a reasonably short time. While the dispatching rules are computationally efficient and easy to use, they are generally myopic in

B. Fleischmann et al.(Eds.), *Operations Research Proceedings 2008*,
DOI: 10.1007/978-3-642-00142-0_26, © Springer-Verlag Berlin Heidelberg 2009

both space and time and may result in poor long term performance [5]. Decomposition methods aim at developing solutions to complex problems by decomposing them into a number of smaller subproblems which are more tractable and easier to understand. The Shifting Bottleneck Heuristic (SBH), originally proposed by Adams et al. [6] and improved by Balas et al. [7], is a powerful decomposition method for the JSSP to minimize makespan ($J_m \parallel C_{max}$). It takes advantage of the disjunctive graph representation introduced by Roy and Sussmann [8] to model interactions between the subproblems. At each iteration, the most critical unscheduled machine is identified and scheduled optimally. Every time a new machine is scheduled, the constraints imposed by these new decisions are propagated through the partial solution using the directed graph representation. In the original SBH, each subproblem is that of minimizing maximum lateness (L_{max}) on a single machine in the presence of release times and dues dates. This problem is NP-hard [9] and it is solved optimally by a version of Carlier's branch-and-bound method [10] for small-size problems. The subproblem formulation and solution has been later improved by Dauzere-Peres and Lassere [11] and Balas et al. [7] with the addition of delayed precedence constraints to insure the feasibility of the single machine subproblem solutions. Ovacik and Uzsoy [12] later extended the SBH to minimize maximum lateness with reentrant product flows, sequence-dependent setup times. Oey and Mason [13] and Mason and Oey [14] studied a complex job shop problem with reentrant flow and batch processing machines, and proposed a modified SBH for generating machine schedules to minimize the total weighted tardiness. Most of the experiments conducted on the SBH have focused on relatively small problems and small sets of benchmark problems available in the literature. As noted in Singer [15], most of the proposed methods are incapable of solving large problem instances with more than 30 jobs. Given that an instance with 30 jobs is a rather small problem in real life, we propose a bottleneck-based heuristic (BBH) for solving the large-scale RJSSPs with 100 or more jobs with the objective of minimizing makespan ($J_m|recrc|C_{max}$), in realistic industrial contexts. The proposed BBH is adapted from the SBH and tailored to the needs of the RJSSP. As in the SBH, the problem is decomposed into a number of single machine subproblems, and additionally a specialized sequencing algorithm is proposed for the solution of subproblems so that the large-scale RJSSPs can be handled conveniently. The following section describes the BBH. In Section 3, a case study in a textile factory is given, which evaluates the performance of the BBH in comparison to the well-known dispatching rules. Finally, concluding remarks are given in Section 4.

2 The Proposed Bottleneck-Based Heuristic (BBH)

The solution procedure for the BBH contains the same main steps as the original SBH. However, there are some differences related to subproblem solution and cycle elimination procedure. A new sequencing algorithm named as SAL is developed for solving the nonpreemptive single-machine maximum lateness subproblem with release times ($1|r_j|L_{max}$). The description of the SAL is given below. Note that O_U represents the set of all operations that are not scheduled yet, whereas O_S is the set of scheduled operations. O'_U is the set of all operations that are not scheduled and have release times smaller than the scheduled time t on the machine.

Step 1. Initialization of the variables used in the algorithm.
Step 2. If set O_U is not empty, go to Step 3, otherwise, go to Step 11.
Step 3. Compute the earliest completion time (release time plus processing time) for all operations that are not scheduled yet. Find the minimum completion time and assign this value to the current scheduled time t.
Step 4. If there is no unscheduled operation with release time smaller than t, go to Step 2, otherwise, go to Step 5.
Step 5. Identify the unscheduled operations whose release times are smaller than t and put them in set O'_U.
Step 6. For each operation in set O'_U that belongs to a job with reentrance property on the associated machine, repeat steps 7 to 9.
Step 7. Identify its preceding reentrant operation. If this operation has been scheduled, then compute a new release time value by adding the processing times of all operations between these reentrant operations to its completion time, otherwise, update set O'_U by removing this operation.
Step 8. If the computed value is greater than its current release time, then replace it with this value, otherwise, do not make any change.
Step 9. If the new release time is greater than t, then update set O'_U by removing this operation, otherwise, set O'_U remains unchanged.
Step 10. If there is no operation left in set O'_U, go to step 2, otherwise, select an operation from set O'_U with minimum due date, release time and index values, respectively in case of ties. Schedule the selected operation next on the machine. Update the sets O_U and O_S. Give a

sequence number for the scheduled operation. Compute its completion time. Assign this value to t. Initialize set O'_U. Go to Step 4.
Step 11. Calculate the maximum lateness for the resultant operation sequence on the machine.
The machine with the largest L_{max} is determined as the "bottleneck machine" after the SAL is applied to all unscheduled machines at each iteration of the BBH. The graph representing the partial schedule is updated by fixing the disjunctive arcs corresponding to the sequence of the operations on the bottleneck machine. The C_{max} is increased by L_{max}(bottleneck). Different from the JSSP, in the RJSSP, while the sequence of operations on the machine is being determined, the precedence relationships among the operations that belong to the same job and use the same machine must be ensured. In the SAL, the release times of reentrant operations may need to be shifted due to the prerequisite operations. While the algorithm identifies the candidate operations to schedule next, the release time of a reentrant unscheduled operation must be updated according to its preceding operation on the same machine, which has been already scheduled. If its completion time has been shifted recently, it must be checked whether it affects the release time of this reentrant operation. This control is performed through Steps 7-9 of the proposed algorithm. In the SBH, delayed precedence constraints, which were proposed by Dauzere-Peres and Lassere [11] and Balas et al. [7], are used to insure the feasibility of the single machine subproblems. Without these constraints the SBH may end up in a situation where there is a cycle in the disjunctive graph and the schedule is infeasible. Instead of imposing delayed precedence constraints, a new cycle elimination procedure is developed to prevent infeasibility in the BBH. This procedure restricts the SAL from selecting an operation that will create a cycle due to the previously scheduled machines.

3 A Real Life Application in Textile Industry

The proposed BBH is applied for solving the RJSSP of the dyeing-finishing facility of a textile factory. This facility is reentrant because the ordered jobs have to be processed on some of the machines more than once. The factory produces a large variety of manufactured fabrics, and works under a make-to-order policy in a complicated processing environment ranging from yarn producing to yarn dyeing, weaving and dyeing-finishing. Currently, the planning department loads the machines using the First Come First Served (FCFS) rule. In this application, a 4-week production schedule has been prepared on a daily basis.

On a typical day, more than one hundred jobs and approximately one thousand operations have to be scheduled using 20 different machines. Since the number of the jobs to be scheduled in a day is very high, the unfinished jobs are delayed to the next day. The delayed jobs are scheduled first on the relevant machines the next day and their planned completion times on the machines are regarded as the release dates of the waiting jobs to be scheduled on that particular day. We compared the BBH with respect to the dispatching rules, Most Work Remaining (MWKR), Longest Processing Time (LPT), Largest Number of Successors (LNS), Shortest Processing Time (SPT), Least Work Remaining (LWKR), FCFS, and RANDOM. The BBH outperforms all dispatching rules based on the C_{max} values. The smallest gap is 11.51 % for the FCFS rule. The solution times are very short for all the dispatching rules, within less than 1 minute. However, the average solution time is less than 5 minutes for the BBH and this duration can be traded off for the increased solution quality.
As a second performance indicator, the average delayed times of the jobs on the last machine through which all of them must go through are measured, which are 883, 887, 853, 979, 903, 738, and 860 minutes respectively for MWKR, LPT, LNS, SPT, LWKR, FCFS, and RANDOM, whereas it is 525 minutes for the BBH. Over the 28-day period only 36.21 % of the jobs are completed the next day for the BBH, while this is 83.01 %, 82.51 %, 65.21 %, 54.51 %, 48.01 %, 56.61 %, and 62.91 % respectively for the above listed dispatching rules. Thirdly, the average completion times of the jobs are analyzed. It is 948 minutes for the BBH, whereas it is 1527, 1525, 1342, 1305, 1238, 1233, and 1357 minutes respectively for the listed dispatching rules. The average completion and waiting times of jobs are reduced, and machine utilization rates are increased. Finally, using the paired t-test we showed that the difference between the dispatching rules and the BBH is statistically significant at 1 % level in solution quality.

4 Conclusion

In this paper, we present a new bottleneck-based heuristic method to minimize the makespan in reentrant job shops. The proposed approach is capable obtaining high-quality solution for large-size JSSPs in very short computing times. A new heuristic algorithm has been developed for the $(1|r_j|L_{max})$ subproblem solution in the adapted SBH. We have evaluated the performance of the BBH through computational experiments on a case study in a textile factory. We have stressed the applicability of our approach and focused on the resulting benefits and savings. The results show that the algorithm has considerable improvements in comparison to the dispatching rules and the average solution time is less than five minutes. Whereas the computational burden of the original SBH increases rapidly as the number of operations increases, the proposed BBH is well-suited for real life applications.

References

1. Wang MY, Sethi SP, Van De Velde SL (1997) Minimizing makespan in a class of reentrant shops. Oper Res 45:702-712
2. Pan JCH (1997) A study of integer programming formulations for scheduling problems. J Oper Res Soc 28:33-41
3. Singer M, Pinedo M (1998) A computational study of branch and bound techniques for minimizing the total weighted tardiness in job shops. IIE Trans 30:109-118
4. Aytug H, Kempf K, Uzsoy R (2002) Measures of subproblem criticality in decomposition algorithms for shop scheduling. Int J Prod Res 41(5):865-882
5. Bhaskaran K, Pinedo M (1991) Dispatching. In: Salvendy G (ed) Handbook of Industrial Engineering. Wiley, New York
6. Adams J, Balas E, Zawack D (1988) The shifting bottleneck procedure for job shop scheduling. Manage Sci 34(3):391-401
7. Balas E, Lenstra JK, Vazacopoulos A (1995) The one machine scheduling with delayed precedence constraints. Manage Sci 41:94-109
8. Roy B, Sussman B (1964) Les problémes d'ordonnancements avec contraintes disjonctives. Proceedings of SEMA, Montrouge
9. Garey MR, Johnson DS (1979) Computer and intractability: A guide to the theory of NP-completeness. WH.Freeman, San Francisco, CA
10. Carlier J (1982) The one-machine sequencing problem. Eur J Oper Res 11:42-47
11. Dauzere-Peres S, Lasserre JB (1993) A modified shifting bottleneck procedure for job shop scheduling. Int J Prod Res 31:923-932
12. Ovacik IM, Uzsoy R (1992) A shifting bottleneck algorithm for scheduling semiconductor testing operations. J of Electronic Manufacturing 2:119-134
13. Oey K, Mason SJ (2001) Scheduling batch processing machines in complex job shops. Proceedings of the 2001 Winter Simulation Conference, pp 1200-1207
14. Mason SJ, Oey K (2003) Scheduling complex job shops using disjunctive graphs: A cycle elimination procedure. Int J Prod Res 41(5):981-994
15. Singer M (2001) Decomposition methods for large job shops. Comput Oper Res 28:193-207

Project Scheduling with Precedence Constraints and Scarce Resources: An Experimental Analysis of Commercial Project Management Software

Norbert Trautmann and Philipp Baumann

Universität Bern, Departement Betriebswirtschaftslehre, AP Quantitative Methoden, Schützenmattstrasse 14, 3012 Bern, Schweiz
{norbert.trautmann,philipp.baumann}@pqm.unibe.ch

Summary. We report on the results of an experimental analysis where we have compared the resource allocation capabilities of recent versions of 5 commercial project management software packages against state-of-the-art solution methods from the literature. For our analysis, we have used 1560 RCPSP instances from the standard test set PSPLIB. The results indicate that using the automatic resource allocating feature of those packages may result in project durations that are considerably longer than necessary, in particular when the resource scarcity is high or when the number of activities is large.

1 Introduction

The resource-constrained project scheduling problem RCPSP can be described as follows (cf. [1]). Given are a set V of n non-interruptible activities $i \in V$, a set E of precedence relationships $(i, j) \in V \times V$ among these activities, and a set R of renewable resources $k \in R$ with constant capacities $R_k > 0$. The execution of an activity $i \in V$ takes a prescribed amount of time $p_i > 0$ and requires a prescribed amount $r_{ik} \geq 0$ of each resource $k \in R$. Sought is a baseline schedule, i.e. a start time $S_i \geq 0$ for each activity $i \in V$, such that [a] the precedence relationships are taken into account (i.e., $S_j \geq S_i + p_i$ for all $(i, j) \in E$), [b] for every resource, at no point in time the total requirement exceeds the available capacity (i.e., $\sum_{i \in V: S_i \leq t < S_i + p_i} r_{ik} \leq R_k$ for all $k \in R$ and $t \geq 0$), and [c] the project duration $\max_{i \in V} S_i + p_i$ is minimized.
For this NP-hard optimization problem, a large variety of exact and heuristic solution methods has been presented in the literature; for a

B. Fleischmann et al.(Eds.), *Operations Research Proceedings 2008*,
DOI: 10.1007/978-3-642-00142-0_27,

survey, we refer to [6]. In practice, however, it is common to use commercial software packages for determining project schedules (cf. [3]). In experimental analyses, [2, 4, 5, 7] have compared the resource allocation capabilities of such software packages with those of specific solution methods. While all these studies are based on rather small test sets, [9] have used the 1560 RCPSP instances with 30, 60, and 120 activities, respectively, from the standard test set PSPLIB (cf. [8]). In this paper, we continue that research using recent versions of the project management software packages Acos Plus.1 (ACOS Projektmanagement GmbH), Turbo Project Professional (OfficeWork Software), CS Project Professional (CREST Software), Microsoft Office Project 2007 (Microsoft Corporation), and PS8 (Sciforma Corporation).
The remainder of this paper is structured as follows. Section 2 outlines the resource allocation features of the individual software packages. Section 3 describes the design and the results of our experimental analysis. Section 4 is devoted to some concluding remarks.

2 Software Packages

For our analysis, we installed Release 8.9a of Acos Plus.1 (ACO), Release 4.00.221.2 of Turbo Project Professional (TPP), Release 3.4.1.20 of CS Project Professional (CSP), Release 12.0.4518.1014 of Microsoft Office Project 2007 (MSP), and Release 8.5.1.9 of PS8 (PS8) on various standard PCs with Windows XP or Windows Vista as operating system. In this section, we briefly explain how resource allocation can be performed manually and automatically with the individual packages.

2.1 Acos Plus.1

In Acos Plus.1 the resource capacities and requirements need to be specified in units per time period. An activity filter highlights activities involved in resource overloads. Overloads can be resolved manually by dragging and dropping activities within a Gantt-chart.
Automatic resource allocation can be performed for all or for selected resources and activities only. Optionally, a project completion time may be prescribed. As activity priorities, one of the options smallest total/free float time, affiliation to the critical path, longest/shortest processing time, number of predecessors/successors, total number of predecessors and successors, or user-defined values can be selected. In our analysis, we tested all these options except user-defined values and

selected the schedule with the shortest project duration. For some instances and options, however, the allocation procedure created cyclic precedence relationships and returned an infeasible schedule; we did not take such schedules into account.

2.2 Turbo Project Professional

In Turbo Project the resource capacities and requirements must be defined in units per day. Activities involved in resource overloads are indicated in an activity table. Turbo Project supports manual resource allocation with a dialog box which allows to locate and display information about activities involved in resource overloads.
For automatic resource allocation the user must decide whether the actual project duration may be increased or not. A resource filter allows for removing overloads of only certain resources. The activity priorities can be chosen manually or set to a constant value. In our analysis, we used the latter option. The time horizon considered for the allocation can be selected either according to the actual schedule or manually. In the first case, automatic resource allocation may result in an infeasible schedule, because resource overloads beyond the current project completion time are not resolved. Therefore, we tried both to use a sufficiently long time horizon and to repeat calling the resource allocation function until we obtained a feasible schedule, and selected the resulting schedule with shorter project duration.

2.3 CS Project Professional

In CS Project the resource capacities and requirements must be specified in hours per day. A resource profile and a resource overload indicator support manual resource allocation via drag-and-drop.
Automatic resource allocation can be performed for all or only for selected activities and resources. It is also possible to freeze the actual project completion time. The user can select one of 65536 different priority rules by combining duration, early start, start-baseline, finish-baseline, late finish, total/free float, and user-defined values, each in ascending or descending order, in a four-level hierarchy. In our analysis, we tested 80 of the possible priority rules, using the first two levels of the hierarchy.

2.4 Microsoft Office Project 2007

In Microsoft Office Project the resource capacities and requirements are modeled as aggregated workloads; correspondingly, the duration of

an activity is in general seen as variable, but may also be fixed. Filters and a resource chart support manual resource allocation.
For automatic resource allocation the user must specify the time horizon and the period length. A resource overload occurs when the total requirement within a period exceeds the total capacity; within a period, however, no individual time points are considered. The activity priorities can be set either manually or to predefined values, which are computed according to precedence relationships and float times. In our analysis we used these predefined values.

2.5 PS8

In PS8 the resource capacities and requirements are specified in units per time period (e.g. a minute or a day). For manual resource allocation, a Gantt-chart and a resource profile with common time line are displayed simultaneously; moreover, the software can search for resource overloads.
For automatic resource allocation the user has to choose a leveling time scale which is equivalent to the period concept in Microsoft Office Project. The allocation can be restricted to certain activities and resources; there are no options for choosing priority values.

3 Experimental Analysis

For our analysis, we have used the 1560 RCPSP instances from the PSPLIB test sets J30, J60, and J120 (cf. [8]). The J30 and J60 test sets contain 480 instances with $n = 30$ and $n = 60$ activities, respectively, and the J120 test set contains 600 instances with $n = 120$ activities. We have downloaded the project durations in the optimal schedules (J30 set) and the best known feasible schedules (J60 and J120 set) on May 18, 2008 from the site `http://129.187.106.231/psplib/`. In our analysis, we have used these project durations as reference values.
Table 1 lists the mean, the maximum, and the variance of the relative makespan deviation from the reference values for the three test sets. With respect to all criteria, ACOS Plus.1 and CS Project performed best, closely followed by PS8. The results of all software packages are getting worse when the number n of project activities increases.
As described in [8], the problem instances have been generated by systematically varying the relative resource scarcity (resource strength RS), the mean number of resources used (resource factor RF), and the mean number of precedence relationships (network complexity NC).

Table 1. Relative makespan deviation

	Mean [%]			Maximum [%]			Variance [%^{2}]		
n	30	60	120	30	60	120	30	60	120
ACO	3.35	4.28	11.06	26.15	24.24	31.29	26.16	45.19	60.95
TPP	8.61	9.92	24.42	48.91	54.43	72.90	110.55	155.31	212.54
CSP	2.88	4.47	12.12	20.59	23.53	29.25	18.49	44.60	62.83
MSP	5.18	6.51	15.19	31.03	32.94	48.70	44.66	68.88	102.71
PS8	4.93	5.25	12.26	37.93	41.41	37.43	48.52	62.83	68.23

Table 2 shows the mean relative makespan deviation for the values of these parameters in set J120. For all packages, the results are as follows:

- When the resource scarcity gets higher (decreasing RS value), the deviation increases considerably.
- Interestingly, the deviation is smaller if the activities require one resource only ($RF = 0.25$) or require all resources ($RF = 1$) than if the activities require several, but not all resources.
- The mean number of precedence relationships NC has no strong influence on the deviation.

Table 2. Mean relative makespan deviation in set J120

	RS					RF				NC		
	0.5	0.4	0.3	0.2	0.1	0.25	0.5	0.75	1	1.5	1.8	2.1
ACO	2.39	6.82	10.79	15.56	19.72	6.02	12.73	13.39	12.09	10.35	10.86	11.96
TPP	9.27	17.67	24.28	31.33	39.53	15.71	30.05	30.09	21.82	23.78	24.47	25.01
CSP	3.38	8.41	12.32	16.56	19.96	6.11	14.17	14.77	13.45	11.59	11.93	12.85
MSP	4.70	9.61	14.22	20.07	27.37	8.41	17.83	18.45	16.07	14.65	15.02	15.90
PS8	3.60	8.12	12.13	16.66	20.79	6.03	14.51	14.96	13.54	11.80	12.01	12.97

4 Conclusions

In this paper, we have reported on the results of an experimental analysis in which we evaluated the resource allocation capabilities of 5 commercial project management software packages. It has turned out that when using any of these packages for resource allocation, a project manager must be aware of the risk that the project takes considerably more time than necessary. This gap increases significantly with the number of project activities and the resource scarcity. In our comparison, the project duration computed by Acos Plus.1, CS Project, or PS8 was generally shorter than for Turbo Project or Microsoft Office Project. A possible reason for this behavior could be the resource allocation algorithms used by the software packages. However, we note that in none of the software packages it is possible to explicitly specify an objective such as e.g. minimization of the project duration. Packages like Acos Plus.1 or CS Project Professional offer the possibility to select a priority-rule for the resource allocation; this may explain their (relative) better performance. In contrast, all packages offer extensive modeling possibilities, e.g. for calendars or time-varying resource requirements.

Acknowledgement. The authors would like to thank Christoph Mellentien for his support in preparing the tests.

References

1. Brucker P, Drexl A, Möhring R, Neumann K, Pesch E (1999) Resource-constrained project scheduling: notation, classification, models, and methods. European Journal of Operational Research 112:3–41
2. Farid F, Manoharan S (1996) Comparative analysis of resource-allocation capabilities of project management software packages. Project Management Journal 27:35–44
3. Herroelen W (2005) Project Scheduling—Theory and Practice. Production and Operations Management 14:413–432
4. Johnson R (1992) Resource constrained scheduling capabilities of commercial project management software. Project Management Journal 22:39–43
5. Kolisch R (1999) Resource allocation capabilities of commercial project management software packages. Interfaces 29(4):19–31
6. Kolisch R, Hartmann S (2006) Experimental investigation of heuristics for resource-constrained project scheduling: An update. European Journal of Operational Research 174:23–37
7. Kolisch R, Hempel K (1996) Experimentelle Evaluation der Kapazitätsplanung von Projektmanagementsoftware. Zeitschrift für betriebswirtschaftliche Forschung 48:999–1018
8. Kolisch R, Sprecher A, Drexl A (1995) Characterization and generation of a general class of resource–constrained project scheduling problems. Management Science 41:1693–1704
9. Mellentien C, Trautmann N (2001) Resource allocation with project management software. OR Spectrum 23:383–394

Part V

Supply Chain and Inventory Management

Interactive Multi-Objective Stochastic Programming Approaches for Designing Robust Supply Chain Networks

Amir Azaron[1], Kai Furmans[1], and Mohammad Modarres[2]

[1] University of Karlsruhe (TH), Karlsruhe, Germany
{Amir.Azaron,Kai.Furmans}@ifl.uni-karlsruhe.de
[2] Sharif University of Technology, Tehran, Iran,
modarres@sharif.edu

1 Introduction

Many attempts have been made to model and optimize supply chain design, most of which are based on deterministic approaches, see for example [3], [8], [4] and many others. In order to take into account the effects of the uncertainty in the production scenario, a two-stage stochastic model is proposed in this paper.
There are a few research works addressing comprehensive (strategic and tactical issues simultaneously) design of supply chain networks using two-stage stochastic models including [6], [9], [1] and [7]. [2] developed a multi-objective stochastic programming approach for designing robust supply chains. Then, they used goal attainment technique, see [5] for details, to solve the resulting multi-objective problem. This method has the same disadvantages as those of goal programming; namely, the preferred solution is sensitive to the goal vector and the weighting vector given by the decision maker, and it is very hard in practice to get the proper goals and weights. To overcome this drawback, we use STEM method in this paper to solve this multi-objective model.

2 Problem Description

The supply chain configuration decisions consist of deciding which of the processing centers to build. We associate a binary variable y_i to these decisions. The tactical decisions consist of routing the flow of each product from the suppliers to the customers. We let x_{ij}^k, z_j^k and

B. Fleischmann et al.(Eds.), *Operations Research Proceedings 2008*,
DOI: 10.1007/978-3-642-00142-0_28, © Springer-Verlag Berlin Heidelberg 2009

e_j denote the flow of product k from a node i to a node j of the network, shortfall of product k at customer center j and expansion amount of processing facility j after building the plants and revealing the uncertain parameters, respectively.
We now propose a stochastic programming approach based on a recourse model with two stages to incorporate the uncertainty associated with demands, supplies, processing/transportation, shortage and capacity expansion costs. Considering $\zeta = (d,\ s,\ q,\ h,\ f)$ as the corresponding random vector, the two-stage stochastic model, in matrix form, is formulated as follows (see [7] for details):

$$Min \quad C^T y + E[G(y, \zeta)] \qquad \text{[Expected Total Cost]} \tag{1}$$

s.t

$$y \in Y \subseteq \{0,\ 1\}^{|P|} \qquad \text{[Binary Variables]} \tag{2}$$

where $G(y, \zeta)$ is the optimal value of the following problem:

$$Min \quad q^T x + h^T z + f^T e \tag{3}$$

s.t.

$$Bx = 0 \qquad \text{[Flow Conservation]} \tag{4}$$

$$Dx + z \geq d \qquad \text{[Meeting Demand]} \tag{5}$$

$$Sx \leq s \qquad \text{[Supply Limit]} \tag{6}$$

$$Rx \leq My + e \qquad \text{[Capacity Constraint]} \tag{7}$$

$$e \leq Oy \qquad \text{[Capacity Expansion Limit]} \tag{8}$$

$$x \in R_+^{|A|*|K|},\ z \in R_+^{|C|*|K|},\ e \in R_+^{|P|} \ \text{[Continuous Variables]} \tag{9}$$

In this paper, the uncertainty is represented by a set of discrete scenarios with given probability of occurrence. The role of unreliable suppliers is implicitly considered in the model by properly way of generating scenarios.

3 Multi-Objective Supply Chain Design Problem

To develop a robust model, two additional objective functions are added into the traditional supply chain design problem, which only minimizes the expected total cost as a single objective problem. The first is the minimization of the variance of the total cost, and the second is the minimization of the downside risk or the risk of loss. The definition of downside risk or the expected total loss is:

$$DRisk = \sum_{l=1}^{L} p_i * Max(Cost_l - \Omega,\ 0) \tag{10}$$

where p_l, Ω and $Cost_l$ represent the occurrence probability of lth scenario, available budget and total cost when lth scenario is realized, respectively. The downside risk can be calculated as follows:

$$DRisk = \sum_{l=1}^{L} p_l * DR_l \tag{11}$$

$$DR_l \geq Cost_l - \Omega \qquad \forall l \tag{12}$$

$$DR_l \geq 0 \qquad \forall l \tag{13}$$

The proper multi-objective stochastic model for our supply chain design problem will be:

$$Min \quad f_1(x) = C^T y + \sum_{l=1}^{L} p_l(q_l^T x_l + h_l^T z_l + f_l^T e_l) \tag{14}$$

$$Min \quad f_2(x) = \sum_{l=1}^{L} p_l[q_l^T x_l + h_l^T z_l + f_l^T e_l - \sum_{l=1}^{L} p_l(q_l^T x_l + h_l^T z_l + f_l^T e_l)]^2 \tag{15}$$

$$Min \quad f_3(x) = \sum_{l=1}^{L} p_l * DR_l \tag{16}$$

s.t

$$Bx_l = 0 \qquad l = 1, \ldots L \tag{17}$$

$$Dx_l + z_l \geq d_l \qquad l = 1, \ldots L \tag{18}$$

$$Sx_l \leq s_l \qquad l = 1, \ldots L \tag{19}$$

$$Rx_l \leq My + e_l \qquad l = 1, \ldots L \tag{20}$$

$$e_l \leq Oy \qquad l = 1, \ldots L \tag{21}$$

$$c^T y + q_l^T x_l + h_l^T z_l + f_l^T e_l - \Omega \leq DR_l \qquad l = 1, \ldots L \tag{22}$$

$$y \in Y \subseteq \{0,\ 1\}^{|P|} \tag{23}$$

$$x \in R_+^{|A|*|K|*L}, \quad z \in R_+^{|C|*|K|*L}, \quad e \in R_+^{|P|*L}, \quad DR \in R_+^L \tag{24}$$

3.1 STEM method

Step 0. Construction of a pay-off table:
A pay-off table is constructed before the first interactive cycle. Let $f_j,\ j = 1, 2, 3$, be feasible ideal solutions of the following 3 problems:

$$Min \quad f_j(x), j = 1, 2, 3 \tag{25}$$

s.t.

$$x \in S \ \text{ (Feasible region of problem } (14 - 24)) \tag{26}$$

Step 1. Calculation phase:
At the mth cycle, the feasible solution to the problem (27-30) is sought, which is the "nearest", in the MINIMAX sense, to the ideal solution f_j^* :

$$Min\ \gamma \tag{27}$$

s.t.

$$\gamma \geq (f_j(x) - f_j^*) * \pi_j, \quad j = 1, 2, 3 \tag{28}$$

$$x \in X^m \tag{29}$$

$$\gamma \geq 0 \tag{30}$$

where X^m includes S plus any constraint added in the previous $(m-1)$ cycles; π_j gives the relative importance of the distances to the optima. Let us consider the jth column of the pay-off table. Let f_j^{max} and f_j^{min} be the maximum and minimum values; then $\pi_j,\ j = 1, 2, 3$, are chosen such that $\pi_j = a_j / \sum_i a_i$, where $a_j = \frac{f_j^{max} - f_j^{min}}{f_j^{max}}$.

Step 2. Decision phase:
The compromise solution x^m is presented to the decision maker(DM). If some of the objectives are satisfactory and others are not, the DM relaxes a satisfactory objective f_j^m enough to allow an improvement of the unsatisfactory objectives in the next iterative cycle. The DM gives Δf_j as the amount of acceptable relaxation. Then, for the next cycle the feasible region is modified as:

$$X^{m+1} = \begin{cases} X^m \\ f_j(x) \leq f_j(x^m) + \Delta f_j, \ \text{if } j = 1, 2, 3 \\ f_i(x) \leq f_i(x^m) \quad \text{if } i = 1, 2, 3, \quad i \neq j \end{cases} \tag{31}$$

The weight π_j is set to zero and the calculation phase of cycle $m + 1$ begins.

4 Numerical Experiments

Consider the supply chain network design problem, as explained in [2]. A drink manufacturing company is willing to design its supply chain. This company owns three customer centers located in three different cities L, M, and N, respectively. Uniform-quality drink in bulk (raw material) is supplied from four plants located in A, B, C and D. There are four possible locations E, F, G and H for building the bottling plants.

The problem attempts to minimize the expected total cost, the variance of the total cost and the downside risk in a multi-objective scheme. Now, we use the STEM method to solve this multi-objective supply chain design problem. First, we construct the pay-off table, which is shown in Table 1.

Table 1. Pay-off table

MEAN	1856986	307077100000	119113
VAR	6165288	0	3985288
DRISK	2179694	1495374000	4467.3

Then, we go to the calculation phase and solve the problem (27-30) using LINGO 10 on a PC Pentium IV 2.1–GHz processor. The compromise solution for the location (strategic) variables is $[1,1,1,0]$. Then, in the decision phase, DM compares the objective vector $f^1 = (f_1^1, f_2^1, f_3^1) = (2219887, 253346, 39887)$ with ideal $f^1 = (f_1^*, f_2^*, f_3^*) = (1856986, 0, 4467.3)$. If f_2^1 is satisfactory, but the other objectives are not, the DM must relax the satisfactory objective f_2^1 enough to allow an improvement of the unsatisfactory objectives in the first cycle. Then, $\Delta f_2 = 999746654$ is considered as the acceptable amount of relaxation. In the second cycle, the compromise solution for the location variables is still $[1,1,1,0]$. This compromise solution is again presented to the DM, who compares its objective vector $f^2 = (f_1^2, f_2^2, f_3^2) = (2132615, 1000000000, 4467.3)$ with the ideal one. If all objectives of the vector f^2 are satisfactory, f^2 is the final solution and the optimal vector including the strategic and tactical variables would be x^2 . The total computational time to solve the problem using STEM method is equal to 18 : 47 (mm:ss), comparing to 02 : 26 : 37 (hh:mm:ss) in generating 55 Pareto-optimal solutions using goal attainment technique (see [2] for details).

5 Conclusion

The proposed model in this paper accounts for the minimization of the expected total cost, the variance of the total cost and the downside risk in a multi-objective scheme to design a robust supply chain network. We used STEM method, which is an interactive multi-objective technique with implicit trade-off information given, to solve the problem. The main advantage of the STEM method is that the prefered solution does not depend on the goal and weight vectors, unlike goal attainment technique. We also avoided using several more binary variables in defining financial risk by introducing downside risk in this paper, which significantly reduced the computational times.

Acknowledgement. This research is supported by Alexander von Humboldt-Stiftung and Iran National Science Foundation (INSF).

References

1. Alonso-Ayuso A, Escudero LF, Garin A, Ortuno MT, Perez G (2003) An approach for strategic supply chain planning under uncertainty based on stochastic 0-1 programming. Journal of Global Optimization 26:97-124
2. Azaron A, Brown KN, Armagan Tarim S, Modarres M (2008) A multi-objective stochastic programming approach for supply chain design considering risk. International Journal of Production Economics 116:129-138
3. Bok JK, Grossmann IE, Park S (2000) Supply chain optimization in continuous flexible process networks. Industrial and Engineering Chemistry Research 39:1279-1290
4. Gjerdrum J, Shah N, Papageorgiou LG (2000) A combined optimisation and agent-based approach for supply chain modelling and performance assessment. Production Planning and Control 12:81-88
5. Hwang CL, Masud ASM (1979) Multiple Objective Decision Making. Springer, Berlin
6. MirHassani SA, Lucas C, Mitra G, Messina E, Poojari CA (2000) Computational solution of capacity planning models under uncertainty. Parallel Computing 26:511-538
7. Santoso T, Ahmed S, Goetschalckx M, Shapiro A (2005) A stochastic programming approach for supply chain network design under uncertainty. European Journal of Operational Research 167:96-115
8. Timpe CH, Kallrath J (2000) Optimal planning in large multi-site production networks. European Journal of Operational Research 126:422-435
9. Tsiakis P, Shah N, Pantelides CC (2001) Design of multiechelon supply chain networks under demand uncertainty. Industrial and Engineering Chemistry Research 40:3585-3604

A MILP Model for Production and Distribution Planning in Consumer Goods Supply Chains

Bilge Bilgen[1] and Hans-Otto Günther[2]

[1] Dept. of Industrial Engineering, Dokuz Eylul University, 35100 Izmir, Turkey,
bilge.bilgen@deu.edu.tr

[2] Dept. of Production Management, Technical University of Berlin, Wilmersdorfer Str. 148, 10585 Berlin,
hans-otto.guenther@tu-berlin.de

Summary. In the consumer goods industry there is often a natural sequence in which the various products are to be produced in order to minimize total changeover time and to maintain product quality standards. For instance, in the production of beverages, the final bottling and packaging lines determine the output rate of the entire production system. This type of production system is called "make-and-pack". In this paper, a so-called block planning approach based on mixed-integer linear optimization modeling is presented which establishes cyclical production patterns for a number of pre-defined setup families.

1 Introduction

In the consumer goods industry the focus in production planning and scheduling is shifting from the management of plant-specific operations to a holistic view of the entire supply chain comprising value adding functions like purchasing, manufacturing and distribution. Consequently, in order to improve the performance of the entire logistic chain, operational planning systems have to be established which allocate the forecasted product demand between plants at various locations, determine the distribution of final products to warehouses, and generate detailed schedules for manufacturing the required quantities of products at the various sites.

The intention of this paper is to develop an integrated production and distribution model for application in the consumer goods industry that addresses the above-mentioned challenges. Its specific contributions are

B. Fleischmann et al.(Eds.), *Operations Research Proceedings 2008*,
DOI: 10.1007/978-3-642-00142-0_29,

the incorporation of cyclical patterns according to the block planning concept proposed by [2] into the production model and the integration of a distribution model that covers the transportation between the plants and the distribution centres in a supply chain. The model is intended for decision support in operative planning with a typical planning horizon of 4 to 12 weeks.

2 Scheduling "Make-and-Pack" Production

In the consumer goods industry, for example in the production of detergents, cosmetics, food and beverages, as well as in the fine-chemicals industry, there is often only one single production stage after which final products are packed and shipped to distribution centres or individual customers. This type of production environment is known as "make-and-pack production". A practical and computationally efficient approach to short-term scheduling for this type of production system can be seen in the block planning approach proposed by [2]. The key idea behind this approach is to schedule blocks, i.e. a predefined sequence of production orders of variable size, in a cyclical fashion. For instance, one block is scheduled per period. The start and ending times of the blocks are variable except that blocks have to be completed before the end of the period they are assigned to. This way block planning provides a higher degree of flexibility regarding the time-phasing and sequencing of production orders compared to classical dynamic lot sizing models. Applications of the block planning approach can be found, for instance, in the production of yoghurt (cf. [3]) or in the production of hair dyes (cf. [2]). Figure 1 illustrates the concept of the block planning approach and its integration with distribution planning.

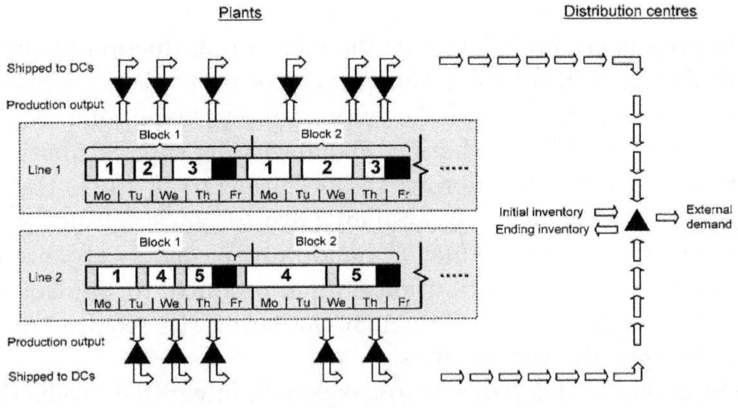

Fig. 1. Block planning and its integration with distribution planning

3 Model Formulation

In the following, a mixed-integer linear programming (MILP) model for production and distribution planning is developed. The formulation is based on a continuous time representation to model the production runs for all items within a setup family and a discrete time representation with micro-periods for the daily assignment of demand elements. The following specific modelling assumptions should be mentioned.

- Since the location of a production line is not essential here, we number production lines consecutively irrespective of the plant location.
- Production lots on each line are consecutively numbered based on the given setup sequence within a block. Hence, constraints which refer to the time phasing of production activities address lots rather than individual products.
- For each lot, a time window is defined which indicates the range of micro-periods during which the production lot is allowed to finish.
- Major setup times of blocks occur at the end of the processing time of a block when the equipment is prepared for the next setup family.
- Due to the unique assignment of blocks to macro-periods, the same time index can be used for both entities.

The notation used in the model formulation is given below. For convenience, we use weeks and days as lengths of macro and micro-periods, respectively.

Indices and sets

$i \in I$	plants
$i \in I_j$	plants which supply DC j
$j \in J$	distribution centers (DCs)
$J \in J_i$	DCs which are supplied by plant i
$p \in P$	products
$l \in L$	production lines
$l \in L_{ip}$	lines at plant i which produce product p
$k \in K_l$	consecutive number of production lots on line l
$k \in K_{lt}$	lots on line l scheduled in block t
$k \in K_{pl}$	lots which produce product p on line l
first(lt)	first lot of block t on line l
$t \in T$	macro-periods: weeks ($t = 1, 2, \ldots, \lvert T\rvert$)
$\tau \in D$	micro-periods: days ($\tau = 1, 2, \ldots, \lvert D\rvert$)
$\tau \in D_{kl}$	time window (micro-periods) for the completion of lot k on line l ($\tau = \underline{d}_{kl}, \ldots, \overline{d}_{kl}$) with d_{kl} = length of the time window

Parameters

c_l^P	operating cost of line l per time unit
c_{kl}^S	minor setup cost for lot k on line l
c_p^I	inventory holding cost per unit of product p per day
c_{pij}^T	transportation cost per unit of product p from plant i to DC j
a_{kl}	unit production time for lot k on line l
s_{kl}	minor setup time for lot k on line l
S_l	major setup time of a block on line l
M_{kl}	maximum size of lot k on line l
$\underline{\alpha}_{lt}$	earliest start-off time (day) of block t on line l
γ_t	end of week t expressed on the daily time scale
$E_{pj\tau}$	external demand of product p at DC j on day τ

Decision Variables

$x_{kl} \geq 0$	size of lot k produced on line l
$y_{kl} \in \{0,1\}$	=1, if lot k is set up on line 1 (0, otherwise)
$z_{kl\tau} \in \{0,1\}$	=1, if lot k has been finished up to day τ on line 1 (0, otherwise)
$\alpha_{lt} \geq 0$	start time of block t on line l
$\Omega_{kl} \geq 0$	end time of lot k on line l
$\delta_{lt} \geq 0$	duration of block t on line l
$q_{kl\tau}$	output from lot k at line l on day τ
$u_{pij\tau}$	quantity of product p shipped from plant i to DC j on day τ
$F_{pj\tau}$	inventory level of product p at DC j at the end of day τ

The objective function aims to minimize total costs comprised of production costs, minor set up costs for the production lots of the individual products, inventory holding costs at distribution centers, and transportation costs.

$$\sum_{l\in L}\sum_{t\in T} c_l^P \cdot \delta_{lt} + \sum_{l\in L}\sum_{k\in K_l} c_{kl}^S \cdot y_{kl} + \sum_{p\in P}\sum_{\tau\in D}\left(\sum_{j\in J} c_p^I \cdot F_{pj\tau} + \sum_{i\in I}\sum_{j\in J_i} c_{pij}^T \cdot u_{pij\tau}\right) \quad (1)$$

Constraints related to blocks

$$\delta_{lt} = S_l + \sum_{k\in K_{lt}} s_{kl} \cdot y_{kl} + \sum_{k\in K_{lt}} a_{kl} \cdot x_{kl} \qquad \forall\, l \in L,\ t \in T \quad (2)$$

$$\alpha_{lt} \geq \alpha_{l,t-1} + \delta_{l,t-1} \qquad \forall l \in L,\ t = 2,\ldots,|T| \text{ with } \alpha_{l1} = 0 \quad (3)$$

$$\alpha_{lt} \geq \underline{\alpha}_{l,t} \qquad \forall l \in L,\ t \in T \quad (4)$$

$$\alpha_{lt} + \delta_{lt} \leq \gamma_t \qquad \forall l \in L,\ t \in T \tag{5}$$

Constraint (2) determines the duration of a block. According to (3) a block is allowed to start as soon as the predecessor block has been completed. Bounds (4) define an earliest start-off day for each block. Constraint (5) ensures that each block is completed before the end of the assigned week.

Constraints related to production lots

$$x_{kl} \leq M_{kl} \cdot y_{kl} \qquad l \in L,\ k \in K_l \tag{6}$$

$$\Omega_{\text{first}(lt)} = \alpha_{lt} + s_{\text{first}(lt)} \cdot y_{\text{first}(lt)} + a_{\text{first}(lt)} \cdot x_{\text{first}(lt)} \qquad l \in L,\ t \in T \tag{7}$$

$$\Omega_{kl} = \Omega_{k-1,l} + s_{kl} \cdot y_{kl} + a_{kl} \cdot x_{kl} \qquad l \in L,\ k \in K_{lt} \backslash (\text{first}(lt),\ t \in T) \tag{8}$$

$$\frac{\tau - \Omega_{kl}}{d_{kl}} \leq z_{kl\tau} \leq 1 + \frac{\tau - \Omega_{kl}}{d_{kl}} \qquad l \in L,\ k \in K_l,\ \tau \in D_{kl} \tag{9}$$

Constraint (6) enforces the lot size k on line l to zero if no corresponding setup takes place. The calculation of the end time of the lots is expressed in (7) for lots which are the first ones in a block and in (8) for the remaining lots. In order to trace the completion of lots on a daily time scale, auxiliary binary decision variables, so-called heaviside variables, are introduced (cf. [1]). These variables indicate that lot k has been finished on line l up to a particular day τ. (9) enforces the heaviside variables to zero for all periods prior to the completion period of the lot and to one for the remaining periods.

Constraints related to production output

$$q_{kl\tau} \leq M_{kl} \cdot z_{kl\tau} \qquad l \in L,\ k \in K_l,\ \tau = \underline{d}_{kl} \tag{10}$$

$$q_{kl\tau} \leq M_{kl} \cdot (z_{kl\tau} - z_{kl\tau-1}) \qquad l \in L,\ k \in K_l,\ \tau \in D_{kl} \backslash \underline{d}_{kl} \tag{11}$$

$$\sum_{\tau \in D_{kl}} q_{kl\tau} = x_{kl} \qquad l \in L,\ k \in K_l \tag{12}$$

The completion of a lot is indicated by a switch from 0 to 1 in the periodic development of the heaviside variables according to (9). Only at the day when the heaviside variable switches from 0 to 1, the q-variable may take a positive value indicating that output is available from lot k produced on line l. Otherwise, its value is enforced to zero. (10) considers the first period in the relevant time window, (11) the remaining periods. Constraint (12) allocates the lot size between the daily output quantities of production lot k at line l.

Constraints related to distribution centers

$$\sum_{l \in L_{ip}} \sum_{k \in K_{pl}} q_{kl\tau} = \sum_{j \in J_i} u_{pij\tau} \qquad p \in P,\ i \in I,\ \tau \in D \tag{13}$$

$$F_{pj\tau} = F_{pj\tau-1} + \sum_{i \in I_j} u_{pij\tau} - F_{pj\tau} \qquad p \in P,\ j \in J,\ \tau \in D \text{ with } F_{pj0} = \text{ given} \tag{14}$$

Constraint (13) expresses the balance between total output quantities achieved from lots producing product p at the production lines in plant i and the shipping quantities from plant i to the connected DCs. Inventory balances are given in (14).

4 Conclusions

Generally, the problem of determining production order sizes, their timing and sequencing in make-and-pack production with setup considerations is equivalent to a capacitated lot size problem. However, the related models presented in the academic literature do not sufficiently reflect the need for scheduling production orders on a continuous time scale with demand elements being assigned to distinct delivery dates. Moreover, issues like definition of setup families with consideration of major and minor setup times and multiple non-identical production lines with dedicated productline assignments are seldom addressed in a realistic way. In this paper an MILP-based block planning concept has been presented that is suited to make-and-pack production system in the consumer goods industry.

Acknowledgement. The first author's research was supported by The Scientific and Technological Research Council of Turkey (TUBITAK).

References

1. Blömer, F, Günther, H-O (2000) LP-based heuristics for scheduling chemical batch processes. International Journal of Production Research, 38:1029–1051
2. Günther, H-O, Grunow, M, Neuhaus, U (2006) Realizing block planning concepts in make-and-pack production using MILP modelling and SAP APO , International Journal of Production Research, 44:3711–3726
3. Lütke Entrup, M, Günther, H-O, van Beek, P, Grunow, M, Seiler, T (2005) Mixed integer linear programming approaches to shelf-life-integrated planning and scheduling in yoghurt production. International Journal of Production Research, 43:5071-5100

An Exact Discrete-Time Model of a Two-Echelon Inventory System with Two Customer Classes

Lars Fischer and Michael Manitz

Universität zu Köln, Seminar für Allgemeine Betriebswirtschaftslehre,
Supply Chain Management und Produktion, 50923 Köln
{lars.fischer,manitz}@wiso.uni-koeln.de

Summary. In this paper, we analyze a two-echelon inventory system with one warehouse, one retailer and a priority customer class with direct supply from the warehouse. The customer demands are arbitrarily distributed. At both stockpoints, the inventory control is according to a reorder-point policy with periodic review and fixed replenishment order sizes. For such a system, we present a discrete-time Markov chain model to describe the inventory positions. With this model, the service levels for both customer classes can be computed. A simple priorization rule such that the priority customer class will be served first as well as a critical-level policy can be considered.

1 Problem Description

The warehouse faces two types of customers: retailers, and important customers who buy directly as end consumers at the warehouse. It is assumed that an incomplete supply to such customers leads to higher backorder costs in comparison to the retailers. Therefore, they will be served with higher priority.
Both, the demand of an arbitrary time period at the retailer, D_R, and the end consumer demand D_W that is directed to the warehouse are generally distributed. To restrict the state space under consideration, we regard $d_R^{\max}$ and $d_W^{\max}$ as the maximum demand sizes. That is, we truncate the demand distributions such that, if the maximum demands are chosen sufficiently large, the probability of demand sizes larger than these values are close to 0.
The replenishment lead time is deterministic on a discrete time axis (e. g. days). It is assumed that the basic time period is chosen according to the length of the replenishment lead time. That is, the lead

B. Fleischmann et al.(Eds.), *Operations Research Proceedings 2008*,
DOI: 10.1007/978-3-642-00142-0_30, © Springer-Verlag Berlin Heidelberg 2009

time is exactly one period (e. g. 1 day). This assumption ensures the Markov property in the models described later on. Otherwise, any transition probability between two inventory positions would require the information on how long the actual replenishment would still take. We assume the following periodic schedule: The inventory is registered after receiving a replenishment order. The new customer orders are collected up to a point of time later on. They may be fulfilled by the available stock on-hand. Finally, after delivery, a replenishment order is triggered if the stock on-hand is reduced to or below the reorder point. Nota bene there is a time shift between the replenishment order releases of the retailer and those of the warehouse such that a retailer order is to be considered as an add-on to the demand directed to the warehouse. Anyway, a replenishment order usually is triggered at the end of a period. It will arrive next period (just before the next inspection point in time).

At both stockpoints, reorder-point policies are established. We assume that the replenishment order quantities q_R and q_W are chosen such that at most only one order can be outstanding at the inspection points. That means, the replenishment order size has to be larger than the maximum demand of one period. To ensure steady-state in the case of a critical-level policy without partial backordering, the restriction is a little harder: $q_W > 2 \cdot (d_W^{\max} + q_R)$. This avoids unlimited waiting times for the retailer.

2 The Basic Model

In case of Poisson demand, the evolution of the inventory position can be described by a Markov model with the state space $\{s+1, \ldots, s+q\}$; see [2] for stockpoints that are controlled in continuous time. It is shown that the inventory position has an uniform distribution in the steady-state. The probability distribution of the on-hand inventory (see also [1]) and of the amount of backorders can be derived from these probabilities. Similar models can be developed for multi-echelon systems. Then, the state space consists of state vectors describing the inventory position at each stage. The state space for the inventory position remains the same also in the case of a discrete time axis if just before an inspection point a replenishment order of an appropriate size can be released.

To model class-dependent service, we take into consideration the evolution of the net inventory at the warehouse. The net inventory is equal to the inventory position except when outstanding orders are on their

way. According to the assumptions above, only one order of size q_W can be outstanding. Hence, the net inventory stock may be reduced by q_W units below the inventory position that includes the outstanding order. The minimum possible value of the net inventory stock is reached if the inventory position was s_W+1 at the last inspection point and if the maximum possible demand occurs. So, the net inventory is between $s_W + 1 - (d_W^{\max} + q_R)$ and $s_W + q_W$. The lower ones of these values can only be reached with a positive q_R. That is, the retailer has placed a replenishment order which was triggered if his net inventory stock had fallen down to s_R or below by the demand d_R during the last period before. With the same amount of d_R units, the retailer's inventory position (that includes the outstanding order of size q_R and that is at least $s_R + 1$ by assumption) has been reduced, too. It takes its minimum possible value if the maximum demand $d_R^{\max}$ has occured, i. e.: $s_R+1-d_R^{\max}+q_R = s_R+q_R-(d_R^{\max}-1)$. Hence, in case of the maximum demand $d_W^{\max}$ and net inventories between $s_W + 1 - (d_W^{\max} + q_R)$ and $s_W + 1 - (d_W^{\max} + 1) = s_W - d_W^{\max}$ at the warehouse (i. e. with positive q_R), the inventory position of the retailer is between its maximum value $(s_R + q_R)$ and at most $(d_R^{\max} - 1)$ units below it.
In general, a state of the system is denoted by (i, j, k), where i is a particular net inventory stock and j the inventory position at the warehouse. The third entry represents a particular inventory position of k items at the retailer. The transitions between these states are registered as changes of these inventory positions during one period. The states will change driven by stochastic demands and according to the pre-determined (s, q) replenishment policies. The probability of a certain transition corresponds to the probability that during one period a certain demand d_W occurs at the warehouse, and that a certain demand d_R occurs at the retailer: $\mathrm{P}(D_W = d_W, D_R = d_R) =: p\,(d_W, d_R)$.
First, let us consider the case $i = j$ when the system has reached a certain state (i, j, k) at which no replenishment order has been placed by the warehouse at this particular time instant. Then, there are four kinds of initial states a transition into (i, j, k) may come from:

$$
\begin{aligned}
&\mathrm{P}(i,j,k)\\
&= \sum_{d_W \in \mathcal{D}_W^1} \sum_{d_R \in \mathcal{D}_R^1} p\,(d_W, d_R) \cdot \mathrm{P}(i + d_W, j + d_W, k + d_R)\\
&+ \sum_{d_W \in \mathcal{D}_W^2} \sum_{d_R \in \mathcal{D}_R^2} p\,(d_W, d_R)\\
&\qquad \cdot \mathrm{P}(i + d_W + q_R, j + d_W + q_R, k + d_R - q_R)\\
&+ \sum_{d_W \in \mathcal{D}_W^{3_1}} \sum_{d_R \in \mathcal{D}_R^{3_1}} p\,(d_W, d_R) \cdot \mathrm{P}(i + d_W - q_W, j + d_W, k + d_R)\\
&+ \sum_{d_W \in \mathcal{D}_W^{3_2}} \sum_{d_R \in \mathcal{D}_R^{3_2}} p\,(d_W, d_R) \cdot \mathrm{P}(i + d_W - q_W, j + d_W, k + d_R)\\
&+ \sum_{d_W \in \mathcal{D}_W^{4_1}} \sum_{d_R \in \mathcal{D}_R^{4_1}} p\,(d_W, d_R)\\
&\qquad \cdot \mathrm{P}(i + d_W + q_R - q_W, j + d_W + q_R, k + d_R - q_R)\\
&+ \sum_{d_W \in \mathcal{D}_W^{4_2}} \sum_{d_R \in \mathcal{D}_R^{4_2}} p\,(d_W, d_R)\\
&\qquad \cdot \mathrm{P}(i + d_W + q_R - q_W, j + d_W + q_R, k + d_R - q_R)\\
&\qquad\qquad \begin{pmatrix} i, j \in [s_W + 1, s_W + q_W] =: \mathcal{I}_1 \text{ with } i = j\\ k \in [s_R + 1, s_R + q_R] =: \mathcal{K}_1 \end{pmatrix} \qquad (1)
\end{aligned}
$$

with

$$
\begin{aligned}
\mathcal{D}_W^1 &= [0, \min\{s_W + q_W - i, d_W^{\max}\}],\ \mathcal{D}_R^1 = [0, \min\{s_R + q_R - k, d_R^{\max}\}]\\
\mathcal{D}_W^2 &= [0, \min\{s_W + q_W - q_R - i, d_W^{\max}\}]\\
\mathcal{D}_R^2 &= [s_R + 1 + q_R - k, \min\{s_R + q_R + q_R - k, d_R^{\max}\}]\\
\mathcal{D}_W^{3_1} &= [\max\{0, s_W + 1 - (d_W^{\max} + q_R) + q_W - i\},\\
&\qquad \min\{s_W - d_W^{\max} + q_W - i, d_W^{\max}\}]\\
\mathcal{D}_R^{3_1} &= [\max\{0, s_R + q_R - (d_R^{\max} - 1) - k\}, \min\{s_R + q_R - k, d_R^{\max}\}]\\
\mathcal{D}_W^{3_2} &= [\max\{0, s_W + 1 - d_W^{\max} + q_W - i\}, \min\{s_W + q_W - j, d_W^{\max}\}]\\
\mathcal{D}_R^{3_2} &= [0, \min\{s_R + q_R - k, d_R^{\max}\}]\\
\mathcal{D}_W^{4_1} &= [\max\{0, s_W + 1 - d_W^{\max} - 2 \cdot q_R + q_W - i\},\\
&\qquad \min\{s_W - d_W^{\max} - q_R + q_W - i, d_W^{\max}\}]\\
\mathcal{D}_R^{4_1} &= [s_R + 2 \cdot q_R - (d_R^{\max} - 1) - k, d_R^{\max}]\\
\mathcal{D}_W^{4_2} &= [\max\{0, s_W + 1 - d_W^{\max} + q_W - q_R - i\},\\
&\qquad \min\{s_W + q_W - q_R - j, d_W^{\max}\}]\\
\mathcal{D}_R^{4_2} &= [s_R + 1 + q_R - k, d_R^{\max}]
\end{aligned}
$$

For states with $i = j - q_W \leq s_W$ (i.e. a replenishment order at the warehouse has been released and still is outstanding), it follows:

$$
\mathrm{P}(i,j,k) = \sum_{d_W \in \mathcal{D}_W^5} \sum_{d_R \in \mathcal{D}_R^5} p\,(d_W, d_R) \cdot \mathrm{P}(i + d_W, j + d_W - q_W, k + d_R)
$$

$$\begin{pmatrix} i \in [s_W + 1 - d_W^{\max}, s_W] =: \mathcal{I}_2 \\ j \in [s_W + 1 - d_W^{\max} + q_W, s_W + q_W] =: \mathcal{J}_2 \\ k \in [s_R + 1, s_R + q_R] =: \mathcal{K}_2 \end{pmatrix} \quad (2)$$

$$\begin{aligned} &\mathrm{P}(i,j,k) \\ &= \sum_{d_W \in \mathcal{D}_W^6} \sum_{d_R \in \mathcal{D}_R^6} p(d_W, d_R) \cdot \mathrm{P}(i + d_W + q_R, j + d_W - q_W + q_R, k + d_R - q_R) \end{aligned}$$

$$\begin{pmatrix} i \in [s_W + 1 - (d_W^{\max} + q_R), s_W - d_W^{\max}] =: \mathcal{I}_3 \\ j \in [s_W + 1 - (d_W^{\max} + q_R) + q_W, s_W + q_W - d_W^{\max}] =: \mathcal{J}_3 \\ k \in [s_R + q_R - (d_R^{\max} - 1), s_R + q_R] =: \mathcal{K}_3 \end{pmatrix} \quad (3)$$

with

$$\mathcal{D}_W^5 = [s_W + 1 - i, d_W^{\max}], \mathcal{D}_R^5 = [0, \min\{s_R + q_R - k, d_R^{\max}\}]$$
$$\mathcal{D}_W^6 = [\max\{0, s_W + 1 - q_R - i\}, d_W^{\max}]$$
$$\mathcal{D}_R^6 = [s_R + 1 + q_R - k, \min\{s_R + 2 \cdot q_R - k, d_R^{\max}\}]$$

The system of steady-state equations becomes complete with the normalization constraint that replaces one of the equations above:

$$\sum_{i \in \mathcal{I}_1} \sum_{j \in \mathcal{I}_1} \sum_{k \in \mathcal{K}_1} \mathrm{P}(i,j,k) + \sum_{i \in \mathcal{I}_2} \sum_{j \in \mathcal{J}_2} \sum_{k \in \mathcal{K}_2} \mathrm{P}(i,j,k) + \sum_{i \in \mathcal{I}_3} \sum_{j \in \mathcal{J}_3} \sum_{k \in \mathcal{K}_3} \mathrm{P}(i,j,k) = 1 \quad (4)$$

3 The Service Levels

Due to priorization, the service levels differ for both customer classes. The customers (denoted as end consumers (E)) that directly buy at the warehouse are served first. The service level that is offered to the retailer (R) is significantly lower. To describe the service levels, we analyze the probability of shortages at the warehouse. For a certain inventory state, backorders may occur due to according demands. Hence, the probability of backorders are:

$$\begin{aligned} \mathrm{P}(B_E) = &\sum_{i \in \tilde{\mathcal{I}}_1} \sum_{j \in \tilde{\mathcal{J}}_1} \sum_{k \in \mathcal{K}_1} \mathrm{P}(i,j,k) \cdot \sum_{d_W \in \mathcal{D}_W} \mathrm{P}(D_W = d_W) \\ &+ \sum_{i \in \tilde{\mathcal{I}}_2} \sum_{j \in \tilde{\mathcal{J}}_2} \sum_{k \in \mathcal{K}_2} \mathrm{P}(i,j,k) \cdot \sum_{d_W \in \mathcal{D}_W} \mathrm{P}(D_W = d_W) \\ &+ \sum_{i \in \tilde{\mathcal{I}}_3} \sum_{j \in \tilde{\mathcal{J}}_3} \sum_{k \in \mathcal{K}_3} \mathrm{P}(i,j,k) \cdot \sum_{d_W \in \mathcal{D}_W} \mathrm{P}(D_W = d_W) \end{aligned} \quad (5)$$

Shortages only occur if the available inventory is not enough, i. e. only for $i \in \widetilde{\mathcal{I}}_\ell$ which are defined as $\mathcal{I}_\ell$ (and $\mathcal{J}_\ell$, respectively) without inventory levels beyond $d_W^{\max}$. On the other hand, only demands from range $\mathcal{D}_W = [\max\{i+1,1\}, d_W^{\max}]$ lead to backorders. For the retailer, a backorder may occur also in the case $D_W = 0$ if $q_R > i$. The retailer orders exactly the quantity q_R if the demand D_R is large enough such that the retailer's net inventory falls down from k to s_R. Using modified state sets (now considering an overall demand $D_W + q_R$ at the warehouse), the probability of a retailer backorder can be calculated as follows:

$$\begin{aligned} \mathrm{P}(B_R) = &\sum_{i\in\widetilde{\mathcal{I}}_1}\sum_{j\in\widetilde{\mathcal{J}}_1}\sum_{k\in\mathcal{K}_1}\mathrm{P}(i,j,k)\cdot\sum_{d_W\in\widetilde{\mathcal{D}}_W}\sum_{d_W\in\widetilde{\mathcal{D}}_R}p(d_W,d_R)\\ &+\sum_{i\in\widetilde{\mathcal{I}}_2}\sum_{j\in\widetilde{\mathcal{J}}_2}\sum_{k\in\mathcal{K}_2}\mathrm{P}(i,j,k)\cdot\sum_{d_W\in\widetilde{\mathcal{D}}_W}\sum_{d_W\in\widetilde{\mathcal{D}}_R}p(d_W,d_R)\\ &+\sum_{i\in\widetilde{\mathcal{I}}_3}\sum_{j\in\widetilde{\mathcal{J}}_3}\sum_{k\in\mathcal{K}_3}\mathrm{P}(i,j,k)\cdot\sum_{d_W\in\widetilde{\mathcal{D}}_W}\sum_{d_W\in\widetilde{\mathcal{D}}_R}p(d_W,d_R) \end{aligned} \tag{6}$$

with $\widetilde{\mathcal{D}}_W = [\max\{i-q_R+1,0\}, d_W^{\max}]$ and $\widetilde{\mathcal{D}}_R = [k-s_R, d_R^{\max}]$.

4 An Extended Model for Critical-Level Policies

With an additional dimension of the state space that registers the number of backordered replenishments of the retailer, the model can be extended to consider a critical inventory level c that should be reserved for the important customer; see for example [4], [3]. Without partial backordering, the retailer is only served from a physical stock equal to or larger than $q_R + c$. This increases the service level that can be offered to important customers.

References

1. Galliher, H. P., P. M. Morse, and M. Simond (1959). Dynamics of two classes of continuous-review inventory systems. Operations Research 7 (3), 362-393.
2. Hadley, G., and T. M. Whitin (1963). Analysis of Inventory Systems. Englewood Cliffs: Prentice-Hall.
3. Möllering, K. T., and U. W. Thonemann (2007). An optimal critical level policy for inventory systems with two demand classes. Working paper, University of Cologne, Department of Supply Chain Management and Management Science, Cologne.
4. Tempelmeier, H. (2006). Supply chain inventory optimization with two customer classes. European Journal of Operational Research 174 (1), 600-621.

Supply Chain Coordination Models with Retailer's Attitudes Toward Risk

Harikrishnan K Kanthen[1] and Chhaing Huy[2]

[1] Faculty of Engineering,Nottingham University Malaysia Campus,43500 Semenyih, Selangor, Malaysia.
Harikrishnan.kk@nottingham.edu.my

[2] Graduate School of Economics,Osaka University, Toyonaka, Osaka 560-0043, Japan

1 Introduction

Supply Chain Management (SCM) to effectively integrate various facilities and partners, such as suppliers, manufacturers, distributors, and retailers is the key to yield the competitive advantage for the companies. Nowadays, integrated supply chain management is possible due to advances in information technology. Despite these advances, it is still difficult to achieve the best supply chain performance without the coordination among members of a supply chain, because different facilities and partners in the supply chain may have different, conflicting objectives (see for instance [4]).

We consider a two level supply chain model in a newsvendor's problem. First, we shall show a property on the supplier's share of the supply chain's expected profit in a buyback contract where the retailer is risk-neutral. Second, we discuss the effect of the attitudes toward risk of the retailer on the coordination in a supply chain. Most of the previous literature assume the retailer is risk-neutral. But recent studies indicate that the risk-averse approach in SCM plays a very crucial role in achieving a better supply chain performance. For example, Wang et al.[5] studied about the risk-averse newsvendor order at a higher selling price. Zhang et al. [7] analyzed the supply chain coordination of loss-averse newsvendor with contract. Keren and Pliskin [2] studied a benchmark solution for the newsvendor problem where the retailer is risk-averse, in which the demand distribution is uniformly distributed. We study this problem in a more general setting; using the utility functions representing the retailer's preferences, the optimal order quantity

DOI: 10.1007/978-3-642-00142-0_31,

placed by a risk-averse retailer is compared with that of a risk-neutral retailer under the wholesale price contract and the buyback contract respectively. We also derive interesting properties between the retailer's order quantity and the Pratt's risk aversion function in the wholesale price contract and the buyback contract in a special case where both the demand function and the retailer's utility function are exponential.

2 Models Where the Retailer is Risk-Neutral

In these models, the retailer must choose an order quantity before the start of a single selling season that has stochastic demand. Let $D \geq 0$ be the demand occurring in a selling season. Let F be the distribution function of demand and f its density function. F is differentiable, strictly increasing and $F(0) = 0$.
The notation we use is as follows: c the production cost per unit; w the supplier's wholesale price per unit; p the retailer's selling price per unit; b the supplier's buyback cost per unit; ν the salvage price for unsold item per unit at the end of the season; D demand;and Q the order quantity.

2.1 Wholesale Price Contract Model

Under wholesale price contract, the supplier charges the retailer w per unit purchased. The retailer gets a salvage value ν per unit unsold. It is reasonable to assume $\nu < w$. When the retailer is risk-neutral, he/she wishes to maximize his/her profit function $\Pi_R^{(W)}(Q)$,

$$\begin{aligned}\Pi_R^{(W)}(Q) &= \int_0^Q [pD + \nu(Q-D) - wQ] f(D)\, dD + \int_Q^\infty (p-w)Qf(D)\, dD \\ &= (p-w)Q - (p-\nu)\int_0^Q F(D)\, dD. \qquad (1)\end{aligned}$$

The retailer's optimal order quantity $Q_R^{(W)*}$ is

$$Q_R^{(W)*} = F^{-1}\left(\frac{p-w}{p-\nu}\right). \qquad (2)$$

On the other hand, the supply chain's expected profit is

$$\Pi_{SC}^{(W)}(Q) = \Pi_R^{(W)}(Q) + \Pi_S^{(W)}(Q) = (p-c)Q - (p-\nu)\int_0^Q F(D)\, dD, \qquad (3)$$

Then the supply chain's optimal order quantity is:

$$Q_{SC}^{(W)*} = F^{-1}\left(\frac{p-c}{p-\nu}\right). \tag{4}$$

Since $Q_R^{(W)*} = F^{-1}\left(\frac{p-w}{p-\nu}\right) < Q_{SC}^{(W)*} = F^{-1}\left(\frac{p-c}{p-\nu}\right)$ when $w > c$, the wholesale price contract does not coordinate the supply chain if the supplier earns a positive profit (see,for more detail[1]).

2.2 Buyback Contract Model

An obvious explanation for a buyback contract is risk-sharing; that is, the retailer returns the unsold products to the supplier or the supplier offers a credit on all unsold products to the retailer [6]. With the buyback contract, the supplier charges the retailer w per unit and pays b per unit unsold item at the end of the season. In our model, we assume that all unsold units are physically returned back to the supplier, and $b \geq \nu$ so the supplier salvages those units. The retailer's expected profit, $\Pi_R^{(B)}(Q)$, under this buyback contract is

$$\begin{aligned}\Pi_R^{(B)}(Q) &= \int_0^Q [pD + b(Q-D) - wQ]f(D)\,dD + \int_Q^\infty (p-w)Qf(D)\,dD \\ &= (p-w)Q - (p-b)\int_0^Q F(D)\,dD. \end{aligned} \tag{5}$$

The retailer's optimal order quantity is,

$$Q_R^{(B)*} = F^{-1}\left(\frac{p-w}{p-b}\right). \tag{6}$$

The supply chain's expected profit is

$$\Pi_{SC}^{(B)}(Q) = \Pi_R^{(B)}(Q) + \Pi_S^{(B)}(Q) = (p-c)Q - (p-\nu)\int_0^Q F(D)\,dD, \tag{7}$$

which is maximized at

$$Q_{SC}^{(B)*} = F^{-1}\left(\frac{p-c}{p-\nu}\right). \tag{8}$$

From Eqs. (6) and (8), there exists $b^* = \frac{p(w-c)+\nu(p-w)}{p-c}$ such that $Q_R^{(B)*} = Q_S^{(B)*} = Q_{SC}^{(B)*}$.
Observe that

$$\Pi_S^{(B)}(Q) = (w-c)Q - (b^* - \nu)\int_0^Q F(D)\,dD = \frac{w-c}{p-c}\,\Pi_{SC}^{(B)}(Q)$$

and

$$\frac{\partial}{\partial Q}\left(\frac{\Pi_S^{(B)}(Q)}{\Pi_{SC}^{(B)}(Q)}\right) = \frac{[QF(Q)-\int_0^Q F(D)\,dD]}{(p-c)[(p-c)Q-(p-\nu)\int_0^Q F(D)\,dD]^2}(b^*-b).$$

Then we have

Theorem 1.

1. *For $b = b^* = \frac{p(w-c)+\nu(p-w)}{p-c}$, the supplier's share of the supply chain's expected profit is the same (equals to $\frac{w-c}{p-c}$), regardless of the order quantity Q.*
2. *if $b < b^*$, the supplier's share of the supply chain's expected profit is increasing in the order quantity Q.*
3. *if $b > b^*$ the supplier's share of the supply chain's expected profit is decreasing in the order quantity Q.*

Note that , for b=b*, the pie (the supply chain's profit) is maximized at $Q_R^{(B)*}$.

3 Models Where the Retailer is Risk-Averse

Let $u(x)$ be the utility function which represents the retailer's preferences. Assume that $u(x)$ is strictly increasing and twice continuously differentiable at all x, then $u'(x) > 0$ for all x.
As is well-known, the decision maker is risk-averse if and only if her utility function is concave. In what follows, assume that the retailer is risk- averse. Thus $u''(x) \leq 0$ for all x. According to [3], a measure of risk aversion to indicate the extent that the decision maker wants to averse the risk is $r(x) = -u''(x)/u'(x)$.

3.1 Wholesale price contract model

The retailer's expected utility is

$$Eu_{R,a}^{(W)}(Q) = \int_0^Q u[pD + \nu(Q-D) - wQ]f(D)\,dD + \int_Q^\infty u[(p-w)Q]f(D)\,dD. \quad (9)$$

Let $Q_{R,a}^{(W)*}$ be the solution which maximizes $Eu_{R,a}^{(W)}(Q)$. Then observe that

$$\begin{aligned}\frac{\partial Eu_{R,a}^{(W)}\left(Q_R^{(W)*}\right)}{\partial Q} &\leq -(p-\nu)u'\left[(p-w)Q_R^{(W)*}\right]\left(\frac{p-w}{p-\nu}\right)\\ &\quad + (p-w)u'\left[(p-w)Q_R^{(W)*}\right] = 0\\ &= \frac{\partial Eu_{R,a}^{(W)}\left(Q_{R,a}^{(W)*}\right)}{\partial Q}. \qquad (10)\end{aligned}$$

Since $\frac{\partial Eu_{R,a}^{(W)}}{\partial Q}$ is a decreasing function, we obtain

Theorem 2. *The optimal order quantity for a risk averse retailer is less than or equal to that of a risk neutral one; that is,* $Q_{R,a}^{(W)*} \leq Q_R^{(W)*}$.

A Special Case

Assume that the retailer's utility function is $u(x) = -e^{-kx}$, where k is a positive number and that the random demand for the item follows an exponential density function, $f(D) = \lambda e^{-\lambda D}$. Therefore, the risk aversion function is: $r(x) = k$. Noting that

$$Q_{R,a}^{(W)*} = \frac{1}{kp - k\nu + \lambda} \ln\left(\frac{(p-\nu)(kp - kw + \lambda)}{\lambda(w-\nu)}\right), \qquad (11)$$

we have

Theorem 3. *The more risk-averse the retailer is, the less the order quantity is. That is, the order quantity* $Q_{R,a}^{(W)*}$ *decreases as k increases.*

3.2 Buyback Contract Model

The retailer's expected utility is

$$Eu_{R,a}^{(B)}(Q) = \int_0^Q u[pD + b(Q-D) - wQ]f(D)\,dD + \int_Q^\infty u[(p-w)Q]f(D)\,dD. \qquad (12)$$

Let $Q_{R,a}^{(B)*}$ be the solution which maximizes $Eu_{R,a}^{(B)}(Q)$. Then we have

Theorem 4. *The optimal order quantity for a risk-averse retailer is less than or equal to that of a risk neutral one; that is,* $Q_{R,a}^{(B)*} \leq Q_R^{(B)*}$.

A Special Case

Under the same assumptions as a special case in 3.1, we have

Theorem 5. *The more risk-averse the retailer is, the less the order quantity is. That is, the order quantity* $Q_{R,a}^{(B)*}$ *is decreases as k increases.*

4 Conclusions

We have derived a property on the supplier's share of the supply chain's profit in a buyback contract. Then we explored some fundamental properties of the order quantity of risk-averse retailer. In particular, we can conclude that the optimal order quantity of a risk-averse retailer is less than or equal to that of a risk-neutral one, and so the supply chain performance in the risk-averse is worse off than that in the risk-neutral case. Furthermore, the more risk-averse the retailer is, the less the performance of the supply chain is. Therefore, if there is a significant risk aversion in a supply chain then our findings would suggest that some care to be given to the retailer's order quantity in order to maximize the system's profit.

References

1. Cachon, G. P.: 2003, Supply chain coordination with contracts, in A. G. de Kok and S. C. Gravers (eds), Handbooks in Operations Research and Management Science, Vol. 11, Elsevier, Boston, pp. 229–340.
2. Keren, B. and Pliskin, J. S.: 2006, A benchmark solution for the risk-averse newsvendor problem, European Journal of Operational Research 174, 1643–1650.
3. Pratt, W. J.: 1964, Risk aversion in the small and in the large, Econometrica 32, 122–136.
4. Simchi-Levi, D., Kaminsky, P. and Simchi-Levi, E.: 2000, Designing and Managing the Supply Chain: Concepts, Strategies, and Case Studies, McGraw-Hill Companies.
5. Wang, X. C , Webster,S and Suresh, N. C: 2008, Would a risk-averse newsvendor order less at a higher selling price?, European Journal of Operational Research (forthcoming)
6. Yue, X. and Raghunathan, S.: 2007, The impacts of the full returns policy on a supply chain with information asymmetry, European Journal of Operational Research 180, 630–647.
7. Zhang, L. C , Song, S and Wu, C: 2005, Supply chain coordination of loss-averse newsvendor with contract, Tsinghuan Science and Technology 10, 133-140.

Zur Erweiterung der Kennlinientheorie auf mehrstufige Lagersysteme

Karl Inderfurth und Tobias Schulz

Fakultät für Wirtschaftswissenschaft, Otto-von-Guericke-Universität Magdeburg: {karl.inderfurth,tobias.schulz}@ovgu.de

1 Einleitung

Zum Design logistischer Systeme sowie zur Planung und Kontrolle logistischer Prozesse ist es notwendig, messbare Kenngrößen zu deren Beurteilung heranzuziehen sowie qualitative und quantitative Relationen zwischen diesen Kenngrößen einschätzen zu können. Für die Darstellung funktionaler Zusammenhänge zwischen wichtigen Kenngrößen und deren graphische Umsetzung in Form von Kurvenverläufen wird im ingenieurwissenschaftlichen Bereich der Begriff der "logistischen Kennlinie" verwendet. In ihrer Monografie "Logistische Kennlinien" beschreiben die Autoren P. Nyhuis und H.-P. Wiendahl [2] unter anderem, wie sich ein solcher Zusammenhang für die Kenngrößen mittlerer Bestand und mittlerer Lieferverzug in einem einstufigen Lagersystem approximativ ableiten lässt. In mehreren Arbeiten erweitern Inderfurth und Schulz (siehe unter anderem in [1]) diesen Ansatz dahingehend, dass sie eine Methodik zur exakten Ableitung der Lagerkennlinie erarbeiten und diese dem approximativen Ansatz nach Nyhuis und Wiendahl gegenüberstellen.
In diesem Beitrag soll die exakte Methodik auf ein mehrstufiges Lagersystem erweitert werden, womit an theoretische Aspekte der mehrstufigen stochastischen Lagerhaltung (siehe unter anderem in Tempelmeier [3]) angeschlossen wird. Dazu soll im folgenden Abschnitt der Modellrahmen kurz erläutert und die Extrempunkte der exakten Kennlinie (Maximalwerte von mittlerem Bestand und mittlerem Lieferverzug) bestimmt werden. Der Verlauf der Kennlinie zwischen den Extrempunkten steht im Mittelpunkt der Analyse des dritten Abschnitts, bevor die Arbeit im letzten Kapitel kurz zusammengefasst und ein Ausblick auf zukünftige Forschungsrichtungen gegeben wird.

B. Fleischmann et al.(Eds.), *Operations Research Proceedings 2008*,

DOI: 10.1007/978-3-642-00142-0_32,

2 Festlegung der Extrempunkte der Kennlinie

In diesem und dem nächsten Unterpunkt soll die exakte Kennlinienanalyse von Inderfurth und Schulz auf ein mehrstufiges Lagersystem erweitert werden. Es wird als Modellrahmen ein serielles zweistufiges Lagersystem ohne Lieferzeiten zugrundegelegt, das aus einer Vor- und einer Endstufe besteht. Das Lager der Endstufe sieht sich einer zufälligen konstanten Nachfragerate gegenüber und wird dabei durch Bestellungen beim Vorstufenlager aufgefüllt. Das Vorstufenlager muss den Bestellungen der Endstufe nachkommen und wird selbst durch einen stets lieferfähigen Lieferanten versorgt. Beide Lager werden mit einer (t,S)-Regel disponiert, bei der nach Ablauf einer Kontrollperiode t das Lager auf den disponiblen Bestand S aufgefüllt wird. Die Bestellgrenzen, welche im Folgenden für Vor- und Endstufe mit S^V bzw. S^E bezeichnet werden, können dabei unabhängig voneinander festgelegt werden. Um die Mehrstufigkeit des Systems angemessen zu berücksichtigen, werden die Bestellintervalle beider Lager miteinander gekoppelt und im Folgenden beispielhaft durch die Relation $t^V = 2 \cdot t^E$ miteinander verknüpft. Durch diese Relation kommt es innerhalb einer Kontrollperiode der Vorstufe t^V zu folgenden Ereignissen. Zu Beginn der Kontrollperiode liegen die physischen Bestände auf beiden Lagerstufen bei S^V bzw. S^E Einheiten, was gleichzeitig bedeutet, dass sämtliche in vorherigen Kontrollperioden verursachte Fehlmengen getilgt wurden. Nach Ablauf der ersten Kontrollperiode der Endstufe t^E, in der eine stetige Kundennachfrage das Lager der Endstufe kontinuierlich verringert, erhöht das Endstufenlager seinen disponiblen Bestand auf S^E Einheiten, in dem es beim Lager der Vorstufe eine entsprechende Bestellung aufgibt. Da nicht immer genügend Einheiten im Vorstufenlager vorhanden sein müssen, kann es durchaus vorkommen, dass zu Beginn der zweiten Kontrollperiode des Endstufenlagers der physische Lagerbestand nicht bei S^E Einheiten liegt.

Zur Berechnung der Extremwerte der Lagerkennlinie werden zur Vereinfachung der Analyse noch folgende Annahmen gemacht. Die Nachfragerate der Kunden in jedem Kontrollintervall t^E kann lediglich zwei Ausprägungen annehmen, eine hohe (mit r_o bezeichnet) und eine niedrige (mit r_u bezeichnet). Zur Auswertung der möglichen Bestandsverläufe müssen stets zwei konsekutive Teilzyklen der Endstufe mit der Länge t^E untersucht werden, die zusammen dem Bestellzyklus der Vorstufe t^V entsprechen. Daraus folgt, dass unter den gewählten Annahmen insgesamt vier verschiedene Nachfrageszenarien ausgewertet werden müssen, die in Tabelle 1 aufgeführt und mit I bis IV gekennzeichnet sind.

Table 1. Nachfrageausprägung in den Teilzyklen der Nachfrageszenarien

	1. Teilzyklus	2. Teilzyklus
Nachfrageszenario I	niedrige Nachfrage	niedrige Nachfrage
Nachfrageszenario II	niedrige Nachfrage	hohe Nachfrage
Nachfrageszenario III	hohe Nachfrage	niedrige Nachfrage
Nachfrageszenario IV	hohe Nachfrage	hohe Nachfrage

Eine Kennlinie weist stets zwei Extremwerte auf, den maximal möglichen mittleren Lieferverzug L_{max} sowie den maximal möglichen mittleren Bestand B_{max}. Analog zur Analyse von Inderfurth und Schulz in [1] können diese beiden Punkte durch eine Gewichtung der Kenngrößen aller möglichen Szenarien errechnet werden. Dabei gilt es zu beachten, dass die Lagerstände beider Stufen für die Berechnung des mittleren Bestandes genutzt werden, während nur die Fehlmengen auf der Endstufe in den mittleren Lieferverzug eingehen. Der maximal mögliche mittlere Lieferverzug L_{max} ist genau dann zu beobachten, wenn in keinem Nachfrageszenario ein Bestand im System vorliegt. Dies wird unter den gegebenen Annahmen für ein mehrstufiges System durch die Bestellgrenzen $S^V = S^E = 0$ erreicht. In Abbildung 1 sind alle vier potentiellen Nachfrageszenarien für die Nachfragedaten $r_u = 5$ und $r_o = 10$ sowie für ein Kontrollintervall $t^E = 1$ beispielhaft dargestellt. Der maximal mögliche mittlere Bestand B_{max}, der gerade den Punkt signalisiert, bei dem das System keinen Lieferverzug aufweist, wird unter den gegebenen Annahmen durch die Bestellgrenzen $S^V = S^E = t^E r_o$ erreicht. Die vier potentiellen Nachfrageszenarien zur Bestimmung von B_{max} sind in Abbildung 2 dargestellt.
Nach Auswertung der einzelnen Nachfrageszenarien für beide Extremwerte können folgende Ergebnisse ermittelt werden, wobei mit $\overline{r}$ die mittlere Kundennachfrage pro Periode bezeichnet wird:

$$L_{max} = t^E \tag{1}$$
$$B_{max} = t^E \cdot \overline{r} + 2 \cdot (r_o - \overline{r}). \tag{2}$$

Vergleicht man die Lage der Extrempunkte mit dem einstufigen Fall aus [1], wird man feststellen, dass sowohl L_{max} als auch B_{max} exakt doppelt so groß sind wie im einstufigen Fall. Nachdem die exakte Ermittlung der Extrempunkte der Lagerkennlinie für ein serielles zweistufiges Lagersystem bei unsicherer Nachfrage in diesem Kapitel vorgestellt wurde, konzentriert sich das folgende Kapitel auf den Verlauf der Kennlinie zwischen diesen beiden Punkten.

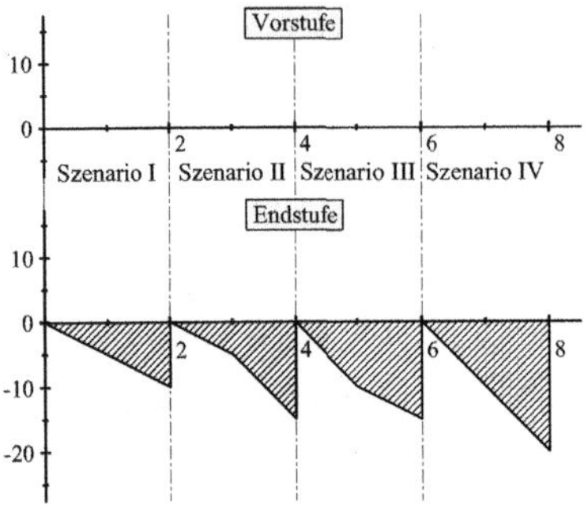

Fig. 1. Bestandsverlauf zur Berechnung von L_{max}

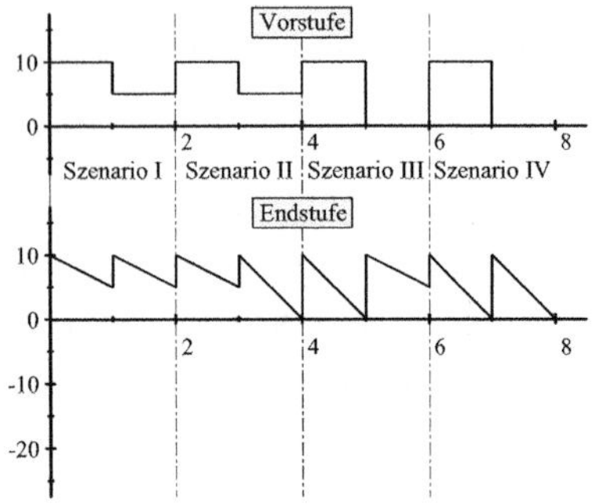

Fig. 2. Bestandsverlauf zur Berechnung von B_{max}

3 Exakter Verlauf der Lagerkennlinie

Im Gegensatz zur Extremwertberechnung weist die Ermittlung des exakten Verlaufs der Lagerkennlinie im mehrstufigen Fall einen gravierenden Unterschied im Vergleich zum einstufigen Fall auf, der die exakte Analyse erheblich erschwert. Hierbei handelt es sich um die Tatsache, dass keine eineindeutige Zuordnung eines mittleren Lieferverzugs zu einem mittleren Bestand vorgenommen werden kann, da nun zwei unterschiedliche Bestellgrenzen das Systemverhalten determinieren. Durch Variation der beiden Bestellgrenzen lässt sich im mehrstufigen Fall für unterschiedliche Parameterkombinationen aus S^V und S^E unter Umständen der gleiche mittlere Bestand beobachten. Allerdings weisen diese Kombinationen trotz eines identischen mittleren Bestands üblicherweise einen anderen mittleren Lieferverzug auf. Da dieses Charakteristikum der mehrstufigen Kennlinienanalyse eine Vielzahl von Kennlinien für einen bestimmten Artikel zulässt, soll im Folgenden lediglich die sogenannte effiziente Kennlinie herausgearbeitet werden. Diese ist dadurch gekennzeichnet, dass für jeden beliebigen mittleren Bestand B (mit $0 \leq B \leq B_{max}$) der minimal mögliche mittlere Lieferverzug bestimmt wird. Bei einer genauen Analyse der effizienten Kennlinie ergeben sich die folgenden, generell gültigen Eigenschaften. Die effiziente Kennlinie kann in zwei Abschnitte unterteilt werden, die im Folgenden mit A und B bezeichnet und in Abbildung 3 für das im letzten Kapitel vorgestellte Beispiel präsentiert werden (mit $\overline{r} = 7,5$). Für alle Punkte auf der effizienten Kennlinie in Abschnitt A gelten die folgenden Bestellgrenzen: S^V=0 und $0 \leq S^E < t^E r_o$. In Abschnitt B liegen die Bestellgrenzen aller effizienten Parameterkombinationen in den Intervallen: $0 < S^V \leq t^E r_o$ und S^E=$t^E r_o$.

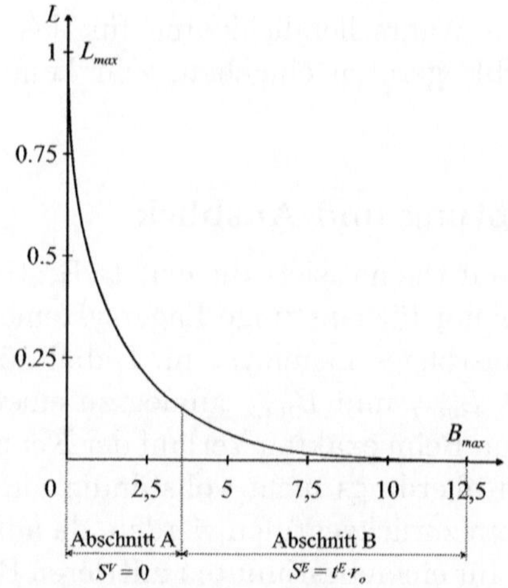

Fig. 3. Die effiziente Lagerkennlinie im mehrstufigen Fall

Nachdem die effiziente Kennlinie näher analysiert wurde, soll nun versucht werden, den Kennlinienverlauf mithilfe einer mathematischen Funktion exakt zu beschreiben. Um den Kennlinienverlauf zwischen den Extrempunkten L_{max} und B_{max} exakt zu bestimmen, muss die Kennlinie in bis zu fünf Teilbereiche (L_1 und L_2 für Abschnitt A sowie L_3 bis L_5 für Abschnitt B) unterteilt werden. Für jeden dieser Teilbereiche lässt sich dann in Abhängigkeit der Systemparameter ein geschlossener Ausdruck ermitteln.

$$L(B) = \begin{cases} L_1 \text{ für } & S^V = 0 & , \quad 0 \le S^E \le t^E r_u \\ L_2 \text{ für } & S^V = 0 & , t^E r_u \le S^E \le t^E r_o \\ L_3 \text{ für } & 0 \le S^V \le t^E(2r_u - r_o) & , \quad S^E = t^E r_o \\ L_4 \text{ für } & \max(0, t^E(2r_u - r_o)) \le S^V \le t^E r_u & , \quad S^E = t^E r_o \\ L_5 \text{ für } & t^E r_u \le S^V \le t^E r_o & , \quad S^E = t^E r_o \end{cases}$$

Die einzelnen Teilabschnitte unterscheiden sich nach dem Auftreten von Fehlmengen in den einzelnen Nachfrageszenarien. So tritt beispielsweise im ersten Teilbereich L_1 in allen vier Nachfrageszenarien eine Fehlmenge auf, wohingegen im fünften Teilbereich L_5 nur im Nachfrageszenario IV eine Fehlmenge zu beobachten ist. Eine Besonderheit bei der Analyse sind der dritte und vierte Teilbereich der exakten Kennlinie. Sollte r_u größer als $r_o/2$ sein, muss der dritte Teilbereich für die exakte Kennlinie ausgewertet werden. Ist diese Bedingung nicht erfüllt, kann bei der Ermittlung des Kennlinienverlaufs direkt vom zweiten zum vierten Teilbereich übergegangen werden. Auf eine beispielhafte Auswertung des mathematischen Zusammenhangs soll an dieser Stelle verzichtet werden, da die einzelnen Teilbereichsfunktionen, in die auch die Wahrscheinlichkeiten für das Auftreten der unterschiedlichen Nachfrageraten eingehen, sehr komplexer Natur sind.

4 Schlussbetrachtung und Ausblick

Die vorliegende Arbeit thematisiert die exakte Bestimmung der Lagerkennlinie, die bisher nur für einstufige Lagersysteme analysiert wurde, für ein serielles zweistufiges Lagersystem. Dabei können die Extremwerte der Kennlinie L_{max} und B_{max} analog zu einem einstufigen System ermittelt werden. Beim exakten Verlauf der Kennlinie zwischen den Extremwerten kann allerdings nicht vollständig auf die Methodik für ein einstufiges System zurückgegriffen werden, da aufgrund der Bestellgrenzen S^V und S^E für einen bestimmten mittleren Bestand je nach Parameterkonstellation mehrere mittlere Lieferverzüge beobachtet werden können. Es ist allerdings möglich, für jeden mittleren Bestand den minimalen Lieferverzug zu ermitteln, um damit die sogenannte effiziente Kennlinie aufzuspannen.

Der in dieser Arbeit vorgestellte Modellrahmen kann in vielerlei Hinsicht erweitert werden. So könnte zum Beispiel in zukünftigen Forschungsarbeiten versucht werden, die angenommene Relation zwischen den Kontrollintervallen der beiden Stufen zu verallgemeinern, so dass gilt: $t^V = n \cdot t^E$. Ebenso ist es möglich, die effiziente Kennlinie bei Vorliegen mehrerer Unsicherheitsquellen (wie stochastische Lieferzeit und Liefermenge) sowie bei stetigen Wahrscheinlichkeitsverteilungen für die einzelnen Quellen der Unsicherheit in einem mehrstufigen Lagersystem zu ermitteln. Außerdem könnte die Untersuchung für andere Dispositionsregeln als die (t,S)-Regel vorgenommen werden. Schließlich besteht eine weitere herausfordernde Aufgabe darin, die vorgenommene Analyse auf komplexere mehrstufige Systeme mit konvergierender oder divergierender Struktur auszudehnen.

References

1. Inderfurth, K. und Schulz, T. (2007) Zur Exaktheit der Lagerkennlinie nach Nyhuis und Wiendahl. In: Otto, A. und Obermaier, R. (Hrsg.) Logistikmanagement. *Gabler*, Wiesbaden, S. 23-49.
2. Nyhuis, P. und Wiendahl, H.-P. (2003) Logistische Kennlinien, 2. Aufl., *Springer*, Berlin.
3. Tempelmeier, H. (2005) Bestandsmanagement in Supply Chains, *Books on Demand*, Norderstedt.

Setup Cost Reduction and Supply Chain Coordination in Case of Asymmetric Information

Karl Inderfurth and Guido Voigt

Otto-von-Guericke University Magdeburg, Faculty of Economics and Management,
{karl.inderfurth,guido.voigt}@ovgu.de

1 Motivation

The model utilized in this paper captures a supply chain planning problem, in which the buyer asks the supplier to switch the delivery mode to Just-in-Time (JiT). We characterize the JiT mode with low order sizes. The buyer faces several multidimensional advantages from a JiT-delivery, which we aggregate to the buyer's holding costs per period. Hence, if the buyer faces high holding costs she is supposed to have high advantages from a JiT-delivery, and vice versa. On the other hand, smaller order sizes can cause an increase of the supplier's setup and distribution costs. In our modelling approach, the supplier's setup costs per period reflect these disadvantages. Yet, it is well known that small order sizes are not sufficient for a successful implementation of the JiT concept. Setup cost reduction, thus, is regarded to be one main facilitator for JiT to be efficient. Our model depicts the need for accompanying process improvements by the supplier's option to invest in setup cost reduction (see [4]). From a supply chain perspective, an implementation of a JiT strategy will only be profitable, if the buyer‘s cost advantages exceed the supplier‘s cost increase. Yet, this is not always the case. The supplier, thus, may have a strong incentive to convince the buyer to abandon the JiT strategy, i.e. to accept higher order sizes. However, the buyer is supposed to be in a strong bargaining position and will not be convinced unless she is offered a compensation for the disadvantages of not implementing the JiT strategy. Yet, as long as pareto improvements are possible, the supplier can compensate the buyer while improving his own performance. Nonetheless, the above-mentioned advantages of a JiT strategy contain to a major extent private information of the

B. Fleischmann et al.(Eds.), *Operations Research Proceedings 2008*,

DOI: 10.1007/978-3-642-00142-0_33,

buyer. Thus, they can not be easily observed and valued by the supplier. The buyer, thus, will apparently claim that switching towards higher order sizes causes substantial costs and that a high compensation is required. Assuming the strategic use of private information, it is in the supplier's best interest to offer a menu of contracts (see [1]). Basically, this menu of contracts aligns the incentives of the supply chain members such that a buyer with low advantages of a JiT delivery will agree upon higher order sizes than a buyer with high advantages of this supply mode. However, the incentive structure provided by this menu of contracts causes inefficiencies, because the resulting order sizes are too low compared to the supply chain's optimal solution. Starting from this insight, our main focus in this study is to analyse the impact of investments in setup cost reduction on this lack of coordination. Summing up, there are basically two streams of research (namely the inefficiencies due to asymmetric information and the optimal set-up cost reduction in an integrated lot-sizing decision) this paper combines.

2 Outline of the Model

This paper analyses a simplified Joint-Economic-Lotsizing model (see [2]). In a dyadic relationship, composed of a buyer (B) and a supplier (S), the buyer decides upon the order lotsize (Q) from her supplier. Let f denote the setup costs for each delivery incurred by the supplier. These setup costs are a decision variable for the supplier's decision problem. The cost for reducing the setup costs from its original level f_{max} by $f_{max} - f, \forall f \geq f_{min} \geq 0$ are captured by the investment function $k(f)$. The investment $k(f)$ leads to a setup cost reduction over the whole (infinite) planning horizon. Hence, the supplier faces costs of $r \cdot k(f)$ in each period, where r denotes the company specific interest rate. The buyer faces holding costs h per item and period. The demand is, without loss of generality, standardized to one unit per period. Hence period costs equal unit costs. As the supplier's full information about all JiT related advantages of the buyer is a very critical assumption we study a situation in which the supplier can only estimate these advantages. We formalize this estimation with a probability distribution p_i $(i = 0, ..., n)$ over all possible holding cost realizations h_i $(h_i < h_j \,\forall\, i < j; i, j = 0, ..., n)$ that is assumed to be common knowledge. For simplifying forthcoming formulas we additionally define: $p_0 = 0$ and $h_0 = 0$. Let the indices AI and FI refer to the situation under asymmetric information and full information, respectively. The supplier minimizes his expected cost by offering a menu of contracts Q_i^{AI}, T_i^{AI} $(i = 1, ..., n)$, where T_i^{AI} denotes the buyer's compensation for accepting higher order sizes. However, for analyzing the impact of setup cost reduction on supply chain coordination it is not necesssary to present details on the amount of these side payments. The incentive scheme provided by the menu of contracts ensures that the buyer facing holding costs h_i chooses the order size Q_i^{AI} as long as $h_i + (h_i - h_{i-1}) \cdot \sum_{t=0}^{i-1} p_t/p_i < h_j + (h_j - h_{j-1}) \cdot \sum_{t=0}^{j-1} p_t/p_j \quad \forall \quad i = 1, ..., n-1; j = 2, ..., n; j > i$ holds. Please refer to [3] for a detailed discussion of the model characteristics and assumptions.

3 Coordinating the Supply Chain Through Setup Cost Reduction?

If the supplier faces no asymmetric information, he knows with certainty the buyer's holding costs h_i. In this situation, the supplier will offer the supply chain optimal order size $Q_i^{FI} = \sqrt{2 \cdot f_i^{FI}/h_i}, \forall i = 1, ..., n$ which is the well known economic order quantity. The optimal order size under asymmetric information is $Q_i^{AI} = \sqrt{2 \cdot f_i^{AI}/(h_i + \phi_i)}, \forall i = 1, ..., n$ with $\phi_i = (h_i - h_{i-1}) \cdot \sum_{t=0}^{i-1} p_t/p_i, \; \forall \; i = 1, ..., n$. In the following, we will show that there is an overinvestment due to asymmetric information, i.e. $f_i^{AI} \leq f_i^{FI}$, $\forall i = 1, ..., n$. As $\phi_i \geq 0$, $\forall i = 1, ..., n$ holds, it follows directly that the well-known downward distortion of order sizes (i.e. $Q_i^{FI} \geq Q_i^{AI}$, $\forall i = 1, ..., n$) is prevalent if setup reduction is possible. We refer for a derivation and discussion of this menu of contracts to [3].

Overinvestment due to asymmetric information: Next, we show that asymmetric information leads to an overinvestment in setup cost reduction. We restrict our analysis to a convex investment function $k(f)$. Let $MR(f_i)^{AI} = \sqrt{(h_i + \phi_i)/(2 \cdot f_i)}$ and $MR(f_i)^{FI} = \sqrt{h_i/(2 \cdot f_i)}$ denote the marginal revenues (i.e. cost savings) for reducing the setup cost level f_i. The marginal costs $MC(f_i) = -r \cdot \frac{dk(f_i)}{df_i}$ for reducing the setup cost level f_i are the same under asymmetric and full information. If the optimal setup cost levels f_i^{AI} and f_i^{FI} are interior solutions, they can be computed from $MC(f_i) = MR(f_i)^{AI}$ and $MC(f_i) = MR(f_i)^{FI}$, respectively. Otherwise, the optimal setup cost level is a corner solution (i.e. f_{min} or f_{max}). Comparing the marginal revenues under asymmetric and under full information, we find that $MR^{AI}(f_i) \geq MR^{FI}(f_i), \forall i = 1, ..., n$ holds. The downward distortion of order sizes due to asymmetric information, thus, increases the benefits from setup cost reduction.

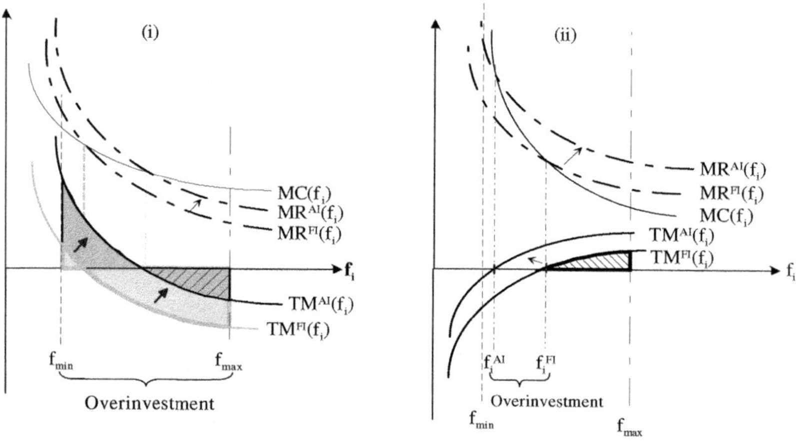

Fig. 1. Overinvestment in case of a convex investment function.

Figure 1 depicts the case of (i) a monotonically decreasing TM-curve of total marginal cost savings and (ii) a monotonically increasing TM-curve, where $TM(f_i) = MR(f_i) - MC(f_i)$. If the TM-curve is not monotonic at all, there are multiple interior solutions. However, the argumentation in this situation can either be reduced to case (i) or (ii). In case (i), all setup cost levels which are higher than the root of the TM-curve cause a loss, as the marginal revenues are smaller than the marginal costs. The root of the TM-curve, thus, is a cost maximum instead of a cost minimum and the optimal setup cost levels are corner solutions, i.e. f^{AI} and f^{FI} are either f_{min} or f_{max}. However, as the marginal revenues of setup cost reduction are higher under asymmetric information (i.e. $MR^{AI} \geq MR^{FI}$) the setup cost level may be distorted, i.e. $f_i^{FI} = f_{max} > f_{min} = f_i^{AI}$. Figure 1(i) depicts this case. Please note, that it can be shown that the analysis for a concave or linear investment function reduces to this case, see [3]. In case (ii), the root of the TM-curve is a cost minimum. As the marginal revenues increase due to asymmetric information we observe again an overinvestment in setup cost reduction (see figure 1(ii)).

As such, one can summarize that there is always the possibility of an overinvestment in setup reduction, regardless of the actual shape of the investment function. Yet, it is not obvious if the coordination deficit (i.e. the performance gap between supply chain optimum and screening

contract) is increasing or decreasing due to an investment in setup cost reduction. In the following, we illustrate the previous analysis for the investment function $k_i(f_i) = a \cdot f_i^{-b} - d$, with $a = 4700, b = 0.094$ and $d = 2500$ (see [4]). We assume that $f_{min} = 0, h_1 = 1, h_2 = 5, p_1 = 0.5$ and $p_2 = 0.5$. The interest rate r is varied for a comparative static analysis. Let $K_i^{SC}(Q_i, f_i) = \frac{f_i}{Q_i} + \frac{h_i}{2}Q_i + r\left(af_i^{-b} - d\right), i = 1, 2$ denote the supply chain costs that result if the buyer faces holding costs h_i. Additionally, $E\left[K^{SC}\right] = \sum_{i=1}^{2} p_i \cdot K_i^{SC}$ denote the expected supply chain costs. To analyse the impact of the option to invest in setup cost reduction we introduce $Q_i^{f_{max}}$ which is the order size that results if no setup cost reduction is possible. As there is no coordination deficit, if the buyer faces holding costs h_1 ($\phi_1 = 0$) we restrict the analysis for the coordination deficit to the cases in which the buyer faces holding costs h_2. The coordination deficit with setup cost reduction, then, results from $CD = K_2^{SC}\left(Q_2^{AI}, f_2^{AI}\right) - K_2^{SC}\left(Q_2^{FI}, f_2^{FI}\right)$. In contrast, $CD^{f_{max}} = K_2^{SC}\left(Q_2^{AI,f_{max}}, f_{max}\right) - K_2^{SC}\left(Q_2^{FI,f_{max}}, f_{max}\right)$ denotes the coordination deficit without the option to reduce setup costs. The change in the coordination deficit, thus, results from $\Delta CD = CD - CD^{f_{max}}$. If $\Delta CD > 0$, the supply chain deficit increases due to the option to invest in setup cost reduction. To analyse the effect on the overall supply chain performance we compute the expected change in supply chain costs that result if setup cost reduction is possible, i.e. $\Delta P = E\left[K^{SC}\left(Q_i^{AI,f_{max}}, f_{max}\right)\right] - E\left[K^{SC}\left(Q_i^{AI}, f_i^{AI}\right)\right]$. If $\Delta P < 0$, the supply chain performance deteriorates in the presence of the setup cost reduction option due to asymmetric information. Figure 2 depicts ΔCD and ΔP in dependence of the interest rate r.
Obviously, the impact of the option to invest in setup cost reduction is ambiguous. Setup cost reduction, thus, is a proper coordination device if the investment is inexpensive (i.e., if r is sufficiently small). Otherwise, the overinvestment caused by asymmetric information leads to an even greater coordination deficit. This coordination deficit may even cause a deterioration of the overall supply chain performance.

4 Conclusion

JiT delivery has received ever-increasing attention in the recent past. Typically, the implementation of JiT strategies is accompanied by setup cost reductions. The supplier's optimal reaction is offering a pareto improving menu of contracts if he only possesses imperfect information

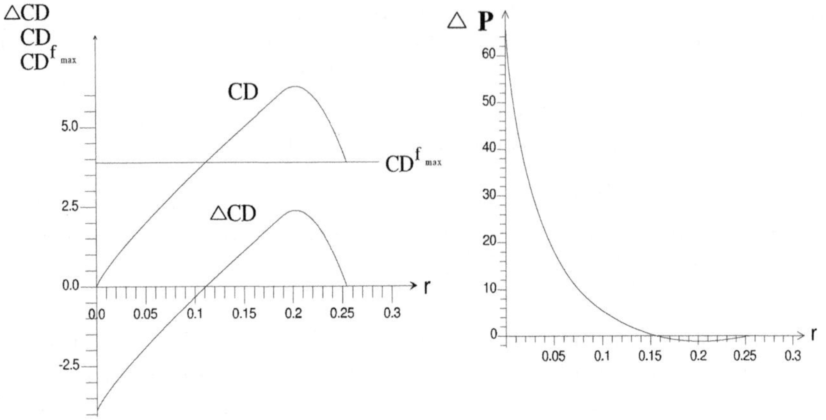

Fig. 2. Changes in supply chain deficit and performance in dependence from interest rate r.

about the buyer's cost position. Obviously, the supplier will not be worse off in terms of expected profits, as the status quo (i.e. f_{max}) is still feasible. Yet, the effect on supply chain coordination and performance is ambiguous. The incentive structure provided by the optimal menu contracts can lead to suboptimally low order sizes. In turn, this can lead to suboptimally high investments in setup cost reduction. This analysis is robust for a wide variety of investment functions. Hence, the coordination deficit caused by asymmetric information can typically not be overcome by setup cost reduction. Particularly, if the investment is expensive the overinvestment can even seriously harm the supply chain performance and increase the coordination deficit.

References

1. C.J. Corbett and X. de Groote. A supplier's optimal quantity discount policy under asymmetric information. Management Science, 46(3):444-450, 2000.
2. S.K. Goyal. An integrated inventory model for a single supplier-single customer problem. Int. J. Product. Res., 15(1):107-111, 1977.
3. K. Inderfurth and G. Voigt. Setup cost reduction and supply chain coordination in case of asymmetric information. FEMM Working Paper Series, No. 16/2008, 2008.
4. E.L. Porteus. Investing in Reduced setups in the EOQ Model. Management Science, 31(8):998-1010, 1985.

A Heuristic Approach for Integrating Product Recovery into Post PLC Spare Parts Procurement

Rainer Kleber and Karl Inderfurth

Otto-von-Guericke University Magdeburg, Faculty of Economics and Management,
{rainer.kleber,karl.inderfurth}@ovgu.de

Summary. Product recovery gains increasing importance in spare parts acquisition, especially during the post product-life-cycle (PLC) service period. Considering product recovery in addition to traditional procurement options (final order, extra production) makes the underlying stochastic dynamic planning problem highly complex. An efficient heuristic for solving the integrated procurement problem is developed, which can be used to exploit the flexibility offered by product recovery.

1 Introduction

The efficient procurement of spare parts is one of the core issues in after-sales service. This task is especially complicated because of the uncertainty surrounding spare parts demand and its inherently dynamic nature (see [5, 1]). An especially challenging situation is present during that part of the service period, which lies after the end-of-production of the parent product, where the conditions for re-supply often fundamentally change.

This research is motivated by an industry project with a large European car manufacturer, which operates a central warehouse for spare parts. Because of the high profitability of the after-sales service but also due to legal obligations, the manufacturer guarantees a service period of 15 years after producing the last car of a certain model and even worse, after having regular production processes available. Providing spare parts in an efficient way therefore poses a large challenge for the manufacturer. Legal take back obligations (e.g. AltfahrzeugG in Germany) yield a stream of returning used cars and thus, also product

DOI: 10.1007/978-3-642-00142-0_34,

recovery is becoming increasingly important for extending the scope of possible sourcing options.
Procurement options include a so-called *final order* for parts at low unit cost which is placed at the time when regular sourcing ends. However, when used to satisfy demand over a long period this is connected with large stocks and a low level of flexibility yielding a high risk of obsolescence of the stored parts. More flexibility can be provided by adding options such as extra production and remanufacturing of used products (for a comprehensive overview on further options see [8]). *Extra production/procurement* incurs high flexibility since this option can react on additional information about demand becoming available during the service period. Because of the loss of scale economies, variable production cost is higher than regular production cost, and often considerable lead times apply. *Remanufacturing of used products* normally causes only moderate variable cost. However, that cost must be adapted to include a price discount for selling a 'refurbished' part instead of a new one. Flexibility of this option is limited by the availability of recoverable products, yielding further uncertainty in timing and quantity of remanufacturable returns.
Quantitative approaches for spare part sourcing focus on the determination of the final lot size (for an overview see, e.g., [7]). Most papers only consider a single additional source and therefore can not efficiently coordinate all three supply sources, but there are a few exceptions. Spengler and Schröter [9] evaluate different strategies to meet spare parts demand in the electronics industry using a System Dynamics approach and Inderfurth and Mukherjee [4] present a formulation of the problem as a stochastic dynamic decision problem.
In this paper we aim to give decision support by providing a heuristic that efficiently coordinates all three supply sources. The paper is organized as follows. In Section 2 we formulate the problem and provide basic ideas of our heuristic approach. Results of a limited performance test are shown in Section 3 and Section 4 provides some conclusions.

2 A Heuristic for Determining Final Ordering, Remanufacturing and Extra Production

2.1 Assumptions and Model

We consider a periodic review system with a finite planning period of length T. Each period t, a random number of spare parts D_t is demanded and a random number of used products R_t returns. In the first period the size of the final order y is set. Furtheron, in each period the stocks of serviceables and recoverables are observed and decisions are made on remanufacturing r_t and extra production p_t quantities. Extra production increases stock level after a lead time l_P. Two state variables are relevant: the net serviceables stock I_t^S and the stock of used products I_t^R. The time invariant cost structure includes:

$c_{F/P/R}$ unit cost for final lot production / extra production / remanufacturing
h holding cost of serviceable spare parts per unit and period
v backorder cost of serviceable spare parts per unit and period
p per unit penalty for unsatisfied demand at the end of horizon

In order to restrict to situations where all options are used we assume that following inequality w.r.t. sourcing cost holds: $c_F \leq c_R \leq c_P$. Holding cost for returned used products are neglected since the capital tied up in returns is more or less zero. We distinguish between backorder cost v for postponing the supply of a demanded item and a penalty p which we incur if it is not possible to meet a demanded part at all. We consider the following model (let $(x)^+$ denote $\max\{x; 0\}$ and $p_t = 0$ $\forall t < 1$)

$$\min TEC = E\Bigg\{ c_F \cdot y + \sum_{t=1}^{T-l_p} c_P \cdot p_t + \sum_{t=1}^{T-1} \left[c_R \cdot r_t + h \cdot \left(I_{t+1}^S\right)^+ + v \cdot \left(-I_{t+1}^S\right)^+ \right] + c_R \cdot r_T + h \cdot \left(I_{T+1}^S\right)^+ + p \cdot \left(-I_{T+1}^S\right)^+ \Bigg\} \quad (1)$$

$$I_{t+1}^S = \begin{cases} y + p_{1-l_p} + r_1 - D_1 & \text{for } t = 1 \\ I_t^S + p_{t-l_p} + r_t - D_t & \text{for } t = 2, ..., T \end{cases} \quad (2)$$

$$I_1^R = 0 \text{ and } I_{t+1}^R = I_t^R - r_t + R_t \quad \text{for } t = 1, ..., T \quad (3)$$

$$y \geq 0, p_t \geq 0, \text{ and } 0 \leq r_t \leq I_t^R \quad \text{for } t = 1, ..., T \quad (4)$$

The objective (1) is to select values of the decision variables that minimize total expected cost TEC over the entire planning period T. It includes sourcing cost, holding cost for serviceable parts, backorder cost when not immediately satisfying demand, and a penalty for unmet demand at the end of the planning period. Constraints (2) and (3) are inventory balance equations, where for ease of presentation initial stocks are set to zero. Restrictions (4) assure validity of decisions.

2.2 A Three Stage Procedure to Determine Policy Parameters

Problem (1)-(4) has the basic structure of a multi-period stochastic inventory control problem with proportional cost. From [2] it is known, that the optimal policy has a simple structure with two order-up-to levels (M_t for remanufacturing and S_t for extra production) only in the case of zero lead time. However, in our heuristic we use this policy structure also for the general lead time case in the following way:

$$r_t = \begin{cases} \min\{(M_1 - I_1^S - y)^+, I_1^R\} & \text{for } t = 1 \\ \min\{(M_t - I_t^S - p_{t-l_p})^+, I_t^R\} & \text{otherwise} \end{cases} \quad (5)$$

$$p_t = \begin{cases} (S_1 - I_1^S - y - I_1^R)^+ & \text{for } t = 1 \\ (S_t - I_t^S - \sum_{i=1}^{l_p} p_{t-i} - I_t^R)^+ & \text{for } 2 \le t \le T - l_p \\ 0 & \text{otherwise} \end{cases} \quad (6)$$

Parameter determination is separated into three stages where in a first step remanufacture-up-to levels M_t are set according to a newsvendor approach. In a second step produce-up-to levels S_t are obtained using an adapted version of the approach put forth in [6]. In the last step the final order size y is calculated under marginal cost considerations and incorporating parameter values derived in the previous steps.
In this way, parameters are fixed as follows. Remanufacture-up-to levels M_t are given by

$$M_t : \Phi_t^D(M_t) = \begin{cases} \frac{v}{v+h} & \text{for } t < T \\ \frac{p-c_R}{p+h} & \text{for } t = T \end{cases} \quad (7)$$

where Φ_t^D is the cumulative density function of demand in period t. Produce-up-to levels S_t are determined by

$$S_t : \Psi_t(S_t) = \begin{cases} \frac{v-(c_P-c_R)\alpha_{t+l_P}}{v+h} & \text{for } t < T - l_P \\ \frac{p-c_P}{p+h} & \text{for } t = T - l_P \end{cases} \quad (8)$$

where α_t is the probability of gathering more returns than required after period t, i.e. $\alpha_t = P\left\{\sum_{i=t}^{T-1} R_i > \sum_{i=t+1}^{T} D_i\right\}$, and Ψ_t is the cumulative density function of relevant net demand $\sum_{i=t}^{t+l_P} D_i - \sum_{i=t}^{t+l_P-1} R_i$. For determining the final order size y^+, we calculate an approximate value for its marginal impact on the objective function. Assuming convexity of total expected cost, we choose y^+ such that

$$y^+ = \arg\min\{c(y) \ge 0\} \quad (9)$$

where $c(y)$ is given by

$$c(y) = c_F + \theta(y) \cdot h - \pi(y) \cdot c_P - \beta(y) \cdot c_R. \quad (10)$$

Here, $\theta(y)$ denotes the expected number of periods a marginal item is in stock: $\theta(y) = 1 + \sum_{t=2}^{T} P\left\{y - \sum_{i=1}^{t-1} D_i > M_t\right\}$. $\pi(y)$ is an approximation of the probability of reducing extra production by one unit, i.e. $\pi(y) = P\left\{y - \sum_{i=1}^{T-l_P-1} D_i + \sum_{i=1}^{T-l_P-1} R_i < S_{T-l_P}\right\}$. $\beta(y)$ stands for the probability of not reducing extra production but instead being able to lower remanufacturing by one unit: $\beta(y) = (1-\pi(y))\cdot P\left\{y - \sum_{i=1}^{T-1} D_i < M_T\right\}$.

3 Numerical Test

In order to assess the performance of our heuristic, a numerical test was conducted where the heuristic was compared with the optimal solution obtained by stochastic dynamic programming. For this test, we considered two demand and return scenarios: a static and a dynamic one with identical total number of expected demands and returns over a planning horizon of ten periods. Demand and returns approximately follow normal distributions with means as given in Table 1 and constant coefficients of variation ρ_D and ρ_R, respectively.

Table 1. Expected demand and returns in the considered scenarios

	static scenario									
Period	1	2	3	4	5	6	7	8	9	10
expected demand	6	6	6	6	6	6	6	6	6	6
expected returns	3	3	3	3	3	3	3	3	3	0
	dynamic scenario									
Period	1	2	3	4	5	6	7	8	9	10
expected demand	2	3	5	7	8	8	6	4	2	1
expected returns	1	2	3	5	5	4	4	3	2	0

With respect to all relevant parameters we used a full factorial design including two values for each parameter except for final order production cost, which was normalized to 10. Parameter values are $c_R \in \{12, 16\}$, $c_P \in \{16, 20\}$, $h \in \{1, 3\}$, $v \in \{25, 75\}$, $p \in \{75, 200\}$, $\rho_D \in \{0.1, 0.4\}$, and $\rho_R \in \{0.1, 0.4\}$. The lead time for extra production ranged between 0 and 2 periods, yielding a total number of 768 combinations.

For each instance we determined the relative costs deviation of our heuristic approach from the optimal solution ΔTEC. Average cost deviations over these instances are depicted in Table 2.

Table 2. Worst case and average performance of the heuristic

ΔTEC [in %]	overall $l_p = 0$	overall $l_p = 1$	overall $l_p = 2$	static scenario $l_p = 0$	static scenario $l_p = 1$	static scenario $l_p = 2$	dynamic scenario $l_p = 0$	dynamic scenario $l_p = 1$	dynamic scenario $l_p = 2$
maximum	2.15	3.86	5.83	0.16	2.66	3.46	2.15	3.86	5.83
average	0.22	0.99	1.59	0.01	0.59	1.16	0.43	1.38	2.02

Results reveal that the average cost deviation of the heuristic is quite small for all scenarios. Since also the worst case performance is within the 6% limit the heuristic seems very promising.

4 Conclusions and Outlook

In this paper we developed an efficient heuristic approach for the coordination of sourcing decision for spare parts procurement after end-of-production. This heuristic can also be used to assess the value of flexibility when using remanufacturing and extra production in addition to the final order. A first numerical study (see [3]) reveals under realistic conditions that the additional flexibility of each of these options results in considerable profitability gains. Further improved and extended versions of our heuristic would allow to extend the scope of flexibility studies.

References

1. Hesselbach J, Mansour M, Graf R (2002) Reuse of components for the spare parts management in the automotive electronics industry after end-of-production. 9th CIRP International Seminar, Erlangen, Germany
2. Inderfurth K (1997) Simple optimal replenishment and disposal policies for a product recovery system with leadtimes. OR Spektrum 19:111–122
3. Inderfurth K, Kleber R (2008) Modellgestützte Flexibilitätsanalyse von Strategien zur Ersatzteilversorgung in der Nachserienphase. FEMM Working Paper No. 26/2008, Faculty of Economics and Management, Otto-von-Guericke University Magdeburg
4. Inderfurth K, Mukherjee K (2008) Decision support for spare parts acquisition in post product life cycle. Central European Journal of Operations Research 16:17–42
5. Kennedy WJ, Patterson JW, Fredendall LD (2002) An overview of recent literature on spare parts inventories. International Journal of Production Economics 76:201–215
6. Kiesmüller GP, Minner S (2003) Simple expressions for finding recovery system inventory control parameters. Journal of the Operational Research Society 54:83–88
7. Kleber R, Inderfurth K (2007) Heuristic approach for inventory control of spare parts after end-of-production. In Otto A, Obermaier R (eds) Logistikmanagement - Analyse, Bewertung und Gestaltung logistischer Systeme. DUV, Wiesbaden
8. Schröter M (2006) Strategisches Ersatzteilmanagement in Closed-Loop Supply Chains. DUV, Wiesbaden
9. Spengler T, Schröter M (2003) Strategic management of spare parts in closed-loop supply chains - a system dynamics approach. Interfaces 6:7–17

The Economic Lot and Supply Scheduling Problem Under a Power-of-Two Policy

Thomas Liske and Heinrich Kuhn

Catholic University of Eichstätt Ingolstadt, Chair of Production and Operations Management, Auf der Schanz 49, 85049 Ingolstadt
{Thomas.Liske,Heinrich.Kuhn}@ku-eichstaett.de

Summary. The paper presents a new problem class, the Economic Lot and Supply Scheduling Problem (ELSSP). The ELSSP deals with the simultaneous vehicle routing and production planning and is an extension of the well known Economic Lot Scheduling Problem (ELSP). To solve the described problem the junction point method of Yao and Elmaghraby (2001) for the solution of the ELSP under a power-of-two policy is modified and extended in order to consider the vehicle routing problem caused by the deliveries of the input materials.

1 Problem Description

The Economic Lot and Supply Scheduling Problem (ELSSP) considers a single capacitated production facility on which several end items $j, j \in P$ have to be produced with an end item specific production rate $p_j, j \in P$. Each of the end items faces a constant demand rate $d_j, j \in P$. These assumptions are identical to the ones from the ELSP. In addition to the ELSP, the ELSSP assumes that the production of each end item requires a specific amount of dedicated input materials $i, i \in S$, $a_{ij}, i \in S, j \in P$. The input materials are purchased from geographically dispersed suppliers by a given fleet of vehicles with limited capacity Q. Furthermore, the input materials have to be available at the production site before the related production of an end item starts. The input materials can be stored between delivery and production in an inbound warehouse which causes inventory holding costs $h_i, i \in S$ per unit and time unit. The end items are stored in an outbound warehouse where the customer demand occurs. This causes inventory holding costs $h_j, j \in P$ per unit and time unit. Similar to the ELSP an infinite planning horizon is assumed. Therefore, the solution of the ELSSP results in a cyclic plan with cycle time T. In addition, we assume that the production plan satisfies the base period approach

B. Fleischmann et al.(Eds.), *Operations Research Proceedings 2008*,

DOI: 10.1007/978-3-642-00142-0_35,

with a power-of-two policy. Thus, a common base period B and for every end item a multiplier k_j have to be determined. The multipliers, however, satisfy the power-of-two condition. Since the product k_jB denotes the end item specific cycle time, the overall cycle time is given by $T = max_{j \in P}\{k_jB\}$.
Figure 1 shows the inventory levels for an ELSP under a power-of-two policy. These inventory levels are adequate to the inventory levels of the end items in the outbound warehouse under the ELSSP assumptions. The displayed example consists of three end items having multipliers $k_1 = 1$, $k_2 = 2$, and $k_3 = 4$. Obviously, the multiplier of end item 3 determines the overall cycle time T, so $T = k_3B$.

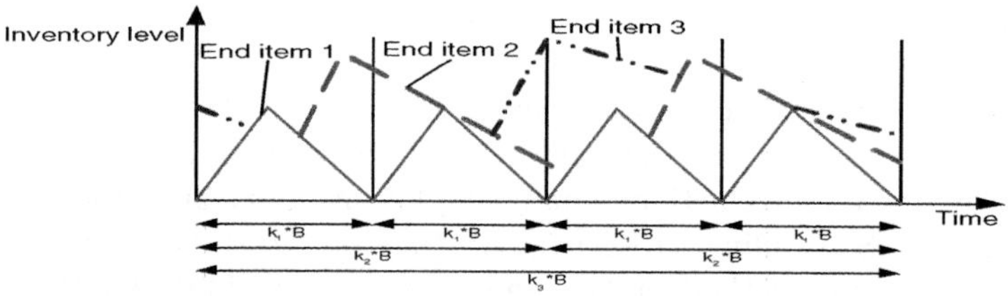

Fig. 1. Inventory levels of an ELSP under a power-of-two policy

The aim of the ELSSP is to determine simultaneously the delivery quantities $q_i, i \in S$ of the input materials, the routes for the deliveries, the production sequence of the end items, and the production lot-sizes in order to minimize the average transportation, inventory holding, and production costs. The objective function $TC(\{k_j\}, B)$ of the ELSSP under a power-of-two policy can be formulated as follows:

$$\sum_{j \in P} \sum_{l \in L_j} \frac{TSP(S_j^{(l)})}{k_jB} + \tag{1}$$

$$+ \sum_{j \in P} \frac{1}{k_jB} \Big(\sum_{i \in S} \frac{1}{2} t_j^{p^2} a_{ij} p_j h_i + \frac{1}{2} h_j k_j B (p_j - d_j) t_j^p + q_i b_{ij} t_j^s + s_j \Big)$$

The overall costs consist of transportation, inventory holding, and setup costs. The first term of the objective function denotes the average transportation costs per unit time. In this term the function $TSP(S_j^{(l)})$ describes the transportation costs of an optimum tour along all suppliers supplying the input materials of end item $j, j \in P$ who are pooled to an identical tour l. The set of suppliers belonging to a certain tour l is denoted by $S_j^{(l)}$. The set L_j, however, denotes all tours necessary to

supply the input materials of end item j. Thus, the set $L_j, j \in P$ contains all tours related to end item $j, j \in P$. We assume that all tours start and end at the production site.
Term $\sum_{i=1}^{S} \frac{1}{2} t_j^{p^2} a_{ij} p_j h_i$ in the objective function represents the costs for storing the input materials. The input materials have to be stored during the production time of end item j, $t_j^p, j \in P$. The inventory holding costs for the end items are given by the term $\frac{1}{2} h_j k_j B(p_j - d_j) t_j^p$ which are identical to the ELSP case.
The inventory holding costs for the input materials occurring during the setup time of the production facility are modeled by the term $q_i b_{ij} t_j^s$. The parameters $b_{ij}, i \in S, j \in P$ indicate whether an input material is needed for producing end item j ($b_{ij} = 1$) or not ($b_{ij} = 0$). And $t_j^s, j \in P$ donate the setup times for end item $j, j \in P$. The last term of the objective function represents the sequence independent setup costs for end item $j, j \in P$, s_j.

2 Literature Review

The ELSSP combines the Economic Lot Scheduling Problem (ELSP) [2] and the Capacitated Vehicle Routing Problem (CVRP) [9]. The literature suggests different approaches to combine production planning and transportation planning. The problem of determining optimal order quantities for several products with respect to group specific ordering costs is described by the Joint Replenishment Problem (JRP) [5]. The linkage between the ELSP and the problem of delivering the produced end items to the customers is treated by the Economic Lot and Delivery Scheduling Problem (ELDSP) [4]. However, the transportation costs considered in an ELDSP model are not given by the solution of an VRP. Therefore, the transportation costs are independent of the location of the customers. The coordination of outgoing deliveries from a central warehouse to geographically dispersed regional warehouses, facing a constant demand rate, is the focus of the Inventory Routing Problem (IRP) [1]. Thus, the overall costs, consisting of transportation costs for the deliveries and inventory costs caused by storing the end items at the regional warehouses, have to be minimized. The IRP, however, neglects production costs.
The ELSSP is the first approach considering lot-sizing, production and supply decisions simultaneously.

3 Solution Method

To solve the ELSSP model under a power-of-two policy we modify the junction point method of Yao and Elmaghraby ([12]). The junction

point method, however, is not restricted to the ELSP under a power-of-two policy and has already been used to solve different lot-sizing and scheduling problems ([7], [11], [10]).
The objective function (1) can be separated in $|P|$ objective functions, one for each end item $j, j \in P$:

$$TC_j(k_j, B) = \sum_{l \in L_j} \frac{TSP(S_j^{(l)})}{k_j B} + \tag{2}$$

$$+\frac{1}{k_j B}\Big(\sum_{i \in S} \frac{1}{2} t_j^{p^2} a_{ij} p_j h_i + \frac{1}{2} h_j k_j B (p_j - d_j) t_j^p + q_i b_{ij} t_j^s + s_j\Big)$$

This leads to a modified formulation of the objective function, which is useful for further investigations.

$$TC_{opt}(\{k_j\}, B) = \sum_{j=1}^{P} min_{k_j} TC_j(k_j, B) \tag{3}$$

From equation (3) it is obvious that for a given base period B a set of optimal multipliers $\{k_j\}, j \in P$ has to be found. The basic idea of the solution method is to find intervals of B, where the set of optimal multipliers is equal for all values of B within the interval. As an implication of equations (1) and (2) it follows that only the functions $TC_j(k_j, B), j \in P$ have to be considered in order to find an optimal solution. It can be shown that $TC_j(k_j, B), j \in P$ are piecewise convex functions which is a result of the intersections of the functions $TC_j^{k_j}(B), j \in P$ with a fixed k_j. Therefore, intervals can be found with optimal multipliers and every junction point indicates a change of one of the optimal multipliers.
In addition, with decreasing B the limited capacity of the vehicles may be reached which again leads to new multipliers k_j.
A third reason for a change in the optimum value of k_j results, since the optimal solution of the CVRP may change; equivalent as described in [6].
The mentioned conditions result in three different junction points which can be calculated quite easily. Obviously, a junction point is equivalent to a change in the set of optimal multipliers. So, the multiplier has to be changed as soon as an intersection, a new feasibility of the CVRP, or a change in the costs of the CVRP occurs. Therefore, it is possible to calculate intervals in which $TC_j(k_j, B)$ is convex and therefore $TC(\{k_j\}, B)$ is convex within these intervals too.

4 Results

The ELSSP under a power-of-two policy is analyzed by a sequential and a simultaneous procedure. The sequential procedure solves the ELSP first under a power-of-two policy using the junction point method ([12]). The solution is then used to calculate the delivery times and quantities. Afterwards the corresponding CVRP is solved. The simultaneous procedure solves both problems in an integrated manner. The CVRP is solved in both approaches with the savings algorithm ([3]) in combination with the 2-opt improvement method ([8]).
The analysis considers 24 instances with different setup costs. The setup costs vary from 0.5 to 50 monetary units. The results of the simultaneous procedure and the sequential solution approach are compared in terms of relative cost savings by the simultaneous procedure. The results are shown in figure 2.

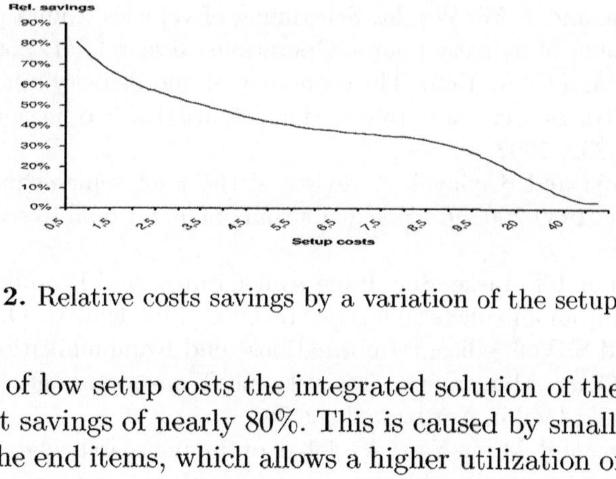

Fig. 2. Relative costs savings by a variation of the setup costs

In the case of low setup costs the integrated solution of the ELSSP results in cost savings of nearly 80%. This is caused by small production cycles for the end items, which allows a higher utilization of the supply tours and savings in transportation costs and inventory holding costs due to smaller delivery quantities. With increasing setup times the cost savings gained from reduced transportation costs and inventory holding costs are not able to compensate the higher setup costs any more. This results in less opportunities to save transportation and inventory holding costs. With higher setup costs the simultaneous solution approach can still realize cost savings of about 4% due to higher utilized transports.

5 Conclusion

The described ELSSP can be found in a lot of industrial sections. The automobile industry as well as the retail industry are examples for

potential applications. The ELSSP offers one possibility to model such systems and the results in Section 4 indicate noticeable cost savings applying the simultaneous planning approach.
The presented model and solution approach is only a first attempt to model and solve integrated transportation and production problems. The results so far justify a more intensive research in this area. In doing so, extensions according to the model and its assumptions are worth to examine. The ELSSP, as presented in this paper, may be an appropriate starting point for further research.

References

1. A. M. Campbell, L. Clarke, A. J. Kleywegt, and M. W. P. Savelsbergh. The inventory routing problem. In T. G. Crainic and G. Laporte, editors, Fleet Management and Logistics, pages 95-113. Kluwer Academic Publishers, Boston, 1998.
2. P. Carstensen. The economic lot scheduling problem-survey and lp-based method. OR Spektrum, (21):429-460, 1999.
3. G. Clarke and J. W. Wright. Scheduling of vehicles from a central depot to a number of delivery points. Operations Research, (12):568-581, 1964.
4. J. Hahm and C. A. Yano. The economic lot and delivery scheduling problem: The single item case. International Journal of Production Economics, (28):235-252, 1992.
5. M. Khouja and S. Goyal. A review of the joint replenishment problem literature: 1989-2005. European Journal of Operational Research, (186):1-16, 2008.
6. H. Kuhn and T. Liske. Simultane Anlieferungs- und Produktionsplanung in der Automobilzulieferindustrie. In D. C. Mattfeld, H.-O. Günther, L. Suhl, and S. Voß, editors, Informations- und Kommunikationssysteme in Supply Chain Managment, Logistik und Transport, pages 39-53. Books on Demand Gmbh, Norderstedt, 2008.
7. F.-C. Lee and M.-J. Yao. A global optimum search algorithm for the joint replenishemnt problem under power-of-two policy. Computers and Operations Research, (30):1319-1333, 2003.
8. S. Lin. Computer solutions of the traveling salesman problem. Bell Systems Technical Journal, (44):2245-2269, 1965.
9. P. Toth and D. Vigo, editors. The Vehicle Routing Problem. Society for Industrial and Applied Mathematics, Philadelphia, 2002.
10. M.-J. Yao. The economic lot scheduling problem without capacity constraints. Annals of Operations Research, (133):193-205, 2005.
11. M.-J. Yao and C.-C. Chiou. On a replenishment coordination model in an integrated supply chain with one vendor and multiple buyers. European Journal of Operational Research, (159):406-419, 2004.
12. M.-J. Yao and S. E. Elmaghraby. The economic lot scheduling problem under power-of-two policy. Computers and Mathematics with Applications, (41):1379-1393, 2001.

Towards Coordination of Decentralized Embedded System Development Processes

Kerstin Schmidt[1], Grit Walther[1], Thomas S. Spengler[1], and Rolf Ernst[2]

[1] Institut für Produktion und Logistik, Technische Universität Braunschweig, Katharinenstr. 3, 38106 Braunschweig
{kerstin.schmidt|g.walther|t.spengler}@tu-bs.de

[2] Institut für Datentechnik und Kommunikationsnetze, Technische Universität Braunschweig, Hans-Sommer-Str. 66, 38106 Braunschweig
ernst@ida.ing.tu-bs.de

1 Introduction

An acceleration in technological change and shorter product life cycles lead among other things to companies' concentration on their core competencies and to a parallelization of formerly sequential processes. This results in an increasing number of collaborations between firms. This trend can also be observed in research and development, where projects are often conducted corporately by various partners. An example for such a cooperation is the development of embedded systems. Embedded systems are characterized by a modular design of different interacting hard- and software components. Thereby, the performance of the overall system depends on the interaction of the individual components' performances. In current system industry's practice, these components are developed by specialized suppliers and are then combined to the embedded system by a system integrator. Since these partners are usually legally and economically independent companies, the cooperation is regularized by contractual agreements. In practice, fixed-price contracts are most commonly used. In these contracts, components' requirements as well as prices are fixed ex-ante by the system integrator. However, uncertainties exist with regard to the outcome of the development process. For instance, the performance of components as well as of the overall system cannot be predicted with certainty. Additionally, partners may follow different objectives. Thus, inefficiencies in the design process as well as in the design of the embedded systems often occur. Some of these uncertainties and inefficiencies might

B. Fleischmann et al.(Eds.), *Operations Research Proceedings 2008*,
DOI: 10.1007/978-3-642-00142-0_36,

be absorbed by making use of existing substitutional dependencies between components. However, this is not possible when inappropriate contracts as well as insufficient incentive structures are applied, since these lead to a decreasing flexibility in the development process and thus to an increase in development costs. Thereby, economic risk for suppliers and integrators increases. Overcoming these difficulties requires improved coordination of the partners ahead of and during the development process. Hence, the aim of this contribution is to improve collaborative development processes of embedded systems by adapting mechanisms from supply chain management to development processes. As in supply chain management cooperation is regularized by contracts. In addition, uncertainties exist in the decentralized development processes as in supply chain management, which lead to inefficiencies in the cooperation. Supply chain management has rich experience in flexible contracting with various incentives, targeting overall flexibilization, risk mitigation, and economic fairness [1]. Unlike in supply chain management there are substitutional dependencies between components' attributes in the development process. However, differences between production and development processes currently prevent an easy adoption of these mechanisms. First approaches to the flexibilization of contracts in embedded system development processes are given by [2, 3]. In the following section a mathematical model for cooperative development processes is described and analyzed with regard to the optimal actions of the partners.

2 A Model for the Development of Embedded Systems

This section studies coordination in embedded system development processes with one omniscient integrator and two independent suppliers. In the following, we first describe the model. Afterwards we analyze a deterministic case in a centralized as well as in a decentralized setting. In the deterministic case both suppliers can determine the result of their development with certainty. The centralized setting is analyzed first, i.e. how the decision would be taken by one actor developing and integrating both components. Then, the deterministic decentralized setting, i.e. decision making by independent actors, is analyzed with regard to coordination ability of a fixed-price contract. Thereby, the omniscient integrator specifies the parameters of the contract to influence the decisions of the suppliers and thus the result of the development. The model is then expanded by taking into account uncertainties of one supplier with regard to the attribute of the component. In the stochastic case a centralized setting is analyzed.

2.1 Model Description

In this (single-period) model, there are three independent actors, one integrator and two suppliers. The two suppliers each develop one component of an embedded system for the integrator. The operability of the embedded system, and thus the utility of the integrator, depends on the attributes a_i of the components to be developed, with $i = 1, 2$ indicating the two different suppliers. There is a substitutional dependency, with regard to the components' attributes. The function $a_2 = s(a_1)$ describes the efficient boundary of combinations of values of the components' attributes in the interval $[a_{i,min}, a_{i,max}]$, for which the overall system is executable. If the system is executable, the utility of the integrator is one, otherwise it is zero. These interdependencies are shown in Figure 1. To model the 0-1-character of the utility function n_I^a of the integrator I in a continuously differentiable manner, it is approximated by a three-dimensional Sigmoid-function. To take the substitutional dependency between the two components into account, the Sigmoid-function is adjusted, such that the function $s(a_1)$ runs exactly through its turning point. This results in the following utility function of the integrator, with d specifying the slope in the turning point, $d \to \infty$:

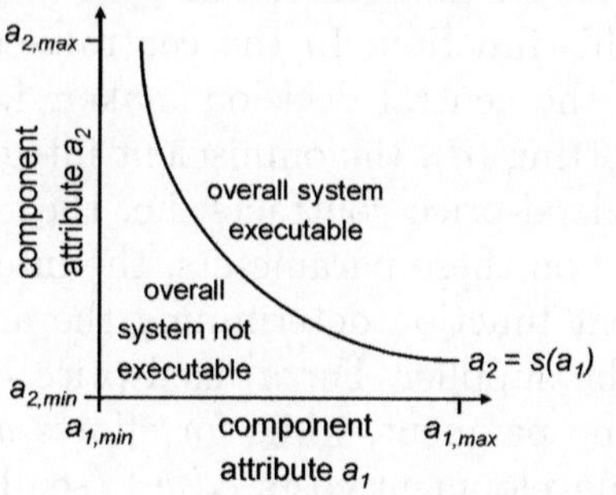

Fig. 1. Interdependencies between a_1 and a_2.

$$n_I^a(a_1, a_2) = 1/(1 + \exp(d \cdot s(a_1)) \cdot \exp(-d \cdot a_2)) \qquad (1)$$

The results of the development process, as realized values of the components' attributes, depend on the efforts w_i of the suppliers. This coherence between effort w_i and value a_i can be described by the function $a_i = k_i(w_i)$. Thus, the utility function of the integrator can be rewritten as follows:

$$n_I^w(w_1, w_2) = 1/(1 + \exp(d \cdot s(k_1(w_1))) \cdot \exp(-d \cdot k_2(w_2))) \qquad (2)$$

The development costs depend on the effort w_i of supplier i and are described by the function $c_i(w_i)$.

2.2 Model Analysis

Deterministic Case. In the central setting (z), one decision maker can control development processes at suppliers as well as the integration. Hence, its aim is to maximize the total profit Π of the development process. This profit results from the utility of the integrator n_I^w minus the development costs c_i of the suppliers as follows:

$$\Pi(w_1, w_2) = n_I^w(w_1, w_2) - c_1(w_1) - c_2(w_2) \tag{3}$$

The optimal efforts $w_{i,z}^*$ of the suppliers, that lead to maximized profit of the overall system, result from the analytical determination of the extreme point of this function. In the central setting, these parameters can be set by the central decision maker, i.e. the integrator. In the decentralized setting (d), the omniscient integrator determines the parameters of the fixed-price contracts, i.e. the required efforts $\hat{w}_i$ of the suppliers. Based on these parameters, the integrator then specifies the transfer payment function determining the amount to be paid by the integrator to the supplier. For a fixed-price contract, an effort of $w_i < \hat{w}_i$ results in no payment, while for efforts $w_i \geq \hat{w}_i$ the payment correspond to the development costs $c_i(\hat{w}_i)$ (see Figure 2). To model

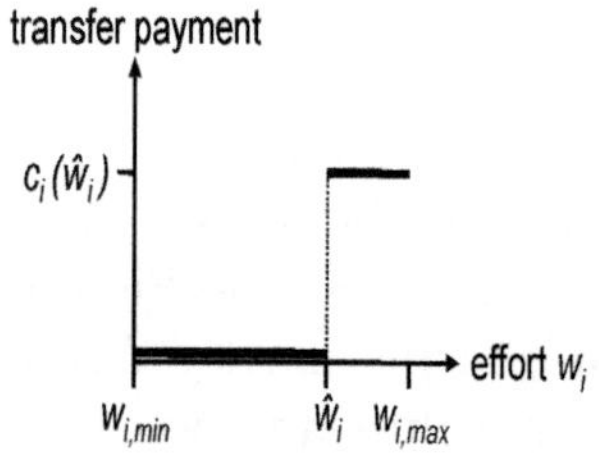

Fig. 2. Fixed-price contract.

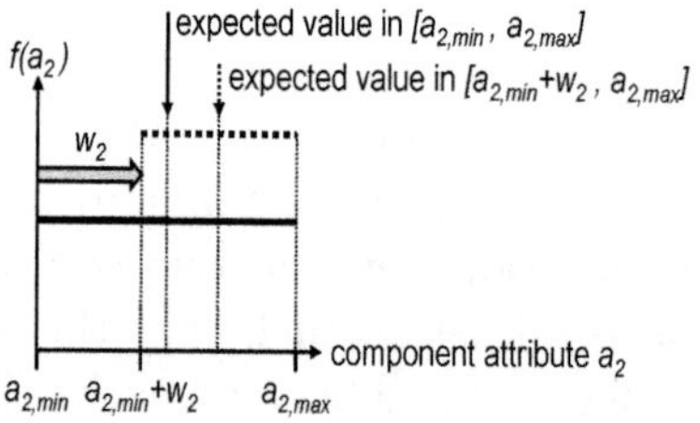

Fig. 3. Interdependencies between a_2 and w_2.

the 0-1-character of the transfer payment function in a continuously differentiable manner, it is approximated by a Sigmoid-function. The turning point of the Sigmoid-function is described by the required effort $\hat{w}_i$. This results in the transfer payment function t_i, where m specifies the slope in the turning point, $m \to \infty$:

$$t_i(\hat{w}_i, w_i) = c_i(\hat{w}_i)/(1 + \exp(m \cdot \hat{w}_i) \cdot \exp(-m \cdot w_i)) \tag{4}$$

Since the integrator is omniscient and he wants to maximize his utility, he chooses the required effort $\hat{w}_i$ in accordance to the optimal efforts $w^*_{i,z}$ in the central case. The supplier then determines the effort w_i, with which he can maximize his profit π_i based on the transfer payments set by the integrator and resulting development costs. The optimal effort $w^*_{i,d}$ of the supplier results from the analytical determination of the extreme point of the following function:

$$\pi_i(w_i) = t_i(\hat{w}_i, w_i) - c_i(w_i) \tag{5}$$

The analysis shows that the optimal effort in the deterministic decentralized setting with fixed-price contract corresponds to the optimal effort in central planning: $w^*_{i,d} = \hat{w}_i = w^*_{i,z}$.
Interpretation: In the deterministic case, the omniscient integrator is able to set the parameters of the fixed-price contract in a manner, that the operability of the overall system is achieved with minimal costs. Due to the transfer payment structure of the fixed-price contract, suppliers choose exactly the required effort. A higher and thus more costly effort would not be profitable for the suppliers, because the transfer payment does not increase, while a lower effort would result in zero transfer payment. Thus, suppliers choose the effort, which is optimal for the overall development process. This means that the fixed-price contract coordinates the cooperation partners during the development process.

Stochastic Case. In the following, the model is enhanced taking uncertainties with regard to development results of one of the suppliers into account. This is done, replacing the functional relationship $a_i = k_i(w_i)$ between effort w_i and resulting value a_i of the components' attributes by a stochastic correlation. It is assumed that the probability of a certain value a_i of the component's attribute is uniformly distributed in the interval $[a_{i,min}, a_{i,max}]$. Such a uniform distribution is typical for new developments, where no experiences of similar projects exist and the development outcome is totally uncertain within a certain interval. However, the supplier can still influence the development results with effort w_i. Every effort $w_i > 0$ shifts the lower interval limit to the right by w_i, so that the existing interval of uncertainty is reduced to the interval $[a_{i,min} + w_i, a_{i,max}]$. Thus, an increase in effort shifts the expected value to the right, as can be seen in Figure 3. Optimal efforts of the suppliers from the perspective of a central planner in the stochastic case (s) are calculated integrating the stochastic correlation between effort and value of the component attribute of one supplier (here supplier 2) as follows:

$$n_I^{EW}(w_1,w_2) = E[n_I^w(w_1,w_2)] = \int_{a_{2,min}+w_2}^{a_{2,max}} \frac{1}{1+\exp\left(d\cdot s(k_1(w_1))\right)\cdot\exp\left(-d\cdot\frac{a_2}{a_{2,max}-(a_{2,min}+w_2)}\right)}d(a_2) \tag{6}$$

The development costs $c_i(w_i)$ of the suppliers remain the same. The optimal efforts $w_{i,s}^*$ of the suppliers result from the analytical determination of the extreme point of the following function:

$$\Pi_s(w_1,w_2) = n_I^{EW}(w_1,w_2) - c_1(w_1) - c_2(w_2) \tag{7}$$

The analysis shows that the optimal efforts $w_{i,s}^*$ of suppliers differ from $w_{i,z}^*$ and lead to an expected combination of components' attributes which is on the right side of the efficient boundary $a_2 = s(a_1)$ of the deterministic case. Hence, uncertainties with regard to the development results of one supplier can lead to configurations of the overall system, which are overdimensioned and thus inefficient compared to the deterministic case.

3 Conclusion and Outlook

In this paper we presented a mathematical model for cooperative development processes of embedded systems. We analyzed a deterministic case in centralized and decentralized setting as well as a stochastic case in centralized setting, each with regard to optimal action of the partners. Future research will concentrate on the decentralized setting in the stochastic case.

Acknowledgement. This work has been funded by the German "Deutsche Forschungsgemeinschaft" (DFG). The authors would like to thank for the support.

References

1. Cachon GP (2003) Supply chain coordination with contracts. In: Graves SC, de Kok AG (eds) Supply chain management: design, coordination and operation. Elsevier, Amsterdam et al.
2. Kruse J, Volling T, Thomsen C, Ernst R, Spengler T (2005) Introducing flexible quantity contracts into distributed SoC and Embedded System design processes. In: DATE 2005. Proceedings.
3. Kruse J, Volling T, Thomsen C, Ernst R, Spengler T (2005) Towards flexible systems Engineering by using flexible quantity contracts. In: AAET 2005. Proceedings.

Collaborative Planning: Issues with Asymmetric Cost and Hold-Up in Automotive Supply Chains

Thomas Staeblein

Daimler AG, Research, 71032 Böblingen,
thomas.staeblein@daimler.com

Summary. Supply chain optimization has emerged as an important topic in several industries. In the automotive industry supply chains are prevalently composed of independent agents with specific preferences, e.g. distinct firms or profit centers in holdings. In either case, one must expect that no single party has control over the entire chain, and no partner has the power to optimize the whole chain by hierarchical planning. Under such decentralized decisions a way to improve supply chain performance is achievable through coordination and synchronization. The field of collaborative planning offers novel coordination concepts for such situations. We characterize issues in such concepts in automotive supply chains under asymmetric cost allocations. Here, and as well in other industries, few assembly sites (mostly owned by OEM's) are linked to multiple suppliers in a converging structure. We find, that under such setting, an iterative negotiation-based process based on counter-proposals is little different to upstream-planning, as local supplier-side savings have comparatively small effects. Further, we study the interaction between collaborative planning and the hold-up problem (i.e. at least one party performs relationship-specific investments), as an additional characteristic in the automotive industry.

1 Introduction

After the introduction of Japanese pull techniques in the 1990's, European car manufacturers and their suppliers are realizing that mastering these techniques alone is no longer enough to achieve competitive advantage. Increased pressure to attain annual cost reduction is driving the industry to re-evaluate their respective supply chain activities and relationships as they strive to improve overall performance. Such strategy, however, requires the ability to consider additional information that is beyond the individual planning domain. To align plans beyond

B. Fleischmann et al.(Eds.), *Operations Research Proceedings 2008*,
DOI: 10.1007/978-3-642-00142-0_37,

the local planning domain, the concept of 'collaborative planning' (CP) can be used. The main idea is to "[...] directly connect planning processes that are local to their planning domain in order to exchange the relevant data [...]" [5, p.259]. There are, however, practical difficulties when creating such collaboration process involving different partners. We study the current state of practice of CP, as a method to align plan between independent organizational units that have agreed on a supplier-buyer relationship using evidence of real-world automotive supply chains. Even after recent supplier base reductions, there remain several suppliers actively managed from a car manufacturer. Further, we study the hold-up problem and its impact in CP. The hold-up situation results as at least one party has to perform specific investments.

2 Collaborative Planning in Automotive Supply Chains

From the functional viewpoint of a car manufacturer, the supply chain is convergent to the point of the final assembly. The bill-of-material is strictly convergent and the final assembly operations are dominant. Sales and distributions networks of final products have a divergent structure, typically from central sales to market organisations responsible for certain regions and a high number of retailers or whole sale agents (see [4, 8]). A collaboration process with suppliers is achieved today by using different concepts, such as Vendor Managed Inventory, kanban, Just-in-Time or Continuous Replenishment. Most of these concepts have been introduced on operational level, based on the idea to create a visible demand pattern that paces the entire chain to enable synchronized supply. A seamless and real-time data exchange has been introduced that enables partners in the supply chain to perform calculations regarding their inventory and capacity on call-off level. Web-based systems are implemented to report mismatches to the car manufacturer. Collaboration needs to be shifted from operational to tactical level (i.e. from execution to planning) to further avoid mismatches and improve overall performance. Here, plans rather then schedules can be coordinated along the supply chain. Several publications propose 'collaborative planning' to combine mathematical programming and distributed decision making to achieve inter-organizational plan alignment [9]. Those concepts are not used from German car manufacturers, yet. Two aspects are barriers to practical introduction: (1) intra-domain decision model(s) and (2) a collaboration mechanism for the inter-relation between decision domains.

The choice of the intra-domain decision model depends on the involvement of the organisational departments, i.e. the objective governing decisions. A typical intra-domain planning model is the multi-level capacitated lot-sizing problem by Stadtler. The model objective is to minimize the total cost of fulfilling given deterministic customer demand. In order to create feasible plans, capacity constraints and multi-level stage manufacturing structure (i.e. consumption of dependent demands such as components, parts, raw materials, and other intermediate goods) are considered. For car manufacturers, however, the availability of components under stochastic customer demand is a core planning objective. This objective is challenging and interacts with the question: "how to plan component demand for complex configurable products under partial information?" Model styles, engine types, and interior options are only the most common categories within a customer can choose to configure a modern car. Due to the combinatorial possible, even a small number of those features amounts to an exponentially growing number of different (and valid!) product configurations, which each result in a distinct component demand pattern. As not all customer orders are received in the relevant time horizon, the component demand pattern remains unknown, and can be only roughly estimated with given methods. It has been reported that further important tasks such as master production scheduling are performed forecast-driven and often only supported by spreadsheet modelling [8].

In recent time, considerable investments in IT systems by European car manufacturers can be observed. While the average total annual spending on IT by manufacturers in Europe is around 2 to 3% of total turnover, a considerable proportion of this is used in maintaining the existing legacy infrastructure. Several manufacturers have to deal with up to 200 individual IT systems per plant, unable to 'switch off' applications that have been built onto existing legacy platforms. Such situation makes it particular challenging to introduce CP based on the use of fast decision support systems.

Concerning the use of coordination mechanisms in CP, issues arise from asymmetric cost situations in the car industy. The dominant supply chain operation is still the final assembly. In this operation more then 10.000 different components are assembled to create a modern car. The customization trend is unbroken and results in an immense variety. Due to recent supplier base reduction about 50 first tier ('key') suppliers contribute to more then 80% of the value of sourced components per car model. However, more then 500 first tier suppliers have to be actively managed at a large car manufacturer, as the final assembly operations

relies on the timely availability of all components in a Just-in-Time manufacturing environment.
The relative power in the buyer-supplier relationship has a strong impact on CP. All phases of the collaboration are tangled by power, e.g. the definition of collaboration conditions, bargaining, exception handling. Power can be theoretical exercised by any of the partners, e.g. in terms of know-how, financial resources, access to customers. For the coordination mechanisms, the largest share in the value creation of the final product is of particular importance and defines power in automotive supply chains. To measure power in this context, the amount of cost contributed by each supplier compared to total manufacturing cost can be used.
In the German car industry, typically no single component sourced from suppliers contributes to more then 10% of total manufacturing cost. Under this situation, available coordination mechanisms using a bilateral negotiation are hardly different to the most basic upstream planning coordination see [1]. Negotiation based CP schemes are based on the idea to iteratively compute plans and exchange them between buyers and suppliers to improve the over supply chain performance cost wise [3]. Each re-negotiation proposal from the buyer impacts all supply requirements, and can also result in cost changes for other suppliers. With costs at the buyer domain being asymmetric high, any supplier-side cost saving due to a new supply proposal to the initial supply requirement is relatively small compared to overall manufacturing cost. Transactional costs and the organisational challenge to coordinate a large supplier base during renegotiations results in the situation that the transactional burden for CP can be higher then saving opportunities in current practice.
How CP interacts with behaviour in buyer-supplier relationships (e.g. moral hazard, adverse selection, hold-up) remains unstudied yet. Although agent's behaviour has a major impact on results, decision models used in CP suppress such behavior. In general, there are only a few exceptions in coordination models that capture both the processing dynamics and the agent's behaviour, e.g. [2].

3 Hold-Up Problem and Collaborative Planning

The hold-up problem plays an important role as a foundation of modern contract and organization theory. The research is based on the case study 'Fisher Body - GM' by Coase and is commonly used for economic studies in supplier-buyer relationships [6]. Hold-up interacts

also with CP, as processes in most coordination schemes are based on the assumption of long-term partnership.
The hold-up situation arises as part of the return on a supplier's (buyer's) relationship-specific investments is ex post expropriable by the buyer (supplier). The main inefficiency results from the situation, that investments are often geared towards a particular buyer-supplier relationship, in which case the returns on them within the relationship exceed those outside it [7]. Once an investment is sunk, the investor has to share part of the gross returns with the other partner. This problem is inherent in many buyer-seller relationships. In the automotive industry, manufacturers and suppliers often customize their production equipment and further processes towards a specific relationship. The risk being held up discourages the capacity investor from making desirable investments. In following, a simple model of the hold-up problem is described to illustrate the main under-investment hypothesis (see [10] for a formal proof).
Consider a manufacturer and his supplier, denoted M and S, trade quanitity: $q \in [0, \overline{q}]$, where $\overline{q} > 0$. The transaction can benefit from the supplier (specific and irreversible) investment, modelled with a binary decision: $I \in \{0, 1\}$, where $I = 1$ stands for 'invest' and $I = 0$ stands for 'not invest'. The investment I cost the supplier $k \cdot I$, where $k > 0$. Given investment I, the manufacturer's gross surplus from consuming q is $v_I(q)$ and the cost of delivering q by the supplier is $c_I(q)$, where both v_I and c_I are strictly increasing and $v_I(0) = c_I(0) = 0$. Let $\phi_I = \max_{q \in Q}[v_I(q) - c_I(q)]$ denote the efficient social surplus given S's investment, and let (q_I^*) be the efficient level of trade. The net surplus is then $W(I) := \phi_I - k \cdot I$. Suppose that

$$\phi_I - k > \phi_0 \tag{1}$$

gives S the interest to invest in the desirable capacity amount. An important assumption is that S's investment is not verifiable (even it might be observable), therefore it can not be contracted upon. Let us assume both parties contract a la Nash, yielding an efficient decision q_I and splitting the gross surplus ϕ_I equally between both partners. One party (in this example, the supplier) thus appropriates only a fraction (here, the half) of the investment returns, while bearing the cost of investment k, so the net payoff will be $U_s(I) := \frac{1}{2}\phi_I - k \cdot I$, following her investment. Suppose that

$$\frac{1}{2}\phi_I - k < \frac{1}{2}\phi_0 \tag{2}$$

then, even though the investment is overall desirable, S will not invest. An underinvestment arises, which influences coordination within the supplier-buyer relationship. Hence, the underinvestment impacts supply chain performance and renegotiation in CP, as the initial capacity portfolio in distributed ownership (buyer-supplier relationship) is less efficient compared to vertical integration (central coordination).

4 Conclusion

The current state-of-practice concerning 'collaborative planning' at car manufacturers has been described. Issues with coordination mechanisms under asymmetric cost allocations have been discussed. Due to the strictly converging supply chain structure of operations, we find that, an iterative negotiation-based process based on counter-proposals is little different to upstream-planning, as local supplier-side savings have comparatively small effects. We study further that renegotiation opportunities in 'collaborative planning' interact with the hold-up problem (i.e. at least one party performs relationship-specific investments), as a further characteristic in the automotive industry.

References

1. Bhatnagar R, Chandra P, Goyal SK (1993) Models for multi-plant coordination. EJOR 67:141–160
2. Cachon G, Laviviere M (2001) Contracting to assure supply: How to share demand forecasts in a supply chain. Management Sci. 47:629–646
3. Dudek G, Stadtler H (2005) Negotiation-based collaborative planning between supply chain partners. EJOR 163:668–687
4. Juergens U (2004) Characteristics of the European automotive system: is there a distinctive European approach? IJATM 4: 112–136
5. Kilger C, Reuter B (2005) Collaborative Planning. In: Stadtler H, Kilger C (eds) Supply Chain Management and Advanced Planning. Springer, Berlin Heidelberg New York
6. Klein B (2000) Fisher-General Motors and the Nature of the Firm. J. of Law and Economics 43:105-141
7. Lau S (2008) Information and bargaining in the hold-up problem. RAND J. of Economics 39:266–282
8. Meyr H (2004) Supply Chain Planning in the German automotive industry. OR Spectrum 26:447–470
9. Stadtler H (2009) A framework for collaborative planning and state-of-the-art. OR Spectrum 31:5-30
10. Tirole J (1986) Procurement and Renegotiation. J. of Political Economy 94:235–259

Negative Default Dependencies: Do Automotive Suppliers Benefit from Their Competitors' Default?

Stephan M. Wagner[1], Christoph Bode[1], and Philipp Koziol[2]

[1] Chair of Logistics Management, Swiss Federal Institute of Technology Zurich
{stwagner,cbode}@ethz.ch
[2] Chair of Finance, University of Gottingen
pkoziol@uni-goettingen.de

1 Introduction

Given the criticality of the supplier network and the increasing number of supplier defaults [3], automotive OEMs are well advised to actively manage their supplier networks, consider supply risks [9], and avoid negative and exploit positive consequences of supplier defaults. We aim to extend the research on supplier defaults along two dimensions:
First, with a few exceptions [1, 10], previous studies have focused on the default consequences of individual suppliers [2]. However, since suppliers do not operate in isolation, their financial situation is often interlinked. We argue that default dependence (also default correlation) exists in supplier networks and that OEMs should manage the entire network as a group and consider interdependencies among suppliers.
Second, all studies that have considered default dependencies have exclusively modeled positive dependence, i.e., they have assumed that, given two suppliers 1 and 2, the default of supplier 2 is positively related to the default of supplier 1. In contrast, we model negative default dependence (i.e. scenarios where supplier 2 benefits from the default of supplier 1).
Our results highlight that in the presence of negative default dependencies between two suppliers the default of one supplier increases the survival probability of the second supplier.

B. Fleischmann et al.(Eds.), *Operations Research Proceedings 2008*,
DOI: 10.1007/978-3-642-00142-0_38, © Springer-Verlag Berlin Heidelberg 2009

2 Model Development and Estimation

The core of our approach consists of copulas which can model the marginal distributions and the dependence structure of the random variables separately [7].

2.1 The Default Model

We use Li's [6] default model as a starting point and apply it to supplier networks. Let $\boldsymbol{\tau} = (\tau_1, \ldots, \tau_n)$ be the random vector of default times of the n suppliers in the network. The default time τ_i of supplier i is exponentially distributed with parameter λ_i in the way that $F_i(t_i) = P(\tau_i \leq t_i) = 1 - e^{-\lambda_i t_i}$. Supplier i has defaulted by time t_i if $\tau_i \leq t_i$, so that the default probability is $F_i(t_i)$. Dependence is now introduced using a copula C according to Sklar's theorem, so that the dependence structure is totally determined by the copula. Thus, the joint default distribution is as follows (applicable to any copula C):

$$P(\tau_1 \leq t_1, \ldots, \tau_n \leq t_n) = C(F_1(t_1), \ldots, F_n(t_n)). \tag{1}$$

For $\tau_i > 0$, there is no closed form solution for this problem. For this reason, we realize our approach by means of Monte Carlo simulation.

2.2 Estimation Methodology

In addition to an appropriate copula C representing the dependence structure, two input factors are necessary: (1) the default intensity of each supplier and (2) the dependence level between the suppliers in the network represented by a *Kendall's tau* value.

To estimate the suppliers' *individual default intensities* λ_i, we applied Jarrow and Turnbull's [5] defaultable claims pricing model, since it is analytically tractable and the most parsimonious intensity-based model. It models independency between the interest rate risk and the default risk and assumes a frictionless, arbitrage-free, and complete market. For the estimation of the default intensities, the Jarrow/Turnbull model requires (1) the default-free zero-coupon bond prices $b(t,T)$, (2) the recovery rate δ, and (3) the default-risky zero-coupon bond (market prices) $v(t,T)$. The default-risky zero-coupon bond is priced under the risk-neutral martingale measure as follows:

$$v(t,T) = b(t,T) \cdot (\delta + (1-\delta) \cdot Q(\tau > T)) \tag{2}$$

Where $Q(\tau > T)$ represents the probability that the underlying bond will survive at least the maturity T. The price of a default-risky coupon

bond with maturity T, $V(t,T)$, is then the linear combination of its cash flows $C(t)$ at time t. We use one market price of the default-risky bond and the default-free bond and can directly calculate the default intensity λ for the considered companies solving the former equation. Similar to [4], we a priori fixed the recovery rate at a reasonable δ of 50% for all bonds.
The second ingredient is the *level of the suppliers' default dependence.* In practice, it is nearly impossible to estimate the default dependence of two companies and calculate a corresponding Kendall's tau because defaults are rare events and there are not enough time-series data on corporate defaults available. For this reason, we will simulate various levels of default dependence ranging from the independence case to a Kendall's tau of -0.60.

3 Negative Supplier Default Dependence: A Model and Simulation Results

3.1 Model Specification

We take the perspective of an automotive OEM that works with two distinct suppliers (dual sourcing): supplier 1 and 2 (subsequently called the OEM's *supplier portfolio*). The suppliers' individual default probabilities follow exponential marginal distributions, the standard distribution for modeling survival times, with default intensities λ_1 and λ_2. We consider, w.l.o.g., the scenario where supplier 1 folds in $\tau_1 \in [0, z_1]$, in order to examine the effect of supplier 1's default on supplier 2's survival probability $P(\tau_2 > z_2 = (z_1 + a)|\tau_1 \in [0, z_1])$ with $a = 1, 2, \dots$ years.
In order to investigate the dependence structure between supplier 1 and 2, we specify the model with the Gauss and t copula, respectively. The choice of the copula reflects the dependence structure and risk profile of the supplier portfolio. The Gauss copula is the standard copula for many applications, but has the main shortcoming that it does not model dependence between extreme events. The t copula, here specifically the t_5 copula, can be regarded as the common extension of the Gauss copula in a way also captures *fat tails.*

3.2 Data

Financial data of automotive supplier firms necessary for specifying and adjusting our model were drawn from the Datastream database.

For the estimation of the default intensities, the prices of the default-risky coupon bonds were considered from April 2006 to March 2007. The price at the coupon payment date was taken in this interval, so that no coupon time transferring problems occurred. For the calculation of the European risk-free bond prices, we used daily estimates of the [8] model from the Deutsche Bundesbank. For the US bonds the zero curve interest rates are applied.

3.3 Simulation Results

For the dependence structure (captured by a copula) and dependence level (captured by a Kendall's tau value), we generated 500,000 random pairs (τ_1, τ_2) of default times for supplier 1 and 2. Based on these numerical results, we approximated the joint default distribution of this supplier portfolio. Our interest focuses on the effect of negative default correlation and on the influence of different dependence structures. We simulated the models with the parameters $z_1 = a := 2$ years. Then, we subjected the obtained data to a comparative static analysis illustrating the impact of negative default correlation in supplier portfolios.

First, we constructed *homogenous* portfolios, i.e., the two suppliers 1 and 2 have identical default intensities. The first portfolio (Figure 1(a)) contains two suppliers with rather low default intensities of $\lambda_1 = \lambda_2 = 0.0073$ (level of Bosch). The second portfolio (Figure 1(b)) consists of two suppliers with significantly higher default intensities of $\lambda_1 = \lambda_2 = 0.1368$ (level of Lear). The comparison of the two Figures plainly shows that the effects are stronger, if the individual default probability is higher. When the dependence structure is characterized by a Gauss copula, a rather low Kendall's tau of -0.2, for instance, increases the survival probability of the second supplier by 2.7% when both suppliers' default intensity is low, and by 15.4% when both suppliers have a high default intensity. Moreover, looking at the imposed dependence structure, we find that the survival probability of supplier 2 increases stronger under the dependence structure determined by the Gauss copula than under the one determined by the t_5 copula.

Second, we built a realistic portfolio of automotive suppliers with *heterogeneous* default intensities. That is, the portfolio contains a supplier with a low and another with a high default intensity. The results show the interesting effect that it is critical which supplier of the portfolio defaults first. If supplier 1 (here Michelin) has a low default intensity and folds, the survival probability of supplier 2 (here Goodyear) increases drastically with increasing (negative) dependences levels (Figure 1(c)).

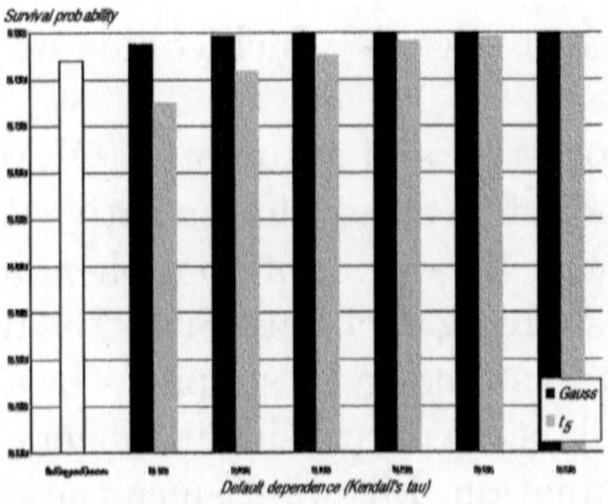

(a) Low default intensity (Bosch/Bosch)

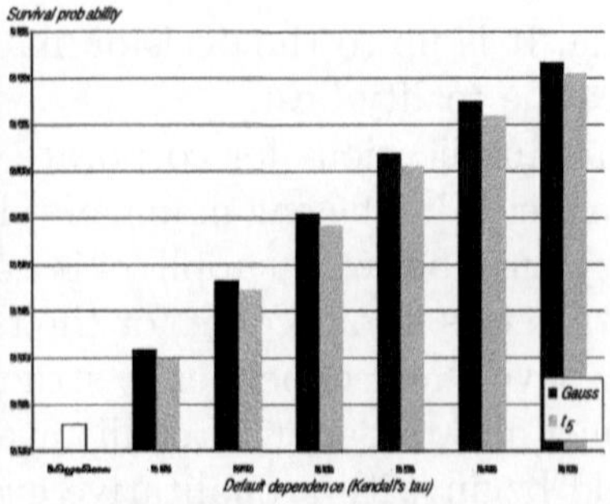

(b) High default intensity (Lear/Lear)

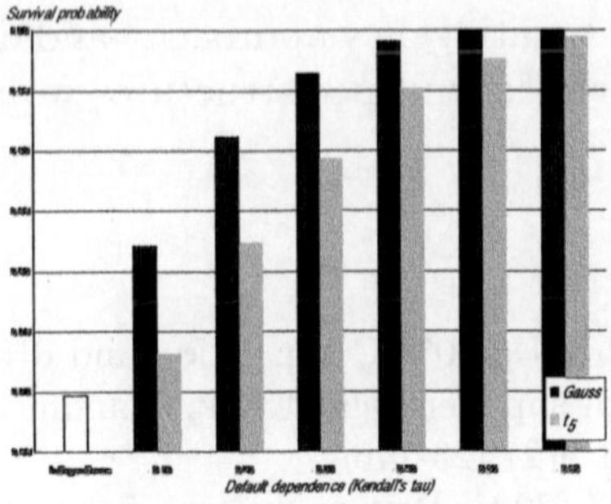

(c) 1. Michelin/2. Goodyear

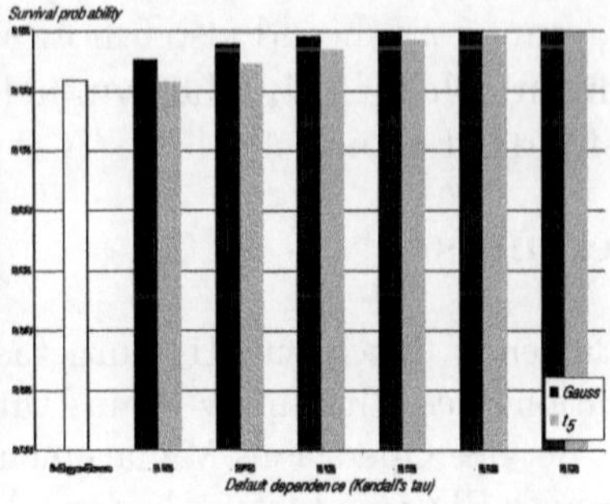

(d) 1. Goodyear/2. Michelin

Fig. 1. Portfolios with two suppliers: Survival probability of supplier 2 $P(\tau_2 > 4 \text{ years} | \tau_1 < 2 \text{ years})$.

In contrast, in case supplier 1 with high default intensity folds first, the survival probability of supplier 2 is not as much affected (Figure 1(d)).

4 Discussion and Implications

Summarizing the results of our simulation and comparative static analysis we can provide several insights. First, the simulation results depict that negative default dependence among suppliers in a supplier network has consequences for the suppliers' survival probabilities. The higher the individual default intensity of a supplier, the stronger the effect of negative default correlation on the survival probability of a second supplier. Second, the dependence structure, reflected by the choice of copula, is decisive. There is a variety of copulas which differ in the dependence structure they represent. Our results demonstrate that the Gauss copula imposes a dependence structure that increases the survival probability of the second supplier more drastically than the t_5

copula. It is up to the decision maker to decide which copula approximates the reality best.

Several implications for corporate practice and automotive OEMs can be derived. Purchasing managers should be aware that negative default dependence between suppliers is not at all an exceptional phenomenon and take this into account for their sourcing decisions. Since the dependence levels and dependence structures among all suppliers in a firm' supplier network are not readily available to purchasing managers, firms should begin with a qualitative evaluation. Likewise, since the dependence structures underlying the supplier default model (i.e., Gauss or t_5 copulas) have a significant impact on the survival of the second supplier, managers should also follow a "contingency approach" and expert judgment. Most likely, the symmetric dependence structures would be the standard choice.

References

1. Babich V, Ritchken PH, Burnetas AN (2007) Competition and diversification effects in supply chains with supplier default risk. Manufacturing & Service Operations Management 9(2):123-146
2. Berger PD, Gerstenfeld A, Zeng AZ (2004) How many suppliers are best? A decision-analysis approach. Omega 32(1):9-15
3. Department of Commerce (2008) U.S. automotive parts industry annual assessment. Washington, DC: U.S. Department of Commerce, Office of Aerospace and Automotive Industries
4. Frühwirth M, Sögner L (2006) The Jarrow/Turnbull default risk model: Evidence from the German market. European Journal of Finance 12(2):107-135
5. Jarrow RA, Turnbull SM (1995) Pricing derivatives on financial securities subject to credit risk. Journal of Finance 50(1):53-85
6. Li DX (2000) On default correlation: A copula function approach. Journal of Fixed Income 9(4):43-54
7. Nelsen RB (2006) An Introduction to Copulas, 2nd edn. Springer, New York, NY
8. Svensson LEO (1994) Estimating forward interest rates with the extended Nelson & Siegel method. Sveriges Riksbank Quarterly Review 3:13-26
9. Wagner SM, Bode C (2006) An empirical investigation into supply chain vulnerability. Journal of Purchasing and Supply Management 12(6):301-312
10. Wagner SM, Bode C, Koziol P (2007) Supplier default dependencies in the automotive industry. CSCMP Annual SCM Educators' Conference, Philadelphia, PA

Part VI

Traffic and Transportation

An Optimization Approach for the Crew Scheduling Problem in Waste Management

Jens Baudach, Annette Chmielewski, and Tobias Mankner

Technische Universität Dortmund
Lehrstuhl für Verkehrssysteme und -logistik
{baudach,chmielewski,mankner}@vsl.mb.tu-dortmund.de

1 Introduction and Problem Description

Difficult financial situations in most German cities and increasing competition in waste management between private and public companies require an efficient allocation of all resources that are involved in the waste disposal process. The two major resources are waste collection vehicles and crews. They can be assigned to the corresponding planning steps of finding so called *waste collection districts* and appropriate crew schedules which are planned manually at the moment. This paper focuses on the optimization of the crew scheduling process where different crews have to be allocated to specific waste collection districts and vehicles respectively. The paper shows first results of a joint research project[1] including the Universities of Braunschweig and Dortmund as well as two companies from the private and public sector.

In the context of our paper waste management means the disposal of household waste which in Germany includes the four so called *fractions* paper, household-like wrappings, biowaste, and residual waste. Starting point of the planning process is a *waste disposal area* which usually is determined by city authorities. Within this area all four fractions of waste have to be disposed in a periodic way. A first planning step – which is prior to the crew scheduling phase and not part of this paper – implies the generation of optimal waste collection districts. Each waste collection district is a part of the waste disposal area and represents a set of households that have to be disposed by one vehicle within a given period, the so called *waste disposal cycle.* Waste collection districts have to be generated for each fraction and therefore can differ in size and length of the associated waste disposal cycle. Figure 1 shows

[1] funded by the German Federal Ministry of Education and Research (BMBF)

B. Fleischmann et al.(Eds.), *Operations Research Proceedings 2008*,

DOI: 10.1007/978-3-642-00142-0_39, © Springer-Verlag Berlin Heidelberg 2009

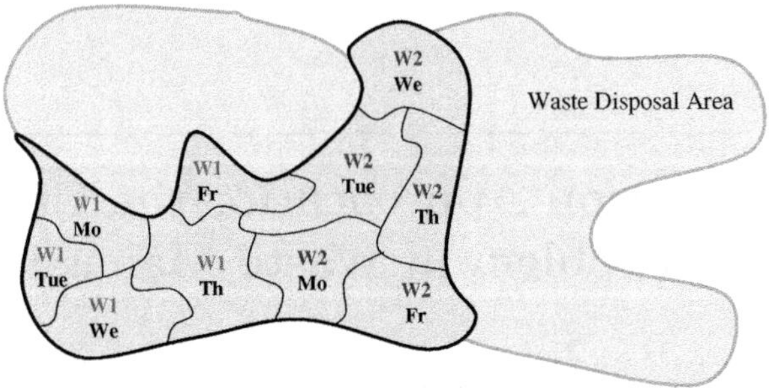

Fig. 1. Waste Collection District with a Waste Disposal Cycle of 2 Weeks

a waste collection district belonging to one specific fraction and vehicle respectively for a waste disposal cycle of two weeks which then can be iterated. The main parameters that influence this part of the optimization are the capacity and speed of the used waste collection vehicles, the applied working time model, and the number of employees belonging to vehicle crews. Working time models control the regular or maximal working time per day or week as well as the (number of) free days within the *deployment cycles* of employees which are usually a multiple of a waste disposal cycle (see Fig. 2) [4]. Free working days are necessary in order to realize a higher utilization of the vehicles by longer working times per day or additional work on saturdays because companies are very skeptic in terms of crew changes which are common practice in many other areas of transportation [2, 3].

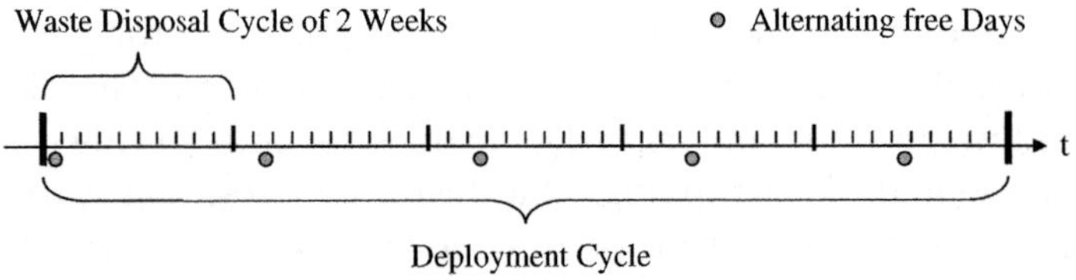

Fig. 2. Deployment Cycle

The above illustrated first phase of waste management planning provides specific information about all generated waste collection districts which have to be covered by crews in the subsequent crew scheduling phase. Thereby, each day of every waste collection district (of all fractions) has to be covered with at least one driver and a determined number of loaders for the waste bins. Additional requirements like district knowledge, crew reserves, or the working time model have to be considered, too. In the following chapter we present a modelling approach that tries to find optimal crew schedules by allocating specific waste disposal cycles to each employee such that all households get disposed. In the current sequential optimization approach the number of required crews is indirectly determined in the previous planning phase of the waste collection districts. Thus, we are focussing on important soft factors in the crew scheduling optimization phase.

2 Mathematical Modelling

Let N be the set of crew members, Z the set of deployment cycles, W the set of waste disposal cycles, D the set of working days in a waste disposal cycle, R the set of waste collection districts, and F the set of fractions. The appertained index to a set is in each case the lower alphabetic character. ($M > 0$ is an auxiliary skalar.)

Input Parameters:

c_{zwdf} $\begin{cases} 1, & \text{fraction } f \text{ is disposed in deployment cycle } z \text{ in waste disposal cycle } w \text{ on (working) day } d \\ 0, & \text{else} \end{cases}$

e_{drf} $\begin{cases} 1, & \text{fraction } f \text{ has to be disposed in waste collection district } r \text{ on working day } d \text{ of a waste disposal cycle} \\ 0, & \text{else} \end{cases}$

b_{drf} number of crew members necessary for the disposal of fraction f in waste collection district r on day d in a waste disposal cycle

l_{drf} number of crew members which have to be familiar in the disposal of fraction f in waste collection district r on day d

i_n $\begin{cases} 1, & \text{if crew member } n \text{ is a reserve pool employee in principle} \\ 0, & \text{else} \end{cases}$

u_n maximum number of waste collection districts a crew member n can be assigned to

q_n $\begin{cases} 1, & \text{crew member } n \text{ has a driving license} \\ 0, & \text{else} \end{cases}$

$p_n^{\min}$ minimum percentage of work crew member n has to be appointed to unfamiliar waste collection districts

$p_n^{\max}$ maximum percentage of work crew member n is allowed to be appointed to unfamiliar waste collection districts

k_{nrf} $\begin{cases} 1, & \text{crew member } n \text{ is familiar with the disposal of fraction } f \text{ in waste collection district } r \\ 0, & \text{else} \end{cases}$

Variables:

$$\alpha_{nz} \begin{cases} 1, & \text{crew member } n \text{ works to the rules of deployment cycle } z \\ 0, & \text{else} \end{cases}$$

$$\beta_{nwdrf} \begin{cases} 1, & \text{crew member } n \text{ is assigned to fraction } f \text{ in waste} \\ & \text{collection district } r \text{ on day } d \text{ in waste disposal cycle } w \\ 0, & \text{else} \end{cases}$$

$$\gamma_{nwdrf} \begin{cases} 1, & \text{crew member } n \text{ is a relief person for fraction } f \text{ in waste} \\ & \text{collection district } r \text{ on day } d \text{ in waste disposal cycle } w \\ 0, & \text{else} \end{cases}$$

$$\omega_{nr} \begin{cases} 1, & \text{crew member } n \text{ is assigned to waste collection district } r \\ & \text{at least once} \\ 0, & \text{else} \end{cases}$$

$$\lambda_{wdrf} \begin{cases} 1, & \text{fraction } f \text{ in waste collection district } r \text{ on day } d \text{ of} \\ & \text{waste disposal cycle } w \text{ is covered by a relief person} \\ 0, & \text{else} \end{cases}$$

Mathematical Program:

$$\max \sum_{w\in W}\sum_{d\in D}\sum_{r\in R}\sum_{f\in F} \lambda_{wdrf}$$

$$\text{s.t.} \quad \sum_{w\in W}\sum_{d\in D}\sum_{f\in F} (\beta_{nwdrf} + \gamma_{nwdrf}) \leq M \cdot \omega_{nr} \quad \forall n, r \qquad (1)$$

$$\sum_{n\in N} \gamma_{nwdrf} \geq \lambda_{wdrf} \qquad \forall w, d, r, f \quad (2)$$

$$\sum_{r\in R} \omega_{nr} \leq u_n \qquad \forall n \quad (3)$$

$$\sum_{n\in N} \beta_{nwdrf} = b_{drf} \cdot e_{drf} \qquad \forall w, d, r, f \quad (4)$$

$$\sum_{n\in N} k_{nrf} \cdot \beta_{nwdrf} \geq l_{drf} \cdot e_{drf} \qquad \forall w, d, r, f \quad (5)$$

$$\sum_{n\in N} q_n \cdot \beta_{nwdrf} \geq e_{drf} \qquad \forall w, d, r, f \quad (6)$$

$$\sum_{w\in W}\sum_{d\in D}\sum_{r\in R}\sum_{f\in F} (1 - k_{nrf}) \cdot \beta_{nwdrf} \geq p_n^{\min} \cdot \sum_{w\in W}\sum_{d\in D}\sum_{r\in R}\sum_{f\in F} \beta_{nwdrf} \qquad \forall n \quad (7)$$

$$\sum_{w\in W}\sum_{d\in D}\sum_{r\in R}\sum_{f\in F} (1 - k_{nrf}) \cdot \beta_{nwdrf} \leq p_n^{\max} \cdot \sum_{w\in W}\sum_{d\in D}\sum_{r\in R}\sum_{f\in F} \beta_{nwdrf} \qquad \forall n \quad (8)$$

$$\sum_{z\in Z} v_{nz} = 1 \qquad \forall n \quad (9)$$

$$\sum_{r\in R} \beta_{nwdrf} + \gamma_{nwdrf} = \sum_{z\in Z} v_{nz} \cdot c_{zwdf} \qquad \forall n, w, d, f \quad (10)$$

Constraints Explanation:

(1) The helping variables ω_{nr} are forced to contain the correct value.
(2) The helping variables λ_{wdrf} are forced to contain the correct value.
(3) The number of waste collection districts crew member n is allowed to work in is bounded by u_n.
(4) The required number of crew members of each day of all waste collection districts has to be covered.
(5) The required number of crew members familiar with the waste disposal district (and a fraction) has to be satisfied for each day.
(6) At least one driver has to be assigned to each (vehicle) crew.
(7) Minimum and maximum boundaries of work of a crew member in
(8) unfamiliar waste collection districts have to be observed.
(9) Exactly one deployment cycle has to be assigned to a crew member.
(10) Each crew member has to work to the rules of its assigned cycle.

Objective Function

The objective function presented in the mathematical program above honours a broad and even distribution of the reserve pool members to different days and fractions of the waste collection districts. We identified and implemented many other of these *soft* objective functions based on different needs of the employees and the companies. They can be formulated considering different aspects of crew planning and usually need additional variables and constraints respectively.

3 Test Results

The process and complexity of crew planning in the two involved waste management companies is quite different. This also results in different input data for the mathematical programm:

	n	z	w	d	r	f	# Variables	# Constraints
Company 1	104	1	1	10	25	1	54954	5056
Company 2	47	10	5	10	11	2	9805	105487

CPlex finds an optimal solution for company 1 in a few seconds, but CPlex is not able to find an optimal solution for company 2. The calculation was aborted after 5 days of computing. If we lower the number of waste collection districts to 8 and the number of cycles to 5, CPlex finds an optimal solution in about 1 hour. As the input size of many other waste disposal areas, for example in huge cities like Berlin or

Hamburg, will be even bigger, additional efforts to increase the performance of our models (and future algorithms) are necessary. Further difficulties may be caused by very complex objective functions (with additional variables or constraints) in order to reproduce one or maybe more of the above mentioned *soft objectives* in crew scheduling.

4 Outlook

Current and future areas of research include an extension of the existing models like consideration of workload, balanced age structures within crews, and meeting different desires of employees. In order to handle the complexity of the given problem we intend to develop problem specific algorithms based on column generation and lagrangean relaxation approaches [1, 3]. In the last phase of the research project we want to realize an integrated optimization of waste collection districts, vehicles, and crews in order to achieve additional optimization potential which actually gets lost in a sequential approach.

References

1. Barnhart C, Johnson, EL, Nemhauser GL, Savelsbergh MWP, Vance PH (1998) Branch-and-Price: Column generation for solving huge integer programs. Operations Research 46(3): 316–329
2. Caprara A, Fischetti M, Toth P, Vigo D, Guida PL (1997) Algorithms for railway crew management. Mathematical Programming 79: 125–141
3. Huisman D (2004) Integrated and dynamic vehicle and crew scheduling, Ph.D. Thesis, Tinbergen Institute, Erasmus University Rotterdam
4. Spengler T (2006) Modellgestützte Personalplanung. FEMM: Faculty of economics and management Magdeburg, working paper series, March, 10

Milk Run Optimization with Delivery Windows and Hedging Against Uncertainty

Carsten Böhle and Wilhelm Dangelmaier

Heinz Nixdorf Institut
{carsten.boehle,whd}@hni.uni-paderborn.de

1 Introduction

Milk runs are an important transportation concept, e.g. in the automotive industry. Trucks start from a depot, pick up goods at different suppliers, and deliver those goods to a single customer. Therefore, milk runs are technically a pick up and delivery problem with multiple pick ups and a single delivery. They make it possible to deliver small lots efficiently and thus lower average inventory levels. Prerequisite is that there are several frequent orders from suppliers that are closely located, otherwise transshipment centers will be used in spite of handling costs. Pickup&Delivery problems calculate routes for single days, sometimes with the additional restriction of time windows for delivery. Models and algorithms for these problems exist and can help in practice as lead-times are usually very short, in most cases only one day. It has been discussed whether it would be better to allow for delivery windows of a few days so that carriers have more leeway for route optimization (cf. [1]). Now the routing problem is extended with the problem of allocating orders to days. The integrated solution requires vehicles to make multiple trips and is formulated as the VRP with Multiple Trips (VRPM)[1]. The VRPM has found only little attention so far: "Although in practice multiple route assignment is common, there is a shortage of papers covering this feature." [2] It is even more interesting to look at the problem from a dynamic point of view, i.e. to iterate through the days and to assign incoming orders to days without having information on future orders.
Some researchers have tackled similar problems. An early work has

[1] The VRPM is sometimes also referred to as the Vehicle Routing Problem with Multiple Use of Vehicles

B. Fleischmann et al.(Eds.), *Operations Research Proceedings 2008*,
DOI: 10.1007/978-3-642-00142-0_40,

been presented by Brandão and Mercer[3]. Gendreau et al. [4],[5] have repeatedly investigated the problem and presented algorithms for optimal solutions. Zhao et al. [6] also stress the fact that the problem of vehicle routing with multiple use of vehicles within the concerned time horizon is of practical relevance but seldomly studied. They present a tabu search algorithm. A similar approach is given by Taillard et al. [7]. A genetic algorithm for the VRPM is presented by Salhi and Petch [8]. An introduction to online VRP can be found in Allulli et al. [9].

2 Problem Formulation

The four-index vehicle flow formulation is based on the one given by Zhao et al. [6] for the MTCDVRP[2]. It adds three restrictions (4, 7, and 8) which will be discussed. The problem will be called VRPPDMTW[3] and DVRPPDMTW when it adresses the online version.
The objective funtion is:

$$min. \sum_{j}^{N} \sum_{l}^{L} \sum_{t}^{T} x_{0jlt} * f + \sum_{i}^{N} \sum_{j}^{N} \sum_{l}^{L} \sum_{t}^{T} c_{ij} * x_{ijlt} \tag{1}$$

N is the set of nodes including the depot and the customer. f are the fix costs for each truck. c_{ij} are the costs for traveling from i to j. x_{ijlt} is binary and indicates if truck l travels from i to j in period t. Node 0 will be regarded as the depot, node 1 as the customer. Nodes 2 to n are suppliers.

$$\sum_{i}^{N} \sum_{l}^{L} \sum_{t}^{T} x_{ijlt} = 1 \qquad \forall j \in N \setminus Depot, Customer \tag{2}$$

$$\sum_{i}^{N} x_{iplt} - \sum_{j}^{N} x_{pjlt} = 0 \qquad \forall p \in N, l \in L, t \in T \tag{3}$$

Restrictions 2 and 3 let every node except depot and customer be visited exactely once. Also, each node that is visited has to be left as well.

---

[2] Multi trip capacity and distance vehicle routing problem

[3] VRP with pickup and delivery and multiple use of vehicles and time windows; note that the solution given in this paper particularly addresses milk runs and is not a generalized pickup and delivery problem

$$\sum_{i}^{N}\sum_{j}^{N} x_{ijlt} \leq x_{1,0,l,t} * M \qquad \forall l \in L, t \in T \tag{4}$$

Restriction 4 assures that every truck in every period travels from the customer to the depot in case it travels to any supplier, thus forming a valid milk run that starts and ends at the depot and visits the customer last.

$$\sum_{i}^{N}\sum_{j}^{N} (d_i * x_{ijlt}) \leq C \qquad \forall l \in L, t \in T \tag{5}$$

$$\sum_{i}^{N}\sum_{j}^{N} (c_{ij} * x_{ijlt}) \leq D \qquad \forall l \in L, t \in T \tag{6}$$

Restrictions 5 and 6 enforce tours to be within capacity limits set by the amount of goods that can be transported by a truck and the distance it can travel.

$$a_j \leq \sum_{i}^{N}\sum_{l}^{L}\sum_{t}^{T} (x_{ijlt} * t) \qquad \forall j \in N \tag{7}$$

$$b_j \geq \sum_{i}^{N}\sum_{l}^{L}\sum_{t}^{T} (x_{ijlt} * t) \qquad \forall j \in N \tag{8}$$

Restrictions 7 and 8 set the time windows for each order. a is the earliest day for pick up, b the latest.

$$\sum_{i \in S}\sum_{j \in S} x_{ijlt} \leq |S| - 1 \qquad S \subseteq N, |S| \geq 2 \qquad \forall l \in L, t \in T \tag{9}$$

Subtour elimination is guaranteed by restriction 9.

3 Test Instance

The data set includes 1,000 orders from 10 suppliers that are within a range of 100 km around the depot. The orders cover 100 days and are served by a fleet of 10 trucks. A truck can transport 300 units over a distance of 500 km. Order sizes are uniformly distributed between 10 and 50 units. 10% of the orders do not have a time window, i.e. the day they have to be served is fixed. If there is a time window, its size is uniformly distributed between 2 and 4 days. 10% of the orders have no leadtime, i.e. they are announced one day in advance. The rest has a leadtime of 2 to 5 days which again is uniformly distributed.

4 Heuristic

The heuristic can run in two different modes. The first mode runs in two stages. In the first stage incoming orders for a certain period are randomly assigned to days. Orders within one day are optimized using the parallel Clark and Wright Savings algorithm. Orders that cannot be served are put on hold. Then, orders are shifted between days using a tabu search algorithm. These stages run in a loop until all orders are assigned to a day or ultimatively have to be rejected. All accepted orders are then fixed which means they cannot be moved in future periods. After that the orders of the next period are looked at. If the period length is the total number of days the solution represents the offline solution. The second mode does not use the tabu search part. It daily receives new orders which have to be fixed on a date. Orders are sorted by descending sizes and then assigned to that day within their time window which currently holds the shortest distance to travel. At this point the method deviates from the tabu search optimization. The latter will only in certain cases shift orders to empty days because this creates tours from and to the depot and customer. This contradicts the myopic goal of shortening routes. The idea is to spread orders over the days to leave buffers in every day so that orders arriving later can still be served. Not being able to serve an order is the worst case.
In contrast to most VRP implementations, the number of vehicles is limited so that the situation may arise that orders have to be rejected. This mimics reality where only a limited number of trucks is available. Further capacity has to sourced from the so-called spot market at a higher price. This resource is currently not included.

5 Results

All results were produced using the heuristic because the size of data sets solvable to optimum by mathematical solvers is limited and less than the size used in this work. The quality of results is expressed by the following formula:

$$\frac{TotalOrderSize - RejectedOrderSize}{TotalDistance} \tag{10}$$

It rewards the amount of orders served and penalizes long tours and rejected orders. The test set described in 3 produced the average results given here:

Table 1. Results

Approach	Quality	Rejected orders
1 day horizon	0.32	105
2 day horizon	0.34	58
5 day horizon	0.49	0
10 day horizon	0.53	0
Offline (= 100 days)	0.57	0
1 day equal distribution	0.56	0

As expected, the offline heuristic performs best and results are getting better the longer the horizon is. The competitive ratio is 1.78 and there are 10.5% missed orders on average when employing a single day planning. The approach of trying to spread the orders equally over all days performs considerably well. Its competitive ratio is 1.02.

6 Conclusion and Outlook

A heuristic was presented which can solve special pickup and delivery problems, so-called milk runs, with time windows and the multiple use of vehicles. It has been shown that methods that deliver good results for the offline version do not necessarily have to deliver good results in the online case, i.e. if decisions have to be made which affect the future which holds considerable uncertainty in terms of additional incoming orders. A simple method, assigning orders to the least booked day, delivered very good results in a test case. However, the impact of certain parameters such as the size of the time window or the leadtime has to be examined in detail. Also, the question of whether the results remain equally good when workload is increased has to be studied closely. Last, it might be interesting to include the spot market to see when orders should be handled externally. The major challenge however is to include other distributions than the uniform distribution regarding the share of orders each supplier holds, the order sizes, and the time windows and leadtimes. This allows for more sophisticated hedging methods and probability calculations.

References

1. Pape, U. (2006) Agentenbasierte Umsetzung eines SCM-Konzepts zum Liefermanagement in Liefernetzwerken der Serienfertigung. HNI-Verlagsschriftenreihe, Paderborn

2. Petch, RJ, Salhi, S. (2004) A multi-phase constructive heuristic for the vehicle routing problem with multiple trips. Discrete Applied Mathematics 133:69–92
3. Brandão, JCS, Mercer, A. (1998) The multi-trip vehicle routing problem. Journal of the Operational Research Society 49:799–805
4. Azi, N., Gendreau, M., Potvin, J.-Y. (2007) An exact algorithm for a single-vehicle routing problem with time windows and multiple routes. European Journal of Operational Research 178:755–766
5. Azi, N., Gendreau, M., Potvin J.-Y. (2007) An Exact Algorithm for a Vehicle Routing Problem with Time Windows and Multiple Use of Vehicles. Tristan VI, Phuket Island, Thailand
6. Zhao, QH, Wang, SY, Lai, KK, Xia, GP (2002) A vehicle routing problem with multiple use of vehicles. Advanced Modeling and Optimization 4:21–40
7. Taillard, ED, Laporte, G., Gendreau, M. (1996) Vehicle Routing with Multiple Use of Vehicles. Journal of the Operational Research Society 47:1065–1071
8. Salhi, S., Petch, R. J. (2007) A GA Based Heuristic for the Vehicle Routing Problem with Multiple Trips. Journal of Mathematical Modelling and Algorithms 6:591–613
9. Allulli, L., Ausiello, G., Laura, L. (2005) On the Power of Lookahead in On-Line Vehicle Routing Problems. Proceedings of the 11th Annual International Conference COCOON 2005, Kunming, China, Springer-Verlag

MEFISTO: A Pragmatic Metaheuristic Framework for Adaptive Search with a Special Application to Pickup and Delivery Transports

Andreas Cardeneo[1], Werner Heid[2], Frank Radaschewski[2], Robert Scheffermann[1], and Johannes Spallek[3]

[1] FZI Forschungszentrum Informatik, Karlsruhe
{cardeneo,scheffermann}@fzi.de
[2] PTV AG, Karlsruhe
{werner.heid,frank.radaschewski}@ptv.de
[3] johannes@spallek.name

Summary. We present MEFISTO, a pragmatic framework for transport optimization algorithms that has been jointly developed by the authors and is integral part of logistics planning solutions provided by PTV AG. We present design aspects that have led to the architecture of the framework using a variant of Granular Tabu Search. For the case of vehicle routing problems with pickup and delivery transports, we present a specialized local search procedure. Summarized results on benchmark and real world problems are given.

1 Introduction

In this paper, we introduce MEFISTO (Metaheuristic Framework for Information Systems in Transport Optimization), a metaheuristic framework developed jointly by the authors of this paper. Throughout this article, we first present the framework in section 2 and continue with an application to vehicle routing problems with pickup and delivery transports in section 3. We present a summary of computational results for both artificial benchmark problems and real world instances and close with directions for further research.

2 Metaheuristic Framework

The development of the framework was motivated by the observation that, until recently, the main concerns of commercial VRP package

B. Fleischmann et al.(Eds.), *Operations Research Proceedings 2008*,
DOI: 10.1007/978-3-642-00142-0_41,

users were the accuracy of the planning model representing their operations as closely as possible and computation time of planning. This has led to quite complex restrictions like asymmetric distances and travel times, multiple customer time-windows and resource operating intervals, heterogeneous fleets, waiting time limits, multiple capacity dimensions, precedence constraints, customer-vehicle type and customer-driver assignment patterns etc. This type of problems is usually known as rich vehicle routing problems and has received some attention in the past (see [2]). Low computation times used to be a key distinctive feature, eventually leading to fast constructive heuristics possibly with a subsequent post-optimization[4] phase.
More recently, customers started to focus more strongly on solution quality without neglecting the above mentioned aspects. At the same time, the advance of computer hardware had led to a situation where the computation time for planning was well below the user accepted threshold. This development opened the path to more elaborated search algorithms and made it possible to introduce metaheuristic search techniques on a general scale beyond single customer specific projects.
Looking at the metaheuristic landscape it became clear that the rapid development of the field would require more than an implementation of a single metaheuristic. In order to being able to change the underlying metaheuristic without modification of the interfaces to other components of the planning system, a more general framework was needed.
MEFISTO was created having the following objectives in mind:

- Offer an application and algorithm independent framework for a large set of metaheuristic techniques,
- provide a simple interface to users of the framework,
- allow the reuse of existing code through a strong separation of concerns,
- include mechanisms for a dynamic control of search.

The resulting system is a C++ framework with a three-layered architecture: The topmost layer is both application and metaheuristic independent and is basically a template of a general metaheuristic. It offers base classes for elementary metaheuristic elements like the metaheuristic itself, a move generator, moves, local searches, and search spaces. The central layer is built by specific metaheuristic classes derived from abstract base classes at the topmost layer and implements

[4] In using the term *optimization* here we refer to the process of generating eventually improving solutions. We do not claim to achieve provably optimal solutions with the methods presented here.

the metaheuristic functionality. The undermost layer is application specific and implements algorithm elements strongly dependent on the problem type at hand, e.g. for vehicle routing applications it implements classes describing node- or arc-exchanging moves.

2.1 Granular Tabu Search

The first[5] metaheuristic we realized within MEFISTO was the Granular Tabu Search (GTS) method by [6]. Figure 1 shows the three-layered architecture with classes derived for GTS.

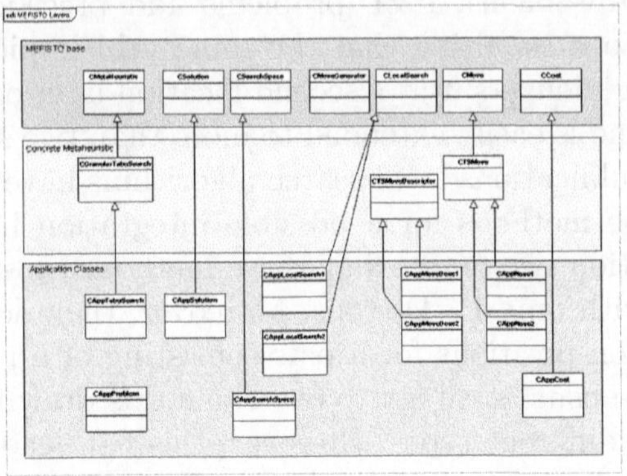

Fig. 1. MEFISTO's three layered-architecture

GTS was chosen because it is an algorithm that is controllable by a small number of parameters and shows good and stable performance in benchmark tests. Moreover, it is a method without any randomization techniques, thus it is able to reproduce results even on different platforms – a behaviour very strongly required in commercial environments. Unlike the original publication, the granularity filter was changed from an arc-length based criterion to the node-based k-closest-neighbours criterion. Furthermore, a set of classical local search algorithms for VRP-type problems (see also [6], namely the crossover-, relocate-, or-exchange and swap-moves, were implemented.
The results obtained when applied to customer problem instances were quite satisfactory, leading to considerably better solution in terms of

[5] Apart from the greedy method which is the default behaviour of the framework and, in a sense, could be seen as a pathological case of a metaheuristic without any mechanism of exiting local optima.

total distance resp. travel time and the number of vehicles used. Encouraged by the progress made using the MEFISTO framework, we decided to address a special case of the VRP, namely the VRP with pickup and delivery-transports. Here, we first had to devise a move type capable of efficiently dealing with the precedence constraints imposed by pickup and delivery restrictions.

3 Pickup & Delivery Transport

Quite a few papers have been published on the pickup and delivery problem (P&D-problem). P&D-problems add precedence constraints between two stops requiring that the same vehicle picks up a load at one location and delivers it at a second location later on the same tour. The P&D-setting is easily extended to more than two stops. In [5] , [4], [1] and other publications local search algorithms have been presented. Analyzing these methods for a possible integration into our MEFISTO-based solution, we found that these methods mostly are designed for problems with only P&D-stops. Moreover, they do not explore all possible insertion positions for a pair consisting of a pickup and a delivery stop. The usual strategy to overcome this drawback is to concat P&D-moves of different types. However, this can lead to an unnecessarily large number of feasibility checks. The complexity of feasibility checks for rich vehicle routing models can generally be considered as the performance bottleneck. It is thus advisable to minimize the number of moves to be evaluated and it is interesting in this context, that the local search algorithms published can generate duplicates, i.e. these methods generate the same stop sequence more than once.

3.1 PD-Relocate

We have developed a simple relocate move for P&D-pairs that avoids the generation of duplicates. Figure 2 shows which arcs are exchanged by the PD-relocate move. In the figure, bold dashed lines indicate newly inserted arcs and light dotted lines represent arcs that are removed. P&D-stop a (with pickup stop a_1 and delivery stop a_2) is to be inserted before stop b (with pickup at b_1 and delivery at b_2). Arcs (π_{b_i}, b_i), (π_{a_i}, a_i) and (a_i, σ_{a_i}) are removed and arcs (π_{b_i}, a_i), (a_i, b_i) and $(\pi_{a_i}, \sigma_{a_i})$ are reinserted.

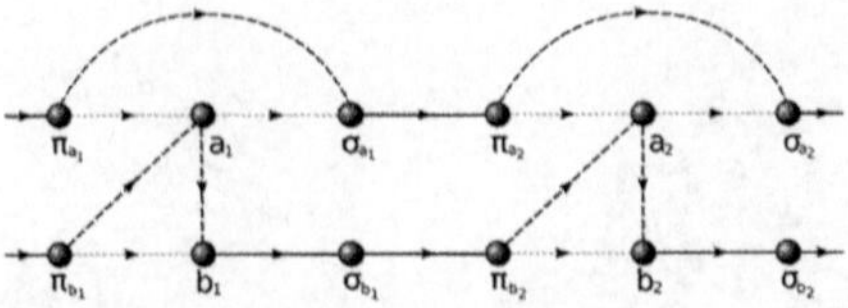

Fig. 2. Principle of the PD-Relocate move

4 Computational Results

We have chosen the benchmark instances from [3] to compare our approach to the existing ones from the literature. In a second step, we have solved a number of practical instances with MEFISTO augmented by the PD-Relocate move. Due to spatial restrictions, we do not report the full set of results here, rather we summarize our findings and draw conclusions from the results.

Our experiments have compared the solution quality of the currently available constructive heuristic with the results obtainable from the metaheuristic approach and the literature benchmark. In general, we were able to get close to the literature results for easy instances and fall short for less clustered instances. Analyzing the results we feel that this is mainly due to deficiencies of the constructive heuristic that produces solutions quite far away from the benchmark. As we currently do not employ a local search method especially designed to reduce the number of tours, the improvement phase is not able to compensate for the bad initial decisions. A second reason is the type of granularity check used that better fits standard VRPs without pairs of stops. Using k-closest neighbours, pickup and delivery stop pairs with stops distant from each other, frequently do not pass the granularity filter.

One of the real world instances we have tested the approach with consists of 26 P&D-orders and 224 deliveries. The stops are distributed in two semi-circles around two major seaports and there is a fleet of 43 vehicles available. A subset of stops has time-windows. Figure 3 shows this example.

The solution of this instance takes roughly the double amount of moves compared to the system without the PD-Relocate move. The total number of vehicles could be reduced from 26 (result after constructive heuristic) to 24 and total distance was reduced from 9364km to 8691km.

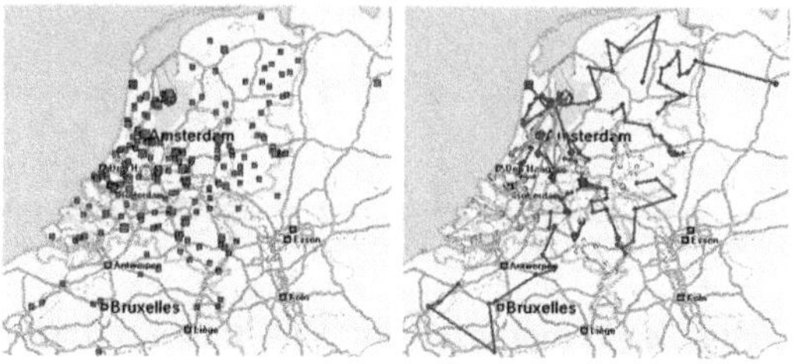

Fig. 3. Real world problem instance

5 Concluding Remarks

The development and introduction of MEFISTO was a successful step that has improved the planning results and has enabled us to quite easily introduce a new dedicated local search procedure into the system. Applying the system to benchmark instances and practical problems has shown the necessity to both improve the construction heuristic and to introduce a local search procedure dedicated to the reduction of the number of tours resp. vehicles.

References

1. O. Bräysy and M. Gendreau. Vehicle routing problem with time windows, part i: Route construction and local search algorithms. Transportation Science, 39(1):104-118, 2005.
2. R. F. Hartl, G. Hasle, and G. K. Janssens (editors). Rich vehicle routing problems. In Special Issue, Central European Journal of Operations Research, Vol. 14(2), 2006.
3. H. Li and A. Lim. A metaheuristic for the pickup and delivery problem with time windows. In 13th IEEE International Conference on Tools with Artificial Intelligence (ICTAI'01), pages 160-169, Los Alamitos, CA, USA, 2001. IEEE Computer Society.
4. W. P. Nanry and J. W. Barnes. Solving the pickup and delivery problem with time windows using reactive tabu search. Transportation Research Part B Methodological, 34(2):107-122, 2000.
5. P. Toth and D. Vigo. Heuristic algorithms for the handicapped persons transportation problem. Transportation Science, 31(1):60-71, Februar 1997.
6. P. Toth and D. Vigo. The granular tabu search and its application to the vehicle-routing problem. INFORMS Journal on Computing, 15(4):333-346, 2003.

Planning in Express Carrier Networks: A Simulation Study

Sascha Dahl and Ulrich Derigs

WINFORS - Seminar für Wirtschaftsinformatik und Operations Research, Universität zu Köln, Albertus-Magnus-Platz, 50923 Köln
{sascha.dahl,ulrich.derigs}@uni-koeln.de

Summary. In this paper we present first results of a simulation study analyzing the effect of alternative compensation schemata in a courier network compared to alternative organizational settings: centralized collaborative planning and individual non-collaborative planning.

1 Motivation

During the last years, transportation firms have faced an increasing cost pressure and revenue erosion at the same time. Large express carriers serving ad-hoc one-way shipping orders can solve the problem of dead-head trips by consolidating and combining orders to efficient roundtrips. Small-sized companies may compensate their competitive disadvantage by allying with partners to a cooperation network to establish a more profitable portfolio of orders. Each partner in the network plans his orders and his vehicle fleet independently with the option to exchange orders with partners. While end customers are charged based on a specific price function, a carrier operating an order for a network partner receives a monetary compensation specified by a generally agreed upon compensation schema.

Our experience with a distributed real-time internet-based collaborative Decision Support System (DSS) for a large courier network (see [1]) has shown, that the success is highly depending on the potential to detect and realize a sufficient number of orders which can be fulfilled efficiently by one of the partners. Figure 1 shows the architecture of our DSS and points out the planning paradigm: decentralized planning at each partner site with a central supporting system that searches and proposes options of profitable exchanges in real-time based on data from the individual dispatching systems and fleet telematics.

DOI: 10.1007/978-3-642-00142-0_42,

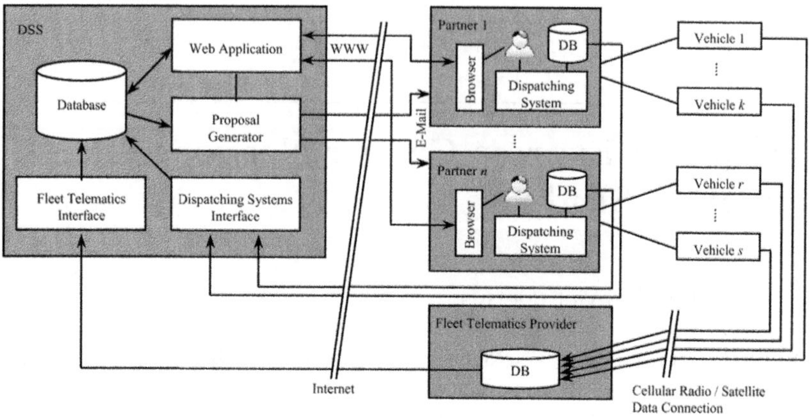

Fig. 1. DSS architecture

The compensation schema determines the profitability of a specific order exchange for both partners involved. Hence, the design of the compensation schema with respect to incentiveness and fairness is a key factor for the overall performance of the network. An analysis of the effect of alternative compensation schemata and a comparison with alternative planning approaches is the foundation for an optimization of the potentials of the collaboration.

2 Planning Situation

Let P be the set of partners in the network. Each partner $p \in P$ is located at a specific depot dep_p and owns a set of vehicles V_p. Each vehicle v is assigned to a vehicle class vc_v which describes physical and technical transportation capabilities. The set of vehicle classes is assumed to be partially ordered with the semantic that a vehicle of a 'larger' class can always transport orders which require a 'smaller' vehicle class. For each vehicle class vc we use two cost rates: price_{vc} the transportation price for the customer and $\text{impCost}_{\text{vc}}$ an internal calculatorial imputed cost rate representing the operational costs. Each order o is defined by its pickup location p_o, its delivery location d_o, time windows for pickup and delivery and its capacity requirement, where vc_o denotes the minimum required vehicle class for the order. Note, that each vehicle can transport more than one order at a time and that it is the task of the partners' local dispatching to combine orders to tours. Under this static view, a vehicle will always return to its depot after serving all orders of a tour, yet, due to the highly dynamic business all plans have to be adapted constantly and a vehicle may receive new orders on tour. For two locations x and y let $l(x, y)$ denote their distance. Then the revenue $\text{rev}(p, o)$, which a partner p obtains from his customer after serving its order o, is calculated as follows, where for picking up the order only the distance exceeding l^{fix} is charged:

$$\text{rev}(p, o) = \text{price}_{\text{vc}_o} \left[\max\left(0; l(\text{dep}_p, p_o) - l^{\text{fix}}\right) + l(p_o, d_o) \right] \quad (1)$$

Now, the carrier p has two options: to serve the order o with his own fleet or to have it served by one of the partners. In the first case, the contribution to profit $\text{ctp}(p, o)$ is calculated as usual, reducing the revenue by the marginal imputed costs. The following formula calculates this contribution in the simple case that the order is served by a dedicated tour on vehicle v:

$$\text{ctp}(p, o) = \text{rev}(p, o) - \text{impCost}_{\text{vc}_v} \left[l(\text{dep}_p, p_o) + l(p_o, d_o) + l(d_o, \text{dep}_p) \right] \quad (2)$$

In the second case, if one of the partners, say q, serves the order, the compensation $\text{comp}(q, o)$ is contractually determined. Here we have to distinguish several cases: If the order is served by a dedicated tour, the calculation is similar to the revenue above, with the only difference that twice the imputed cost rate is applied:

$$\text{comp}(q, o) = 2 \cdot \text{impCost}_{\text{vc}_o} \left[\max\left(0; l(\text{dep}_q, p_o) - l^{\text{fix}}\right) + l(p_o, d_o) \right] \quad (3)$$

If the order can be inserted into a tour already served, and thus can be combined efficiently with other orders, then the load distance $l(p_o, d_o)$ plus the marginal cost caused by the service of o applies. This more complicated calculation is formalized and illustrated in figure 2 for the case that o is combined with two other orders r and s, respectively (bold arrows indicate legs actually driven, while dashed arrows symbolize empty trips):

$$\begin{aligned} \text{comp}(q, o) = \text{impCost}_{\text{vc}_o} & \left[l(p_o, d_o) + l(p_r, p_o) + l(p_o, d_r) - l(p_r, d_r) \right. \\ & \left. + l(p_s, d_o) + l(d_o, d_s) - l(p_s, d_s) \right] \quad (4) \end{aligned}$$

Note, that in the two latter cases the cost rate of the required vehicle class is applied, not the rate of the vehicle actually serving the order.

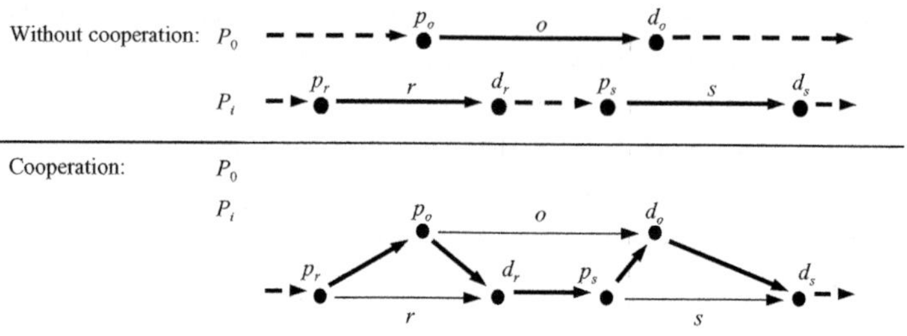

Fig. 2. Compensation in case of combined order execution

Based on this compensation model, the DSS checks all partners' vehicles for the best slot for an acceptable insertion. Here, an insertion is acceptable if $\text{comp}(q, o) > \Delta\text{impCost}(q, o)$, i.e. the partner takes an advantage from serving o. Now, a recommendation is communicated to p and that partner q, offering the acceptable insertion with lowest marginal cost, if $\Delta\text{impCost}(p, o) > \text{comp}(q, o)$.

3 Design of the Simulation Study

3.1 Simulation Data

For the simulation study we have used real data from a European-wide operating cooperative logistic network consisting of 49 carriers collected over seven randomly chosen weeks in 2008. The whole network has a capacity of 8905 vehicles assigned to five vehicle classes. This high number of vehicles reflects the fact that in this courier network each carrier has access to a virtually unlimited number of subcontractors. This carrier and vehicle data has been considered as static, while the order data is highly dynamic. To gather real order data we have timestamped and logged every change of order data for all carriers. All data changes reflecting order exchanges between carriers have been marked by use of an order-fingerprint and have been eliminated from the simulation input, because such decisions are subject to the planning approach in the simulation.

3.2 Planning Scenarios and Methods

We have analyzed the impact of alternative organizational settings: the collaborative planning approach supported by our DSS, a non-collaborative scenario where each carrier plans his own orders only and a centralized approach where vehicles and orders of all carriers are planned simultaneously as if owned by only one single carrier. The non-collaborative scenario requires to solve a set of Dynamic Pickup and Delivery Problems with Time Windows (DPDPTW), while the centralized approach requires to solve a single Dynamic Pickup and Delivery Problem with Time Windows and Multiple Depots (DPDPTWMD) (see [3]). We have solved the non-collaborative scenario problems modifying a heuristic PDPTW-solver called ROUTER, which had been developed at our institute (see [2]). ROUTER first constructs good feasible solutions using a cost-based cheapest insertion strategy and then applies a local search based metaheuristic for improvement in a second phase. In order to realistically mimic the carriers' dispatching in the extremely dynamic environment we have omitted the time consuming improvement phase. To cope with the dynamic problems we use a single event optimization paradigm with the static solver, i.e at each time a data change occurs a static problem is solved with all orders fixed which are already in execution. We found that the real-world DPDPTWMD resulting from centralized planning approach are solved best with a very simple cheapest insertion heuristic considering all orders already planned to be fixed and continuously evaluating the cheapest insertion position for each incoming or updated order.

3.3 Compensation Schemata

We have analyzed the impact of two different compensation concepts: the one presently used in our real-world DSS and another simple marginal cost-based scheme:

$$\mathrm{comp}_{\mathrm{alt}}(q,o) = \frac{1}{2}\left[\Delta\mathrm{impCost}(p,o) + \Delta\mathrm{impCost}(q,o)\right] \qquad (5)$$

i.e. we calculate the marginal imputed costs obtained from cheapest insertion of o into the transportation plans of p and q respectively. $\mathrm{comp}_{\mathrm{alt}}$ is incentive compatible in the sense that whenever there is the potential for decreasing the total network-wide costs through an exchange then there is a positive gain from it for each of the involved carriers.

4 Simulation Results

Table 1 lists the network-wide total imputed costs as percentage of those costs arising in the isolated planning scenario. Here it is interesting to note that applying the marginal cost-based compensation schema $comp_{alt}$ yields the same plans and costs as simulating centralized planning, which under an optimization aspect would always yield the smallest operational costs. The results indicate that cooperative planning pays off in general and that the choice of the compensation schema has an enormous impact on the magnitude of cost savings with 5.2% at average for comp and 18.3% at average with $comp_{alt}$. All values have only small variances over the calendar weeks and their relative order is stable.

Table 1. Simulation results

	CW2	CW4	CW5	CW6	CW9	CW10	CW16	Average
#Orders	1248	1421	1292	1355	1636	1641	1499	1442
Isolated planning [%]	100.0	100.0	100.0	100.0	100.0	100.0	100.0	100.0
Simulation comp [%]	94.1	93.7	96.0	95.6	95.9	94.3	93.7	94.8
Simulation $comp_{alt}$/ Centralized planning [%]	81.9	79.9	83.0	82.1	81.8	80.4	82.5	81.7

5 Conclusion and Further Research

In this paper we have presented a simulation-based analysis for studying the advantage of collaboration over independent planning. We have shown that the selection of the compensation scheme has an enormous impact on the cost savings. Future work will extend the analysis of the different compensation schemata and focus on the impact on fairness, which is an essential property to ensure sustainability of the network.

References

1. Dahl S, Derigs U (2007) Ein Decision Support System zur kooperativen Tourenplanung in Verbünden unabhängiger Transportdienstleister. In: Koschke R, Otthein H, Rödiger KH, Ronthaler M (eds) Informatik 2007: Information trifft Logistik. Bd. 1. Gesellschaft für Informatik, Bonn
2. Derigs U, Doehmer T (2008) Indirect Search for the Vehicle Routing Problem With Pickup and Delivery and Time Windows. OR Spectrum 30:149–165
3. Ropke S, Pisinger D (2007) A General Heuristic for Vehicle Routing Problems. Computers and Operations Research 34:2403–2435

Stability of Airline Schedules

Viktor Dück, Natalia Kliewer, and Leena Suhl

Decision Support & Operations Research Lab, University of Paderborn,
Warburger Str. 100, 33098 Paderborn
{vdueck,kliewer,suhl}@upb.de

Summary. Stability describes the ability of a plan to remain feasible and to give acceptable results under different variations of the operating environment without any modifications of the plan. In this paper we explore the effect of crews changing aircrafts on the stability of airline schedules using a simulation model for delay propagation, considering the interdependence between aircraft rotations and crew pairings.

1 Introduction

Airline schedules are created several months in advance to the day of operations. A detailed *flight schedule*, i.e. given departure and arrival times for all flights, provides the basis for the operational scheduling tasks. The first step is typically the assignment of an aircraft fleet type to each flight, considering the demand forecasts, the capacity and the availability of the aircrafts. After the *fleet assignment* a specific aircraft is assigned to each flight subject to maintenance constraints (*aircraft routing*). *Crew scheduling* is typically decomposed into two processes. In the crew pairing phase anonymous crew itineraries are generated regarding general regulations, such as maximal allowed working or flying time per duty. In the second phase, individual crew members are assigned to these itineraries. The main goal of this operational scheduling is cost optimality. For a detailed description of the airline scheduling process and the objective functions see [2].
On the day of operations disruptions lead to infeasible aircraft and crew schedules, due to absence of resources or violation of crew rules. The process of finding new schedules is called recovery or disruption management. [1] provide a comprehensive review of concepts and models used for recovery of airline resources.

B. Fleischmann et al.(Eds.), *Operations Research Proceedings 2008*,
DOI: 10.1007/978-3-642-00142-0_43, © Springer-Verlag Berlin Heidelberg 2009

Besides disruption management on the day of operations the possibility of disruptions can be also considered during the schedule construction. This is called robust scheduling. The goal of robust scheduling is to construct schedules, which remain feasible and near cost optimal under different variations of the operating environment. Lot of literature about robust scheduling exist in the airline field, see for example [6], [3] and [5].

[4] distinguish two aspects of robustness: *stability* and *flexibility*. Stable schedules are likely to remain feasible and near cost-optimal under changing operating environment, there against flexible schedules can be adapted to changing operating environment efficiently. Especially in the airline field *robust* schedules have to be stable and flexible at the same time. Recovery options are in many cases very complex and challenging for the operations control of the airlines, but at the same time often necessary to operate efficiently.

The natural way to measure the robustness of a schedule is to measure the additional operational costs caused by disruptions on the day of operations. Such costs can be specified for reserve crews and crew overtime as well as for loss of revenue due to canceled flights and disrupted passengers. To get an idea of real costs the consequences of delays should also be tracked at other operational fields, like gate and runway schedules at airports.

To predict the operational costs of schedules in advance an integrated model of all airline operations and information about gate and runway schedules at airports and schedules of other airlines would be necessary. But this is hardly possible due to lack of information. Instead, the adherence to the original schedule can be measured as an indicator for the recovery effort. One possibility to measure the punctuality performance of a schedule is the on-time performance. This describes the percentage of flights arriving or departing at the gate on-time. An arrival or departure is on-time if it is within x minutes of the originally scheduled time. x is usually 0, 5, 15 and 60 minutes.

In this paper we measure the stability of schedules in a schedule simulator. We distinguish between primary and propagated delays. Primary delays arise due to external influences, i.e. technical errors or missing passengers. Propagated delays result from delayed aircrafts or crews during a duty. We do not propagate delays due to violations of pairing rules, such as maximal flying time or minimal rest time. In such cases propagation is not allways realistic and would lead to long propagated delays and thus falsify our results. Instead, we count the number of violations of pairing rules. The consideration of sophisticated recovery actions for crew and aircrafts in the simulation model is one important goal for future development. Such aircraft and crew recovery make possible to guarantee the legality of aircraft maintenance and pairing rules restrictions and is therefore necessary to measure the flexibility of schedules.

2 Generating Stable Crew Pairings

One important aspect of generating efficient crew schedules is to plan crews to change aircrafts during a duty. On the other side, such aircraft changes are considered to be less robust. An arrival delay of an incoming flight before an aircraft change can lead to propagated delays of two different outgoing flights, the next flight of the crew and the next flight of the aircraft. To explore the effect of aircraft changes on the stability of the schedules we construct crew pairing schedules with different penalties for aircraft changes.
For crew pairing optimization we use a method based on column generation. The master problem is a set-partitioning model and the pricing problem is a resource constrained shortest path model. Aircraft changes can be easily detected and penalized during the pairing generation, without changing the structure of the problem. The easiest way to get less aircraft changes is to penalize each aircraft change with a fixed penalty factor c_γ. We call this strategy ac^{fix}. Another possibility is to penalize aircraft changes according to the buffer time. Therefore for each aircraft change the propagation factor $\pi(f_1, f_2)$ is computed, see eq. (1). An aircraft change between the flights f_1 and f_2 is then penalized with the product $c_\gamma \cdot \pi(f_1, f_2)$. This strategy is called ac^{prop}.

$$\pi(f_1, f_2) = 1 - \left(\frac{\text{connection time} - \text{minimal connection time}}{\text{minimal connection time}} \right)^2 \quad (1)$$

Table 1 shows the average costs of ten different schedules for different settings for crew pairing optimization. Each schedule has a period of three days and contains between 340 and 470 flights. The first setting, called "‘cost optimal"’ reduces the number of production hours and does not penalize aircraft changes. The amount of production hours and aircraft changes of the cost optimal solutions represents the 100% level. The results for other settings are described as relative increase of costs and relative decrease of aircraft changes compared to the cost optimal solution. The penalty costs for aircraft changes are defined relative to the cost of one production hour (c_h).

Table 1. Solution properties of crew pairing schedules

	Production hours	Aircraft changes
Cost optimal	100% (1572)	100% (56.5)
ac^{fix}, $c_\gamma = 5 \cdot c_h$	+1.5% (1596)	-44% (31.5)
ac^{fix}, $c_\gamma = 10 \cdot c_h$	+1.8% (1601)	-44% (31.6)
ac^{prop}, $c_\gamma = 5 \cdot c_h$	+0.3% (1576)	-14% (48.4)
ac^{prop}, $c_\gamma = 10 \cdot c_h$	+0.8% (1584)	-18% (46.2)
ac^{prop}, $c_\gamma = 15 \cdot c_h$	+1.7% (1599)	-17% (46.7)

With the strategy ac^{fix} the amount of aircraft changes is significantly lower, although the increase in costs is relatively small. The reduction of aircraft changes with the strategy ac^{prop} is noticeable low, although the increase in cost for the higher penalty factors is comparable with the strategy ac^{fix}. Therefore in the next section we discuss among others the question how the amount of aircraft changes coheres with the on-time performance.

3 Evaluation of Robustness

Each crew schedule described in section 2 is simulated using the following model of airline operations, desribed by equations (2) - (3). For a set F of flights we distinguish between scheduled and actual times of departure and arrival. The scheduled departure time is given for each flight $f \in F$ by τ_f^{SD} and the actual departure and arrival times are given by τ_f^{AD} and τ_f^{AA}. The actual time of arrival τ_f^{AA} mainly depends on the actual time of departure τ_f^{AD} and the block time τ_f^{B} as shown in equation (2).

$$\tau_f^{AA} = \max\{\tau_f^{SA}, \tau_f^{AD} + \tau_f^{B}\} \tag{2}$$

Before a flight can depart the turn process of the aircraft as well as the ground task of the crew have to be finished. Equation (3) shows this dependency. $\tau_{((r(f),f)}^{AG}$ and $\tau_{((p(f),f)}^{CG}$ represent the ground times for aircrafts and crews. The predecessor flights are given by $r(f)$ for aircraft rotations and by $p(f)$ for crew pairings.

$$\tau_f^{AD} = \max\{\tau_f^{SD}, \max\{\tau_{r(f)}^{AA} + \tau_{(r(f),f)}^{AG}, \tau_{p(f)}^{AA} + \tau_{(p(f),f)}^{CG}\}\} \tag{3}$$

In the experiments in this paper we consider only departure delays of flights due to disruptions of aircraft ground tasks. Given the actual and

Table 2. Operational performance of airline schedules

	Flights with propagated delays of			Violations of
	1 - 5 min	6 - 15 min	$\geq$ 16 min	pairing rules
No crew	7.9%	7.1%	5.9%	–
Cost optimal	10.1%	9.2%	7.6%	7.7
ac^{fix}, $c_\gamma = 5 \cdot c_h$	9.1%	8.3%	6.8%	7.8
ac^{fix}, $c_\gamma = 10 \cdot c_h$	9.1%	8.3%	6.9%	7.1
ac^{prop}, $c_\gamma = 5 \cdot c_h$	9.3%	8.5%	7.1%	7.8
ac^{prop}, $c_\gamma = 10 \cdot c_h$	9.1%	8.3%	6.9%	7.6
ac^{prop}, $c_\gamma = 15 \cdot c_h$	8.9%	8.2%	6.8%	7.7

scheduled times the computation of the departure delay x_f^D for flight f is shown in equation (4).

$$x_f^D = \tau_f^{AD} - \tau_f^{SD} \tag{4}$$

We assume that some flights are more liable to be affected by a delay than other. Therefore, for each flight $f \in F$ an individual disruption value $\delta_f = (a \in \{0 \dots 1\}, b > 0)$ describes the combination of the probability that a delay occurs and an expected length of the delay. The length of the delays is exponentially distributed. The experiments in this paper were performed with following flight delay settings: 40% of flights with $\delta_f = (30\%, 30\text{min})$, 10% of flights with $\delta_f = (20\%, 120\text{min})$ and 50% without any delays.

Table 2 shows the amount of flights with propagated delays greater than 0, 5 or 15 minutes and the amount of violations of pairing rules. The first row describes the average on-time performance of the aircraft rotation schedules without crew pairings. The following rows show the results for crew pairing schedules simulated together with the aircraft rotation schedules. Thus, the on-time performance of the aircraft rotation schedule represents a lower bound for any crew pairing schedule simulated together with the aircraft rotation schedule.

There are several noticeable results. First, all described strategies to penalize aircraft changes reduce the amount of flights with propagated delays, but do not reduce the amount of violations of pairing rules, significantly. The strategy ac^{prop} with the penalty factor $c_\gamma = 10 \cdot c_h$ leads to a comparable on-time performance as the strategy ac^{fix}, but at less production hours. The on-time performance of the strategy ac^{prop} does not correlate with the amount of aircraft changes, but with the height of the penalty factor. The increase of the penalty factor for the strategy ac^{fix} does not show any effect on the punctuality.

4 Conclusion and Outlook

The results show, that penalizing aircraft changes generally leads to less propagated delays during a daily duty, but penalizing aircraft changes with little or no buffer time is more efficient. Nevertheless this does not help to reduce the number of violations of pairing rules. Therefore the next important step is to extend the pairing optimization model to reduce the number of violations of pairing rules.
Another important task is the consideration of sophisticated aircraft and crew recovery actions in the simulation model to allow measuring the flexibility of schedules.
The results also show that most propagated delays result from the aircraft rotation schedule. Therefore, it is important to research how the integration of aircraft routing and crew scheduling can improve the overall robustness of the schedules.

References

1. Clausen, J., A. Larsen, and J. Larsen (2005). Disruption management in the airline industry - concepts, models and methods.
2. Klabjan, D. (2005). *Large-Scale Models in the Airline Industry*, pp. 163–196. Kluwer Scientific Publishers.
3. Lan, S., J.-P. Clarke, and C. Barnhart (2006, February). Planning for robust airline operations: Optimizing aircraft routings and flight departure times to minimize passenger disruptions. *Transportation Science 40*(1), 15–28.
4. Scholl, A. (2001). *Robuste Planung und Optimierung: Grundlagen - Konzepte und Methoden - Experimentelle Untersuchungen*, Chapter Robuste Planung, pp. 89 –172. Physica-Verlag Heidelberg.
5. Shebalov, S. and D. Klabjan (2006, August). Robust airline crew pairing: Move-up crews. *Transportation Science 40*(3), 300–312.
6. Weide, O., D. Ryan, and M. Ehrgott (2007, January). Iterative airline scheduling. Technical report, Department of Engineering Science, The University of Auckland, Private Bag 92019, Auckland, New Zealand.

Waiting Strategies for Regular and Emergency Patient Transportation

Guenter Kiechle[1], Karl F. Doerner[2], Michel Gendreau[3], and Richard F. Hartl[2]

[1] Vienna Technical University, Karlsplatz 13, 1040 Vienna, Austria
guenter.kiechle@salzburgresearch.at

[2] University of Vienna, Department of Business Administration, Bruenner Strasse 72, 1210 Vienna, Austria
{karl.doerner,richard.hartl}@univie.ac.at

[3] Centre interuniversitaire de recherche sur les réseaux d'entreprise, la logistique et le transport (CIRRELT), C.P. 6128, succursale Centre-ville, Montréal, Canada H3C 3J7
michelg@crt.umontreal.ca

1 Introduction

Many emergency service providers, especially ambulance departments and companies who provide non-public maintenance services, face the problem that their fleet of vehicles has to provide two different types of services:

1. Cover a certain region and provide immediate service when an emergency occurs;
2. Provide some regular service (e.g., the pick-up and delivery of patients, predetermined service tasks, periodic pick-ups ...).

This is the current situation for the largest Austrian emergency service providers (e.g., the Austrian Red Cross), where the same fleet is used to provide both types of services. Dynamic aspects thus directly influence the schedule for the regular service. When an emergency occurs and an ambulance is required, the vehicle with the shortest distance to the emergency is assigned to serve the emergency patient. Therefore, it may happen that an ambulance vehicle that has to carry out a scheduled transport order of a patient, which has not started yet, is used to serve the emergency request and the schedule for the regular services has to be re-optimized and another vehicle has to be reassigned

B. Fleischmann et al.(Eds.), *Operations Research Proceedings 2008*,
DOI: 10.1007/978-3-642-00142-0_44, © Springer-Verlag Berlin Heidelberg 2009

to the regular patient. Ambulances that carry out emergency transport become available at the hospital after the emergency service and can be then used to carry out regular transport orders. Again, the schedule for regular services has to be re-optimized.

From the perspective of managing the regular services, the objective is minimizing the total traveling distance subject to certain restrictions (e.g., be on time). From the perspective of minimizing the response time for servicing an emergency request, it is necessary to locate and schedule the vehicles in such a way that each possible location where an emergency case may occur can be reached within a given time (see [4]). These two objectives are not totally contradictory, but they certainly conflict: on the one hand, for the emergency transport requests, one has to account for the fact that vehicles are changing positions and are blocked for some time due to some regular service assignment; on the other hand, when regular orders are assigned to vehicles, it is important to keep a certain coverage level to ensure a satisfactory service of the emergency cases, which may require relocating some of the vehicles.

Emergency service providers thus want to find a robust solution for a specific day of the week in order to minimize routing costs, as well as the response time for servicing emergency patients, taking into account that the two types of the transportation services have to be realized with a single fleet.

Some related work has been published where pickup and delivery requests occur dynamically (see [1, 3, 6]). Note that in our case only emergency orders occur dynamically. The nearest empty vehicle is used to serve the emergency immediately. Since subsequent scheduled regular transport requests on the redirected vehicle are not covered any more, a re-optimization step is initiated.

In order to study different dispatch strategies for the real world patient transportation problem, we need a simple, fast and effective solution procedure. Hence, we implemented a constructive heuristic approach. In the construction phase, we exploit the temporal structure of the requests and use a nearest neighbor measure inspired by [5]. The main challenge is to deal with the dynamic nature of the problem, which implies that new solutions can be calculated in very short time. Every time a new emergency request occurs, the distance information for the empty vehicles has to be updated in order to identify very quickly the next empty vehicle to the emergency request. Then, one has to resolve the remaining problem with one less vehicle available to serve the regular orders. When an emergency order is fulfilled and the patient has been unloaded at the hospital, an additional vehicle is made available

and it can be used to fulfill regular patient transportation orders. When this situation occurs, the schedule for the regular transport orders has also to be re-optimized to take advantage of this additional vehicle to improve the objective function.
To evaluate the solution quality various scenarios have to be calculated. In the evaluation of the different scenarios, vehicle movements and the spatial distribution of empty vehicles are monitored over time. The distance matrix is also updated whenever an emergency request occurs.

2 Problem Description

In our study we consider a combination of two problems

- the Dial-a-Ride Problem (DARP) for regular patient transportation and
- dispatching ambulance vehicles for emergency cases.

In the classical DARP, a set of requests announced beforehand are served from a single depot. These requests have hard time windows and a preferred pickup or delivery time. In the problem of dispatching ambulance vehicles for emergency cases, one must ensure short response times in a dynamic environment. The regular patient transportation problem can be considered to be a variation of DARP with additional real world constraints regarding customer preferences or requirements. A comprehensive description of the DARP is given in [2].
Each transport order or request incorporates a pickup location and a delivery location. For each transportation request a time window and a service or loading time for each pickup and delivery location is given. Concerning time windows, we have two different situations - on the one hand, patients should be picked-up as late as possible from their home when they are being transported to hospitals (outbound request); on the other hand, patients should be picked-up as early as possible when they are transported from the hospital back home (inbound request). Time window violations are not allowed and deviations from the desired pick-up and drop-off times within the specified time window are considered in the objective function.
Time windows for each request are defined either by a desired delivery time for outbound requests or a desired pickup time for inbound requests. The resulting time window is calculated by allowing an early pickup or an early delivery by 30 minutes. The pickup or delivery has to be performed within this time window. We allow also a maximum ride time for each passenger. This maximum ride time is defined for

each customer by calculating the shortest path from the pickup to the delivery location and allowing an additional ride time of 30 minutes. In our computational tests the vehicle capacity is two, therefore the maximum number of patients that can be transported at the same time is two. Waiting is not allowed with a passenger on board.
In our real world problem for the regular patients it is desired to minimize transportation costs and to maximize quality of service for patients. And for the emergency patients it is desired to minimize response time for emergency requests.
In the objective function for the constructive procedure the regular patients (transportation costs, quality of service) are considered whereas the response time minimization is considered in the different waiting strategies. Quality of service for the regular transport orders is measured by the deviation of the preferred delivery time for the outbound request and the deviation of the preferred pickup time for the inbound request respectively.

3 Dynamic Aspects

After computing a solution for the problem at hand using one of the proposed solution procedures, improving the coverage criterion is possible without changing the solution structure. More precisely, movements of vehicles may be delayed or pushed forward in time while the sequence of transport requests on each tour remains unchanged.
The distribution is measured at discrete points in time for all empty vehicles, e.g., at equal intervals from the first pick-up to the last drop-off during the day or optimization period. Besides the empty vehicle positions, we use a set of equidistant raster points in the respective area. The covered raster points by the empty vehicles within a certain radius are calculated and used as a proxy for the response time.
Therefore, we distinguish four different waiting strategies to influence movements of empty vehicles on their way from a delivery location to the next pickup location. The first two strategies are static ones called 'early arrival' and 'late arrival'. Early arrival means, that a vehicle departs to its next pickup location immediately after delivering the last patient and waits at the pickup location until the next pickup starts. On the contrary, late arrival means, that a vehicle waits at the last delivery location just as long as possible to arrive at the next pickup location in time.
The last two strategies are dynamic variants of the former ones, which use late arrival or early arrival as default, but may change the strategy

for each single request if an individual coverage calculation indicates an improvement. For each single request both coverage values (late and early arrival) are calculated. The strategy with the better global coverage is performed. Note that potential movements of other vehicles while the respective vehicle drives from delivery to pickup location are not considered in this proxy evaluation.

4 Evaluation of the Approach

In order to test the solution procedures with a comprehensive set of parameters and emergency instances, a simulation environment for the problem at hand was implemented. The efficiency of the solution procedures was evaluated with real world data from the city of Graz provided by the Austrian Red Cross. We used 15 problem instances with a number of regular transport requests ranging from 152 to 286 and grouped them into three classes of five instances each. The small instances encounter from 152 to 192, the medium instances from 200 to 220 and the large instances from 260 to 286 transport requests. All requests occur after 6:00 a.m. and before 5:00 p.m. and for each outbound request a corresponding inbound request occurs a few hours later.
In the computational studies, we evaluated the impact of different waiting strategies as described above. Hence, we evaluated response times to artifical emergency transport requests for a large number of runs. For each instance and each waiting strategy we measured response times for three randomly distributed emergency requests over 2000 independent runs. In Table 1 we report average response time, maximum response time and coverage for each instance class as well as for each waiting strategy. The improvements of the dynamic strategies compared to the static counterpart are reported in the last two rows of the tables (imp.). The results show that the dynamic strategies outperform the static ones, whereas early arrival generally performs better than late arrival. The average response time using early arrival strategy can be reduced from 5.04 to 5.01 minutes. Also the maximum response time in the case of early arrivals can be reduced from 7.87 minutes to 7.83 minutes. The improved coverage enables these improvements. By using a dynamic waiting strategy the coverage can be improved e.g. for the early arrival strategy by 1.65 %. Improvements from the static to the dynamic strategies are remarkably higher for late arrival.

Acknowledgement. Financial support from the Austrian Science Fund (FWF) under grant #L286-N04 is gratefully acknowledged.

Table 1. Response times and coverage subject to waiting strategies

	average response time				maximum response time			
instance class	small	medium	large	average	small	medium	large	average
static late arrival	5.38	5.19	5.12	5.23	8.32	8.09	7.93	8.11
static early arrival	5.20	5.00	4.94	5.04	8.07	7.85	7.69	7.87
dynamic late arrival	5.17	5.00	4.97	5.05	8.03	7.83	7.72	7.86
dynamic early arrival	5.14	4.97	4.93	5.01	8.01	7.81	7.67	7.83
imp. late arrival	4.06%	3.65%	2.97%	3.57%	3.49%	3.39%	2.74%	3.21%
imp. early arrival	1.18%	0.54%	0.21%	0.65%	0.78%	0.50%	0.20%	0.50%

	coverage			
instance class	small	medium	large	average
static late arrival	30.0%	32.4%	31.0%	31.2%
static early arrival	30.6%	33.2%	32.2%	32.0%
dynamic late arrival	32.0%	34.4%	33.1%	33.2%
dynamic early arrival	32.2%	34.6%	33.3%	33.4%
imp. late arrival	6.16%	5.81%	6.43%	6.13%
imp. early arrival	4.90%	3.93%	3.56%	4.12%

References

1. Attanasio, A., Cordeau, J.-F., Ghiani, G., and Laporte, G. (2004) Parallel Tabu search heuristics for the dynamic multi-vehicle dial-a-ride problem. Parallel Computing 30:377–387.
2. Cordeau, J.-F., and Laporte, G. (2003) "The Dial-a-Ride Problem (DARP): Variants modelling issues and algorithms". *4OR* **1**, pp. 89–101.
3. Gendreau, M., Guertin, F., Potvin, J.-Y., and Séguin, R. (2002): "Neighborhood Search heuristic for a Dynamic Vehicle Dispatching Problem with Pickups and Deliveries". Technical Report, Centre de recherche sur les transports, Université de Montréal. Forthcoming in *Transportation Research C.*
4. Gendreau, M., Laporte, G., and Semet, F. (2001): "A dynamic model and parallel tabu search heuristic for real-time ambulance relocation". *Parallel Computing* **27**, 1641–1653.
5. Jaw J.-J., Odoni, A., Psaraftis, H., Wilson, N. (1986): "A heuristic algorithm for the multi-vehicle advance request dial-a-ride problem with time windows". *Transportation Research B* **20** (3), pp. 243-257.
6. Mitrović-Minić, S., and Laporte, G. (2004): "Waiting Strategies for the Dynamic Pickup and Delivery Problem with Time Windows". *Transportation Research B* **38**, pp. 635-655.

Eine heuristische Methode zur Erhöhung der Robustheit von Mehrdepot-Umlaufplänen im ÖPNV

Stefan Kramkowski, Christian Meier und Natalia Kliewer

Decision Support & Operations Research Lab, Universität Paderborn
Warburger Str. 100, 33098 Paderborn, Germany
{kramkowski,meier,kliewer}@dsor.de

1 Einleitung

Die Einsatzplanung für Busse, die sogenannte Umlaufplanung, ist eine Hauptaufgabe des operativen Planungsprozesses eines Verkehrsbetriebes im Öffentlichen Personennahverkehr (ÖPNV). In ihr werden Fahrzeuge einer vorgegebenen Menge von Servicefahrten zugewiesen, so dass die Kosten minimal sind. Die Kosten setzen sich zusammen aus Fixkosten pro eingesetztem Fahrzeug, variablen Kosten pro gefahrenem Kilometer und variablen Kosten pro Zeiteinheit, die ein Bus außerhalb des Depots verbringt. Ein Umlaufplan ist genau dann zulässig, wenn jede Servicefahrt mindestens einem Fahrzeug zugewiesen ist, alle Servicefahrten nur Fahrzeugen zulässigen Typs zugewiesen sind und jeder Fahrzeugumlauf in einem zulässigen Depot startet und am Ende des Planungshorizontes in dieses wieder zurückkehrt. Für eine genauere Beschreibung des Umlaufplanungsproblems und der verschiedenen Lösungsmethoden siehe [1].
Die Umlaufpläne werden im ÖPNV traditionell mehrere Wochen vor dem Tag der Ausführung erstellt. Deshalb können in dieser Planung nicht die tatsächlichen Fahrtzeiten berücksichtigt werden. Stattdessen wird mit Erfahrungswerten für verschiedene Strecken und Tageszeiten geplant. Somit gehen die Störungen im Fahrbetrieb und die daraus resultierenden Verspätungen nicht in diese *offline* Umlaufplanung ein. In den letzten Jahren sind die Umlaufpläne durch den Einsatz spezialisierter Planungssoftware und Verbesserungen bei den verwendeten Methoden immer kostengünstiger geworden. Mit der Senkung der geplanten Kosten geht aber eine Erhöhung der Störanfälligkeit einher, da die Wartezeit der Busse verringert wird, die zum Auffan-

B. Fleischmann et al.(Eds.), *Operations Research Proceedings 2008*,
DOI: 10.1007/978-3-642-00142-0_45, © Springer-Verlag Berlin Heidelberg 2009

gen von Verspätungen zur Verfügung steht. Dies führt bei Auftreten von Störungen zu einem außerplanmäßigen Mehreinsatz an Fahrzeugen und zu Strafzahlungen, die der Verkehrsbetrieb an die staatliche Verwaltung entrichten muss, wenn eine vertraglich vereinbarte minimale Servicefahrt-Pünktlichkeit unterschritten wird (siehe [2]). Deshalb steigen manchmal die tatsächlichen Kosten entgegen der ursprünglichen Zielsetzung an, obwohl die geplanten Kosten gesunken sind.

Um diesen unerwünschten Effekt zu vermeiden, gibt es verschiedene Möglichkeiten: Die Umlaufpläne können während der Ausführung erstellt bzw. laufend an die aktuellen Gegebenheiten angepasst werden. Methodisch wird dies umgesetzt durch den Einsatz von *online Algorithmen* oder dem Lösen von *Recovery Problemen*. In diesen Bereich fällt auch der in [2] vorgestellte Ansatz der *Dynamischen Umlaufplanung*.

Wir verfolgen hier einen anderen Ansatz: Die Umlaufplanung wird wie bisher *offline* ausgeführt, wobei jetzt aber auch potenzielle Störungen berücksichtigt werden. So wird eine zu starke Verringerung der Wartezeit vermieden. Ein so berechneter Umlaufplan ist in einem gewissen Maß robust, oder genauer gesagt stabil, gegenüber Störungen. Er ist in der Lage, eine bestimmte Menge an Verspätungen aufzufangen d.h. zu tolerieren. Die soeben verwendeten Begriffe der Robustheit und Stabilität werden in [3] in einem allgemeineren Kontext als der Umlaufplanung definiert.

In Abschnitt 2 präsentieren wir eine Methode, die diesen *offline* Ansatz in Form eines Verbesserungsverfahrens umsetzt. Wir nutzen dabei eine Heuristik auf Basis des in [4] vorgestellten *Simulated Annealing for Noisy Environments*, abgekürzt SANE. Ausgehend von einem z.B. kostenoptimalen Umlaufplan wird die Zuordnung von Fahrzeugen zu Servicefahrten iterativ abgeändert. So wird die Störungstoleranz schrittweise verbessert, ohne hierbei die Kosten zu vernachlässigen. In Abschnitt 3 werden einige Testläufe und Ergebnisse präsentiert.

2 Heuristik zur Erhöhung der Robustheit

Vor der Beschreibung unseres Verfahrens muss der Begriff der "Verspätung" genauer definiert werden. In der Literatur (z.B. [5]) wird zwischen *primären* und *sekundären* Verspätungen unterschieden. Primäre Verspätungen sind aus Sicht der Planung exogen und direkt durch Störungen hervorgerufen. Sekundäre Verspätungen dagegen sind endogen und entstehen aus den primären durch die Abhängigkeiten der Fahrten eines Fahrzeugs untereinander. Nur die sekundären Verspätungen können durch Änderungen des Planes beeinflusst werden.

Wir setzen hier sekundäre Verspätungen mit *propagierten Verspätungen* synonym, weil bezüglich der Messung der Stabilität von Umlaufplänen der Einsatz von Recovery-Maßnahmen nicht sinnvoll ist. D.h. Verspätungen werden immer propagiert bis sie durch Wartezeiten aufgefangen werden oder der Fahrzeugumlauf endet. Außerdem werden propagierte Verspätungen nur auf Servicefahrten gemessen, da zu spät startende Leerfahrten für Kunden und Verkehrsbetriebe nicht von Bedeutung sind. Nur so ist die Bewertung der Stabilität unverfälscht.

2.1 Konzept

Simulated Annealing for Noisy Environments ist eine mono-kriterielle Meta-Heuristik, die genutzt werden kann, wenn der Zielfunktionswert einer Lösung stochastischer Unsicherheit unterliegt. Unsere Zielfunktion z ist in Anlehnung an die in [2] verwendete definiert als:

$$z = \text{geplante Kosten} + \text{Verspätungskosten} \tag{1}$$

$$\text{Verspätungskosten} = \sum (\text{Länge prop. Verspätung})^2 \cdot \frac{\text{Bus Fixkosten}}{1920^2} \tag{2}$$

Dabei sind die geplanten Kosten deterministisch für jeden Umlaufplan. Die propagierten Verspätungen dagegen sind nur anhand einer Menge von primären Verspätungen berechenbar. Um eine repräsentative Menge primärer Verspätungen zu erhalten, verwenden wir Monte-Carlo Simulation und eine Wahrscheinlichkeitsfunktion, die steuert, ob und wieviel jede Fahrt primär verspätet ist. Die Länge (in Sekunden) jeder so berechneten propagierten Verspätung einer Servicefahrt wird in der Zielfunktion quadriert und mit den Fixkosten des jeweiligen Fahrzeugs gewichtet (für eine Begründung dieser Berechnung siehe [2]). Dadurch sind die Verspätungskosten stochastisch beeinflußt und somit auch der Zielfunktionswert. Deshalb verwenden wir SANE als Basis für unsere Methode und kein herkömmliches *Simulated Annealing*.

Unsere Methode läuft wie in Algorithmus 2.1 dargestellt ab: Ausgehend von einer Initiallösung und einer Nachbarschaftslösung wird $\sigma^2_{\Delta E}$ geschätzt. Als nächstes wird die Initialtemperatur zufällig gesetzt mithilfe des *Temperature Equivalent* für eine Stichprobe (siehe [4]) und einer im Intervall $[0;1[$ gleichverteilten Zufallsvariable. In einer Schleife werden nun solange Nachbarschaftslösungen generiert, bewertet und entweder akzeptiert oder verworfen, bis die Abbruchbedingung erfüllt ist. Dabei werden anders als in [4] mindestens 50 Stichproben des Zielfunktionswertes pro Nachbarschaftslösung gezogen, weil wir nicht nur

Algorithmus 1 : Verbesserungsheuristik robuster Umlaufplan

Eingabe : Initiallösung Umlaufplan x_I
Ergebnis : Umlaufplan x_C

Aktueller Umlaufplan $x_C \leftarrow x_I$
Generiere Nachbarschaftslösung $x_N \in N(x_C)$
Verspätungen in x_C und x_N 50-mal simulieren
Erwartungswert der Zielfunktion für x_C und x_N berechnen
$\hat{\delta} \leftarrow E(x_N) - E(x_C)$
$\sigma^2_{\Delta E}$ schätzen als Varianz von $\hat{\delta}$
Iterationsanzahl $n \leftarrow 0$
Aktuelle Temperatur $T_n \leftarrow (1 + U[0;1[) \cdot \sigma_{\Delta E} \cdot \sqrt{\pi/8}$
repeat
    repeat
        $n \leftarrow n + 1$
        Generiere Nachbarschaftslösung $x_N \in N(x_C)$
        Verspätungen in x_N 50-mal simulieren
        Erwartungswert der Zielfunktion für x_N berechnen
        $\hat{\delta} \leftarrow E(x_N) - E(x_C)$
        Schätzung von $\sigma^2_{\Delta E}$ aktualisieren
        if $T_n \geq \sigma_{\Delta E} \cdot \sqrt{\pi/8}$ **then**
            Akzeptanzprüfung nach Ceperley und Dewing
        else
            Sequentielles Sampling mit Akzeptanzprüfung nach Glauber
        end
    until $n \bmod 20 = 0$
    $T_n \leftarrow T_{n-1} \cdot 0,98$
until $n > 5000$

die Fitnesswertdifferenz schätzen, sondern auch die Varianz der Fitnesswertdifferenz.
Wie die Nachbarschaftsgenerierung abläuft, wird in Abschnitt 2.2 beschrieben. Die *Akzeptanzprüfung nach Ceperley und Dewing* und das *Sequentielle Sampling mit Akzeptanzprüfung nach Glauber* laufen wie in [4] dargestellt ab. Der im Algorithmus 2.1 beschriebene *Annealing Schedule* verringert die Temperatur alle 20 Iterationen um 2%. Dies hat sich in unseren Tests als gut herausgestellt, weshalb wir hier nur diesen behandeln.

2.2 Nachbarschaftsgenerierung

In der Nachbarschaftsgenerierung muss ausgehend vom aktuellen Umlaufplan ein leicht veränderter Umlaufplan, die sogenannte Nachbarschaftslösung, erzeugt werden. Der von uns verwendete Nachbarschaftsoperator wählt zufällig eine Servicefahrt aus dem aktuellen Umlaufplan und ordnet diese einem anderen Fahrzeug zu, das diese Servicefahrt typgerecht und ohne zeitliche Kollision bedienen kann. Falls kein solches Fahrzeug existiert, wird dem Umlaufplan ein Fahrzeug hinzugefügt, dem die Servicefahrt zugeordnet wird. Dieses neue Fahrzeug ist vom selben Typ wie das Fahrzeug, dem die Servicefahrt bisher zugeordnet war. In jedem Fall muss die Gültigkeit der beiden geänderten Umläufe bzw. des geänderten und des neuen Umlaufs durch Hinzufügen und Entfernen von Leerfahrten hergestellt werden. In Abbildung 1 ist dieser Vorgang für zwei geänderte Umläufe A und B grafisch als Time-Space Netzwerk dargestellt, wobei die neu zugeordnete Servicefahrt mit T bezeichnet ist.

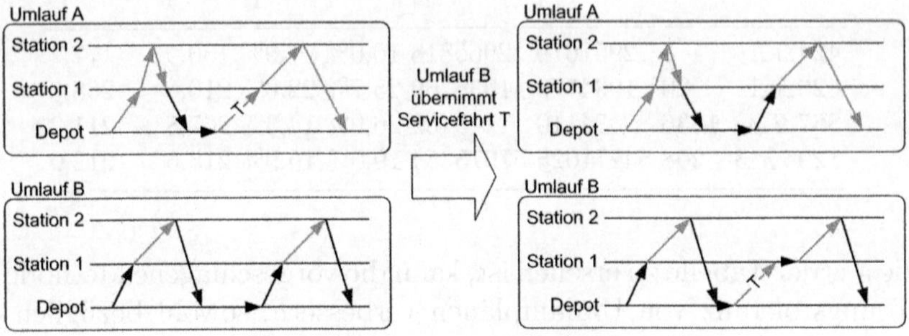

Fig. 1. Nachbarschaftsoperator

3 Ergebnisse

Wir haben unsere Methode mit realen Fahrplänen vier deutscher Städte getestet. Als Initiallösung wurden kostenoptimale Umlaufpläne verwendet, weil wir davon ausgehen, dass ein kostengünstiger robuster Umlaufplan sich nicht stark von einem kostenoptimalen Umlaufplan unterscheidet. Für die Simulation der primären Verspätungen wurde für alle Fahrten folgende zusammengesetzte Wahrscheinlichkeitsfunktion benutzt: Anhand eines Bernoulli-Experiments wurden die 20% der Fahrten identifiziert, die primär verspätet sind. Die Länge jeder primären Verspätung in Sekunden wurde aus einer Dreiecksverteilung im Intervall [1;600] mit Dichtemaximum bei 1 gezogen. Alle Berechnungen wurden auf einem Windows-PC mit einer Pentium M 2 GHz CPU und 2 GB RAM ausgeführt.

In Tabelle 1 sind die geplanten Kosten, die Wahrscheinlichkeit einer sekundären Verspätung pro Servicefahrt und die erwartete Länge einer sekundären Verspätung pro Servicefahrt in Sekunden dargestellt, wobei jeweils die Initiallösung (I) dem Ergebnis der Verbesserungsheuristik

(H) gegenübergestellt ist. Um besser vergleichen zu können, wurden diese Werte nicht aus der Heuristik übernommen, sondern neu mit jeweils den gleichen primären Verspätungen simuliert. Für eine genaue Schätzung wurden jeweils 500 Simulationsläufe durchgeführt. Außerdem sind jeweils die Laufzeit der Verbesserungsheuristik in Minuten und der Name der Instanz angegeben[1].

Table 1. Testergebnisse

Instanz	CPU	geplante Kosten		$P(SD>0)$		$E(SD\|SD>0)$	
		I	H	I	H	I	H
424_1_1	5	2951679	2966818	10,0%	7,9%	180,9	172,2
426_1_1	20	1934170	1936769	25,7%	22,6%	210,0	208,0
867_2_3	1336	74110	105469	26,0%	14,3%	267,8	211,2
1296_1_3	208	54271025	57975354	19,0%	16,9%	215,6	212,0

Wie aus der Tabelle zu ersehen ist, kann die vorgeschlagene Methode die Störungstoleranz von Umlaufplänen verbessern, sowohl bezüglich der Häufigkeit als auch der Länge der propagierten Verspätungen. Gleichzeitig steigen die geplanten Kosten nur wenig an.

References

1. Bunte S, Kliewer N, Suhl L (2006) An Overview on Vehicle Scheduling Models in Public Transport. Proceedings of the 10th International Conference on Computer-Aided Scheduling of Public Transport, Leeds.
2. Huisman D, Freling R, Wagelmans APM (2004) A Robust Solution Approach to the Dynamic Vehicle Scheduling Problem. Transportation Science 38:447–458.
3. Scholl A (2001) Robuste Planung und Optimierung. Physica-Verlag, Heidelberg.
4. Branke J, Meisel S, Schmidt C (2007) Simulated Annealing in the Presence of Noise. Accepted for Journal of Heuristics, Springer.
5. Zhu P, Schneider E (2001) Determining Traffic Delayes through Simulation. Computer Aided Scheduling of Public Transport, 387–399.

[1] Der Name ist wie folgt codiert: #Servicefahrten_#Depots_#Fahrzeugtypen. Die Testinstanzen können heruntergeladen werden:
`http://dsor.upb.de/bustestset`

The Berth Allocation Problem with a Cut-and-Run Option

Frank Meisel and Christian Bierwirth

Martin-Luther-University Halle-Wittenberg
{frank.meisel,christian.bierwirth}@wiwi.uni-halle.de

Summary. The Berth Allocation Problem consists of assigning berthing times and berthing positions to container vessels calling at a seaport. We extend this problem by a so-called cut-and-run option, which is used in practice to ensure that vessels meet their liner schedules. A mathematical problem formulation, meta-heuristics, and computational tests are provided.

1 Introduction

A main reason for liner schedule disturbances is unexpected waiting time of vessels before their service at a container terminal (CT) starts, see [4]. Hence, deciding on berthing times and positions of vessels within the Berth Allocation Problem (BAP) is a main determinant of service quality of terminals, which motivates intensive scientific investigations, see [5]. One assumption in the vast majority of BAP studies is that all vessels are served completely, where tardy departures are allowed. In practice, however, vessels with a time-critical liner schedule must depart at their due date even if served partially only. This policy is known as *cut-and-run* and incorporated into the BAP in our paper. We explain basic principles in Section 2, provide a mathematical model in Section 3, and describe and test meta-heuristics in Sections 4 and 5.

2 Cut-and-Run

Cut-and-run bases on three principles. First, strict due dates, i.e. latest times of departure, are given for the vessels. Due dates follow from liner schedules or from tide-dependent accessibility of a CT, see [4].
Second, vessels that cannot be served completely up to their latest time of departure receive at least a partial service, where remaining

B. Fleischmann et al.(Eds.), *Operations Research Proceedings 2008*,
DOI: 10.1007/978-3-642-00142-0_46,

containers have to wait for a next vessel. This potential of cut-and-run makes it a preferable option compared to the complete rejection of such vessels as proposed by [2] and [6].
Third, minimizing the total waiting and handling time of vessels, as is typically pursued in BAP studies to achieve high service quality, is an improper objective in presence of the cut-and-run option. Since cut-and-run may lead to a partial service of vessels, maximum service quality depends also on minimizing the amount of unprocessed workload. For a precise measurement of (un)processed workload of vessels, quay crane (QC) operations must be considered within the BAP, as done in [3]. The model of [3] is taken up to incorporate cut-and-run into the BAP.

3 Modeling the BAP with a Cut-and-Run Option

3.1 Notation

A terminal with a quay length L, measured in segments of 10 meters length, is considered. A number of Q QCs is available to serve vessels. A set of n vessels $V = \{1, 2 \ldots n\}$ is projected to be served. The planning horizon is H hours, where T is a corresponding set of 1-hour time periods, i.e. $T = \{0, 1 \ldots H-1\}$.
For each vessel $i \in V$ its length l_i, measured in segments of 10 meters length, an expected time of arrival $ETA_i \geq 0$, and a latest time of departure $LTD_i \leq H$ are given. The workload of vessel i, i.e. the container volume to transship, is represented by a crane capacity demand m_i, measured in QC-hours. The minimum and maximum number of QCs to assign to the vessel are denoted by $r_i^{\min}$ and $r_i^{\max}$.
The decisions to make are to assign a berthing time s_i ($s_i \geq ETA_i$), a berthing position b_i, and a number of cranes q_{it} in period $t \in T$ to vessel $i \in V$. q_{it} is either 0 or taken from the range $[r_i^{\min}, r_i^{\max}]$. s_i and e_i point to the begin of the first period and to the end of the last period with cranes assigned to vessel i. Furthermore, binary variable r_{it} is set to 1 if cranes are assigned to vessel i at time t, i.e. $q_{it} > 0$, and binary variables y_{ij} and z_{ij} indicate the relative positioning of vessels i and j in time and space.
To evaluate a solution, the waiting time $w_i = s_i - ETA_i$ of vessel i is penalized by cost c_i^1 per time period. Unprocessed workload of vessel i, as caused by cut-and-run, is penalized by cost c_i^2 per missing QC-hour. The objective is to minimize the total penalty cost Z of a solution.

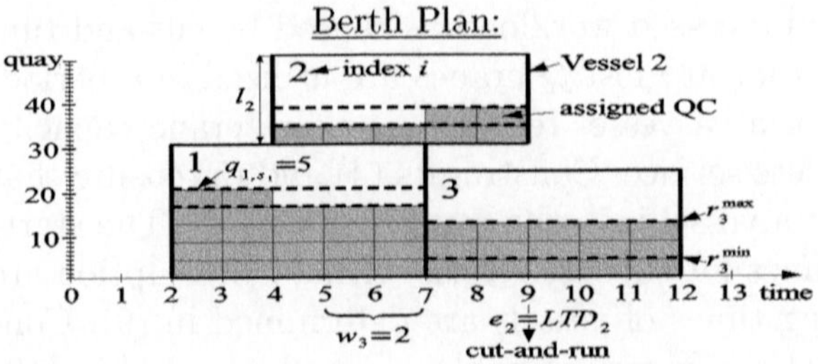

Terminal data:

$L = 50, H = 14, Q = 5$

Vessel data:

i	1	2	3
l_i	30	20	25
m_i	22	10	15
$r_i^{\min}$	4	1	1
$r_i^{\max}$	5	2	3
ETA_i	2	4	5
LTD_i	10	9	14

$c_i^1 = 2, c_i^2 = 1 \quad \forall i \in V$

Fig. 1. Example solution (left) and instance data (right).

A solution of an example instance with $n = 3$ vessels and $Q = 5$ QCs is depicted in a space-time diagram in Fig. 1. In this solution, Vessels 1 and 2 are served immediately upon arrival. Merely Vessel 3 shows a waiting time of $w_3 = 2$ hours, leading to penalty cost of 4 with $c_3^1 = 2$. The service of Vessel 2 is ended by cut-and-run at time $e_2 = LTD_2 = 9$. Although the vessel requires $m_2 = 10$ QC-hours for a complete service, only 7 QC-hours are assigned in the solution. The unprocessed workload of 3 QC-hours leads to cost of 3 with $c_2^2 = 1$. Total cost are $Z = 7$.

3.2 Optimization Model

$$\text{minimize } Z = \sum_{i \in V} \left(c_i^1 \cdot w_i + c_i^2 \cdot \left(m_i - \sum_{t \in T} q_{it} \right) \right) \tag{1}$$

$$\text{subject to } \sum_{i \in V} q_{it} \le Q \qquad \forall t \in T, \tag{2}$$

$$\sum_{t \in T} q_{it} \le m_i \qquad \forall i \in V, \tag{3}$$

$$q_{it} \ge r_i^{\min} \cdot r_{it} \qquad \forall i \in V, \forall t \in T, \tag{4}$$

$$q_{it} \le r_i^{\max} \cdot r_{it} \qquad \forall i \in V, \forall t \in T, \tag{5}$$

$$\sum_{t \in T} r_{it} = e_i - s_i \qquad \forall i \in V, \tag{6}$$

$$(t+1) \cdot r_{it} \le e_i \qquad \forall i \in V, \forall t \in T, \tag{7}$$

$$t \cdot r_{it} + H \cdot (1 - r_{it}) \ge s_i \qquad \forall i \in V, \forall t \in T, \tag{8}$$

$$w_i = s_i - ETA_i \qquad \forall i \in V, \tag{9}$$

$$b_j + M \cdot (1 - y_{ij}) \ge b_i + l_i \qquad \forall i, j \in V, i \neq j, \tag{10}$$

$$s_j + M \cdot (1 - z_{ij}) \ge e_i \qquad \forall i, j \in V, i \neq j, \tag{11}$$

$$y_{ij} + y_{ji} + z_{ij} + z_{ji} \ge 1 \qquad \forall i, j \in V, i \neq j, \tag{12}$$

$$s_i, e_i \in \{ETA_i, \ldots LTD_i\} \quad \forall i \in V, \tag{13}$$

$$b_i \in \{0, 1, \ldots L - l_i\} \qquad \forall i \in V, \tag{14}$$

$$w_i, q_{it} \ge 0, integer \qquad \forall i \in V, \forall t \in T, \tag{15}$$

$$r_{it}, y_{ij}, z_{ij} \in \{0, 1\} \qquad \forall i, j \in V, \forall t \in T. \tag{16}$$

The Objective (1) is the minimization of total penalty cost for waiting of vessels and for unprocessed workload as caused by cut-and-run. Constraints (2) enforce that at most Q cranes are utilized in a period. Constraints (3) ensure that a vessel receives no more crane capacity than needed for a complete service. Constraints (4) and (5) ensure that the number of cranes at a vessel is in the range $[r_i^{\min}, r_i^{\max}]$. The starting times and ending times for serving vessels without preemption are set in (6) to (8). Waiting times of vessels are determined in (9). Constraints (10) and (11) set the binary variables y_{ij} and z_{ij}, used in (12) to avoid overlapping in the space-time diagram. Constraints (13) ensure that the service of a vessel takes place within the time interval spanned by ETA_i and the latest time of departure LTD_i. Note that LTD_i effects a strict due date, i.e. cut-and-run is applied instead of accepting tardiness. From (14) vessels are positioned within the quay boundaries.

4 Solution Methods

A construction heuristic that uses a *priority list* of vessels to generate a solution for the combined problem of berth allocation and crane assignment is proposed in [3]. Moreover, two meta-heuristics are presented, namely *Squeaky Wheel Optimization* (SWO) and *Tabu Search* (TS). Both explore changes of the priority list in order to improve the quality of berth plans. SWO and TS are briefly described in the following with particular focus on the necessary adaptations to cope with cut-and-run.

4.1 Squeaky Wheel Optimization

The idea of SWO is to identify weak performing elements in an existing solution to a combinatorial optimization problem, see [1]. These elements receive higher priority in the solution process by moving them towards the top of a priority list. The new list serves to build a new solution using a base heuristic of the problem. SWO attracts by its ability to eliminate multiple weaknesses of a given solution at one strike without a time consuming exploration of a neighborhood.
SWO is applied to the BAP with a cut-and-run option as follows. First, the individual cost contribution of each single vessel to the total cost of a given solution is calculated. Next, weak performing vessels, i.e. those that show high individual cost, receive higher priority by partially re-sorting the priority list. This is realized by swapping consecutive vessels,

if the cost incurred by the first vessel is lower than the cost incurred by the second vessel. Then, the new priority list is used to construct a new solution. Afterwards, a new iteration is started by analyzing the new solution in the first step and so on. SWO terminates after a preset number of iterations without improvement of the best known solution. As an example for an SWO iteration consider the solution in Fig. 1, as derived from the priority list $P = (1, 2, 3)$. The corresponding cost of Vessels 1, 2, and 3, are 0, 3, and 4, respectively. From partially resorting the priority list, positions of Vessels 1 and 2 are swapped ($P = (2, 1, 3)$) and, then, positions of Vessels 1 and 3 are swapped ($P = (2, 3, 1)$). The derived list is used to construct a new solution for the next iteration.

4.2 Tabu Search

The used TS implementation explores the pairwise-exchange neighborhood of the priority list completely. New solutions are constructed from the modified priority lists. The new solution with least total penalty cost for waiting and unprocessed workload of vessels replaces the current solution. The tabu list management is described in [3]. TS terminates after a given number of iterations without finding a new best solution.

5 Computational Study

We assess the performance of SWO and TS using thirty test instances generated by [3], with ten instances of size $n = 20$, $n = 30$, and $n = 40$, respectively. The instances are modified by increasing the crane capacity demand m_i of vessels to simulate a highly congested terminal situation where the use of cut-and-run is inevitable. Within the instances, vessels are distinguished into feeder, medium, and jumbo class with waiting cost rates c_i^1 set to 1000\$, 2000\$, and 3000\$, respectively. The penalty cost rate for unprocessed workload is set to $c_i^2 = 1000\$$ per QC-hour for all vessels. SWO and TS terminate after 200 iterations without gaining improvement. For comparison, ILOG CPLEX 9.1 is applied to the optimization model with a runtime limit of one hour per instance.

Table 1 reports the total cost Z (in thousand \$) of the solutions delivered by CPLEX, SWO, and TS. The best known solution of an instance is shown bold face. As can be seen, CPLEX delivers a best known solution for instance #3 only. Unfortunately, even for this instance CPLEX does not terminate within the runtime limit and, therefore, optimality of the solution is not proven. SWO delivers best known solutions for 24 of the instances, whereas TS delivers best known solutions for 10 instances. The cost of SWO solutions are on average 49% below the cost of CPLEX solutions, whereas the cost of TS solutions are on average 45% below the cost of CPLEX solutions. The average runtime per instance on a PC P4 2.4 GHz is 279 seconds for SWO and 493 seconds for TS. Summarizing, SWO delivers the best solution quality for the BAP with a cut-and-run option within shortest computation time.

Table 1. Performance comparison of solution methods.

$n = 20$				$n = 30$				$n = 40$			
#	CPLEX	SWO	TS	#	CPLEX	SWO	TS	#	CPLEX	SWO	TS
1	173	**12**	16	11	279	**161**	170	21	401	**282**	303
2	37	**0**	**0**	12	154	**20**	33	22	359	**285**	294
3	**82**	90	94	13	564	**76**	86	23	533	**345**	362
4	102	**43**	**43**	14	166	**60**	79	24	548	451	**446**
5	68	**46**	50	15	283	**93**	133	25	1366	351	**334**
6	49	**32**	**32**	16	206	**82**	94	26	405	**388**	420
7	39	30	**29**	17	217	**86**	**86**	27	1003	370	**353**
8	18	**4**	6	18	182	**66**	99	28	442	**412**	452
9	54	48	**43**	19	1049	**124**	146	29	497	**282**	**282**
10	40	**16**	25	20	306	**113**	125	30	453	**382**	399

6 Conclusions

A cut-and-run option has been incorporated into the Berth Allocation Problem to plan the service of vessels with a time-critical liner schedule. A mathematical model has been formulated that respects the principles of cut-and-run. Meta-heuristics have been presented, where Squeaky Wheel Optimization outperforms Tabu Search and the CPLEX solver.

References

1. D. P. Clements, J. M. Crawford, D. E. Joslin, G. L. Nemhauser, M. E. Puttlitz, and M. W. P. Savelsbergh. Heuristic optimization: A hybrid ai/or approach. In *Proceedings of the Workshop on Industrial Constraint-Directed Scheduling*, 1997.
2. A. Imai, E. Nishimura, and S. Papadimitriou. Berthing ships at a multi-user container terminal with a limited quay capacity. *Transportation Research Part E*, 44:136 – 151, 2008.
3. F. Meisel and C. Bierwirth. Heuristics for the integration of crane productivity in the berth allocation problem. *Transportation Research Part E*, 2008, accepted for publication, doi:10.1016/j.tre.2008.03.001.
4. T.E. Notteboom. The time factor in liner shipping service. *Maritime Economics & Logistics*, 8:19–39, 2006.
5. R. Stahlbock and S. Voß. Operations research at container terminals: a literature update. *OR Spectrum*, 30(1):1–52, 2008.
6. F. Wang and A. Lim. A stochastic beam search for the berth allocation problem. *Decision Support Systems*, 42:2186–2196, 2007.

A Model for the Traveling Salesman Problem Including the EC Regulations on Driving Hours

Herbert Kopfer and Christoph Manuel Meyer

Chair of Logistics, University of Bremen, Wilhelm-Herbst-Strasse 5, 28359 Bremen;
{kopfer,cmmeyer}@uni-bremen.de

1 Introduction

Since April 2007 the new EC Regulation No 561/2006 concerning driving hours in road transport is effective. This regulation restricts the length of time periods for driving and requires minimum breaks and rest periods for drivers [2]. An analysis of the EC Regulation with respect to vehicle routing can be found in [3]. In this paper the restrictions on driving times and the need for breaks are formalized and integrated in an optimization model of the TSPTW. The solution space of the extended traveling salesman problem with time windows and EU-constraints (TSPTW-EU) contains all Hamiltonian circuits which fulfil the given time windows and restrictions of the Regulation relevant for a time period up to one week. The presented approach for extending the TSPTW to the TSPTW-EU is also applicable for the extension of the VRPTW and PDPTW, thus offering a possibility to include the EC Regulations in vehicle routing and scheduling.

2 Integration of the EC Regulation into the TSPTW

For the integration of the Regulation a position-based formulation of the TSPTW is used since it allows the calculation of driving times of sub-routes. It represents the position q, at which a location j will be visited within a route (see e.g. [1]). For its formulation the following sets and variables are introduced.
I: set of locations $i \in \{0, 1, \ldots, n\}$ with 0 as starting point of the route
P: $I \setminus \{0\}$

B. Fleischmann et al.(Eds.), *Operations Research Proceedings 2008*,
DOI: 10.1007/978-3-642-00142-0_47,

Q(j): set of positions q, at which location j can occur
I(j,q): set of locations i, from which can be traveled to j, so that j occurs at position q
J(i): set of locations j, to which can directly be traveled from i
J(i,q): set of locations j, to which can be traveled from i, so that j occurs at position q
Q(i,j): set of positions q, at which j can occur, if j is reached from i
We assume that each driver is assigned to a fixed vehicle. In this case the driving times of a driver can simply be modeled by the wheel turning times of the assigned vehicle. The goal of the TSPTW-EU consists in minimizing the total time used for the route including breaks. This results in the objective function (1) where t_{rueck} denotes the time when the route is completed.

$$\min f = t_{rueck} \tag{1}$$

Let w_{ijq} be a binary variable with $w_{ijq} = 1$ iff the route leads from location i to location j so that j is reached at position q. Let tt_{ij} denote the traveling time needed to travel from i to j including breaks. Using the above definition of sets the following restrictions for the TSPTW can be formulated.

$$\sum_{i \in I} \sum_{q \in Q(i,j)} w_{ijq} = 1 \;\; \forall j \in I \tag{2}$$

$$\sum_{j \in P} w_{0j2} = 1 \tag{3}$$

$$\sum_{i \in P} w_{i0,n+1} = 1 \tag{4}$$

$$\sum_{i \in I(s,q)} w_{isq} = \sum_{j \in J(s,q+1)} w_{sj,q+1} \;\; \forall s \in P, q \in Q(s) \tag{5}$$

Restriction (2) requires that each location is reached exactly once. By (3) and (4) it is required that the depot is left at the beginning of the route and is reached again at position n+1. Restriction (5) means: If a location s is reached at position q of the route it must be left so that the location which is visited next will be reached at position q+1. If $t_i > 0$ denotes the arrival time at location $i \in P$, the following restrictions (6) and (7) guarantee that the arrival times at all locations of the route are conform to the traveling times tt_{ij}. The restrictions (8) and (9) postulate that the time windows $[a_i, e_i]$ for each location i are met.

$$t_j \geq t_i + tt_{ij} - M(1 - \sum_{q \in Q(i,j)} w_{ijq}) \ \forall i \in I, j \in J(i) \setminus \{0\} \quad (6)$$

$$t_{rueck} \geq t_i + tt_{i0} - M(1 - \sum_{q \in Q(i,0)} w_{i0q}) \ \forall i \in P \quad (7)$$

$$t_i \geq a_i \ \forall i \in P \quad (8)$$

$$t_i \leq e_i \ \forall i \in P \quad (9)$$

For modeling the breaks included in the traveling times tt_{ij} the following variables and data are introduced. Let $P15_{ijq}$ be the integer variable for the number of those 15 minute breaks which are taken between the locations i and j, while j is at position q of the route. Let $P30_{ijq}$ be the integer variable for the number of second parts of regular breaks with a length of 30 minutes, also between i and j with j at position q. The integer variable p_{ijq} denotes the number of regular breaks separating different driving periods between i and j. Of course, each of these regular breaks can consist of a single break of 45 minutes or two parts with 15 and 30 minutes belonging together to a regular break. Let the variable TR_{ij} denote the duration of a daily rest period taken between the locations i and j. Then, the traveling times tt_{ij} and tt_{i0} in (6) and (7) can be calculated by (10) and (11).

$$tt_{ij} = d_{ij} + \sum_{q \in Q(i,j)} (P15_{ijq} * 0.25 + P30_{ijq} * 0.5) + TR_{ij}$$
$$\forall i \in I, j \in J(i) \setminus \{0\} \quad (10)$$

$$t_{i0} = d_{i0} + \sum_{q \in Q(i,0)} (P15_{i0q} * 0.25 + P30_{i0q} * 0.5) + TR_{i0} \ \forall i \in P \quad (11)$$

The equations (10) and (11) presume that driving between any two customer locations will not require more than one daily rest period. The total driving time necessary to travel from the starting point 0 to the location at the position q of the route amounts to $GZ_q = \sum_{q'=1}^{q} \sum_{j \in I} \sum_{i \in I(j,q)} d_{ij} w_{ijq'}$. The following constraints (12) to (15) refer to the positioning of the breaks during the route. Constraints (12) and (13) ensure that breaks can only be taken at connections between locations which are part of the route. Constraint (14) requires that there are enough breaks before reaching the location at position q, and (15) prevents that breaks are taken in advance in order to use them later on.

$$p_{ijq} + P15_{ijq} \leq M * w_{ijq} \;\; \forall i, j \in I, q \in Q(i,j) \quad (12)$$

$$P30_{ijq} \leq p_{ijq} \;\; \forall i, j \in I, q \in Q(i,j) \quad (13)$$

$$GZ_q \leq 4.5((\sum_{q'=1}^{q} \sum_{j \in I} \sum_{i \in I(j,q)} p_{ijq'}) + 1) \;\; \forall q \in Q(l), l \in I \quad (14)$$

$$GZ_q \geq 4.5(\sum_{q'=1}^{q} \sum_{j \in I} \sum_{i \in I(j,q)} p_{ijq'}) \;\; \forall q \in Q(l), l \in I \quad (15)$$

The conditions (16) to (18) arrange the combination of parts of breaks to regular breaks. The constraint (16) causes that up to an arbitrary position q of the route the number of 15-minute-breaks must be greater or equal than the number of regular breaks. Constraint (17) postulates that up to q the number of 30-minute-breaks must be equal to the number of regular breaks and (18) requires that at any position q at most one 15-minute-break is countable towards input for a regular break. Altogether, the effect is that couples of short breaks of 15 and 30 minutes duration are combined to regular breaks and that a 15-minute-break has always to be taken earlier than its corresponding 30-minute-break.

$$\sum_{q'=1}^{q} \sum_{j \in I} \sum_{i \in I(j,q)} p_{ijq'} - \sum_{q'=1}^{q} \sum_{j \in I} \sum_{i \in I(j,q)} P15_{ijq'} \leq 0 \;\; \forall q \in Q(l), l \in I \quad (16)$$

$$\sum_{q'=1}^{q} \sum_{j \in I} \sum_{i \in I(j,q)} p_{ijq'} - \sum_{q'=1}^{q} \sum_{j \in I} \sum_{i \in I(j,q)} P30_{ijq'} = 0 \;\; \forall q \in Q(l), l \in I \quad (17)$$

$$\sum_{q'=1}^{q} \sum_{j \in I} \sum_{i \in I(j,q)} p_{ijq'} + 1 \geq \sum_{q'=1}^{q} \sum_{j \in I} \sum_{i \in I(j,q)} P15_{ijq'} \;\; \forall q \in Q(l), l \in I \quad (18)$$

The following constraints (19) to (25) formulate the conditions for the positioning and length of the daily rest periods. Let tp_{ijq} be the binary variable with $tp_{ijq} = 1$ iff there is a daily rest period between i and j at position q. The binary variable $dred_{ij} = 1$ iff the daily rest period between the locations i and j is cut down to a reduced daily rest period. Let $bdrive_{ijq}$ be the binary variable indicating whether on the way from location i to j at position q the daily driving time has exceptionally been extended from 9 to 10 hours.

$$bdrive_{ijq} + tp_{ijq} \leq 2w_{ijq} \;\; \forall i,j \in I, q \in Q(i,j) \tag{19}$$

$$TR_{ij} \geq 11 - 2 * dred_{ij} - M(1 - \sum_{q \in Q(i,j)} tp_{ijq}) \;\; \forall i,j \in I \tag{20}$$

$$\sum_{i \in I} \sum_{j \in I} dred_{ij} \leq 3 \tag{21}$$

$$dred_{ij} \leq \sum_{q \in Q(i,j)} w_{ijq} \;\; \forall i,j \in I \tag{22}$$

$$\sum_{q=q'}^{q''} \sum_{i \in I} \sum_{j \in I} d_{ij} w_{ijq} \leq 9 + 9(\sum_{q=q'}^{q''} \sum_{i \in I} \sum_{j \in I} tp_{ijq}) + \sum_{q=q'}^{q''} \sum_{i \in I} \sum_{j \in I} 1 * bdrive_{ijq}$$
$$\forall q', q'' \in Q(i,j), q' < q'' \tag{23}$$

$$\sum_{i \in I} \sum_{j \in I} \sum_{q \in Q(i,j)} bdrive_{ijq} \leq 2 \tag{24}$$

$$\sum_{q=q'}^{q''} bdrive_{ijq} - \sum_{q=q'}^{q''} tp_{ijq} \leq 1 \;\; \forall i,j \in I, q', q'' \in Q(i,j), q' < q'' \tag{25}$$

The condition (19) specifies that daily rest periods and driving time extensions are only possible on used connections (i,j). In (20) it is required that a regular daily rest period must at least last 11 hours and can be reduced by two hours in dependence of the value of $dred_{ij}$. Constraint (21) ensures that the possibility to reduce the daily rest period is used at most three times. Restriction (22) states that reductions of daily rest periods are only allowed on used connections. The condition (23) guaranties that for an arbitrary sub-route from any position q' to q" the accumulated driving times are not greater than the total time of the maximal allowed daily driving times that are situated between the positions q' and q". The number of daily driving times situated between them depends on the number of daily rest periods on the considered

sub-route and on the possible prolongation of single daily driving times by means of $bdrive_{ijq}$. The inequality (24) states that the extension of the daily driving time can only be applied twice a week. Constraint (25) enables that the daily driving time can be extended up to 10 hours by allowing only one variable $bdrive_{ijq}$ to equal 1 between two daily rest periods.
The weekly driving time must not exceed 56 hours per single week and 45 hours on average for any two consecutive weeks. Let Wlz_{w-1} be the driving time in the previous week, then ΔWlz denotes the deviation from the average driving time, i.e. $\Delta Wlz = 45 - Wlz_{w-1}$. In constraints (26) and (27) the restriction of the weekly driving time is applied to the weekly planning period.

$$\sum_{i \in I} \sum_{j \in I} \sum_{q \in Q(i,j)} d_{ij} w_{ijq} \leq 56 \tag{26}$$

$$\sum_{i \in I} \sum_{j \in I} \sum_{q \in Q(i,j)} d_{ij} w_{ijq} \leq 45 + \Delta Wlz \tag{27}$$

The above objective function (1) and the constraints (2) to (27) yield a complete model for the TSPTW-EU incorporating all rules of the EC Regulation which are statutory for the planning of a single weekly planning period, except the rule that a new daily rest period has to be started at least 24 hours after the end of a daily rest period.

Acknowledgement. This research was supported by the German Research Foundation (DFG) as part of the Collaborative Research Centre 637 "Autonomous Cooperating Logistics Processes - A Paradigm Shift and its Limitations" (subproject B9).

References

1. Dethloff J (1994) Verallgemeinterte Tourenplanungsprobleme: Klassifizierung, Modellierung, Lösungsmöglichkeiten. Logistik und Verkehr, Göttingen
2. Regulation (EC) No 561/2006 of the European Parliament and of the Council of 15 March 2006, Official Journal of the European Union L 102/1, 11.4.2006
3. Meyer CM, Kopfer H (2008) Restrictions for the operational transportation planning by regulations on drivers' working hours. In: Bortfeldt A, Homberger J, Kopfer H, Pankratz G, Strangmeier R (eds): Intelligent Decision Support. Gabler, Wiesbaden

Crew Recovery with Flight Retiming

Felix Pottmeyer, Viktor Dück, and Natalia Kliewer

Decision Support & Operations Research Lab, University of Paderborn,
Warburger Str. 100, 33098 Paderborn {pottmeye,vdueck,kliewer}@upb.de

Summary. This paper presents a method for the crew recovery problem with the possibility to retime flights to avoid large modifications of the original schedule. In the presented method a new neighborhood search for the crew scheduling problem is embedded in a tabu search scheme. Dynamic programming on leg based networks is used for generation of pairings. We test our method on different crew schedules and disruption scenarios and show the effects of flight retiming.

1 Introduction

On the day of operations disruptions like aircraft breakdowns or bad weather conditions may disturb the planed schedules. In such situations an airline has to recover many resources, i.e. aircrafts, crews and airport gates, considering different constraints. To recover from the disruption a plan of crew and aircraft swapping, reserve utilization and flight retiming is needed. A decision support system for this purpose consists of many modules, including simulation and optimization, i.e. see [1]. Delaying the departing times as recovery action can be essential to find solutions of good quality, but most optimization modules used for crew recovery regard the scheduled departing times as hard constraints. We integrate a method for flight retiming into a meta heuristic for crew recovery. This allows the method to find good solutions quickly. We consider the main objective of the recovery problem being to cover all flights and to minimize the use of reserve crews. Further objectives are to minimize the use of additional deadheads, hotel stays and delay minutes.
[3] provide a comprehensive review of concepts and models used for recovery of airline resources. Here we give a short overview about a small

B. Fleischmann et al.(Eds.), *Operations Research Proceedings 2008,*
DOI: 10.1007/978-3-642-00142-0_48, © Springer-Verlag Berlin Heidelberg 2009

selection of methods for airline recovery only. [7] present a heuristic that aims at quickly finding a solution that covers all flights and has as few changed crew pairings as possible. The problem is modeled as an integer multi-commodity network flow problem. A branch and bound algorithm is used to solve the model. A node in the search tree is represented by a set of uncovered flights and a list of modified pairings. Then an uncovered flight is picked and for each possible assignment of a crew to this flight a new branch is created. [6] use an integer multi-commodity network flow problem too. The model is solved by a column generation method. In order to generate new pairings a duty based network is build for every crew member. The objective is to cover all flights at minimum costs and with minimum changes to the crew schedules. In order to solve the model quickly, a recovery period is used and only a subset of crews is considered in the optimization. [4] propose another method to reduce the size of the recovery problem. Therefore a maximum number of candidate crews per misconnected flight is defined before the recovery. Then a set covering model with additional variables for canceled flights and deadheads is solved by a column generation method. The aim is to minimize the changes in the crew schedule. [5] present a duty based network model for the airline crew recovery problem. The model is solved with a branch and price method. The set of feasible duties is obtained through enumeration. To reduce the size of the recovery problem only a subset of crews is considered in the optimization.

[2] proposes tabu search and simulated annealing meta heuristics for the aircraft recovery problem. In the underlying local search procedure new rotations for two aircrafts are computed using the uncovered flights and the flights from both rotations. The heuristics use a tree-search algorithm to find new aircraft rotations. Flights can be delayed for recovery, but no maintenance constraints are considered.

In this paper we transfer the method of [2] to crew pairing recovery. We use a similar local search procedure with a tabu search method. For pairing generation we use a network algorithm and propose an alternative method of delaying flights for recovery.

2 Method for Crew Recovery

The main idea of this recovery method is to define the neighborhood in the local search as a set of solutions, which can be reached by rescheduling the pairings of any two crews. To generate new pairings we use connection-based networks and incorporate the ability to delay flights

in this network. This local search method is used inside a tabu search method to find near optimal solutions. To reduce the size of the recovery problem we implement a preprocessing method. Algorithm 2 outlines the main method.
The first step at line 1 is to reduce the number of considered crews for recovery. Active crews are those, which are assigned to disrupted flights or are at an airport when other disrupted flights are arriving or departing. This method is based on the preprocessing method of [5]. Another important aspect for problem reduction, the recovery period, is not stated in the algorithm explicitly. The recovery period determines which period of each pairing is allowed to change and is a parameter to the algorithm.

Algorithm 1: Recovery method

Data: set of uncovered flights N
Data: set of tabu solutions $T = \emptyset$
Data: best solution s_{best}
1 Define the set of active pairings P
`/* meta search loop */`
2 **repeat**
  Data: best local solution s_{local}
  `/* local search loop */`
3 **foreach** $(p_1, p_2), p_1 \in P, p_2 \in P$ **do**
4    $N_{tmp} = N \cup p_1 \cup p_2$
5    $P_1 =$ GenerateAllPairings(N_{tmp}, p_1)
6    $P_2 =$ GenerateAllPairings(N_{tmp}, p_2)
7   Choose two new pairings from $P_1 \cup P_2$,
8   so that the new solution $s_{tmp} \notin T$
9   minimizes the objective function
10   **if** $s_{tmp} < s_{local}$ **then** $s_{local} = s_{tmp}$
11 **end**
12 $T = T \cup s_{local}$
13 **if** $s_{local} < s_{best}$ **then** $s_{best} = s_{local}$
14 **until** *cancelation criteria apply*

The neighborhood $N(s)$ of a solution s consists of all solutions that can be reached from s by applying a local search move. Let a solution be given by a set P of pairings and a set N of uncovered flights. Then for the local search move two crews p_1 and $p_2 \in P$ are chosen and all feasible new pairings for each of them generated, using only the flights of the two crews and the uncovered flights N. Then the best combination of one new pairing for p_1 and one new pairing for p_2 is chosen. As long as the numbers of generated pairings are not too big, a simple pair-wise evaluation is sufficient. This step is repeated for every combination of two distinct crews. Afterwards the solution with the best objective value is saved. You find the local search loop in algorithm 2 at line 3.

In order to generate new pairings we use a connection based network with nodes for each flight and arcs connecting flights, which can be flown in succession. For recovery we have to take into account the scheduled parts before and after the recovery period. Therefore a distinct network for each crew with corresponding source and sink nodes is generated. A path from the source to the sink node represents a sequence of legs that a crew can fly. New pairings are generated with a simple depth first enumeration algorithm. All constraints concerning the connection of two flights in a pairing are considered during the network is build. Other constraints, like maximal working and flying time, are checked during the generation of pairings.

In order to be able to use flight delays as a recovery action, we extended the network with additional nodes for delayed flights. Adding all possibilities for delays into the network would increase the network size and the computation time to forbidden ranges. Hence we allow only delays for successor flights in aircraft rotations of flights, which were disrupted or modified during aircraft rotation recovery. Moreover we restrict the length of secondary delays, so that all aircraft rotations remain feasible.

The presented local search procedure is extended to a tabu search meta heuristic by introducing a tabu list. The tabu list saves the last n solutions that have been visited, line 4 in algorithm 2. During the search a new solution is only allowed to be visited if it is not part of the tabu list. The outer loop at line 2 in algorithm 2 is the meta search loop. We use a time limit as termination criteria for the outer loop.

3 Computational Results

We present results for our method for two different aircraft fleets and three different disruption scenarios applied to the crew schedule of each fleet. Table 1 gives an overview about the size and the properties of the crew schedules. The three disruptions scenarios are:

Scenario 1: delay three random flights which depart close in time at different airports for 120 minutes,

Scenario 2: forbid departures at an airport for one hour delaying all those flights,

Table 1. Problem instances

	Fleet A	Fleet B
Planning horizon	4 weeks	2 weeks
Airports	23	62
Flights	649	1250
Aircraft Rotations	649	1250
Crew Pairings	54	120

Scenario 3: delay one flight for 30 minutes where the successor flight of the aircraft is flown by another crew.

The proposed method was implemented using Microsoft C#.NET and tested on a Windows PC with 1.6 GHz and 2 GB Ram. Before applying this recovery method, the aircraft rotations are recovered by simply propagating all delays.

We found a recovery period of 48 hours to be the best choice regarding both, solution time and solution quality. Therefore we present in table 2 the results for this recovery period. The results are average values for ten random disruptions for each disruption scenario. The column *solution time* shows the seconds until the best solution was found. The first feasible solution that covered all flights was always available within 10 seconds. The last column shows the average length of secondary delays. In 4 of 10 test cases for fleet A and in 3 of 10 test cases for fleet B secondary delays between 3 and 30 minutes were used. Each time a secondary delay prevented the use of an otherwise needed reserve crew. The other columns show the use of additional deadheads using only regular flights of the own airline, additional hotel stays and reserve crews.

Table 2. Results for recovery period of 48 hours

	Scenario	solution time (seconds)	deadheads	hotel stays	reserve crews	delays (minutes)
	1	7	0.4	0.4	0.4	0
Fleet A	2	55.8	0	0	0	0
	3	15.2	0.2	0.3	0.1	0.4
	1	11.6	0	0	0.4	0
Fleet B	2	20.3	0	0.4	0.2	0
	3	9.1	0	0.1	0.2	0.3

4 Conclusion and Outlook

The presented local search move for the crew recovery problem enables meta heuristics to find good solutions quickly. Our experiments show that secondary delays are indispensable for obtaining economical disruption management decisions.
Despite the already very good results the current implementation offers many possibilities for improvement. The pairing generation process can be accelerated by using a time-space network representation. During the local search it is very time consuming to compute all neighborhood solutions; hence there is a need for a scheme how to find good local search moves heuristically

References

1. Abdelghany, K. F., A. F. Abdelghany, and G. Ekollu (2008, March). An integrated decision support tool for airlines schedule recovery during irregular operations. *European Journal of Operational Research 185*(2), 825–848.
2. Andersson, T. (2006, January). Solving the flight perturbation problem with meta heuristics. *Journal of Heuristics 12*(1-2), 37–53.
3. Clausen, J., A. Larsen, and J. Larsen (2005). Disruption management in the airline industry - concepts, models and methods.
4. Lettovský, L., E. L. Johnson, and G. L. Nemhauser (2000). Airline crew recovery. *Transportation Science 34*(4), 337–348.
5. Nissen, R. and K. Haase (2006). Duty-period-based network model for crew rescheduling in european airlines. *Journal of Scheduling 9*, 255–278.
6. Stojkovic, M., F. Soumis, and J. Desrosiers (1998). The operational airline crew scheduling problem. *Transportation Science 32*(3), 232–245.
7. Wei, G., G. Yu, and M. Song (1997, October). Optimization model and algorithm for crew management during airline irregular operations. *Journal of Combinatorial Optimization 1*(3), 305–321.

A Branch-and-Cut Approach to the Vehicle Routing Problem with Simultaneous Delivery and Pick-up

Julia Rieck and Jürgen Zimmermann

Institute of Management and Economics, Operations Research Group,
Clausthal University of Technology, Germany
{julia.rieck,juergen.zimmermann}@tu-clausthal.de

Summary. The vehicle routing problem with simultaneous delivery and pick-up (VRPSDP) is an extension of the capacitated vehicle routing problem in which products have to be transported from the depot to customer locations and other products have to be trucked from customer locations to the depot. Each customer requires a simultaneous delivery and pick-up of goods by the same vehicle. The VRPSDP is a basic problem in reverse logistics: In addition to the distribution process to the customers, re-usable goods have to be transported in the reverse direction. We implement a branch-and-cut approach and study how it can be applied to the solution of the VRPSDP. The computational tests have been performed on known benchmark instances. Some benchmark problems are solved to optimality for the first time.

1 Introduction

Many Companies are forced by environmental laws to take back their products, i.e. batteries, household chemicals or paints, as well as waste electrical or electronic equipment. For these companies, the recovery of used or unused products is normally more profitable than disposal. In addition, competition, marketing and the growing environmental awareness of customers have pushed companies into product recovery activities. Figure 1 shows the supply chain from production through delivery to the customer (*forward flow*) and from the customer to remanufacturing or the disposal facilities (*backward flow*). It is obvious that in reverse logistics networks, each customer requires two types of service: a delivery and a pick-up. To avoid serving customers twice during the planning horizon, the delivery and pick-up must be made simultaneously. The resulting problem faced by companies in the reverse logistics field is an extension of the capacitated vehicle routing

B. Fleischmann et al.(Eds.), *Operations Research Proceedings 2008*,

DOI: 10.1007/978-3-642-00142-0_49,

problem (CVRP). In addition to the distribution activities, re-usable products have to be trucked in the reverse direction.

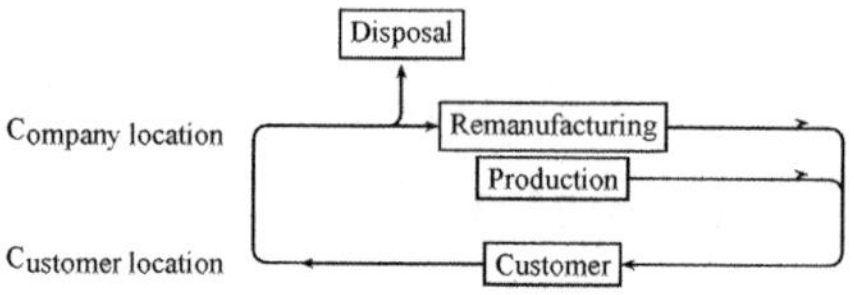

2 Problem Description and Mathematical Formulation

In this paper we consider the vehicle routing problem with simultaneous delivery and pick-up (VRPSDP). Firstly, we start by describing a new mixed integer linear programming formulation for the VRPSDP. The model is a two-index flow formulation that contains $|V|^2$ binary and $2|V| + |C|$ auxiliary variables. Let $G = (V, A)$ be a complete digraph, where $V = \{0, \ldots, n\} = C \cup \{0\}$ is the node set and $A = \{\langle i, j\rangle \mid i, j \in V\}$ is the arc set. Each node $i \in C$ represents a customer, while node 0 corresponds to the depot. Two nonnegative weights c_{ij} and t_{ij} are associated with each arc $\langle i, j\rangle \in A$ which represent the travel costs and the travel time from node i to j, respectively. We assume that $c_{ii} := 0$ for all $i \in V$ and the cost and travel time matrices satisfy the triangle inequality, i.e. $c_{ik} \leq c_{ij} + c_{jk}$ for all $i, j, k \in V$. A set of M identical vehicles, each with capacity Cap, is available at the depot. Each customer $i \in C$ is associated with a demand $d_i \geq 0$ that should be delivered and a demand $p_i \geq 0$ that should be picked-up, whereas $d_i + p_i > 0$. For the depot, we set $d_0 := p_0 := 0$. We measure the demand of customers in abstract transport units which can be calculated from the dimension and weight of the individual product. To ensure that the feasible region is not empty, we assume that $d_i, p_i \leq Cap$ for all $i \in V$. The service time $s_i > 0$ is the time which is necessary for the loading and unloading activities at node $i \in C$. For the depot, we define $s_0 := 0$. The capacity of the vehicles as well as a total travel and service time T_{max} may not be exceeded. With the decision variables

$$x_{ij} := \begin{cases} 1, & \text{if arc } \langle i, j\rangle \text{ belongs to the solution} \\ 0, & \text{otherwise} \end{cases}$$

$i, j \in V$ and the auxiliary variables

$f_i \geq 0$ required travel and service time to node $i \in V$,

$ld_i \geq 0$ amount of load that has to be delivered to node $i \in V$ and to all other following nodes,

$l_i \geq 0$ amount of load after visiting customer $i \in C$

we formulate the VRPSDP as a mixed integer linear (MIP) program.

$$\text{Minimize} \qquad \sum_{i\in V}\sum_{j\in V} c_{ij}x_{ij} \tag{1}$$

subject to

$$\sum_{\substack{i\in V\\ i\neq j}} x_{ij} = 1 \qquad j\in C \tag{2}$$

$$\sum_{\substack{i\in V\\ i\neq j}} x_{ji} = 1 \qquad j\in C \tag{3}$$

$$\sum_{i\in C} x_{0i} \leq M \tag{4}$$

$$f_0 = 0 \tag{5}$$

$$f_j \geq f_i + s_i + t_{ij} - T_{max}(1 - x_{ij}) \qquad i\in V, j\in C \tag{6}$$

$$f_i + s_i + t_{i0} \leq T_{max} \qquad i\in C \tag{7}$$

$$ld_i \geq ld_j + d_i - Cap(1 - x_{ij}) \qquad i\in V, j\in C \tag{8}$$

$$l_i \geq ld_i - d_i + p_i - Cap(1 - x_{0i}) \qquad i\in C \tag{9}$$

$$l_j \geq l_i - d_j + p_j - Cap(1 - x_{ij}) \qquad i,j\in C \tag{10}$$

$$d_i \leq ld_i \leq Cap \qquad i\in V \tag{11}$$

$$p_i \leq l_i \leq Cap \qquad i\in C \tag{12}$$

$$x_{ij} \in \{0,1\} \qquad i,j\in V \tag{13}$$

Objective function (1) represents the transportation costs which are to be minimized. The indegree and outdegree constraints (2) and (3) ensure that each customer is visited exactly once by a vehicle. Constraint (4) states that no more than M routes are created. Inequalities (5) – (7) guarantee that the prescribed travel and service time T_{max} may not be exceeded. With constraints (8) we specify the delivery quantity that has to be loaded at the depot. Constraints (6) and (8) force an order for customer visits in the routes. Therefore, they ensure that no subtours without the depot are generated. Constraints (9) and (10) declare the amount of load of vehicles after the visit of the first customer and the other customers in the routes, respectively. Inequalities (11) and (12) guarantee that the capacity of the vehicles may not be exceeded. Since all decision and auxiliary variables are bounded the feasible region of the VRPSDP is a convex polytope. The model (1) – (13) can be used as input to a MIP solver (e.g. CPLEX). But before solving the VRPSDP, it is appropriate to restrict the domains of auxiliary variables and "big-M-constraints" in the model. For the required travel and service time to node $i \in C$ and for the amount of load after visiting customer $i \in C$ we obtain

$$t_{0i} \leq f_i \leq T_{max} - s_i - t_{i0}$$

$$l_i \leq Cap - \max\{d_i - p_i, 0\}.$$

A common possibility for linearizing disjunctive constraints is the so called "big-M-form", where a "big-M" is introduced to satisfy the wrong way constraints. For example, the inequalities

$$\begin{aligned} l_i &\geq ld_i - d_i + p_i - \mathrm{M}_{0i}(1 - x_{0i}) \quad && i \in C && (8) \\ l_j &\geq l_i - d_j + p_j - \mathrm{M}'_{ij}(1 - x_{ij}) \quad && i, j \in C. && (10) \end{aligned}$$

are "big-M-constraints" with $\mathrm{M}_{0i} = \mathrm{M}'_{ij} = Cap, i, j \in C$. It is well known that big-M-constraints have lousy computational behavior. Therefore, it is important that each big-M is as small as possible. In (8) and (10), we are able to displace the big-Ms as follows

$$\begin{aligned} \mathrm{M}_{0i} &= Cap - d_i && i \in C \\ \mathrm{M}'_{ij} &= Cap - \max\{d_j, d_i - p_i + d_j\} && i, j \in C. \end{aligned}$$

We solved the VRPSDP with CPLEX 10.0 using a branch-and-cut approach with problem-specific preprocessing techniques and cutting planes.

3 Branch-and-Cut Approach

Branch-and-cut is a generalization of the branch-and-bound principle where the problem under consideration is divided into subproblems. At each subproblem additional cutting planes (cuts) are inserted, in order to determine an integral solution for the current subproblem or to strengthen the lower bound given by an optimal solution of the underlying linear programming (LP) relaxation. If one or more cuts are identified by solving the so called separation problem, they are added to the subproblem and the LP is solved again. If no cuts are found, the branching is executed. The features of CPLEX allow the generation of general cuts for mixed-integer linear programs during optimization. In our branch-and-cut approach we consider Gomory fractional, clique, cover, flow cover, mixed integer rounding (MIR), and implied bound cuts. Additionally, we take into consideration the rounded capacity (RC) cuts which are special inequalities for the CVRP. To insert RC cuts, we use the "CVRPSEP package" implemented by [3]. The package is created especially for the symmetric CVRP, and therefore we have to make some necessary adjustments. If we transfer the LP solution $\tilde{x}$ to the separation algorithm, we identify the decision variable $\tilde{x}_{[i,j]}$ (specifies if edge $[i,j]$ is in the solution) with $\tilde{x}_{[i,j]} := \tilde{x}_{ij} + \tilde{x}_{ji}$. The rounded capacity inequalities

$$\sum_{i \in Q} \sum_{j \in V \setminus Q} x_{ij} + x_{ji} \geq 2r(Q) \qquad Q \subseteq V \setminus \{0\}, Q \neq \emptyset \tag{11}$$

impose the vehicle capacity restrictions and also ensure that the routes are connected. Let $r(Q)$ be the minimum number of vehicles needed to serve a customer set $Q \subseteq C$. For the VRPSDP we obtain

$$r(Q) = \left\lceil \max\{\textstyle\sum_{i \in Q} d_i, \sum_{i \in Q} p_i\} / Cap \right\rceil .$$

The separation problem of the rounded capacity cuts is known to be $\mathcal{NP}$-hard. For the computational tractability, we add RC cuts only at 400 nodes of the branch-and-bound tree.

4 Discussion and Computational Results

The computational tests have been performed on the instances of [1] which are derived from the extended Solomon instances R121, R141, R161, R181 and R1101. The benchmark is composed of 25 instances, ranging from 15 to 20 customers. Each instance is solved twice: once by inserting only the RC cuts (test run 1) and once by inserting all explained CPLEX cuts together with RC cuts (test run 2).

Table 1. Computational Results

Problem	n	Cap	costs	m	t_{cpu}	Problem	n	Cap	costs	m	t_{cpu}
R121	15	80	610,80	3	5,20	R121	15	120	542,2	2	1,22
R141	15	80	750,06	3	0,39	R141	15	120	669,72	2	0,14
R161	15	80	1166,82	2	0,13	R161	15	120	1162,58	2	12,94
R181	15	80	1968,38	2	49,09	R181	15	120	1755,95	2	0,09
R1101	15	80	2033,88	2	1,92	R1101	15	120	1809,69	2	0,34
R121	17	80	726,06	4	159,34	R121	17	120	564,39	2	0,42
R141	17	80	791,10	3	0,38	R141	17	120	757,74	2	0,45
R161	17	80	1211,22	3	0,25	R161	17	120	1192,99	2	1,13
R181	17	80	1991,59	3	110,38	R181	17	120	1786,75	2	0,48
R1101	17	80	2295,88	2	100,77	R1101	17	120	2052,03	2	0,36
R121	20	120	623,67	2	6,41	R181	20	120	1865,87	2	6,00
R141	20	120	798,39	2	2,38	R1101	20	120	2119,54	2	1,25
R161	20	120	1279,51	2	5,20						

Table 1 shows the computational results for the benchmark problems, whereas in column t_{cpu} the best CPU time (in seconds) is given. The five instances in bold type are solved to optimality for the first time. $m \leq M$ is the resulting number of routes required to cover all customers. Additionally, we examine the real-life instance of [4] that contains 22 customers. We solved the instance within 6,59 seconds to optimality. The objective function value of 88 monetary units is less than the solution of 89 obtained by [2].

Table 2 shows the average numbers of cuts added in test run 2 during optimization. For instances of [1], we indicate in the first column the number of cuts of instances with 15, in the second with 17 and in the third with 20 customers. Especially, implied bound and rounded capacity cuts are created in the subproblems.

Table 2. Average numbers of cuts

Cuts	Chen & Wu (2006)			Min (1989)
Gomory	3,6	3,2	3,2	2
Clique	5,6	5,6	7,0	6
Cover	1,1	1,8	0,2	0
Flow Cover	6,1	9,3	7,6	4
MIR	0,4	0,1	0,2	0
Implied Bound	317,0	567,5	216,0	236
Rounded Capacity	18,9	20,7	26,4	55
Sum	333,8	588,6	234,2	303

5 Conclusion

The paper considers a new mixed integer linear model for the VRPSDP. We solved the model using CPLEX 10.0 for small and medium instances of the problem in reasonable time. The selective use of preprocessing techniques and cuts improves the solver performance during the branch-and-cut approach. The computational tests shows that the approach is fast enough to be used for practical applications. Future work will concentrate on extensions to the described model (e.g. time windows).

References

1. Chen J-F, Wu T-H (2006) Vehicle Routing Problem with Simultaneous Deliveries and Pickups. Journal of the Operational Research Society 57:579–587
2. Dethloff J (2001) Vehicle Routing and Reverse Logistics: The Vehicle Routing Problem with Simultaneous Delivery and Pick-up. OR Spektrum 23:79–96
3. Lysgaard J (2004) CVRPSEP: A Package of Separation Routines for the Capacitated Vehicle Routing Problem. Working Paper 03–04, Department of Business Studies, Aarhus School of Business, University of Aarhus, Danmark
4. Min H (1989) The Multiple Vehicle Routing Problem with Simultaneous Delivery and Pick-up Points. Transportation Research, Series A 23A:377–386

Vehicle and Commodity Flow Synchronization

Jörn Schönberger[1], Herbert Kopfer[1], Bernd-Ludwig Wenning[2], and Henning Rekersbrink[3]

[1] University of Bremen, Chair of Logistics
{jsb,kopfer}@uni-bremen.de
[2] University of Bremen, Communication Networks
wenn@comnets.uni-bremen.de
[3] University of Bremen, BIBA - Bremer Institut für Produktion und Logistik GmbH
rek@biba.uni-bremen.de

1 Introduction

Network freight flow consolidation organizes the commodity flow with special attention to the minimization of the flow costs but the efficiency of transport resources is not addressed. In contrast, vehicle routing targets primarily the maximization of the efficiency of transport resources but the commodity-related preferences are treated as inferior requirements. This article is about the problem of synchronizing simultaneously the use of vehicles and the flow of commodities in a given transport network. Section 2 introduces the investigated scenario. Section 3 proposes a multi-commodity network flow model for the representation of the flow synchronization problem and Section 4 presents results from numerical experiments.

2 The Flow Synchronization Problem

Related and Previous Work. The synchronization of flows along independently determined paths in a network is investigated under the term *multi-commodity network flow problem* [3]. While most of the contributions aim at minimizing the sum of travel length (or similar objectives) only little work has been published on the assignment of commodity path parts to transport resources with intermediate resource change [1, 2].

B. Fleischmann et al.(Eds.), *Operations Research Proceedings 2008*,
DOI: 10.1007/978-3-642-00142-0_50,

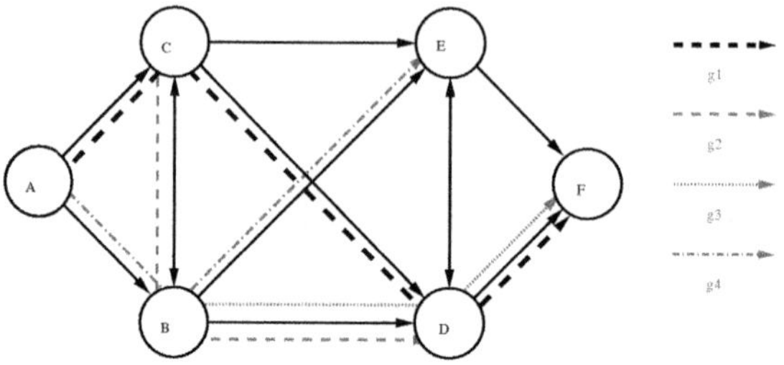

Fig. 1. Example Network with six nodes.

An Example of the Synchronization Challenge. Fig. 1 presents an example with three commodities $\gamma \in \{1, 2, 3\}$ in a network $\mathcal{G} = (\mathcal{V}, \mathcal{A}, c, \delta)$ (node set $\mathcal{V}$, arc set A, arc cost function c and travel time function δ. The path $PP(\gamma)$ was selected by commodity γ connecting its start node (site of availability of commodity γ) with its target node (site of demand for commodity γ): $PP(1) = (C, B, D)$, $PP(2) = (B, D, F)$ and $PP(3) = (A, B, E)$.
A *package hop* $H^p(i, j)$ describes the movement along the arc $(i, j) \in \mathcal{G}$. The hop $H^p(i, j)$ cannot leave from i towards j before time $a_{H^p(i,j)}$ and $dt_{H^p(i,j)}$ is its actually determined departure time. Transport resources hopping along the arcs in $\mathcal{G}$ execute the hops. A transport resource carries several commodities on a **service hop** $H^S(i, j)$. **Internal service hops** have a unchangeable departure time. In our example, an own vehicle travels along $PV(1) = (A, C, D, F)$. It starts at A at time 0 and visits the subsequent stops C, D, F using three internal service hops $H^S(A, C)$, $H^S(C, D)$ and $H^S(D, F)$. An **external service hop** can be booked at every time. External service hops are provided by logistic service providers (LSP) which are paid according to a previously known tariff for executing hops at the determined time (subcontraction).
A hop sequence $H^p(i_1, i_2), H^p(i_3, i_4), \ldots, H^p(i_{k-3}, i_{k-2}), H^p(i_{k-1} . i_k)$ is concatenated if $i_2 = i_3$, $i_4 = i_5$ and so on. Two concatenated hop sequences $H^1(i_1, i_2), \ldots, H^1(i_{l-1} . i_l)$ and $H^2(j_1, j_2), \ldots, H^2(j_{k-1} . j_k)$ are compatible if and only if $i_1 = j_1$ and $i_l = j_k$. In Fig. 1 the second hop of $PP(2)$ is compatible with the last hop in $PV(1)$ and the concatenated package hop sequence $H^p(C, B), H^p(B, D)$ in $PP(1)$ is compatible with the internal service hop $H^s(C, D)$.
Solving the synchronization problem requires the assignment of each hop sequence associated with $PP(1), \ldots, PP(K)$ to exactly one compatible sequence of (internal and/or external) service hops, so that the following requirements are met: If a package departs from a node then it has been brought to this node before (C1). The hop associated with

the first arc in $PV(m)$ ($m \in \{1, \ldots, M\}$) departs at time 0 (C2). The arriving and the departure time of a vehicle at a node coincide and neither a loading nor an unloading operation consumes time (C3). The initial hop in $PP(\gamma)$ starts not before time 0 (C4). The difference between the departure times belonging to two consecutively visited nodes i and j is at least $\delta(i,j)$ (C5). If a package hop sequence is assigned to a service hop sequence the first package hop $H^p(i_i, i_2)$ must be available before the first service hop $H^s(j_1, j_2)$ starts, e.g. $a_{H^p(i_i,i_2)} \leq dt_{H^s(j_1,j_2)}$ (C6). The sum of flow costs is minimal (C7).

3 Construction of the Matching Network

The nodes in the vehicle paths $PV(\cdot)$ and in the package paths $PP(\cdot)$ are re-labeled pair wise distinctly by 1, 2, ... In $\sigma(i)$ the original label of node i is stored. Only one departure time must be managed for each node. The relabeled nodes are shown in Fig. 2. The vehicle path $PV(1) := (A, C, D, F)$ is now $(1, 2, 3, 4)$, the package path $PP(1) := (C, B, D)$ is $(5, 6, 7)$ and so on.

The set $\mathcal{N}^{veh}$ contains all relabeled nodes from $PV(1), \ldots, PV(M)$, $\mathcal{N}^{pac}$ consists of the nodes appearing in $PP(1), \ldots, PP(K)$ and $\mathcal{N}^* := \mathcal{N}^{veh} \cup \mathcal{N}^{pac}$. All arcs from the vehicle paths form the set $\mathcal{A}^{veh}$ and all arcs forming the package paths are collected in the set $\mathcal{A}^{pac}$. Vehicle paths and the package paths are connected by *transfer arcs*. A transfer arc connects a node i in a vehicle path with a node j in a package path if and only if the two nodes i and j are the same in graph $\mathcal{G}$ ($\sigma(i) = \sigma(j)$). The set $\mathcal{A}^{transfer} := \{(i,j) \in \mathcal{N}^{veh} \times \mathcal{N}^{pac} \cup \mathcal{N}^{pac} \times \mathcal{N}^{veh} \mid \sigma(i) = \sigma(j)\}$ contains all transfer arcs. The solid arcs in Fig. 2 represent the internal service hops, the dotted arcs give the external service hops and the dashed arcs are the transfer arcs for changing the transport resource. The *flow synchronization graph* is defined as $\mathcal{G}^* := (\mathcal{N}^*, \mathcal{A}^*, c_f^*, \delta)$ where δ represents the travel times between the nodes along the arcs in $\mathcal{A}^* := \mathcal{A}^{veh} \cup \mathcal{A}^{pac} \cup \mathcal{A}^{transfer}$. The cost for transferring an arc depends upon the type of the arc. It is assumed that loading and unloading operations do not produce any costs so that travelling along a transfer arc is free. The cost function c_f^* assigning flow costs to the arcs in $\mathcal{A}^*$ is defined by $c_f^* := 0$ if $(i,j) \in \mathcal{A}^{veh} \cup \mathcal{A}^{trans}$, $c_f^* := f \cdot c(i,j)$ if $(i,j) \in \mathcal{A}^{pac}$. If $f > 0$ then the LSP incorporation is more expensive than the usage of the own vehicle but if $f = 0$ then internal and external service hops cause equal costs.

The cost reducing effect of reassigning package hop sequences from a sequence of external service hops to a sequence of internal service hops

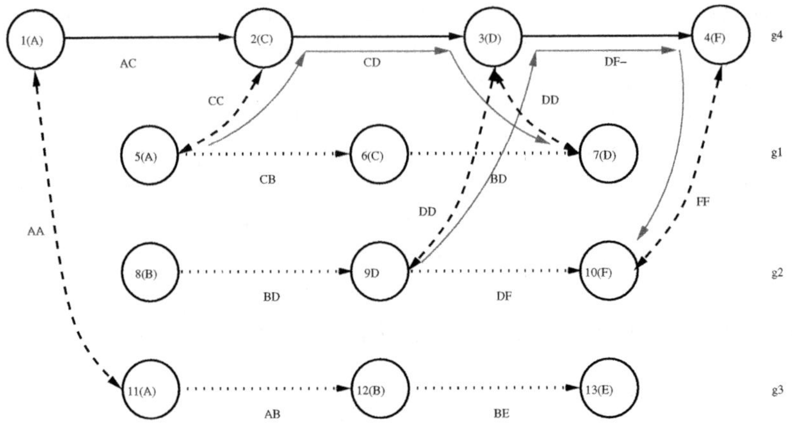

Fig. 2. Matching-Network (original labels of the nodes in brackets)

is demonstrated by the grey arcs in Fig. 2: If commodity $\gamma = 1$ uses the external service hop sequence $H(5,6), H(6,7)$ then the journey from 5 to 7 ($C \to B \to D$) causes costs at 2+2.8=4.8 money units. In case that the commodity uses the transfer hop $H(5,2)$, then the internal service hop $H(2,3)$ and finally the transfer hop $H(3,7)$ then no costs are accounted.

Multi Commodity Network Flow Model. Let K denote the number of commodities merged in the set $\mathcal{R}$. With $D(i,\gamma)$, we denote the offer ($D(i,\gamma) > 0$) of commodity γ at node i and the demand ($D(i,\gamma) < 0$) respectively. The initial node in the path of vehicle $m \in \{1,\ldots,M\}$ is denoted by v_m^+ and the initial node in the path of commodity γ is named i_γ^+. $\mathcal{M}$ is a sufficiently large number ("Big M").

In order to code the necessary flow decisions, we introduce three families of decision variables. The earliest starting time of hops originating from node i is saved in the decision variable d_i. If the arc $(i,j) \in \mathcal{A}^*$ is used then the binary variable u_{ij} is 1. The portion of the overall demand of commodity γ that flows along the arc (i,j) is represented by the continuous decision variable $x_{ij\gamma}$.

$$\sum_{i \in \mathcal{N}^*} \sum_{j \in \mathcal{N}^*} \sum_{\gamma=1}^{K} c_f^*(i,j) \cdot x_{ij\gamma} \rightarrow \min \tag{1}$$

$$\sum_{j \in \mathcal{N}^*} x_{ji\gamma} - \sum_{j \in \mathcal{N}^*} x_{ij\gamma} = D(i,\gamma) \; \forall i \in \mathcal{N}^* \; \forall \gamma \in \mathcal{R} \tag{2}$$

$$d_{v_m^+} = 0 \; \forall m \in \{1, \ldots, M\} \tag{3}$$

$$d_i + \delta(i,j) = d_j \; \forall (i,j) \in \mathcal{A}^{veh} \tag{4}$$

$$d_{i_\gamma^+} \geq 0 \; \forall \gamma \in \mathcal{R} \tag{5}$$

$$d_i + \delta(i,j) \leq d_j \; \forall (i,j) \in \mathcal{A}^{pac} \cup \mathcal{A}^{trans} \tag{6}$$

$$u_{ij} \geq \sum_{\gamma=1}^{M} x_{ij\gamma} \; \forall i,j \in \mathcal{N}^* \tag{7}$$

$$M \cdot (u_{ij} - 1) + d_j + \delta(i,j) \leq d_j \forall (i,j) \in \mathcal{A}^*. \tag{8}$$

Eq. (1) represents C7, (2) addresses C1, (3) corresponds to C2, and (4) to C3. Similarly, (5) ensures C4 and (6) does the same for C5. Finally, (7) and (8) addresses C6.

Test Cases. A set of 36 artificial test cases has been generated, each consisting of a transport graph $\mathcal{G}$, $\alpha = 1, 2$ or 3 vehicles, $\beta = 1, 2, 3$ or 4 commodities and vehicle as well as path proposals $PV(\cdot)$ and $PP(\cdot)$. In the original network $\mathcal{G} := (\mathcal{N}, \mathcal{A}, c, \delta)$ the first 25 nodes from the Solomon instance R104 (dropping the time windows) form the node set $\mathcal{N}$. A minimal spanning tree connecting the 25 nodes generated by Kruskal's algorithm determines the arc set $\mathcal{N}$. The Euclidean Distance $\delta(i,j)$ gives the distance matrix and the costs $c(i,j)$ for traversing the arc (i,j) are set to one money unit for each distance unit. For each commodity γ, a demand and an offer location are randomly generated and the shortest path $PP(\gamma)$ in $\mathcal{G}$ is calculated using the Dijkstra-algorithm. The path $PV(\cdot)$ starts at an arbitrarily selected node, it continues on a shortest path to a randomly selected stop in $PP(1)$, follows $PP(1)$ for a randomly generated number of hops. Then, it continues on a shortest path to a node in the $PP(2)$ and so on.

4 Numerical Experiments

Setup of the Experiments. The flow synchronization graph *calG** is set up for each of the 36 test cases and the model (1)-(8) is solved once with $f = 0$ (internal and external service hops cause the same costs) and once with $f = 1$ (external service hops are more expensive). The *lp_solve* mixed-integer linear program solver is deployed for the derivation of the optimal solution for the model (1)-(8). For each scenario (α, β) we have calculated the averagely observed number $l_f(\alpha, \beta)$ of used external service hops and the increase $l(\alpha, \beta) := \frac{l_1(\alpha,\beta)}{l_0(\alpha,\beta)} - 1$ of this number has been calculated. Similarly, we have determined the increase $v(\alpha, \beta)$ of used internal service hops and the increase $t(\alpha, \beta)$ of used transfer hops.

Results. The observed values for $l(\alpha, \beta)$, $v(\alpha, \beta)$ and $t(\alpha, \beta)$ are compiled in Table 1. If the number of vehicles or the number of commodities is increased then the number of used external service hops decreases ($l(\alpha, \beta)$ decreases). At the same time, the number of used internal service hops is lifted ($v(\alpha, \beta)$ grows up). The utilization of the transfer arcs also increases ($t(\alpha, \beta)$ increases).

Table 1. Variation of the number of used hops.

	$\alpha = 1$			$\alpha = 2$			$\alpha = 3$		
β	$l(\alpha,\beta)$	$v(\alpha,\beta)$	$t(\alpha,\beta)$	$l(\alpha,\beta)$	$v(\alpha,\beta)$	$t(\alpha,\beta)$	$l(\alpha,\beta)$	$v(\alpha,\beta)$	$t(\alpha,\beta)$
3	0%	0%	14%	-33%	18%	43%	-63%	20%	13%
4	-25%	25%	33%	–	–	–	-77%	-19%	133%
5	-30%	35%	67%	-64%	50%	113%	–	–	–
6	-46%	26%	138%	–	–	–	–	–	–

– not solved within 15 minutes

5 Conclusions

We have introduced and modeled the problem of synchronizing path proposals of transport resources and commodities. A mixed integer linear program is proposed. The experimental results demonstrate the intricacy of the flow synchronization. Even for a small number of involved resources and commodities the identification of the best synchronization decisions is impossible. Future research is dedicated to the development of hybrid algorithms that combine a heuristic search with the linear programming in order to shift the border of solvability to larger number of vehicles, commodities and nodes.

References

1. Borndörfer R, Grötschel M, Pfetsch ME (2004) Models for Line Planning in Public Transport. ZIB-Report 04-10, Berlin
2. Pickl S, Mues C. (2005) Transhipment and Time-WIndows in Vehicle Routing. In: Proceedings of International Symposium of Parrallel Architectures, Algorithms and Networks
3. Assad AA (1978) Multicommodity Network Flow – A Survey Networks 8:37–91

Multi-Criteria Optimization for Regional Timetable Synchronization in Public Transport

Ingmar Schüle, Anca Diana Dragan, Alexander Radev, Michael Schröder, and Karl-Heinz Küfer

Fraunhofer Institut für Techno- und Wirtschaftsmathematik (ITWM), Fraunhofer-Platz 1, 67663 Kaiserslautern,
{schuele,dragan,radev,schroeder,kuefer}@itwm.fraunhofer.de

1 Introduction

The need for synchronizing timetables occurs when several public transportation companies interact in the same area, e.g. a city with busses, trains and trams. Existing approaches that optimize a global, waiting time oriented objective function reach their limit quite fast concerning real-life applications. While optimizing public transportation timetables, "winners" and "losers" will inevitably come about and the traffic planner has to keep control of this process. Thus, we propose an optimization approach that enhances the overall situation without losing sight of the deficiencies arising at particular stations.

2 Concept of Timetable Synchronization in Public Transport

In our concept for optimizing public transportation timetables, the goal is to synchronize a set of lines (e.g. busses, trains, ...) by changing their starting times such that passengers can transfer more conveniently. For this, we do not take the often used approach of simply minimizing waiting times, but we use a concept of trying to achieve transfers that can be called convenient. For these, the time should not be too long, but also not too short, in order to reduce the risk of missing a transfer if the arriving vehicle is delayed. For optimization purposes, we assign to all possible waiting times at a transfer a corresponding penalty-value. One of our goals is to minimize the sum of these penalties over all transfers. For a detailed introduction, see [4].

B. Fleischmann et al.(Eds.), *Operations Research Proceedings 2008*,
DOI: 10.1007/978-3-642-00142-0_51, © Springer-Verlag Berlin Heidelberg 2009

We introduce the following notation: In a public transportation network we have a set of *lines* $L = \{l_1, ..., l_m\}$ with the index set $M = \{1, ..., m\}$. For each line we define the set of allowed *shifts* $S_i = \{s_{i1}, ..., s_{in_i}\}$, $s_{ij} \in \mathbb{Z}$, with the index set $N_i = \{1, ..., n_i\}$ for $i \in M$. The task is to assign to each line l_i a shift out of S_i, which can be modelled as a Quadratic Semi-Assignment Problem (QSAP) - see [5]. The shift that is assigned to line l_i is called s_i and this means that all vehicles serving this line start their tour s_i minutes later.
Whenever two lines meet, a theoretical transfer possibility occurs, but in reality most passengers change between lines only at certain important points in the network. These points, called *network nodes*, do not necessarily consist of just a single station, but also of several stations within walking distance.

3 MIP Formulation

For a better mathematical handling of the concept, we give a graph theoretical representation of the problem that was first proposed by Voss [5]. For this, we build a multipartite graph $G = (V, E)$ with all elements $s_{ij} \in S_i$ for each $i \in M$ as vertices (the corresponding vertex to s_{ij} is called v_{ij}). We set edges between every two vertices whose corresponding elements s_{ij} and s_{kl} belong to different lines $l_i, l_k \in L$, i.e. $i \neq k$.
Therefore, we get

$$V = \{v_{ij} \mid i \in M, j \in N_i\} \quad \text{and}$$

$$E = \{\langle v_{ij}, v_{kl}\rangle \mid v_{ij}, v_{kl} \in V, i \neq k\}.$$

Figure 1 shows such a complete m-partite graph. A shift pattern corresponds to an m-clique (a complete subgraph with m vertices) and our goal is to find the m-clique with minimal edge-weights, which is an NP-hard problem.
With this graph we get the following MIP formulation for the given QSAP:

$$\begin{aligned}
\text{(MIP)}\ \min & \sum_{e\in E} b_e \cdot y_e \\
\text{s.t.} & \sum_{j\in N_i} x_{v_{ij}} = 1 \quad \forall i \in M \\
& - x_{v_{ij}} + \sum_{l\in N_k} y_{\langle v_{ij}, v_{kl}\rangle} = 0 \quad \forall v_{ij} \in V\ \forall k \in M \setminus \{i\} \\
& y_e \geq 0 \quad \forall e \in E \\
& x_v \in \{0,1\} \quad \forall v \in V.
\end{aligned}$$

This problem can be solved by a MIP solver (e.g. CPLEX), but for larger problem instances the solution process takes an unreasonable amount of time. These long calculation runs are already known from QASPs that have many local minima. Figure 2 shows the surface of the penalty function for two lines and this simple example already illustrates how complex the given problem is.

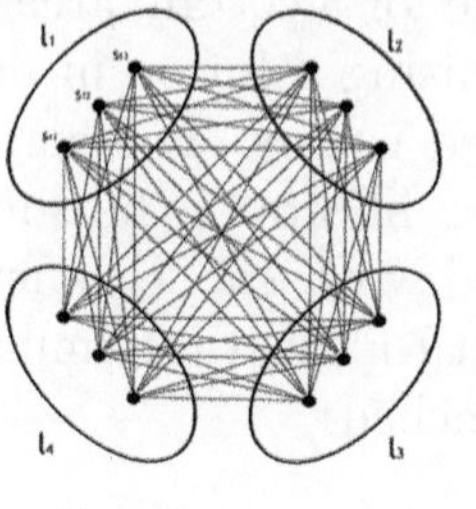

Fig. 1. Graph representation of a QSAP.

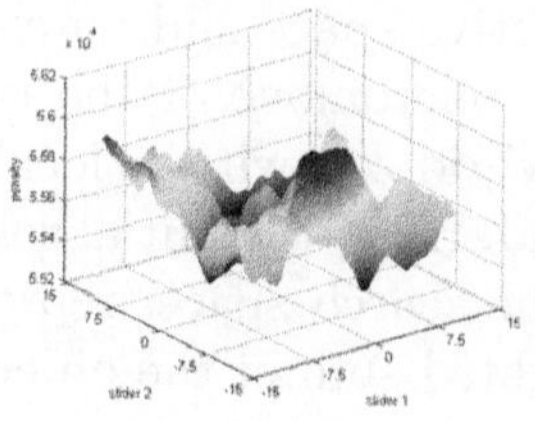

Fig. 2. plot of the surface of the first objective function

A lower bound for the QSAP can be achieved by a relaxation of the integrality of the variables x_{ij}, but the optimality gap of the linear relaxation of MIP is too large for useful solutions.

4 Metaheuristics for Multi-criteria Optimization

The MIP approach has the disadvantage that it models the problem in a single-criteria way, so it cannot reflect the multi-dimensional character of the problem. In this section we choose three objective functions to model the intentions of the traffic planners:

1. Minimize the overall penalty,
2. Minimize the worst result at a network node,

3. Minimize the changes made to the current timetable.

Because the problem is NP-hard even for a single objective, we use different metaheuristics to approximate the pareto front. These are presented in the following sections.

4.1 Ant Colony Optimization

In Ant Colony Optimization (ACO), a colony of virtual ants explores the search space of a given problem to find good solutions. For the timetable synchronization problem, each ant tries to construct a max-clique on the graph from section 3. The ants share information about promising edges via stigmergy, an indirect way of communicating by changing the environment with pheromones. ACO is considered for our problem because it has proven to give promising results on quadratic assignment problems. For more information about ACO, see [1].
We deal with the multi-dimensional character of our problem by using several ant colonies. Each colony is responsible for a certain area in the objective space and an ant that finds a promising solution in an area adds pheromone of the type of the associated colony. The algorithm is hybridized with a local search that further improves the solutions found. Here, the initial solutions are optimized w.r.t. single dimensions in the search space, represented by lines, until a local minimum of a weighted sum of the objective functions is reached.

4.2 Evolutionary Algorithm

Currently, Multiobjective Evolutionary Algorithms (MOEA) are one of the standard approaches for multiobjective combinatorial optimization. It is a population based technique inspired by biological evolution. It starts with initial seeding, followed by iterative application of fitness assignment, and three genetic operators which emulate reproduction: selection (choosing parents for the next generation with preference towards higher fitness by binary tournament), recombination (choosing different genotype portions from selected parents uniformly) and mutation (binomial random novelty). Instead of using binary strings, we encode the genetic information in a manner which utilizes the structure of the problem at hand.
Much of the research in MOEA is concentrated on fitness assignment and parent selection. We have implemented the legacy VEGA as a reference and the state-of-the-art SPEA-2 and NSGA-II strategies. The later are elitist, e.g. they mix old population with new individual and

get the best with respect to non-dominance and a specific population density measure. Detailed presentation of MOEA ranging from basics to latest technical and theoretical details can be found in [2].

4.3 Simulated Annealing

Simulated Annealing (SA) models the way a metal cools into a minimum energy state. Starting from a certain point, the search space is analyzed by choosing a neighbor and moving to that position either if the energy is lower or if the temperature is still high enough to allow transitions to worse states. For more information about SA, see [3].
We define a state as a vector of shifts $(s_1, s_2, ..., s_n)$ and its energy as the weighted sum of the three objective functions used. To find a neighbor, we pick a random line and consider a percentage p of the possible shifts this line can have. Out of those, the one that minimizes the energy is applied and all other lines are kept the same. p increases throughout the run, making the probability to choose a worse neighbor smaller, since more shifts are analyzed. Therefore, when the system reaches a low temperature and does not allow transitions to higher energy states, p will already be large enough to guarantee that most selected neighbors are better. This way, the time wasted on choosing worse states and rejecting them is reduced.
Each run receives a random starting point and a direction in the objective space, consisting of a vector of weights that establish how much each objective influences the energy. The SA solver uses such runs iteratively with different directions to approximate the pareto front.

5 Results

The example we use arises from the current timetable of the city of Kaiserslautern (100.000 inhabitants), including city-busses, regional-busses and trains. There are 31 lines meeting at four network nodes and creating approx. 10.000 transfers in a time period of four hours.
On this example, the computation of the exact MIP solution (single-criterion), CPLEX 11.1 needed slightly more than 25 hours. For the metaheuristics, we used 30 runs for each solver and computed the volumes of the pareto fronts after three time milestones, of about one minute, ten minutes and one hour. The results are presented in Figure 3, which shows how close the average pareto fronts volumes were to the one obtained from the set of all best solutions found. The time milestones are represented here by the number of evaluations of the objective function (which is the most computationally expensive).

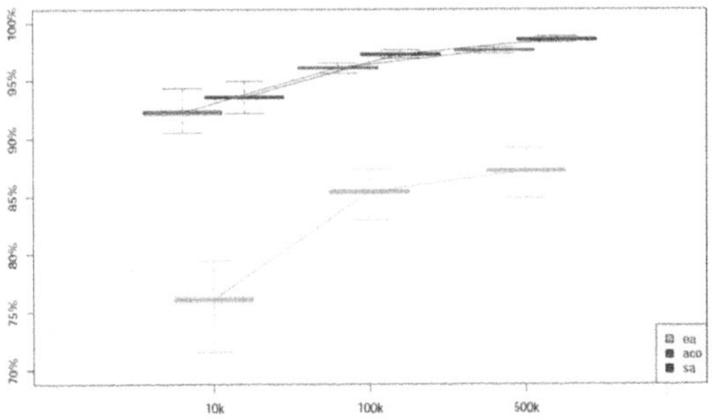

Fig. 3. Results that measure how close the meta-heuristic algorithms got the the pareto front volume (represented by 100%) after the three time milestones. EA (bottom), ACO and SA (top).

6 Conclusions

For the problem at hand, metaheuristics are preferable, as they reach quality solutions within the first few seconds of optimization. Among these, SA and ACO performed best on the tested scenario. However, a lot of interesting research topics are still to be addressed. The traffic planner will be allowed to influence the optimization process by setting constraints (e.g. to force a single transfer to become better or to forbid the transfer possibilites between two lines to become worse). Furthermore, a hybridization of metaheuristics and the MIP model seems promising. Finally, the methods should be tested on more benchmark files that include such constraints.

Acknowledgement. This project is funded by the Fraunhofer Society, project-no. 662984.

References

1. Dorigo, M. and Stützle, T. Ant Colony Optimization. The MITPress, Cambridge, Massachusetts, 2004.
2. Eiben, A. E. and Smith, J. E. Introduction to Evolutionary Computing. Springer, 2003.
3. Kirkpatrick, S., Gelatt Jr., C. D. and Vecchi, M. P. Optimization by simulated annealing. Science, 220(4598), May 1983.
4. Schröder, M. and Schüle, I. Interaktive mehrkriterielle Optimierung für die regionale Fahrplanabstimmung in Verkehrsverbünden. Straßenverkehrstechnik, (6):332-340, June 2008.
5. Voß, S. Network design formulations in schedule synchronization. In: M. Desrochers and J.-M. Rousseau (eds.), Computer-Aided Transit Scheduling, Lecture Notes in Economics and Mathematical Systems 386, Springer, Berlin, 137 - 152, 1992.

Transportation Planning in Dynamic Environments

René Schumann, Thomas Timmermann, and Ingo J. Timm

Information Systems and Simulation, Institute of Computer Science, Goethe University, Robert-Mayer-Str. 10, 60325 Frankfurt am Main
{reschu,timmerm,timm}@cs.uni-frankfurt.de

1 Introduction

Transportation planning in realistic scenarios has to deal with a dynamic uncertain environment. The actual state of all information relevant to planning can be estimated and be the base of a static planning environment which is often the base for classic planning algorithms. If the environment differs from the one used for planning, either by a wrong estimation or due to the dynamics in the environment the actual plan has to be adapted. Dynamic transportation planning is currently an active research area; nevertheless it is not a new problem in the literature. A good overview of the development of this field can be found in [1]. Different technologies like agents or meta-heuristics and their possible integration are in discussion [2]. Even if a lot of research has been done on the question how these problems can be solved, the dynamics has been rarely tried to measure. Commonly, dynamic transportation planning is characterized by the occurrence of disturbing events that invalidate the current plan. What kinds of events occur depends on the planning problem that was assumed. In the current state of research different technologies claim to be fruitful in more dynamic environments. Larsen [3] points out the importance of measuring dynamics for efficient and sufficient performance comparison. Therefore he defines the degree of dynamism (DoD).Moreover, from an engineering perspective it would be very useful to have metrics and measurements at hand that can help to choose one technology for a given problem. The vision is that based on the characteristics of the given application one could decide which is the most appropriate technology. Therefore, different questions have to be answered, like how can dynamics be measured,

B. Fleischmann et al.(Eds.), *Operations Research Proceedings 2008*,

DOI: 10.1007/978-3-642-00142-0_52,

to what DoD classical approaches are competitive, and how different technologies perform on a varying DoD.
In this article we tackle the first question. Therefore, we discuss here different existing approaches to measure dynamics in transportation planning. Based on the well-known multiple depot vehicle routing problem with time windows (MDVRPTW) problem, we present an extended version of the measurement of the DoD. In section 3, we present empirical results and finally we discuss our findings.

2 Metrics for Dynamics in Transportation Planning

In the following we look at abstract problems like the MDVRPTW, but our argumentation is valid for the VRPTW as well, as we did not argue about the existence of different depots so far. The MDVRPTW describes the problem to deliver uniform goods to a set of customers from a set of depots with vehicles with heterogeneous capacities. The delivery has to be done within a customer-specified time window and the vehicles need to return to the same depot where they have started. Each customer has to be delivered once. A customer i is represented as quintuple $(id_i, g_i, ds_i, e_i, l_i)$ with

- id_i identifies a customer i,
- g_i describes the demand of the customer,
- ds_i the needed service time of customer i,
- e_i earliest arrival for service and
- l_i latest arrival for service

The definition of time windows in this case refers to the definition of so-called soft time windows from Chiang and Russell [4]. If a time window is missed, the plan is still valid but the resulting delay will be measured by a penalty P.
Larsen [3] presents a DoD for the dynamic VRP problem with and without time windows which he calls the *Effective Degree of Dynamism (EDOD(TW))*. The EDODTW is defined on dynamic vehicle routing problem (DVRP)-instances where only new customers are liable to a time window restriction but not the customers known from the beginning. This makes a slight modification of the EDODTW formula necessary. Thereby n_d defines the number of new customer requests. The following formula makes sure that only the time window restriction of new customers in the MDVRPTW influences $\phi_{mdvrptw}$:

$$\phi_{mdvrptw} = \frac{\sum_{i=1}^{n_d} [T-(l_i-t_i)]/T}{n_d}.$$

But it has to be mentioned that Larsen only regards one event, the occurrence of new customers. This limits the measurement of dynamism as a lot of other events may occur depending on the given application. Within the MDVRPTW one could think of ten different events that can occur. These are: the set of customers can change, the demand of a customer can change (g_i), the earliest delivery time can change (e_i), the latest delivery time can change (l_i) and the distance between two points can change (d_{ij}). By change we mean it can decrease or increase. It is an open question if events that did not invalidate the current plan have to be regarded, measuring the DoD. One could argue that due to these events improvement potential arises which could be relevant in a comparison between online and offline algorithms. As we do not want to address this question here, we restricted the regarded events in this work to those events that can invalidate or at least decrease the solution quality of a plan. Those events are:

1. Decrease of l_i
2. Increase of e_i
3. Increase of distance d_{ij}
4. Increase of demand g_i of customer i
5. New customer requests.

We extend the DoD presented by Larsen [3] incorporating these five events. The question if and how the remaining events can be incorporated in the measurement of the DoD has to be subject of further research.

Let Γ be the set of the regarded events. Then Γ is composed of the following types of γ: It is always necessary to find the latest point in time *pit* when an event has to be integrated into the route plan. *pit* can be calculated using a function h regarding event γ with $h(\gamma)$. The event type of γ is represented as binary quintuple with $\gamma = (c_1, c_2, c_3, c_4, c_5)$: $c_i = 1$ for event type i and $c_j = 0\ \forall j \neq i$. Thus the formula by Larsen will be modified as follows:

$$\phi_{mdvrptw} = \frac{\sum_{\forall\gamma\in\Gamma} [T-(h(\gamma)-t_\gamma)]/T}{|\Gamma|}$$

The function $h(\gamma)$ examines whether or not an event $\gamma \in \Gamma$ and thus should take effect on $\phi_{mdvrptw}$. In case of an event of type 1 (decrease of l_i to l'_i, represented by $\gamma = (1,0,0,0,0)$), l'_i describes the desired point in time *pit*. In case of an increasing e_i to e'_i causing a delay at one customer j that follows customer i in this route l_j has to be used. If the distance d_{ij} increases between two customers i and j, the *latest-time* restriction l_k of the first following customer k in this route where

the time window is missed. If there is an increase of g_i, the value *pit* is given by the departure time to the first following customer x in the considered route whose demand g_x cannot be satisfied by the remaining goods of the vehicle. *pit* in case of event type new customer request i remains l_i as hitherto.
Let us assume a_i is the planned arrival time of the vehicle at customer i, a route $Tour$ represents an ordered list of customers, CAP is the capacity of the vehicle used on the regarded route and Δ describes the modification of the changed information l_i, e_i and d_{ij} to l'_i, e'_i and d'_{ij} with $\Delta_{e_i} = e'_i - e_i$, $\Delta_{d_{ij}} = d'_{ij} - d_{ij}$ and $\Delta_{g_i} = g'_i - g_i$. Then the function $h(\gamma)$ can be defined as follows:

$$h(\gamma) = \begin{cases} l_i, & \text{if } c_1 = 1 \wedge a_i > l_i \\ l_j, & \text{if } c_2 = 1 \wedge \exists\ min\ j \geq i : a_j + \Delta_{e_i} > l_j \\ l_k, & \text{if } c_3 = 1 \wedge \exists\ min\ k > i : a_k + \Delta_{d_{ij}} > l_k \\ a_x - d_{x-1,x}, & \text{if } c_4 = 1 \wedge \Delta_{g_i} + \sum_{\forall j \in Tour} g_j > CAP: \\ & x \text{ is first customer holds } \sum_{k=n}^{x} g_k > \Delta_{g_i} \\ l_i, & \text{if } c_5 = 1 \\ 0 & \text{else, can't be reached by the definition of } \Gamma \end{cases}$$

Of course, the formula holds only under the ceteris paribus assumption, that is here that there exists no event in the future that changes parameters of that formula. It has to be mentioned that the extended DoD depends on the computed route plans, and therefore on the solution techniques. To compute this DoD two ways are possible. One could compute the DoD after an initial plan has been generated (offline DoD) or one compute it during the plan execution respecting the plan modifications that were done in earlier repair steps (online DoD).

3 Measuring Dynamics: A Case Study

We implemented a tabu-search meta-heuristic with an incorporated repair technique based on local search, details are provided in [5]. We use the large problem instances pr06, pr10, pr16 and pr20 presented by Cordeau et al. [6]. We used an event generator that creates all possible events randomly but uniformly distributed along the planning horizon varying the number of events from 40 up to 150. In figure 1 one can see the results for the offline and online DoD. These results are mean values and base on 179 tests. The offline DoD overestimates the online DoD except for a large number of disturbing events. The reason for

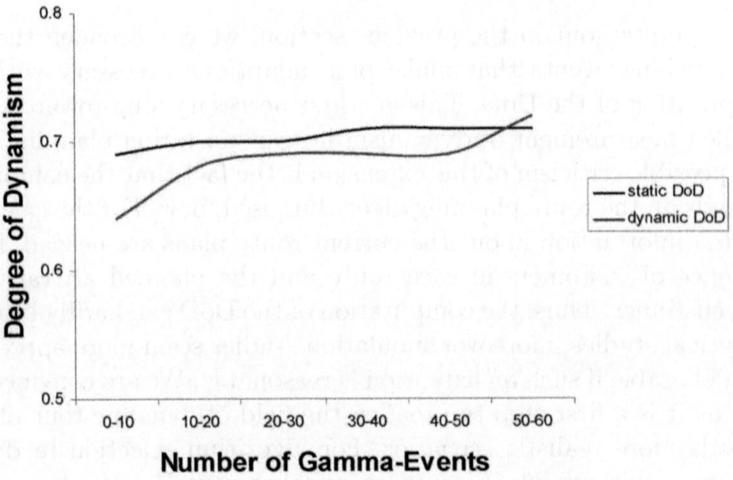

Fig. 1. DoD for different numbers of disturbing events

this is that it cannot incorporate the consequences of earlier plan modifications. As the events are uniformly distributed among the planning horizon, one could expect a linear curve for the offline DoD which is nearly met in our experiments. The curve of the online DoD is more interesting. As expected the DoD increases with the number of events. But one can see three phases of this increasing. We interpret this as follows: First, when only a few events occur the DoD increases faster. This implies that the initial plan has to be adapted. In the second phase, the DoD remains quit constant, even if the number of disturbing events increases. In this phase, the existing plan is quite robust, among other this is a consequence of the repair actions in the first phase. If the number of disturbing events increases more, the DoD increases faster again, as the plan is no longer robust and the need for further plan adaptions occurs. The fraction of events that can possibly invalidate the current plan is 71.71% of all events that occur. Thereby the fraction of events that really causes a repair action is 30.71%. Whereas in the offline computation only 20.97% of all events where assumed to cause a repair action. So the offline computation underestimates the number of critical events by nearly 10%. The fraction of the event "new customer" is 16.42% of all events. Thus, the fraction of events that can be monitored with our extension of the DoD that really forces a plan repair increases by 87.02%. Thus we can broaden our view of dynamic events in tour planning significantly.

4 Discussion of our Findings

As we pointed out in the previous section, we can broaden the varity of disturbing events that make plan adaptions necessary within the computation of the DoD. This is a first necessary step towards a more detailed measurement of dynamism in transportation planning.
One possible criticism of this extension is the fact that the computation depends on the route planning algorithm used, now. For the calculation of $h(\gamma)$ information about the current route plans are needed, like the sequence of customers in each route and the planned arrival time at each customer. Thus, the computation of the DoD can hardly be done in analytical studies, moreover simulation studies seem more appropriate. It is debatable if such an extension is reasonable. We are convinced that it is, as it is a first step to broaden the field of dynamic tour planning towards more realistic scenarios. For algorithm selection in dynamic planning environments it is important to quantify the dynamism within different applications, therefore the extension of the DoD could be a reasonable way.

References

1. Powell WB (2003) Dynamic Models of Transportation Operations. In: de Kok AG, Graves SC (eds) Handbooks in Operations Research and Management Science 11 Supply Chain Management: Design, Coordination and Operation. Elsevier, Amsterdam
2. Langer H, Timm IJ, Schönberger J, Kopfer H (2006) Integration von Software-Agenten und Soft-Computing-Methoden für die Transportplanung. In Nissen V, Petsch M (eds) 9. Symposium Soft Computing Softwareagenten und Soft Computing im Geschäftsprozessmanagement. Cuviller Verlag, Göttingen
3. Larsen A (2001) The Dynamic Vehicle Routing Problem. PhD Thesis, Technical University of Denmark
4. Chiang WC, Russell RA (2004) A Metaheuristic for the Vehicle Routing Problem with Soft Time Windows. Journal of the Operational Research 55:1298–1310
5. Timmermann T, Schumann R (2008) An approach to solve the Multi Depot Vehicle Routing Problem with Time Windows (MDVRPTW) in static and dynamic scenarios. Presented at the 22. PuK Workshop at KI 2008
6. Cordeau JF, Laporte G, Mercier A (2004) Improved Tabu Search Algorithm for the Handling of Route Duration Constraints in Vehicle Routing Problems with Time Windows. Journal of the Operational Research 55:542–546

The CaSSandra Project: Computing Safe and Efficient Routes with GIS

Luca Urciuoli[1] and Jonas Tornberg[2]

[1] Lund Industrial Engineering and Management, Ole Romers Vag 1 Box 118, 22100 Lund, Sweden
luca.urciuoli@tlog.lth.se

[2] Chalmers University of Technology, City and Mobility, SE-41296 Gothenburg, Sweden
jonas@chalmers.se

1 Introduction

Large amounts of dangerous goods are kept constantly on the move in Europe because of their significant impact on economic growth and to support quality of life. According to available statistics [3], road transportation accounts for the movement of the major part of dangerous goods within Europe (58% in 2002). The access to a well built and distributed road infrastructure gives higher flexibility and door to door capabilities [7]. Consequently, transport purchasers perceive this transportation mode as highly effective and economically advantageous. However, the same factors stated above oblige material flows to travel through highly-populated areas or highly-trafficed road segments. As a consequence the exposure of civilians to accident risks increases drastically [3].

History shows that accidents which take place during the transportation of hazardous material can have the same magnitude as those occurring in industrial plants [13]. Possible consequences may include fatality of human beings or ecological disaster if the cargo is dispersed in water catchment areas [8],[6]. The ramifications on private stakeholders may include delayed shipment, undelivered shipment, wasted cargo and higher transportation costs (i.e. bridge collapse) [8],[2].

Dangerous goods or hazardous materials (hazmat) are any solid, liquid or gas substances that can have harmful effects for living organisms, property or environment [11]. Laws and regulations for the transportation of dangerous goods in Europe are first collected by the United Na-

B. Fleischmann et al.(Eds.), *Operations Research Proceedings 2008*,
DOI: 10.1007/978-3-642-00142-0_53, © Springer-Verlag Berlin Heidelberg 2009

tions Economic Commission for Europe, UNECE and then extended to all transportation means (road, rail, sea and air) through specific organizations. The transportation of dangerous goods over European roads is regulated by the Agreement concerning the International Carriage of Dangerous Goods by Roads (ADR) that is enforced in Sweden by the Swedish Rescue Services Agency (SRSA)[11]. The SRSA publishes yearly dangerous goods recommended and restricted road segments. Recommended roads are classified as primary, for throughway traffic and secondary, for local transportation from and to the primary network. The restricted roads are road tunnels and segments in proximity of water catchment areas [10].

Restricted roads are defined in the ADR regulations, while primary and secondary roads are defined independently by each Swedish municipality to grossly avoid densely populated areas. Since municipalities don't perform quantitative risk assessments based on the OD of transportation assignments, there could be further route alternatives that may minimize population exposure to accident risks. In addition, drivers seeking to optimize shipments' efficiency may prefer to minimize travelling time by avoiding only the forbidden segments instead of taking into account both forbidden and recommended segments.

Previous research shows that is possible to model transportation routes by applying heuristic techniques (shortest path algorithms) to minimize a cost function including risk, costs, safety in terms of vehicle collisions, potential exposure of population, road users affected, and travel delays [8],[4],[6]. This investigation is part of a research project co-financed by Volvo Trucks, Volvo Logistics, Ericsson Microwave systems and the Swedish Governmental Agency for Innovation Systems (VINNOVA). The aim of this study is to compare 1) fastest routes computed on the entire transportation network exclusive of restricted roads, with 2) fastest routes on DGR restricted and recommended routes and finally 3) minimum night and day risk routes computed on the whole transportation network excluding the DGR restricted segments. The main objective is to demonstrate how the Swedish recommendations function to minimize societal exposure to accident risks. Additionally, the analysis of the routes will also show how efficiency factors in terms of travel time are affected when avoiding densely populated areas. The analysis is performed on a case study based on Volvo Logistics' real transport operations of material flows containing 2-propanol to be delivered to the Port of Gothenburg, in the region of Vastra Gotaland, Sweden. The results include the development of a Decision Support System, based on Geographic Information Systems (GIS), capable of calculating safer and efficient routes. In addition, this paper discusses the importance of developing a cooperative platform for emergency preparedness and resiliency management including dynamic routing (based on real-time information) and evacuation planning to optimally direct rescue service operations.

2 Method

Shortest path algorithms are exploited to compute minimum cost routes between two points (o/d). These algorithms are based on the representation of the transportation network in oriented graphs, which are structures composed by nodes (roads intersections) and arcs (road segments connecting the nodes). Graphs are mathematically represented with $G = (V, E)$, where V is the set of nodes and E the set of arcs of the graph. To compute shortest paths it is important to assign an impedance function $i : E \rightarrow \Re$ (the amount of resistance required to traverse a line) to each arc $(u, v) \in E$, with $u, v \in V$. The total impedance of a path between two generic nodes u and v is given in the equation below [1].

$$i(p) = \sum_{j=1}^{k} i(v_{j-1}, v_j) \tag{1}$$

A path between the nodes u and v can be formalized as a sequence of nodes $p = (v_0, v_1, ..., v_j, ..., v_k)$ where $v_j \in V, (v_j, v_{j+1}) \in E, j = 1, ...k, v_0 = u$ and $v_k = v$. As a consequence, the shortest path can be defined as in equation 2 [1].

$$\delta(u,v) = \begin{cases} min\{i(p)u \xrightarrow{p} v, \; if \; a \; path \; exists \\ \infty, \; otherwise \end{cases} \tag{2}$$

A commercial database including geometrical and operational characteristics of the transport network (TeleAtlas MultiNet) was configured and used in the GIS environment. In addition, to compute the shortest paths, this study exploits a hierarchical routing algorithm and two impedance functions. The first function is based on travel time to traverse the road segments and the second on a risk index as defined in equation 3. Risk is traditionally defined as a combination of three factors: a scenario, the likelihood of the scenario, and its consequences [5]. However in the analysis of hazard transportation other factors to be considered are accident frequency (α), release probability (ρ), consequences (γ) and the risk preferences of the decision maker (φ) [6],[4],[8].

$$Risk = \alpha \cdot \rho \cdot (\gamma)^{\varphi} \quad (3)$$

The accident rate statistics have been extracted from the Swedish Traffic Accident Data Acquisition database (STRADA). Yearly severity and frequency of traffic accidents is defined, identified and localised by means of geographical coordinates. It is assumed the shipment taking place on urban multi-lane roads in tank trailers with a material release frequency of 0.067 [9]. Accident consequences are measured in terms of exposed population at night and day time within an area of 300 meters around every travelled segment (buffer zone). The Day and Night population density are based on parcel point coordinates provided by Statistics Sweden (SCB). Buffer zone techniques are then exploited to determine the population density per square meter in the proximity of the segments. The damage to society, in form of human fatality, has been estimated by the Swedish Road Administration to about EUR1.5 million [12]. Risk preferences are assumed to be neutral, thus $\varphi = 1$. Finally, the material shipped by Volvo Logistics contains 2-propanol and it is part of the production process of the vehicle industry. 2-propanol is a highly toxic and flammable substance. For this reason, the ADR has classified it as a class III flammable liquids dangerous good (flash point at a temperature below 60.5 °C) [11].

3 Results

This section presents the results obtained by running a shortest path algorithm minimizing travel time and risks on different network configurations. Table 1 shows how travel time (TT), travel distance (TD), accident frequency (AF), Night and Day Population Exposure (NPE and DPE) and risks (RISK), according to equation 3, are affected. The same paths shown in the table are depicted in figure 1. The first path

Table 1. Routing analysis impact measures

Path	Parameter to minimize	TT (min)	TD (km)	NPE	DPE	AF	RISK
1	Travel Time	26	39	0,20	1,02	0,256	32236,2
2	SRSA Travel Time	33	47	0,08	0,17	0,317	8307,1
3	Night Risk	59	62	0,02	0,05	0,04	378,2
4	Day Risk	68	62	0,05	0,03	0,04	403

is computed by minimizing travel time on the whole transportation network excluding the ADR restricted roads. In the table it is possible to notice the elevated risk level of this route. Thus, drivers considering only the forbidden segments place the general population in considerable danger. The second path is instead obtained by minimizing the travel time on the forbidden and recommended roads by the SRSA. This path shows a relatively slight increase of the travel time (+27%) and length (21%) and a noticeable decrement of the risks (-74%), highlighting an optimal trade-off between efficiency and risk factors. The third and fourth routes are computed by minimizing night and day risks on the transportation network deprived of the ADR restricted roads. Comparing them with the second path, travel times (+78% path 3 and +106% path 4) and distances (+32.7% path 3 and 32.6% path 4) increase considerably. However, the exposure to traffic accidents so as of night and day population decreases significantly, determining a fundamental reduction of the risks (-95.5% path 3 and -95.2% path 4).

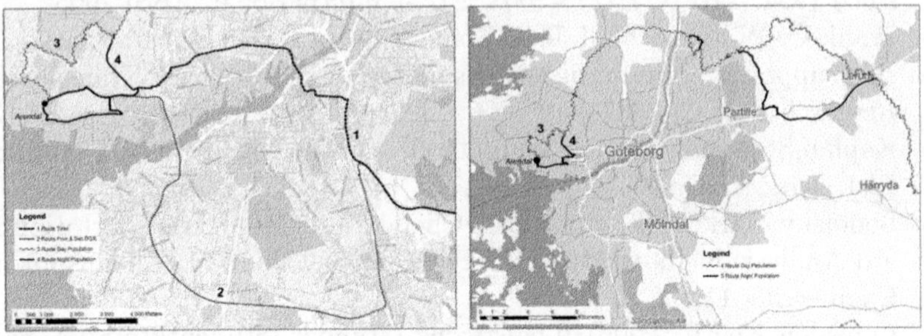

Fig. 1. Routes computations results

4 Discussion

The results of this investigation show that the SRSA recommended and forbidden routes provide an optimal balance of efficiency and risk factors. However, if transport operators behave in an opportunistic way by avoiding only forbidden routes, accident risks increase significantly. Finally, the optimization on day and night population exposure shows the possibility to decrease risks at the cost of efficiency factors (time and length). This study highlights the capability of geographic information techniques to easily handle complicated dangerous goods transportation problems by combining information about the transportation network, class of chemicals transported, population distribution, and traffic statistics into an integrated environment. Additionally, the importance of real-time routing and monitoring of dangerous goods is highlighted as a means to decrease societal risk exposure. Thus, future research has to be oriented towards the integration of the developed methodology into an extended Decision Support System in which vehicles can communicate and exchange information with a central server, where different actors can interact and make decisions.

References

1. Cormen Thomas H., Leiserson Charles E., Rivest Ronald L., Stein Clifford (2001) Introduction to algorithms. The MIT Press, Second Edition
2. DNV Consulting (2005) Study on the impacts of possible European Legislation to improve Transportation Security. Report for European Commission DG-TREN, October 2005
3. EU (2005) Evaluation of EU policy on the transport of Dangerous Goods since 1994. TREN/E3/43-2003, PIRA International, April 2005
4. Frank C. William, Thill Jean-Claude, Batta Rajan (2000) Spatial Decision support system for hazardous material truck routing. Transportation Research Part C 8:337–359
5. Kaplan Stan and Garrick John (1980), On the Quantitative Definition of Risk. Risk Analysis 1
6. Lepofsky Mark, Abkowitz Mark, Cheng Paul (1993) Transportation Hazard Analysis in Integrated GIS Environment. Journal of Transportation Engineering 119
7. Lumsden Kenth (2006), Logistikens Grunder. Studentliteratur, Second Edition
8. Parentela Emelinda M., Cheema Gulraiz (2002) Risk Modeling for Commercial Goods Transport. Report METRANS Transportation Center, June 2002
9. Pet-Armacots Julia J., Sepulveda Jose, Sakude Milton (1999), Monte Carlo Sensitivity Analysis of unknown Parameters in Hazardous Materials Transportation Risk Assessment. Risk Analysis 19:1173–1184
10. Raddningsverket (2006) Raddningsverket vaginformation om Farligt gods. January 2006
11. UNECE (2007) Dangerous Goods and special cargos section homepage. October 2007
12. Vagverket (1997), Vagverkets samhallsekonomiska kalkylmodell. Ekonomisk Teori och Värderingar, Publikation 1997:130, Borlange
13. Vilchez J.A., Sevilla S.H., Montiel H., Casal J. (1995) Historical analysis of accidents in chemical plants and in the transportation of hazardous material. Journal of Loss Prevention in the Process Industries 8

Design and Optimization of Dynamic Routing Problems with Time Dependent Travel Times

Sascha Wohlgemuth[1] and Uwe Clausen[2]

[1] Department of Mechanical Engineering, Technische Universität Dortmund, 44227 Dortmund, Germany,
sascha.wohlgemuth@udo.edu

[2] Fraunhofer-Institute for Material Flow and Logistics, 44227 Dortmund, Germany,
uwe.clausen@iml.fraunhofer.de

1 Introduction

The worldwide transportation of cargo is steadily growing and forwarding agencies handling less-than-truckload freight are no exception. The performance of these companies is influenced strongly by varying transport times between two consecutive points. Surprisingly, traffic information is hardly used within the forwarding industry, even though vehicle location is available in real-time. Numerous unknown customer orders, increasingly received shortly before the actual pickup, are impacting the performance, too.

Typical forwarding agencies perform the pickups and the deliveries conjoined. They have to cope with hundreds of pickups and deliveries each day and a few tens of vehicles are necessary to service the customers in the short-distance traffic region. Furthermore, inquiries of business customers cannot be neglected. In the following we focus on one part of the problem dealing with the integration of varying travel times, often resulting in late deliveries and penalties. Especially in urban areas for many roads rush hour traffic jams are known, but this information is hardly used within forwarding agencies. In particular, real-time approaches solving pickup and delivery problems (PDP) with inhomogeneous goods, capacities of general cargo, time windows, varying travel times, and unknown customer orders, which cannot be neglected, are missing. Thus, the objective is to develop a customized dynamic routing model capable of handling all requirements and assisting forwarding agencies in routing vehicles efficiently in real-time.

B. Fleischmann et al.(Eds.), *Operations Research Proceedings 2008*,
DOI: 10.1007/978-3-642-00142-0_54, © Springer-Verlag Berlin Heidelberg 2009

2 Literature Review

Primarily, there are two ways to solve dynamic problems: A priori models which account in their optimization strategy for changes which might occur and dynamic models which restart computing new solutions, every time new information is available. Often the assumption of constant travel times is unrealistic; therefore, Malandraki and Daskin distinguish two types of variations in travel time [1]. The first type compromises stochastic and unforeseen events like, accidents, weather conditions, vehicle brake downs, or other random events. The second type compromises temporal events like hourly, daily, weekly, or seasonal variations (e.g., heavy traffic during rush hours). Only temporal events can be included by using time dependent functions.

Several papers deal with average speeds for specific areas, a method similar to directly assign a travel time to a link. Therefore, often the FIFO ("first-in-first-out") property does not apply ([2], [3]). Ichoua et al. work with travel speeds and ensure the FIFO property, but they do not assume constant travel speeds other the entire length of a link [4]. The time spent in an interval is calculated by dividing the speed of that interval by the distance covered during the interval. If the planning horizon is divided into three time intervals with different speeds, then the entire travel time for one edge is the sum of the time spent in each interval. This is one of the first approaches to use time dependent travel speeds which satisfy the FIFO property. The time windows used are soft ones, except for the depot, though no capacities are considered.

Fleischmann et al. criticize Ichoua et al., because the drawback of models with varying speeds and constant distances is that they do not pay attention to potential changes of the shortest paths, which might change the distances [5]. Besides, the forward and backward algorithms go through all time slots. With constant travel times the direct link between each pair of visits is taken as the shortest path, but with time varying speeds it might happen that taking other links requires less travel time. Therefore, an approach using Euclidean distances and no real roads is only valid, if the speed distribution preserves the triangle inequality. Fleischmann et al. describe the derivation of travel time data and built a vehicle routing model working with time-varying travel times. The problem is solved using savings and insertion heuristics and a 2-opt improvement heuristic.

Dynamic models can be differentiated in models which anticipate and consider possible incoming orders and models which simply start a recalculation, if new information becomes available. In the following the time dependent modeling is depicted, thus only a small overview of dynamic approaches is given. Fu, for example, analyzes scheduling a dial-a-ride paratransit with tight service time windows and time-varying, stochastic congestion and applies successfully a parallel insertion algorithm [6]. Several other authors analyze time dependent routing problems and different solution techniques (e.g., [7]). Fleischmann et al. also consider a dynamic routing system using forecasted data, where customer orders arrive at random during the planning period [8]. Rarely approaches dealing with dynamic or varying travel times consider all relevant requirements of forwarding agencies, which motivated us to develop an integrative approach suited to these requirements.

3 Optimization Approach

The objective within a first step was to determine how far and under what premises freight forwarding agencies might benefit (i.e., reduction of expenses) from a real-time intelligent planning system, due to the fact that within the freight transportation industry enlarging the low profit margins becomes increasingly important (e.g., [9]).
A mixed integer problem developed earlier is modified to account for time dependent travel times [10]. Besides other modifications, the travel time matrix associated with E is modified according to different time zones (z), which give the corresponding travel times tt^z_{ij}, where $z \in Z$ and $Z = \{1, 2, 3\}$. That is an arc has not only one travel time, instead the travel time depends on the time of the day ($T(0)$, $T(1)$, $\ldots$, $T(n)$), resulting in a step function. Hence, each vehicle k travels a link (i, j) in a certain time zone z: x^z_{ijk}, but still has to comply with all other restrictions (e.g., (1)). In this context the FIFO ("first-in-first-out") property, a method originally applied in inventory management, is important. In vehicle routing it means a vehicle entering a road first, will leave the road first. The FIFO property is violated by using a step function, but not by working with a non step function with a slope of at most minus. We maintain FIFO in allowing vehicles to wait (cf. [1]); accepting the disadvantage of unnecessary waiting.

$$a_{ik} + s_i + tt^z_{ij} - R * \left(1 - x^z_{ijk}\right) \le a_{jk} \forall\, i, j \in V \setminus \{v_1\}, k \in K, z \in Z \quad (1)$$

In the state of North Rhine-Westphalia (NRW) numerous sensors are constantly recording traffic data on the interstates. The regional authorities for central police services provided us with data for one year for detailed analyzes and testing. Subsequently, we determined time zones reflecting the traffic situation in NRW. Figure 1 shows the average number of traffic jams per hour within one week on all interstates

in NRW. Every morning around 6:30 am the number for traffic jams increases clearly and drops around 9 am. Likewise in the afternoon between 3 pm and 6 pm the number of traffic jams increases. An interesting point is that at the beginning of a week the morning spikes are higher than the afternoon ones, but as the weekend nears the afternoon spikes grow in height and width.
Time zones valid for all interstates are established to reduce computational complexity, increase reliability and preserve the triangle inequality for all time zones ($d_{i,j} \leq d_{i,p} + d_{p,j} \; \forall \; i, j, p \in N$). Figure 2 visualizes the three groups of travel times our analysis yielded. In the morning, after 6 am, the average travel time increases on most interstates. Around 7 am the zone with the slowest average travel time is reached, but after 9 am it drops already back to the intermediate zone. Similar characteristics can be observed in the afternoon.

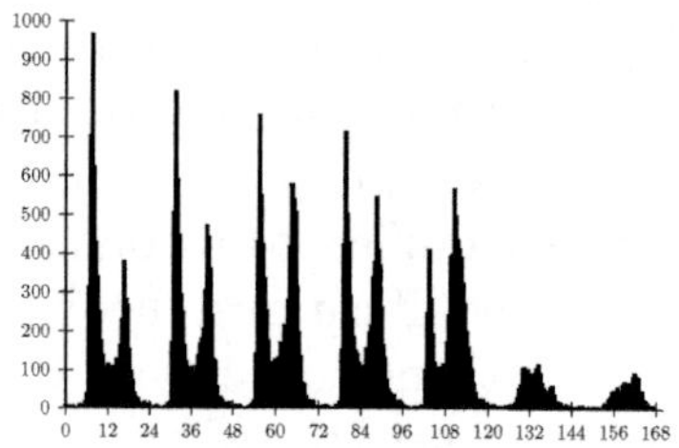

Fig. 1. Number of traffic jams over 168 hours (Monday to Sunday)

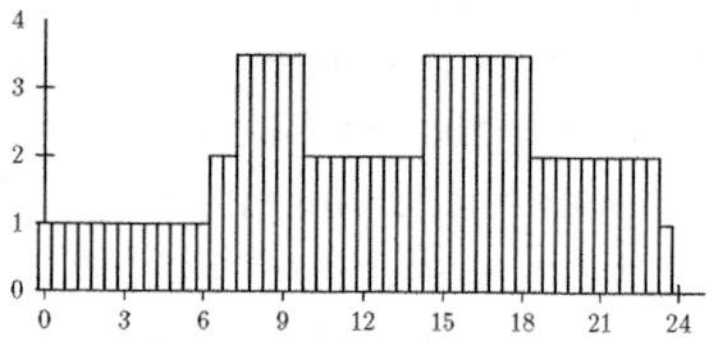

Fig. 2. Travel times zones in North Rhine-Westphalia over 24 hours

The objective is to find a set of routes with minimal total travel time using the previously derived travel time zones, starting and ending at a single depot and serving demands of all customers. Furthermore, vehicle capacity, driving time restrictions and time windows are considered. In addition, the number of vehicles used is minimized, because this is a crucial cost driver. The mixed integer model is implemented in GAMS (22.2145) and is solved with the Branch-and-Cut of CPlex 10.0 for small test instances, but even an explicit column generation approach could not improve computational times for industrial size scenarios. Additionally, the approach is designed for a future real-time routing and dispatching system, requiring a quick response.
Considering objective function, degree of dynamism, and demand rate, we develop a modified tabu search heuristic, an evolved and established solution technique. A tabu search explores only parts of the solution space by moving at each iteration to the most promising neighbor of the

current solution, even if this requires a decrease of the objective function. The objective function $z(x^\eta)$ associated with a particular solution η of an iteration is characterized by the vector $x^\eta = x_{ijk}^{z,\eta}$, denoting the used edges (i, j) for every vehicle k and time zone z. Cycling is avoided by using a tabu list, where recently considered solutions are blocked out for a number of iterations. The initial solution is made up by an insertion algorithm inserting all $n-1$ customers according to their proximity. The neighborhood $N(i, j, \eta)$ of a solution η contains only solutions, where the removed customer j can be inserted approximately to customer i complying all restrictions under time dependent travel times. For a moved vertex a reinsertion or displacement is tabu until the iteration $\eta + \theta$.

4 Computational Experiments

The algorithm is tested using Solomon instances of type R1 modified to fit industrial pickup and delivery scenarios for equally distributed customers. Table 1 compares the results of static and time dependent optimization. Time dependent optimization results in longer overall travel times (t), but still the number of vehicles (k) remains the same. The obtained routes for static and time dependent travel times are tested using the average rush hour slow downs in NRW. At least two customers cannot be served on time using static optimization, whereas with time dependent routing delays (κ) through rush hour traffic jams can be avoided. Anyhow, using real travel times (i.e., including random events) would increase the lateness of both, but with stronger impact on statically planned tours.

Table 1. Static versus time dependent solutions

Instance	Static			Time dependent			
	t	k	κ	t	k	κ	Δt
R1A	1858.18	21	2	2050.60	21	0	192.42
R1B	1999.09	22	2	2123.17	22	0	124.08
R1C	1462.49	15	3	1627.60	15	0	165.11
R1D	1766.36	17	2	1887.81	18	0	121.45

5 Conclusions

In this paper we successfully derived travel time zones and modeled a PDP incorporating practical complexities which received only little attention in literature. The presented algorithm is capable of industrial size problems and computational experiments showed that the algorithm performs well with time dependent travel times and is very helpful in decreasing the probability of lateness due to rush hours slow downs within PDP customized for forwarding agencies.

References

1. C. Malandraki and M. S. Daskin. Time dependent vehicle routing problems: Formulations, properties and heuristic algorithms. *Transportation Science*, 26(3):185–200, 1992.
2. A. V. Hill and W. C. Benton. Modelling intra-city time-dependent travel speeds for vehicle scheduling problems. *The Journal of the Operational Research Society*, 43(4):343–351, 1992.
3. Y.-B. Park and S.-H. Song. Vehicle scheduling problems with time-varying speed. *Computers & Industrial Engineering*, 33(3-4):853–856, 1997.
4. S. Ichoua, M. Gendreau, and J.-Y. Potvin. Vehicle dispatching with time-dependent travel times. *European Journal of Operational Research*, 144:379–396, 2003.
5. B. Fleischmann, M. Gietz, and S. Gnutzmann. Time-varying travel times in vehicle routing. *Transportation Science*, 38(2):160–173, 2004.
6. L. Fu. Scheduling dial-a-ride paratransit under time-verying, stochastic congestion. *Transportation Research Part B*, 36:485–506, 2002.
7. B. Brodén, M. Hammar, and B. J. Nilsson. Online and offline algorithms for the time-dependent tsp with time zones. *Algorithmica*, 39(4):299–319, 2004.
8. B. Fleischmann, S. Gnutzmann, and E. Sandvoß. Dynamic vehicle routing based on online traffic information. *Transportation Science*, 38(4):420–433, 2004.
9. M. Krajewska, H. Kopfer, G. Laporte, S. Ropke, and G. Zaccour. Horizontal cooperation of freight carriers: request allocation and profit sharing. *Journal of the Operational Research Society*, pages 1–9, advance online publication, 12 September 2007.
10. S. Wohlgemuth and U. Clausen. On Profitableness of Considering Dynamics in Forwarding Agencies. In S. I. Ao, O. Castillo, C. Douglas, D. D. Feng, and J.-A Lee, editors, *IMECS*, Lecture Notes in Engineering and Computer Science, pages 1725–1731, Hong Kong, 2008. Newswood Limited & International Association of Engineers.

An Ant Colony Algorithm for Time-Dependent Vehicle Routing Problem with Time Windows

Umman Mahir Yıldırım and Bülent Çatay

Sabancı University, Orhanlı, Tuzla, 34956 Istanbul, Turkey
mahiryldrm@su.sabanciuniv.edu, catay@sabanciuniv.edu

Summary. In this paper, we address the Vehicle Routing Problem with Time Windows, both time-independent and -dependent cases. In the time-independent case, our objective is to minimize the total distance. To solve this problem, we propose an Ant Colony Optimization algorithm. Then we implement the algorithm to solve the time-dependent case where the objective is to minimize the total tour time. The time dependency is embedded in this model by using a deterministic travel speed function which is a step function of the time of the day. An experimental evaluation of the proposed approach is performed on the well-known benchmark problems.

1 Introduction

Optimizing a distribution network has been and remains an important challenge both in the literature and in real-life applications and the routing of a fleet of vehicles is the most widely addressed problem in a distribution network. The Vehicle Routing Problem (VRP) determines a set of vehicle routes originating and terminating at a single depot such that all customers are visited exactly once and the total demand of the customers assigned to each route does not violate the capacity of the vehicle. The objective is to minimize the total distance traveled. An implicit primary objective is to use the least number of vehicles. The Vehicle Routing Problem with Time Windows (VRPTW) is a variant of VRP in which lower and upper limits are imposed to the delivery time of each customer. The arrival at a customer outside the specified delivery time is either penalized (soft time windows) or strictly forbidden (hard time windows). The interested reader is referred to [1] for more details on VRPTW.

In the Stochastic Vehicle Routing Problem, the customer demands and/or the travel times between the customers may vary. Although

B. Fleischmann et al.(Eds.), *Operations Research Proceedings 2008*,

DOI: 10.1007/978-3-642-00142-0_55,

stochastic travel times and demand distributions have been frequently used in the literature, time-varying travel speeds and time-dependent VRPTW (TDVRPTW) have seldom been addressed. In the literature, time dependency is taken into consideration in two ways: stochastic travel times and deterministic travel times. First introduced by [2], stochastic travel times are mainly examined by [4] and [3]. [5] proposed a deterministic travel time based model in which the important non-passing property is introduced. [6] and [7] also use deterministic travel times in a setting where the day is divided into time intervals.

Many exact and heuristic solution approaches were presented for solving VRP and its extensions. One recent approach, Ant Colony Optimization (ACO) is a population-based metaheuristic that can be used to find approximate solutions to difficult optimization problems. A detailed study of ACO and its variants can be found in [8].

In this study, an ACO approach is developed to efficiently solve VRPTW and TDVRPTW with hard time windows. In the next section, we provide a description of the two problems. In Section 3, the proposed algorithm is presented. Section 4 is devoted to the computational study and concluding remarks are given in Section 5.

2 Problem Description

In VRPTW, N geographically dispersed customers are serviced by a homogenous fleet of K vehicles with capacity Q. All vehicle routes start and end at a central depot visiting each customer i, i=1,...,N, exactly once. Each customer has a demand q_i, service time s_i and time window $[e_i, l_i]$. The service time is the loading or unloading service times at the customer i where the terms e_i and l_i denote the earliest and latest available service start time for customer i. The time window may prohibit the visit of certain customer pairs one after the other.

VRPTW is in fact a special case of TDVRPTW. In TDVRPTW the travel time between any source and destination pair on the road network is not a function of the distance alone and is subject to variations due to accidents, weather conditions or other random events. Speed limitations imposed by the road type and the traffic density distribution of the road which is also affected by the time of the day are two main components that cause fluctuations in travel speeds. That is, the travel time between two customers is not constant during the entire scheduling horizon and changes with the changing sub-divisions of the horizon, called time-intervals.

In TDVRPTW, the feasible and infeasible customer pairs are not necessarily same as in the time-independent case. A dynamic travel time calculation is required to check the feasibility in the route construction phase. The arrival time to the next customer may be realized earlier or later compared to the time-independent case.

3 The Proposed Ant Colony Optimization Approach

3.1 ACO for VRPTW

Our ACO approach is inspired from the rank-based Ant System[1] introduced by [10] and is outlined as follows.

Route Construction - Initially, N ants are placed at the N nearest customers to the depot. After a vehicle has returned to the depot, it starts from the customer with the largest attractiveness value. To put a limit on the exploration and to speed up the algorithm, we use a candidate list which consists of the nearest CL (candidate list size) neighbors of the customer. The customers are added to the list by taking their feasibility and distance into account. The next customer is selected from the candidate list using the probabilistic action choice rule as described in [10].

Local Search - In this study, two types of local search procedures, namely Swap and Move, are utilized to improve the solution quality. These procedures are applied at the end of each iteration and pheromone trails are updated afterwards. The simple idea behind the Swap procedure is to exchange two customers in a single route (intra-route) or between routes (inter-route) until no further improvements are available. The Move procedure attempts to improve the solution by removing a customer and inserting it between two other customers, intra-route or inter-route.

Pheromone Update - The pheromone levels are initialized as N/L_0, where L_0 is total distance obtained using the nearest neighbor heuristic. After all ants have constructed their tours, first the pheromone trails are evaporated at the rate ρ then k elitist ants are allowed to reinforce the trails. In our pheromone reinforcement strategy, we utilize *k-1* best-ranked ants for the first P iterations (referred to as preliminary iterations) and in the remainder of iterations we allow *best-so-far* ant along with the *k-1* best-ranked ants to deposit pheromone. Our aim in adopting this strategy is to avoid a quick stagnation.

[1] (AS_{rank})

3.2 Extensions to TDVRPTW

In TDVRPTW, the objective function and travel speeds are adapted accordingly. In addition, the local search and pheromone update procedures are modified in line with the new objective function of minimizing the total travel time. Since the scheduling horizon is divided into multiple time-intervals, the pheromone network also comprises multiple dimensions. An ant in time-interval t deposits pheromone on the corresponding dimension in the network.

4 Computational Study

The performance of the algorithm is tested on the time-independent benchmark problems of [9] using real numbers (float precision). The time-dependent versions are obtained by dividing the scheduling horizon into three time intervals and utilizing different road types which are randomly assigned to each arc. Time-dependent travel speeds are embedded in the algorithm by utilizing a travel time matrix similar to the approach of [6]. In the preliminary runs, we observed that heuristic information such as "1/distance" or "savings" does not improve the solution quality much due to the high performance of local search. Thus, no visibility function is implemented. For each problem, 10 runs are performed with the following parameter setting: $\alpha = 1$, $\beta = 0$, $\rho = 0.15$, number of iterations $= 100$, $P = 25$, number of ants $= N$, $k = 6$, $CL = 50$. Both one dimensional and three dimensional pheromone networks are tested in the time-independent case. The algorithm is shown to be efficient and then applied to the time-dependent case. The primary objective of the time-independent models is to minimize the total distance (TD) whereas total tour time (TT) is minimized in the time-dependent case. The algorithm is coded in C# and executed on a Pentium 2.40 GHz processor.
Table 1 gives the average distance of each problem type for 10 runs as well as the average number of vehicles (VN). In the class column, C and R refer to problem types where the customers are clustered and uniformly randomly distributed, respectively. RC refers to the problem type which includes a combination of clustered and randomly distributed customers. Since a three-dimensional network is needed for time-dependent case we also experimented the effect of such network structure on the time-independent problem. For C problem sets, both one- and three-dimensional pheromone networks exhibit the same performance as a result of the clustered network structure which narrows the feasible region. One-dimensional network outperforms the

Table 1. Average Results

	Time-independent						Time-dependent		
	ACO-Average		ACO-Best		Best-Known				
class	TD	NV	TD	NV	TD	NV	TD	NV	TT
C1	828.380	10.00	828.380	10.00	828.380	10.00	1093.140	10.47	9945.998
C2	590.259	3.00	589.859	3.00	589.859	3.00	941.016	4.16	9854.871
R1	1191.432	13.73	1187.282	13.58	1181.453	13.08	1499.805	12.72	2298.308
R2	913.806	5.78	901.507	5.55	898.067	5.55	1627.551	3.69	2352.808
RC1	1368.344	13.54	1357.403	13.25	1339.235	12.75	1645.410	12.64	2405.311
RC2	1044.907	6.55	1027.401	6.50	1015.738	6.38	1988.114	4.25	2672.617

three-dimensional network in type 1 problems where time windows are narrower and the vehicle capacities are smaller whereas a three-dimensional network is more suitable for type 2 problems. Overall, we observed that a three-dimensional network slightly outperforms the one-dimensional network on the overall average solution quality. Therefore, we report the average and best results belonging to the three-dimensional network setting. For comparison, we also provide best-known solutions as reported in [11]. As seen in Table 1, our distances are comparable to the best-known distances and the average gap is only 0.66 %. Furthermore, we note that we have been able to improve the best-known distance of six instances.

For the time-dependent problems, we report the average tour times as well as the corresponding average distances. The distances are 48.1 % longer on the average compared to the time-independent case. This is an expected result since the two problems have different objective functions. On the other hand, the average number of vehicles is 8.89 % less in time-dependent case. We also observe that type 2 problems are more sensitive to the time-dependent travel times. The distances for type 2 problems increase dramatically due to the existence of tighter constraints.

5 Conclusion

In this paper, we propose an ACO algorithm for solving the VRPTW and TDVRPTW. Our preliminary experimental results show that the proposed algorithm provides good quality results; however, the computation times are rather long. We have observed that the local search procedure enhances the solution quality of ACO significantly. On the other hand, a large portion of the computational time is consumed by the local search procedure. Further research may focus on a selective local search policy to reduce the computational effort. To improve the performance of the algorithm, a visibility function using the time window information can be implemented and a more detailed analysis on the trade-off between the solution quality and computational effort may be conducted.

References

1. Cordeau J, Desaulniers G, Desrosiers J, Solomon MM, Soumis F (2001) VRP with Time Windows. In: Toth P and Vigo D (Eds) The vehicle routing problem. SIAM Monographs on Discrete Mathematics and Applications. SIAM Publishing, Philadelphia, PA, pp. 157-193
2. Laporte G, Louveaux F, Mercure H (1992) The vehicle routing problem with stochastic travel times. Transportation Science 26(3):161-170
3. Potvin J, Xu Y, Benyahia I (2006) Vehicle routing and scheduling with dynamic travel times. Computers & Operations Research 33:1129-1137
4. Kenyon AS, Morton DP (2003) Stochastic vehicle routing with random travel times. Transportation Science 37(1):69-82
5. Ahn BH, Shin JY (1991) Vehicle-routeing with time windows and time-varying congestion. Journal of Operations Research Society 42(5):393-400
6. Ichoua S, Gendreau M, Potvin JY (2003) Vehicle dispatching with time-dependent travel times. European Journal of Operations Research 144:379-396
7. Donati AV, Montemanni R, Casagrande N, Rizzoli AE, Gambardella LM (2008) Time-dependent vehicle routing problem with a multi ant colony system. European Journal of Operational Research 185:1174-1191
8. Dorigo B, Stützle T (2004) Ant colony optimization. MIT Press, Cambridge Massachusetts
9. Solomon MM (1987) Algorithms for the vehicle routing and scheduling problems with time window constraints. Operations Research 35(2):254-265
10. Bullnheimer B, Hartl RF, Strauss C (1999) An Improved Ant System Algorithm for the Vehicle Routing Problem. Annals of Operations Research 89:319-328
11. Alvarenga GB, Mateus GR, Tomi G (2007) A genetic and set partitioning two-phase approach for the vehicle routing problem with time windows. Computers & Operations Research 34:1561-1584

Part VII

Applied Probability and Stochastic Programming

The Comparative Analysis of Different Types of Tax Holidays Under Uncertainty

Vadim Arkin, Alexander Slastnikov and Svetlana Arkina

Central Economics and Mathematics Institute, Nakhimovskii pr. 47, Moscow, Russia
{arkin,slast,s_arkina}@mail.ru

Summary. Tax holidays, that exempt firms (fully or partially) from tax payment for a certain period of time, have been widely used over the world as one of the most effective stimuli for investment attraction (see [1]).
Below, we present the model of investment attraction by means of tax holidays (for other mechanisms of investment attraction see, e.g., [2]). Within the framework of this model, we compare two alternative mechanisms of tax holidays: tax holidays of deterministic (fixed) duration and tax holidays based on the payback period of the initial investment.

1 The Model of Investment Waiting

In this section we study the general investment waiting model. This model describes the behavior of the potential investor who wishes to invest in a project of creating a new enterprise (firm), which produces certain goods, consuming certain resources.
Investment necessary for the project of creation and start of the new firm, is considered to be instantaneous and irreversible so that they cannot be withdrawn from the project anymore and used for other purposes (sunk cost). We also assume that the firm starts to produce immediately after investing.
The most important feature of the proposed model is the assumption that at any moment, the investor can either accept the project and start with the investment, or delay the decision before obtaining new information on its environment (prices, demand etc). In other words investor waits for the most appropriate moment to invest, that is why such a model is called *an investment waiting model.*
Let us suppose that the investment starts at time τ, and that I_τ be the amount of required investment. Let $\pi^\tau_{\tau+t}$ be the "operational profit"

B. Fleischmann et al.(Eds.), *Operations Research Proceedings 2008*, 345
DOI: 10.1007/978-3-642-00142-0_56, © Springer-Verlag Berlin Heidelberg 2009

of the firm at time $\tau + t$, i.e. the difference between the value of the produced goods and its production cost.
The economic environment can be influenced by different stochastic factors (uncertainty in market prices, demand, etc.). For this reason, we will consider that the cost of necessary investment $(I_t, t\geq 0)$ evolves as a stochastic process, and operational profit $(\pi^{\tau}_{\tau+t}, t\geq 0)$ is modeled by a family (in $\tau\geq 0$) of random processes, given on some probability space $(\Omega, \mathbb{F}, \mathbf{P})$ with the flow of σ-fields $\mathcal{F}=(\mathcal{F}_t, t\geq 0)$ (the observable information about the system), and random processes are assumed to be $\mathcal{F}$-adapted.
The lifetime of the project is considered as infinite in our model, capital funds have also a very long useful lifetime (nearly infinite) that is why depreciation charges are not taken into account. Moreover, the model does not take into account maintenance costs of the capital funds.
The corporate profit taxation system will be characterized by the two-level corporate income tax rates: the federal rate γ_f (paid into the federal budget), the regional tax rate γ_r (paid into the regional budget), and the duration of tax holidays ν (during the period of tax holidays, the firm is exempted from the regional part of the profit tax).
The after-tax expected present value of the firm can be described by the following formula

$$V_\tau=\mathbf{E}\left(\int_0^\nu (1-\gamma_f)\pi^{\tau}_{\tau+t}e^{-\rho t}dt + \int_\nu^\infty (1-\gamma_f-\gamma_r)\pi^{\tau}_{\tau+t}e^{-\rho t}dt \,\middle|\, \mathcal{F}_\tau\right), \quad (1)$$

where ρ is the discount rate, and $\mathbf{E}(\cdot|\mathcal{F}_\tau)$ stands for the conditional expectation provided by information about the system up to time τ.
For simplicity we limit our study solely to corporate profit tax. More complicated model with various taxes one can find in [4].
The behavior of the investor, which consists in the choice of investment time, is assumed to be rational in the sense that he chooses time τ (investment rule), in order to maximize his expected net present value (NPV):

$$\mathbf{E}\left(V_\tau - I_\tau\right)e^{-\rho\tau} \to \max_\tau, \quad (2)$$

where the maximum is considered over all Markov times τ (with respect to the flow of σ-fields $\mathcal{F}$).
Simultaneously, the expected tax payments from the future firm into both federal and regional budgets, discounted to investment time τ, are equal to:

$$T_f^\tau = \mathbf{E}\left(\int_0^\infty \gamma_f\pi^{\tau}_{\tau+t}e^{-\rho t}dt \,\middle|\, \mathcal{F}_\tau\right), \quad T_r^\tau = \mathbf{E}\left(\int_\nu^\infty \gamma_r\pi^{\tau}_{\tau+t}e^{-\rho t}dt \,\middle|\, \mathcal{F}_\tau\right) \quad (3)$$

respectively.

Mathematical Assumptions

The dynamics of *operational profits* ($\pi^{\tau}_{\tau+t}$, $t \geq 0$) is specified by a family of stochastic equations $\pi^{\tau}_{\tau+t}=\pi_{\tau}+\int_{\tau}^{\tau+t} \pi^{\tau}_{s}(\alpha_1 ds+\sigma_1 dw^1_s)$, $t \geq 0$, where (w^1_t, $t\geq 0$) is a Wiener process, α_1 and σ_1 are real numbers.
We assume that at any moment τ, observing the current prices on both input and output production one can calculate $\pi_{\tau} = \pi^{\tau}_{\tau}$, which is the profit at the "initial moment" of creation of firm, and, hence, can evaluate the future profits from the project before the actual creation of the firm. We suppose that the process π_{τ} is a geometric Brownian motion with parameters (α_1, σ_1).
The amount of *required investment* I_t is described by geometric Brownian motion $I_t=I_0+\int_0^t I_s(\alpha_2 ds+\sigma_2 dw^2_s)$, $t\geq 0$, where ($w^2_t, t\geq 0$) is a Wiener process, α_2 and σ_2 are real numbers, and I_0 is a given initial state of the process. The pair (w^1_t, w^2_t) is two-dimensional Wiener process with correlation r.

2 Deterministic (Fixed) Tax Holidays

For this type of tax holidays their duration ν is a given deterministic non-negative value.
Let β be a positive root of the quadratic equation

$$\tfrac{1}{2}\sigma^2\beta(\beta-1)+(\alpha_1-\alpha_2)\beta-(\rho-\alpha_2)=0, \tag{4}$$

where $\sigma^2=\sigma_1^2-2r\sigma_1\sigma_2+\sigma_2^2$ is "total" volatility of the investment project. The following theorem specifies an optimal rule for investing under the fixed tax holidays with duration ν.

Theorem 1. *Let* $\sigma>0$, $\alpha_1-\frac{1}{2}\sigma_1^2 \geq \alpha_2-\frac{1}{2}\sigma_2^2$, *and* $\rho>\max(\alpha_1, \alpha_2)$. *Then the optimal investment time for the problem (2) is*

$$\tau^*=\min\{t \geq 0 : \pi_t\geq p^* I_t\}, \quad \textit{where}\; p^* = \frac{\beta}{\beta-1}\cdot\frac{\rho-\alpha_1}{1-\gamma_f-\gamma_r e^{-(\rho-\alpha_1)\nu}}. \tag{5}$$

Knowing the optimal investment time one can find the expected NPV as well as expected present tax revenues into the budgets at different levels under the optimal behavior of the investor:

$$\mathcal{N} = \mathbf{E}(V_{\tau^*} - I_{\tau^*})e^{-\rho\tau^*} = \frac{I_0}{\beta-1}\left(\frac{\pi_0}{I_0 p^*}\right)^{\beta}; \tag{6}$$

$$\mathcal{T}_f = \mathbf{E}T_f^{\tau^*} e^{-\rho\tau^*} = \frac{\gamma_f I_0 p^*}{\rho-\alpha_1}\left(\frac{\pi_0}{I_0 p^*}\right)^{\beta}; \tag{7}$$

$$\mathcal{T}_r = \mathbf{E}T_r^{\tau^*} e^{-\rho\tau^*} = \frac{\gamma_r I_0 p^*}{\rho-\alpha_1} e^{-(\rho-\alpha_1)\nu}\left(\frac{\pi_0}{I_0 p^*}\right)^{\beta}, \tag{8}$$

where p^* is defined in (5).

In [3] we proposed an optimization approach for such type of tax holidays. Thus, for a given duration of deterministic tax holidays ν, one can calculate (using (5)) the optimal investment threshold $p^* = p^*(\nu)$ and the corresponding expected present tax payments in the regional budget $\mathcal{T}_r = \mathcal{T}_r(\nu)$. Region determines tax holidays ν^*, which maximize expected present tax payments in the regional budget. We obtained the following explicit formula for optimal tax holidays ([3]):

$$\nu^* = \begin{cases} 0, & \text{if } \beta \leq (1-\gamma_f)/\gamma_r \\ (\rho - \alpha_1)^{-1} \log\left[\beta\gamma_r/(1-\gamma_f)\right], & \text{if } \beta > (1-\gamma_f)/\gamma_r. \end{cases} \tag{9}$$

3 Tax Holidays Based on Payback Period

In the present paper, as the payback period (PP) of the project we will consider the duration of the time interval (starting from the first occurrence of balance sheet profit), at the end of which the expected profit of the firm (discounted to the investment time) will equal the initial cost. Since the profit of the firm is a stochastic process, the payback period is a random variable.

The generalization of PP is the "modified payback period" (MPP) ν_θ. It is defined as the duration of time interval at the end of which the ratio of the expected discounted net profit (accumulated on this interval) to the initial investment, equals the predefined payback coefficient θ:

$$\nu_\theta = \min\{\nu \geq 0 : \mathbf{E}\left(\int_0^{\nu} \pi_{\tau+t}^{\tau} e^{-\rho t} dt \,\middle|\, \mathcal{F}_\tau\right) \geq \theta I_\tau\},$$

(if the minimum is not attained then $\nu_\theta = \infty$). ν_θ is $\mathcal{F}_\tau$-measurable random variable, but not necessary finite. The standard payback period is a particular case of MPP (with θ=1).

The optimal investment rule, given the MPP tax holidays (with coefficient θ) is characterized by the following theorem.

Theorem 2. *Suppose that* $\sigma > 0$, $\alpha_2 - \frac{1}{2}\sigma_2^2 \geq \alpha_1 - \frac{1}{2}\sigma_1^2$, $\rho > \max(\alpha_1, \alpha_2)$, *and* $\theta < [1 - \gamma_f - (1 - \gamma_f - \gamma_r)/\beta]^{-1}$. *Then the optimal investment time for the problem (2) is*

$$\tau^* = \min\{t \geq 0 : \pi_t \geq p_\theta^* I_t\}, \quad \text{where } p_\theta^* = \frac{\beta}{\beta - 1} \cdot \frac{(\rho - \alpha_1)(1 - \gamma_r \theta)}{1 - \gamma_f - \gamma_r}. \tag{10}$$

On the basis of (10), we can obtain explicit formulas for the expected NPV, the expected present tax payments from the created firm in federal and regional budgets (given optimal investment behavior), as in (6)-(8):

$$\mathcal{N} = \frac{I_0(1 - \gamma_r \theta)}{\beta - 1}\pi^*, \quad \mathcal{T}_f = \frac{\gamma_f I_0 p_\theta^*}{\rho - \alpha_1}\pi^*, \quad \mathcal{T}_r = \gamma_r I_0 \left(\frac{p_\theta^*}{\rho - \alpha_1} - \theta\right)\pi^*, \quad \pi^* = \left(\frac{\pi_0}{I_0 p_\theta^*}\right)^\beta$$

where p_θ^* is defined in (10).
As in the case of deterministic tax holidays, we can propose an optimization approach, through which the region chooses the payback coefficient θ^*, which maximizes expected present tax payments in the regional budget. One can show that :

$$\theta^* = \begin{cases} 0, & \text{if } \beta \leq (1 - \gamma_f - \gamma_r)/\gamma_r \\ \dfrac{1}{\gamma_r} \cdot \dfrac{\beta\gamma_r - (1 - \gamma_f - \gamma_r)}{\beta\gamma_r + (\beta - 1)(1 - \gamma_f - \gamma_r)}, & \text{if } \beta > (1 - \gamma_f - \gamma_r)/\gamma_r. \end{cases} \tag{11}$$

Let us note that the dependence of the optimal payback coefficient on the parameters of the project and the discount rate has been completely specified by β characteristic of the project (4).

4 Comparative Analysis of the Effectiveness of Various Types of Tax Holidays

On the basis of the obtained explicit formulas for optimal tax holidays (9) and optimal payback coefficient (11), it is possible to compare the following types of tax holidays, depending on their impact on the present tax payments in federal $\mathcal{T}_f$ and regional $\mathcal{T}_r$ budgets, as well as expected investor's NPV $\mathcal{N}$:

- optimal MPP tax holidays – ν_{θ^*};
- tax holidays based on standard payback period – ν_1;
- optimal deterministic tax holidays – ν^*.

The value of the optimal payback coefficient θ^* (when β is not too large) is less than 1. This means that the use of the standard payback period as the duration of tax holidays brings, in most cases, decreased expected tax payments from the created firm in the regional budget. If β is not too large ($\beta<\beta_0$) the following relations hold:

$$\mathcal{T}_r(\nu_1) < \mathcal{T}_r(\nu^*) < \mathcal{T}_r(\nu_{\theta^*}).$$

$$\mathcal{T}_f(\nu^*) < \mathcal{T}_f(\nu_{\theta^*}) < \mathcal{T}_f(\nu_1), \quad \mathcal{N}(\nu^*) < \mathcal{N}(\nu_{\theta^*}) < \mathcal{N}(\nu_1).$$

So, for the regional budget, the most effective tax holidays are optimal MPP holidays, and tax holidays based on the standard payback period are the least effective. As for both federal budget and investor, the tax holidays based on the standard payback period are the most effective. As our calculations showed, the parameters of the most "reasonable" investment projects are situated inside the "appropriate" area $\{\beta<\beta_0\}$. Let us notice as well that optimal MPP tax holidays stimulate an earlier arrival of the investor than optimal deterministic tax holidays. But increasing of uncertainty (volatility of the project) generates a loss of efficiency of the various types of optimal tax holidays.

The financial support of RFBR (project 08–06–00154) and of the grant # NSh-929.2008.6, School Support, is gratefully acknowledged.

References

1. Tax Incentives and Foreign Direct Investment. A Global Survey (2000) ASIT Advisory Studies, No. 16. UNCTAD. NY and Geneva: United Nations
2. Arkin V, Slastnikov A (2007) The effect of depreciation allowances on the timing of investment and government tax revenue. Annals of Operations Research 151: 307–323
3. Arkin V, Slastnikov A, Shevtsova E (1999) Tax incentives for investment projects in the Russian economy. Working Paper No 99/03. Moscow, EERC
4. Arkin V, Slastnikov A (2004) Optimal stopping problem and investment models. In: Dynamic Stochastic Optimization. Lecture Notes in Economics and Mathematical Systems 532: 83–98
5. Dixit AK, Pindyck RS (1994) Investment under Uncertainty. Princeton University Press, Princeton

Stochastic Programming Problems with Recourse via Empirical Estimates

Vlasta Kaňková

Institute of Information Theory and Automation
Academy of Sciences of the Czech Republic
Pod Vodárenskou věží 4, 182 08 Praha 8, Czech Republic
kankova@utia.cas.cz

1 Introduction

Let $\xi := \xi(\omega)$ $(s \times 1)$ be a random vector defined on a probability space $(\Omega, \mathcal{S}, P)$; F, P_F the distribution function and the probability measure corresponding to the random vector ξ. Let, moreover, $g_0(x, z)$, $g_0^1(y, z)$ be functions defined on $R^n \times R^s$ and $R^{n_1} \times R^s$; $f_i(x, z)$, $g_i(y)$, $i = 1, \ldots, m$ functions defined on $R^n \times R^s$ and R^{n_1}; $h := h(z)\,(m \times 1)$ a vector function defined on R^s, $h'(z) = (h_1(z), \ldots, h_m(z))$; $X \subset R^n$, $Y \subset R^{n_1}$ be nonempty sets. Symbols $x\,(n \times 1)$, $y := y\,(x, \xi)\,(n_1 \times 1)$ denote decision vectors. (R^n denotes the n–dimensional Euclidean space, h' a transposition of the vector function h.) .

Stochastic programming problems with recourse (in a rather general setting) can be introduced as the following problem:

Find

$$\varphi(F) = \min_{x \in X} \mathsf{E}_F\{g_0(x, \xi) + \min_{\{y \in Y: g_i(y) \leq h_i(\xi) - f_i(x, \xi), i=1,\cdots,m\}} g_0^1(y, \xi)\}, \quad (1)$$

where E_F denotes the operator of mathematical expectation corresponding to F.

A special case of the problem (1) is a stochastic programming problem with linear recourse, where $Y = R^{n_1}$ and, furthermore,

$$\varphi(F) = \min_{x \in X} \mathsf{E}_F\{g_0(x, \xi) + \min_{\{y \in R^{n_1}: Wy = h - Tx, y \geq 0\}} q'y\} \quad (2)$$

with $q := q(\xi)(n_1 \times 1)$, $T := T(\xi)(m \times n), W := W(\xi)(m \times n_1), m \leq n_1, m \leq n$ (generally) random vectors and matrices.

If we denote

B. Fleischmann et al.(Eds.), *Operations Research Proceedings 2008*,
DOI: 10.1007/978-3-642-00142-0_57, © Springer-Verlag Berlin Heidelberg 2009

$$
\begin{aligned}
Q(x,\xi) &= \min_{\{y\in Y: g_i(y)\le h_i(\xi)-f_i(x,\xi), i=1,\dots,m\}} g_0^1(y,\xi)\},\\
f_0(x,\xi) &= g_0(x,\xi) + Q(x,\xi),
\end{aligned} \tag{3}
$$

then evidently the problem (1) is covered by a more general problem: Find

$$\varphi(F) = \inf\{\mathsf{E}_F f_0(x,\xi) | x \in X\}, \tag{4}$$

with $f_0(x,z)$ arbitrary real valued function defined on $R^n \times R^s$.
In applications very often the "underlying" distribution function F has to be replaced by an empirical distribution function F^N. Evidently, then the solution is sought with respect to the "empirical" problem: Find

$$\varphi(F^N) = \inf\{\mathsf{E}_{F^N} f_0(x,\xi) | x \in X\}. \tag{5}$$

If $\mathcal{X}(F), \mathcal{X}(F^N)$ denote the optimal solution sets of the problems (1) and (5), then under rather general assumptions $\varphi(F^N), \mathcal{X}(F^N)$ are "good" stochastic estimates of $\varphi(F), \mathcal{X}(F)$ (see e.g. [1], [4], [5], [12], [13]). There were introduced assumptions guaranteing the consistency, asymptotic normality and convergence rate. Especially, it means in the last case that

$$P\{\omega : N^\beta |\varphi(F) - \varphi(F^N)| > t\} \to_{(N\to\infty)} 0 \quad \text{for} \quad t > 0, \beta \in (0, \frac{1}{2}). \tag{6}$$

To obtain the relation (6), the Hoeffding inequality (see e.g. [2], [5]), large deviation (see e.g. [4]), Talagrand approach (see e.g. [10]) and the stability results (see e.g. [11]) have been employed. To obtain new assertions, we employ stability results [8] based on the Wasserstein metric determined by $\mathcal{L}_1$ norm in R^s. Consequently, our results are based on the assumption of thin tails of one–dimensional marginal distribution functions $F_i(z), i = 1,\dots,s$ corresponding to $F(z)$.

2 Some Auxiliary Assertions

Let $\mathcal{P}(R^s)$ denote the set of all Borel probability measures on R^s, $s \ge 1$; $\mathcal{M}_1(R^s) = \{P \in \mathcal{P}(R^s) : \int_{R^s} \|z\|_s^1 P(dz) < \infty\}$, $\|\cdot\|_s^1$ the $\mathcal{L}_1$ norm in R^s.
First, we recall a little generalized result of [7].
Proposition 1. Let G be an arbitrary s–dimensional distribution function such that $P_G \in \mathcal{M}_1(R^s)$. Let, moreover, $P_F \in \mathcal{M}_1(R^s)$, $f_0(x,z)$ be defined on $R^n \times R^s$. If for every $x \in X$, $f_0(x,z)$ is a Lipschitz function of $z \in R^s$ with the Lipschitz constant $L(x)$ (corresponding to $\mathcal{L}_1$ norm), then

$$|\mathsf{E}_F f_0(x,\xi) - \mathsf{E}_G f_0(x,\xi)| \le L(x) \sum_{i=1}^{s} \int_{-\infty}^{+\infty} |F_i(z_i) - G_i(z_i)| dz_i \quad \text{for} \quad x \in X.$$

(Symbols $F_i, G_i, i = 1, \ldots, s$ denote one–dimensional distribution functions corresponding to F, G.)
Evidently, Proposition 1 reduces (from the mathematical point of view) stability results considered with respect to s–dimensional distribution functions to one–dimensional case. The next assertion has been proven in [8].
Proposition 2. Let $s = 1, t > 0, \bar{R} > 0$. If

1. P_F is absolutely continuous with respect to the Lebesgue measure on R^1,
2. there exists $\psi(N,t) := \psi(N,t,\bar{R})$ such that the empirical distribution function F^N fulfils for $N = 1, 2, \ldots$ the relation

$$P\{\omega : |F(z) - F^N(z)| > t\} \le \psi(N,t) \quad \text{for every} \quad z \in (-\bar{R}, \bar{R}),$$

then for $\frac{t}{4\bar{R}} < 1, N = 1, 2, \ldots$ it holds that

$$\begin{aligned} &P\{\omega : \int_{-\infty}^{\infty} |F(z) - F^N(z)| dz > t\} \le \\ &(\tfrac{12\bar{R}}{t} + 1)\psi(N, \tfrac{t}{12\bar{R}}, \bar{R}) + P\{\omega : \int_{-\infty}^{-\bar{R}} F(z) dz > \tfrac{t}{3}\} + \\ &P\{\omega : \int_{\bar{R}}^{\infty} (1 - F(z)) dz > \tfrac{t}{3}\} + 2NF(-\bar{R}) + 2N(1 - F(\bar{R})). \end{aligned} \quad (7)$$

To recall the next auxiliary assertion (proven in [9]), let $\bar{\xi}, \bar{\eta}$ be random values defined on $(\Omega, \mathcal{S}, P)$. We denote by $F_{(\bar{\xi},\bar{\eta})}, F_{\bar{\xi}}, F_{\bar{\eta}}$ the distribution functions of the random vector $(\bar{\xi}, \bar{\eta})$ and marginal distribution functions of $\bar{\xi}$ and $\bar{\eta}$.
Lemma. Let $\bar{\zeta} = \bar{\xi}\bar{\eta} := \bar{\xi}(\omega)\bar{\eta}(\omega)$, $F_{\bar{\zeta}}$ denote the distribution function of $\bar{\zeta}$. If

1. $P_{F_{\bar{\xi}}}, P_{F_{\bar{\eta}}}$ are absolutely continuous with respect to the Lebesgue measure on R^1 (we denote by $f_{\bar{\xi}}, f_{\bar{\eta}}$ the probability densities corresponding to $F_{\bar{\xi}}, F_{\bar{\eta}}$),
2. there exist constants $C_1^{\bar{\xi}}, C_2^{\bar{\xi}}, C_1^{\bar{\eta}}, C_2^{\bar{\eta}} > 0$ and $T' > 0$ such that

$$f_{\bar{\xi}}(z) \le C_1^{\bar{\xi}} \exp\{-C_2^{\bar{\xi}}|z|\} \quad \text{for} \quad z \in (-\infty, -T') \bigcup (T', \infty),$$
$$f_{\bar{\eta}}(z) \le C_1^{\bar{\eta}} \exp\{-C_2^{\bar{\eta}}|z|\} \quad \text{for} \quad z \in (-\infty, -T') \bigcup (T', \infty),$$

then, there exist constants $C_1^{\bar{\zeta}}, C_2^{\bar{\zeta}} > 0, \bar{T} > 1$ such that for $z > \bar{T}$

$$F_{\bar{\zeta}}(-z) \le \frac{C_1^{\bar{\zeta}}}{C_2^{\bar{\zeta}}} \exp\{-C_2^{\bar{\zeta}}\sqrt{z}\}, \quad (1 - F_{\bar{\zeta}}(z)) \le \frac{C_1^{\bar{\zeta}}}{C_2^{\bar{\zeta}}} \exp\{-C_2^{\bar{\zeta}}\sqrt{z}\}.$$

3 Convergence Rate

Let $\{\xi^i\}_{i=1}^{\infty}$ be a sequence of independent s–dimensional random vectors with a common distribution function F, F^N be determined by $\{\xi^i\}_{i=1}^{N}$.

3.1 General Case

Theorem 1. [8] Let $t > 0$, X be a compact set. If

1. $P_{F_i}, i = 1, \ldots, s$ are absolutely continuous with respect to the Lebesgue measure on R^1 (we denote by $f_i, i = 1, \ldots, s$ the probability densities corresponding to F_i),
2. there exist constants $C_1, C_2 > 0$ and $T > 0$ such that for $i = 1, \ldots, s$

$$f_i(z_i) \le C_1 \exp\{-C_2|z_i|\} \quad \text{for} \quad z_i \in (-\infty, -T) \bigcup (T, \infty),$$

3. $f_0(x, z)$ (defined by the relation (3)) is a uniformly continuous, Lipschitz (with respect to $\mathcal{L}_1$ norm) function of $z \in R^s$, the Lipschitz constant L is not depending on $x \in X$,

then

$$P\{\omega : N^{\beta}|\varphi(F^N) - \varphi(F)| > t\} \to_{(N \to \infty)} 0 \quad \text{for} \quad \beta \in (0, \frac{1}{2}). \qquad (8)$$

Remarks.

1. Some cases, under which $f_0(x, z)$ (defined by (3)) fulfils the assumption 3 of Theorem 1, are introduced e.g. in [6].
2. If $Q(x, z)$ corresponds to the case (2) (with q and simultaneously with at least one of h or T random), then evidently, the assumption 3 of Theorem 1 has not to be fulfilled (for more details see e.g. [3]).

3.2 Stochastic Programming Problems with Linear Recourse

Considering the linear case (2), we assume:

A.1 a. W is a deterministic matrix,
 b. W is a complete recourse matrix (for the definition of the complete recourse matrix see e.g. [3]),

A.2 there exists $u \in R^m$ such that $u'W \leq q$ a.s.

Theorem 2. [8] Let $t > 0$, X be a compact set, the assumptions A.1, A.2 and the assumptions 1, 2 of Theorem 1 be fulfilled. If

1.
$$f_0(x, \xi) = g_0(x, \xi) + Q(x, \xi)$$
$$Q(x, z) = \min_{\{y \in R^{n_1} : Wy = h - Tx, y \geq 0\}} q'y\},$$

2. $g_0(x, z)$ is a uniformly continuous, Lipschitz (with respect to $\mathcal{L}_1$ norm) function of $z \in R^s$, the Lipschitz constant L is not depending on $x \in X$,

then

$$P\{\omega : N^{\beta}|\varphi(F) - \varphi(F^N)| > t\} \rightarrow_{(N \rightarrow \infty)} 0 \quad \text{for} \quad t > 0, \beta \in (0, \frac{1}{2}).$$

Proof. Employing the assertion of Propositions 1, 2, Lemma and the technique employed in [8] we obtain the assertion of Theorem 2. □

4 Conclusion

The paper deals with the convergence rate of the optimal value of the empirical estimates in the case of the stochastic programming with recourse. It is known that if X is a convex, nonempty, compact set and either $f_0(x, z)$ a strongly convex (with a parameter $\rho > 0$) function on X or some growth conditions ([8], [12]) are fulfilled, then also

$$P\{\omega : N^{\beta}\|x(F^N) - x(F))\|^2 > t\} \rightarrow_{(N \rightarrow \infty)} 0 \text{ for } t > 0, \beta \in (0, \frac{1}{2}). \quad (9)$$

To see the conditions under which $Q(x, z)$ is a strongly convex function on X see e.g. [11].

Acknowledgement. The research was supported by the Czech Science Foundation under Grants 402/07/1113, 402/08/0107 and 402/06/0990.

References

1. Dupačvá J, Wets RJB (1984) Asymptotic behaviour of statistical estimates and optimal solutions of stochastic optimization problems. Ann Statist 16: 1517–1549
2. Hoeffding W (1963) Probability inequalities for sums of bounded random variables. Journal of Americ Statist Assoc 58: 13–30
3. Kall P (1976) Stochatic linear programming. Springer, Berlin
4. Kaniovski YM, King AJ, Wets RJB (1995) Probabilistic bounds (via large deviations) for the solutions of stochastic programming problems. Annals of Oper Res 56: 189–208
5. Kaňková V (1978) An approximative solution of stochastic optimization problem. In: Trans 8th Prague Conf Academia, Prague: 349–353
6. Kaňková V (1999) Convexity, Lipschitz property and differentiability in two–stage stochastic nonlinear programming problems. Aportaciones Matematicas Serie Comunicaciones 24
7. Kaňková V, Houda M (2006) Empirical estimates in stochastic programming. In: Proceedings of Prague Stochastics 2006. (M. Hušková and M. Janžura, eds.). MATFYZPRESS, Prague: 426–436.
8. Kaňková V (2007) Empirical estimates via stability in stochastic programming. Research Report UTIA 2007, No. 2192
9. Kaňková V (2008) A remark on nonlinear functionals and empirical estimates. In: Proceedings of Quantitative Methods in Economics (Multiple Criteria Decision Making XIV). The Slovak Society for Operations Research and University of Economics in Bratislava 2008 (to appear)
10. Pflug GCh (2003) Stochastic optimization and statistical inference. In: Stochastic Programming (A. Ruszczynski and A. A. Shapiro, eds.). Handbooks in Operations Research and Managemennt Science, Vol 10. Elsevier, Amsterodam
11. Römisch W, Schulz R (1993) Stability of solutions programs with complete recourse. Matematics of Operations Research 18:
12. Shapiro A (1994) Quantitative stability in stochastic programming. Math Program 67: 99–108
13. Wets RJB (1974) A statistical approach to the solution of stochastic programs with (convex) simple recourse. Research Report University Kentucky USA

Sorting and Goodness-of-Fit Tests of Uniformity in Random Number Generation

Thomas Morgenstern

Hochschule Karlsruhe, Moltkestraße 30, D-76133 Karlsruhe
thomas.morgenstern@hs-karlsruhe.de

Summary. Empirical testing of random number generators includes goodness-of-fit tests with many numbers. The tests involve sorting and classification of random numbers. We study the effects of sorting routines on the computation time of tests of uniformity and propose improvements.

1 Introduction

Unbiased tests must be applied using statistical quantitative measures to judge whether a sequence of numbers is random or not [1].

1.1 Goodness-of-Fit Tests

Two regularly applied statistical goodness-of-fit tests are the χ^2-test and the Kolmogorov-Smirnov test (KS test).
The χ^2-test uses a fairly large number, n, of independent observations. The number of observations falling in each of k categories (we may call buckets) is counted and a test variable χ^2 is computed.
If the test value χ^2 is below the 1% percentile of the χ^2-distribution or above the 99% percentile we reject the numbers as *not sufficiently random.* Values between 1% and 5% or 95% and 99% are *suspect.* The χ^2 test detects global non-random behavior [1].
For a continuous distribution function $F(x)$ the KS test is based on the maximal difference D of the hypothetic distribution function $F(x)$ and the empirical distribution function $F_n(x)$ for observations $x_1, x_2, \ldots, x_n$. We consider the uniform distribution $F(x) = x$ on the unit interval [0, 1] here.
With the observation, that $F(x)$ is increasing and $F_n(x)$ increases only in finite steps at values x_i one derives the procedure in Table 1. A C++

B. Fleischmann et al.(Eds.), *Operations Research Proceedings 2008*,
DOI: 10.1007/978-3-642-00142-0_58, © Springer-Verlag Berlin Heidelberg 2009

Table 1. Algorithm K ([1])

Step 1	`Input the independend observations` $x_1, x_2, \ldots, x_n$
Step 2	`Sort the observations` $x_{\pi(1)} \leq x_{\pi(2)} \leq \ldots \leq x_{\pi(n)}$
Step 3	$d = \max_{i=1,\ldots,n} \{ \frac{i}{n} - F(x_{\pi(i)}),\ F(x_{\pi(i)}) - \frac{i-1}{n} \}$
Step 4	$p = P_{KS}(D \leq d)$.

implementation using quicksort can be found in [2].
We propose to apply the distribution function $F(x)$ before Step 2 and to use linear time sorting algorithms for keys with known range [0, 1]. In addition partial sorting of the data can be sufficient to calculate d.

Remark 1. Both statistical tests require counting rather than sorting.

1.2 Sorting

Good references for the theory of sorting are [3] and [4] and for implementations [5]. We consider *internal sorting* of a *given array* of (random) floating point numbers with a *given (uniform) distribution.*
We sort by counting and distribution using bucket sort, esp. radix sort. Radix sort uses the random numbers as keys and distributes them in order of their digits into buckets giving raise to two principal methods *last significant digits radix sort* (LSD) and *most significant digits radix sort* (MSD). Radix sort can outperform quicksort (see [5, Sect. 10.7]). LSD radix sort requires an auxiliary field to hold a copy of the data, decreasing the number of data that can be sortet internally.

2 Implementation

We generate double floating point random numbers using `ran.h` [2]. We use a simple function (Table 2) to extract the kth digit $d_k \in [0, R-1]$ from $a = 0.d_0d_1d_2 \ldots d_{b-1} \in [0, 1]$ (`factor` $= R^{k+1}$).
For LSD radix sort we take [5, program 10.4] followed by insertion sort [5, program 6.3].

2.1 Counting

Bin sorting algorithms are based on the basic counting procedure in Table 2. The field `count[j]` holds the cumulative absolute frequency of the value j for digit d_k $\texttt{count[j]} = \text{card}\{d_k(\texttt{a[i]}) \mid d_k(\texttt{a[i]}) < j, \texttt{actl} \leq i \leq \texttt{actr}\}$, shiftet by actl, the lower left index of the part actually being sorted, (i.e. the index range [actl, actr] is sorted during the iterations of MSD radix sort).

Table 2. Counting

```
inline int digit(double a, double factor, int R)
      { return ((int)(floor(a * factor ))  \% R ) ;}
for (j = 0; j < R+1; j++) count[j] = 0;
for (i = actl; i <= actr; i++)
      count[digit(a[i],factor,R) + 1]++;
      count[0] = actl;
for (j = 1; j < R+1; j++) count[j] += count[j-1];
```

This counting information can be used to compute the χ^2 statistic:

$$\chi^2 = \sum_{j=0}^{R-1} \frac{(\texttt{count[j+1]} - \texttt{count[j]} - n/R)^2}{n/R} .$$

Remark 2. Sorting can be stopped as soon as the numbers are rejected being non-sufficiently random when the χ^2 statistic is too big.

2.2 Lower Bounds

For subsets $K \subseteq \mathbb{R}$ we get lower bounds for the KS distance $\max_K | x - F_n(x) | \leq \sup_{\mathbb{R}} | x - F_n(x) | = d$. In MSD radix sort $K := \{\texttt{actxl}_j \mid j = 0, \ldots, R-1\}$, the lower bounds of values in a bucket (Table 3).

Table 3. Lower bound on d

```
xj = actxl + j/factor; H = (double)(count[j])/n;
d = max(d, abs(xj-H));
```

Remark 3. Sorting can be stopped as soon as the random numbers are rejected because the KS distance is too big.

2.3 Priority Queue

Like in branch-and-bound algorithms we do not need to inspect all the buckets but only those where the KS distance can exceed the already found lower bound d. Maximal increases are obtained when all the data x_i in bin j have the lowest possible value `actxl+j/factor` or the highest possible value `(actxl+(j+1)/factor)-eps`, giving a bound for bin j as in Table 4.

Table 4. Inspection value of bin j

```
value = max(actxl+(j+1)/factor-(float)(count2[j])/n ,
            (float)(count2[j+1])/n-(actxl+j/factor) );
```

The algorithm is controlled by a priority queue. The bin in the queue with highest value is analysed by counting (the lower bound on d might increase due to the counting results). New bins with values higher than d are inserted into the queue for (possible) later analysis. The exact KS distance d is computed when no bins with values higher than d are in the queue.

2.4 Adaptive, In Place Sorting

Our implementation is based on the MSD radix sort program in [5, program 10.2]. Using permutation loops, MSD radix sort can sort *in place*, i.e. using the same array to store and sort the data [6].
Adaptive radix sort algorithms use different numbers of buckets for different sizes n of data [7]. Counting requires random access to a memory array, possibly slowing down the program [9] considerably. Fine tuning of the memory allocation process can make MSD radix sort approx. three times faster compared to quicksort [10].

3 Experimental Results

Our tests are performed on a Intel Core2 E6600 CPU with 2.4 GHz processor core frequency, 2 GB RAM, 4 MB L2 cashe and 1066 MHz system bus frequency.
We use Microsoft Visual Studio 2008 C++ as compiler with the only additional option /Ot and the Microsoft Windows XP x64 operating system – probably not the best execution and storage control.

We generated sequences of 2^e, $e = 20, \ldots, 27$, random numbers and recorded the execution times in seconds. Table 5 shows the mean of 30 runs and their standard deviation.

Table 5. Computational Results

Quicksort	20	21	22	23	24	25	26	27
mean	0.161	0.338	0.706	1.476	3.079	6.402	13.28	27.53
sdv	0.007	0.008	0.006	0.008	0.010	0.015	0.018	0.032
LSD	20	21	22	23	24	25	26	27
mean	0.100	0.196	0.395	0.797	1.62	3.33	6.84	2823.
sdv	0.009	0.008	0.007	0.003	0.007	0.008	0.016	284.4
MSD	20	21	22	23	24	25	26	27
mean	0.130	0.260	0.537	1.075	2.255	4.590	10.93	24.91
sdv	0.007	0.008	0.008	0.007	0.007	0.007	0.185	0.013
MSD KS	20	21	22	23	24	25	26	27
mean	0.063	0.136	0.269	0.582	1.183	2.591	6.527	16.54
sdv	0.008	0.011	0.010	0.041	0.055	0.206	0.169	0.610

4 Conclusions

The Kolmogorov-Smirnov statistic can be computed significantly faster when the probability function is applied before sorting and special sorting routines are used. Various χ^2-statistics can be computed simultaneously.

Tests using LSD radix sort on the upper digits followed by insertion sort turn out to be approximately two times faster than computations using quicksort (as long as internal sorting is possible) and the LSD radix sort information can be used to compute χ^2-statistics.

In place MSD radix sort needs less memory and can be adapted to the hardware to run faster than quicksort. Using the information generated by MSD radix sort, many χ^2- and KS-statistics can be computed simultaniously.

Requiring exact computation of the KS-statistic only and using MSD radix sort is on average nearly three times faster than tests with complete sorting by quicksort and is also faster than using LSD radix sort. The increase of variance in computation time is due to the fact that for less uniformly distributed random numbers approximate sorting is significantly faster than for perfectly distributed numbers. Computing the KS statistics for non-uniform data is approx. two times faster than for uniform data, also solving worst case considerations for radix sort.

Remark 4. Approximative calculations are even faster.

References

1. Knuth D E (1997) The art of computer programming. Vol. 2: Seminumerical algorithms. third edition, Addison-Wesley, Reading, Mass.
2. Press W H [et. al.] (2007) Numerical Recipes: The art of scientific computing. third edition, Cambridge University Press, Cambridge New York
3. Knuth D E (1997) The art of computer programming. Vol. 3: Sorting and Searching. second edition, Addison-Wesley, Reading, Mass.
4. Mahmoud H M (2000) Sorting: A Distribution Theory. Wiley, New York
5. Sedgewick R (2003) Algorithms in C++. Parts 1-4: Fundamentals, Data Structures, Sorting, Searching. third edition, Addison-Wesley, Reading, Mass.
6. McIlroy P, Bostic K, McIlroy M D (1993) Engineering Radix Sort. Computer Systems 6(1):5–27
7. Andersson A, Nilsson S (1998) Implementing Radixsort. J. Exp. Algorithms 3:Art. No. 7
8. LaMarca A, Landner R E (1999) The influence of cashes on the performance of sorting. J. Algorithms 31:66–104
9. Schutler M E, Sim S W, Lim W Y S (2008) Analysis of Linear Time Sorting Algorithms. The Computer J. 51:451–469
10. Al-Badarneh A, El-Aker F (2004) Effective Adaptive In-Place Radix Sorting. Informatica 15(3):295–302

Constrained Risk-Sensitive Markov Decision Chains

Karel Sladký

Institute of Information Theory and Automation, Academy of Sciences of the Czech Republic, Pod Vodárenskou věží 4, 182 08 Praha 8, Czech Republic
sladky@utia.cas.cz

1 Introduction and Notation

We consider a Markov decision chain $X = \{X_n, n = 0, 1, \ldots\}$ with finite state space $\mathcal{I} = \{1, 2, \ldots, N\}$ and finite set $\mathcal{A}_i = \{1, 2, \ldots, K_i\}$ of possible decisions (actions) in state $i \in \mathcal{I}$. Supposing that in state $i \in \mathcal{I}$ action $a \in \mathcal{A}_i$ is selected, then state j is reached in the next transition with a given probability $p_{ij}(a)$ and one-stage transition rewards $r_{ij}(a)$ and $s_{ij}(a)$ will be accrued to such transition. We shall suppose that the streams of transition rewards are evaluated by an exponential utility function, say $u^\gamma(\cdot)$, with risk aversion coefficient $\gamma < 0$ (the risk averse case) or $\gamma > 0$ (the risk seeking case). Then the utility assigned to the (random) reward ξ is given by $u^\gamma(\xi) := \operatorname{sign}(\gamma) \exp(\gamma\xi)$. Hence the expected utility

$$U^{(\gamma)}(\xi) = \operatorname{sign}(\gamma)\, \mathsf{E}[\exp(\gamma\xi)], \quad (\mathsf{E} \text{ is reserved for expectation}) \tag{1}$$

and for the corresponding certainty equivalent $Z^\gamma(\xi)$ we have

$$u^\gamma(Z^\gamma(\xi)) = \mathsf{E}\,[u^\gamma(\xi)] \Longleftrightarrow Z^\gamma(\xi) = \gamma^{-1} \ln\{\mathsf{E}\,[\exp(\gamma\xi)]\}. \tag{2}$$

A (Markovian) policy controlling the chain identified by $\pi = (f^n)$ where $f^n \in \mathcal{A} \equiv \mathcal{A}_1 \times \ldots \times \mathcal{A}_N$ for every $n = 0, 1, 2, \ldots$ and $f_i^n \in \mathcal{A}_i$ is the decision at the nth transition when the chain X is in state i. Policy which takes at all times the same decision rule, i.e. $\pi \sim (f)$, is called stationary. Stationary policy $\pi \sim (f)$ is randomized, if in state i selects decision $f_i^{(j)}$ with a given probability $\alpha_i^{(j)} \ge 0$ (where $\sum_j \alpha_i^{(j)} = 1$); $\bar{\mathcal{A}}$ denotes the set of all stationary randomized policies. Let

$$\xi_{X_0}^n(\pi) = \sum_{k=0}^{n-1} r_{X_k, X_{k+1}}(f_{X_k}^k), \quad \zeta_{X_0}^n(\pi) = \sum_{k=0}^{n-1} s_{X_k, X_{k+1}}(f_{X_k}^k), \tag{3}$$

B. Fleischmann et al.(Eds.), *Operations Research Proceedings 2008*,
DOI: 10.1007/978-3-642-00142-0_59,

be the sum of transition rewards received in the n next transitions of the considered Markov chain X, and similarly let $\xi_{X_m}^{(m,n)}(\pi)$, $\zeta_{X_m}^{(m,n)}(\pi)$ be reserved for the total (random) additive rewards obtained from the mth up to the nth transition if policy $\pi = (f^n)$ is followed. Since $u^\gamma(\cdot)$ is separable and multiplicative we immediately conclude by (3) that

$$u^\gamma(\xi_{X_0}^n(\pi)) = \exp[\gamma\, r_{X_0,X_1}(f_{X_0}^0)] \cdot u^\gamma(\xi_{X_1}^{(1,n)}(\pi)) \tag{4}$$

$$u^\gamma(\zeta_{X_0}^n(\pi)) = \exp[\gamma\, s_{X_0,X_1}(f_{X_0}^0)] \cdot u^\gamma(\zeta_{X_1}^{(1,n)}(\pi)). \tag{5}$$

Supposing that the chain starts in state $X_0 = i$ and policy $\pi = (f^n)$ is followed, then for expected utility in the n next transitions, the corresponding certainty equivalent, and for mean value of the certainty equivalent we have (E_i^π denotes expectation if policy π is followed and the starting state $X_0 = i$)

$$U_i^\pi(\gamma, n) := \mathsf{E}_i^\pi[u^\gamma(\xi_{X_0}^n(\pi))], \quad \bar{U}_i^\pi(\gamma, n) := \mathsf{E}_i^\pi[u^\gamma(\zeta_{X_0}^n(\pi))], \tag{6}$$

$$Z_i^\pi(\gamma, n) := \frac{1}{\gamma} \ln [U_i^\pi(\gamma, n)], \quad \bar{Z}_i^\pi(\gamma, n) := \frac{1}{\gamma} \ln [\bar{U}_i^\pi(\gamma, n)], \tag{7}$$

$$J_i^\pi(\gamma) := \limsup_{n\to\infty} \frac{1}{n} Z_i^\pi(\gamma, n), \quad \bar{J}_i^\pi(\gamma) := \limsup_{n\to\infty} \frac{1}{n} \bar{Z}_i^\pi(\gamma, n). \tag{8}$$

Similarly, for $m < n$ if the starting state $X_m = i$ we write

$$U_i^\pi(\gamma, m, n) := \mathsf{E}_i^\pi[u^\gamma(\xi_{X_m}^{(m,n)}(\pi))], \quad \bar{U}_i^\pi(\gamma, m, n) := \mathsf{E}_i^\pi[u^\gamma(\zeta_{X_m}^{(m,n)}(\pi))]. \tag{9}$$

We introduce the following matrix notation:
$\boldsymbol{U}^\pi(\gamma, n)$, resp. $\boldsymbol{U}^\pi(\gamma, m, n)$, is reserved for the (column) vector whose ith element equals $U_i^\pi(\gamma, n)$, resp. $U_i^\pi(\gamma, m, n)$,
$\boldsymbol{Q}(f) = [q_{ij}(f_i)]$, resp. $\bar{\boldsymbol{Q}}(f) = [\bar{q}_{ij}(f_i)]$, is an $N \times N$ nonnegative matrix with elements $q_{ij}(f_i) := p_{ij}(f_i) \cdot e^{\gamma r_{ij}(f_i)}$, resp. $\bar{q}_{ij}(f_i) := p_{ij}(f_i) \cdot e^{\gamma s_{ij}(f_i)}$,
$\boldsymbol{P}(f) = [p_{ij}(f_i)]$ is the transition probability matrix, and $\boldsymbol{e}$ is the (column) vector of all ones.
In this note we focus attention on the asymptotic properties of the expected utility and the corresponding certainty equivalents, if the optimal values considered with respect to transition rewards $r_{ij}(\cdot)$ must fulfill certain additional constraint on the expected utility or the certainty equivalent generated by transition rewards $s_{ij}(\cdot)$. Obviously, this additional constraint can also interpreted as fulfilling guaranteed growth rate under a different value of the risk aversion coefficient. Our analysis is based on properties of a collection of nonnegative matrices arising in the recursive formulas for the growth of expected utilities.

2 Risk-Sensitive Optimality and Nonnegative Matrices

Since the considered utility functions are separable and the control policy is Markovian on taking expectations we conclude that for $X_0 = i$

$$\begin{aligned}\mathsf{E}_i^\pi u^\gamma(\xi_{X_0}^n) &= \mathsf{E}_i^\pi\{\exp[\gamma\, r_{X_0,X_1}(f_i^0)]\cdot \mathsf{E}_{X_1}^\pi[u^\gamma(\xi_{X_1}^{(1,n)})|X_1]\}\\ &= \sum_{j\in\mathcal{I}} p_{ij}(f_i^0)\cdot \mathrm{e}^{\gamma r_{ij}(f_i^0)}\cdot \mathsf{E}_j^\pi u^\gamma(\xi_{X_1}^{(1,n)}) \qquad (10)\end{aligned}$$

that can be also written as (recall that $q_{ij}(f_i) = p_{ij}(f_i)\cdot \mathrm{e}^{\gamma r_{ij}(f_i)}$)

$$U_i^\pi(\gamma,0,n) = \sum_{j\in\mathcal{I}} q_{ij}(f_i^0)\cdot U_j^\pi(\gamma,1,n) \quad \text{with } U_i^\pi(\gamma,n,n) = \operatorname{sign}(\gamma) \quad (11)$$

or in vector notation as

$$\boldsymbol{U}^\pi(\gamma,0,n) = \boldsymbol{Q}(f^0)\cdot \boldsymbol{U}^\pi(\gamma,1,n) \quad \text{with } \boldsymbol{U}^\pi(\gamma,n,n) = \operatorname{sign}(\gamma)\,\boldsymbol{e}. \quad (12)$$

Iterating (12) we get if policy $\pi = (f^n)$ is followed

$$\boldsymbol{U}^\pi(\gamma,n) = \boldsymbol{Q}(f^0)\cdot \boldsymbol{Q}(f^1)\cdot \ldots \cdot \boldsymbol{Q}(f^{n-1})\cdot \operatorname{sign}(\gamma)\,\boldsymbol{e}. \quad (13)$$

Since $\boldsymbol{Q}(f)$ is nonnegative by the Perron–Frobenius theorem (see e.g. [1, 3]) spectral radius $\rho(f)$ of $\boldsymbol{Q}(f)$ is equal to the eigenvalue of $\boldsymbol{Q}(f)$ with the largest modulus and the corresponding left and right eigenvectors, say $\boldsymbol{y}(f)$, resp. $\boldsymbol{v}(f)$, can be selected nonnegative, i.e.

$$\rho(f)\,\boldsymbol{y}(f) = \boldsymbol{y}(f)\,\boldsymbol{Q}(f), \qquad \rho(f)\,\boldsymbol{v}(f) = \boldsymbol{Q}(f)\,\boldsymbol{v}(f). \quad (14)$$

Moreover, if every $\boldsymbol{Q}(f)$ is irreducible, then $\boldsymbol{y}(f)$, $\boldsymbol{v}(f)$ can be selected positive. Let for $f, f' \in \mathcal{A}$, $\varphi(f,f') := [\boldsymbol{Q}(f) - \boldsymbol{Q}(f')]\,\boldsymbol{v}(f')$.

Proposition. *If every $\boldsymbol{Q}(f)$ is irreducible, then there exist $\hat{f}, \tilde{f} \in \mathcal{A}$, $\hat{\rho} := \rho(\hat{f})$, $\hat{\boldsymbol{v}} := \boldsymbol{v}(\hat{f})$, $\tilde{\rho} := \rho(\tilde{f})$, $\tilde{\boldsymbol{v}} := \boldsymbol{v}(\tilde{f})$ such that for every $f \in \mathcal{A}$*

$$\hat{\rho}\,\hat{\boldsymbol{v}} = \max_{f\in\mathcal{A}} \boldsymbol{Q}(f)\cdot\hat{\boldsymbol{v}} = \boldsymbol{Q}(\hat{f})\cdot\hat{\boldsymbol{v}} \ge \boldsymbol{Q}(f)\cdot\hat{\boldsymbol{v}} \quad \text{and} \quad \varphi(f,\hat{f}) \le \boldsymbol{0} \quad (15)$$

$$\tilde{\rho}\,\tilde{\boldsymbol{v}} = \min_{f\in\mathcal{A}} \boldsymbol{Q}(f)\cdot\tilde{\boldsymbol{v}} = \boldsymbol{Q}(\tilde{f})\cdot\tilde{\boldsymbol{v}} \le \boldsymbol{Q}(f)\cdot\tilde{\boldsymbol{v}} \quad \text{and} \quad \varphi(f,\tilde{f}) \ge \boldsymbol{0}. \quad (16)$$

The above facts can be easily verified by policy iterations based on

Lemma. *Let f, $f' \in \mathcal{A}$ and $\varphi(f,f') > \boldsymbol{0}$, resp. $\varphi(f,f') < \boldsymbol{0}$. Then $\rho(f) > \rho(f')$, resp. $\rho(f) > \rho(f')$.*

The assertion of Lemma follows immediately since for $f, f' \in \mathcal{A}$ by (14) $\boldsymbol{Q}(f)[\boldsymbol{v}(f) - \boldsymbol{v}(f')] + \varphi(f,f') = \rho(f)\,[\boldsymbol{v}(f) - \boldsymbol{v}(f')] + [\rho(f) - \rho(f')]\,\boldsymbol{v}(f')$ and on premultiplying the above equality by $\boldsymbol{y}(f) > \boldsymbol{0}$ we immediately get $\boldsymbol{y}(f)\,\varphi(f,f') = [\rho(f) - \rho(f')]\,\boldsymbol{y}(f)\,\boldsymbol{v}(f')$. Hence if $\varphi(f,f') > \mathbf{0}$, resp. $\varphi(f,f') < \mathbf{0}$, then $\rho(f) > \rho(f')$, resp. $\rho(f) < \rho(f')$ (for details and slight extensions see e.g. [4, 6, 7]). □

In virtue of Lemma we can easily construct finite sequence $\{f^{(k)} \in \mathcal{A}\}$ such that $\{\rho(f^{(k)}), k = 0, 1, \ldots, K\}$ is increasing, resp. decreasing, and $f^{(K)} = \hat{f}$ fulfills (15), resp. $f^{(K)} = \tilde{f}$ fulfills (16).

Since eigenvectors are unique up to multiplicative constant, on condition that $\hat{\boldsymbol{v}} > \mathbf{0}$ we can choose $\hat{\boldsymbol{v}} \geq \boldsymbol{e}$ or $\hat{\boldsymbol{v}} \leq \boldsymbol{e}$. Then in virtue of (15) we immediately conclude that if $\hat{\boldsymbol{v}} \geq \boldsymbol{e}$ for any policy $\pi = (f^k)$

$$\prod_{k=0}^{n-1} \boldsymbol{Q}(f^k) \cdot \boldsymbol{e} \leq \prod_{k=0}^{n-1} \boldsymbol{Q}(f^k) \cdot \hat{\boldsymbol{v}} \leq (\boldsymbol{Q}(\hat{f})^n \cdot \hat{\boldsymbol{v}} = (\hat{\rho})^n\, \hat{\boldsymbol{v}} \tag{17}$$

and if $\hat{\boldsymbol{v}} \leq \boldsymbol{e}$ then for policy $\hat{\pi} \sim (\hat{f})$ it holds $(\boldsymbol{Q}(\hat{f}))^n \cdot \boldsymbol{e} \geq (\hat{\rho})^n\, \hat{\boldsymbol{v}}$. Quite similarly, by (16), we can show that for suitably selected $\tilde{\boldsymbol{v}} \leq \boldsymbol{e}$

$$\prod_{k=0}^{n-1} \boldsymbol{Q}(f^k) \cdot \boldsymbol{e} \geq \prod_{k=0}^{n-1} \boldsymbol{Q}(f^k) \cdot \tilde{\boldsymbol{v}} \geq (\boldsymbol{Q}(\tilde{f})^n \cdot \tilde{\boldsymbol{v}} = (\tilde{\rho})^n\, \tilde{\boldsymbol{v}} \tag{18}$$

and if $\tilde{\boldsymbol{v}} \geq \boldsymbol{e}$ then for policy $\tilde{\pi} \sim (\tilde{f})$ it holds $(\boldsymbol{Q}(\tilde{f}))^n \cdot \boldsymbol{e} \leq (\tilde{\rho})^n\, \tilde{\boldsymbol{v}}$.

Conclusions. *Under irreducibility of all $\boldsymbol{P}(f)$'s, and hence also of all $\boldsymbol{Q}(f)$'s and $\bar{\boldsymbol{Q}}(f)$, the growth rate of $\boldsymbol{U}^\pi(\gamma, n)$ as well as the mean values of the certainty equivalent (cf. (6) – (8)) are independent of the starting state and bounded from above, resp. from below, by $\hat{\rho}$, resp. by $\tilde{\rho}$, that can be obtained as a solution of (15), resp. (16). Similarly, the growth rate of $\bar{\boldsymbol{U}}^\pi(\gamma, n)$ and the mean value of the certainty equivalent are bounded from above, resp. from below, by $\bar{\rho}$, resp. by $\breve{\rho}$, that can be obtained (along with $\bar{\varphi}(f,f') := [\boldsymbol{Q}(f) - \boldsymbol{Q}(f')]\,\boldsymbol{v}(f')$) as a solution of*

$$\bar{\rho}\,\bar{\boldsymbol{v}} = \max_{f \in \mathcal{A}}\{\bar{\boldsymbol{Q}}(f)\cdot\bar{\boldsymbol{v}}\} = \bar{\boldsymbol{Q}}(\bar{f})\cdot\bar{\boldsymbol{v}} \geq \bar{\boldsymbol{Q}}(f)\cdot\bar{\boldsymbol{v}}, \text{ hence } \bar{\varphi}(f,\bar{f}) \leq \mathbf{0}. \tag{19}$$

$$\breve{\rho}\,\breve{\boldsymbol{v}} = \min_{f \in \mathcal{A}}\{\bar{\boldsymbol{Q}}(f)\cdot\breve{\boldsymbol{v}}\} = \bar{\boldsymbol{Q}}(\breve{f})\cdot\breve{\boldsymbol{v}} \leq \bar{\boldsymbol{Q}}(f)\cdot\breve{\boldsymbol{v}}, \text{ hence } \bar{\varphi}(f,\breve{f}) \geq \mathbf{0}. \tag{20}$$

Unfortunately, if we impose some constraints on the growth of $\bar{\boldsymbol{U}}^\pi(\gamma, n)$ policy optimizing the growth rate of $\boldsymbol{U}^\pi(\gamma, n)$ need not be feasible. In what follows we present policy iterations method yielding stationary policy $\pi^* \sim (f^*)$, in general randomized, maximizing the growth of $\boldsymbol{U}^\pi(\gamma, n)$ (i.e. the spectral radius of $\rho(f)$ of $\boldsymbol{Q}(f)$ or mean value of the corresponding certainty equivalent) on condition that the growth rate of $\bar{\boldsymbol{U}}^\pi(\gamma, n)$ (i.e. the spectral radius $\bar{\rho}(f)$ of $\bar{\boldsymbol{Q}}(f)$ or mean value of the certainty equivalent) is nonsmaller than a given value d (of course, we suppose that there exists such $f \in \mathcal{A}$ that $\bar{\rho}(f) \geq d$).

Quite similarly we can proceed for finding minimal growth rate of $\bar{\boldsymbol{U}}^{\pi}(\gamma, n)$ (or mean value of the corresponding certainty equivalent) under constraints on the growth rate of $\bar{\boldsymbol{U}}^{\pi}(\gamma, n)$ (or mean value of the corresponding certainty equivalent), what may happen if the risk aversion criterion is considered.

3 Finding Optimal Solutions Under Constraints

Employing policy or value iterations (see e.g. [2, 4, 6, 7]) we can find policy $\hat{\pi} \sim (\hat{f})$ maximizing the growth rate of $\boldsymbol{U}^{\pi}(\gamma, n)$ or mean value of the certainty equivalent $J_i^{\pi}(\gamma)$, i.e. the value $\rho(\hat{f})$, cf. (15). In case that $\bar{\rho}(\hat{f}) < d$ for finding optimal policy under a given constraint the following algorithmic procedure, generating decreasing (resp. increasing) sequence $\rho(f^{(n)})$ (resp. $\bar{\rho}(f^{(n)})$) can be used; similar approach works for minimizing growth rate under constraints, and after some modifications also for constrained average Markov decision processes, cf. [5].

Algorithm. (Maximizing growth rate under constraint)

Step 0. Set $f^{(0)} := \hat{f}$.

Step 1. For $n = 0, 1, \ldots$ on employing $f^{(n)} \in \mathcal{A}$ find $\tilde{f}^{(n)} \in \mathcal{A}$ such that $\tilde{f}_i^{(n)} = f_i^{(n)}$ except of one $\tilde{f}_{i_n}^{(n)}$ such that for $\varphi_{i_n}(\tilde{f}^{(n)}, f^{(n)}) < 0$

$$\frac{\bar{\rho}(\tilde{f}^{(n)}) - \bar{\rho}(f^{(n)})}{|\rho(\tilde{f}^{(n)}) - \rho(f^{(n)})|} = \max_{f \in \mathcal{A}} \frac{\bar{\rho}(f) - \bar{\rho}(f^{(n)})}{|\rho(f) - \rho(f^{(n)})|}$$

(to this end check $f_{i_n} = \operatorname{argmax}_{i \in \mathcal{I}, f \in \mathcal{A}} \{\bar{\varphi}_i(f, f^{(n)}) / |\varphi_i(f, f^{(n)})|\}$, and calculate $\bar{\rho}(\tilde{f}^n)$.

Step 2. If $\bar{\rho}(\tilde{f}^{(n)}) < d$ set $f^{(n+1)} := \tilde{f}^{(n)}$ and go to Step 1, else to Step 3.

Step 3. Randomize policies $\tilde{f}^{(n)}, f^{(n)}$ to obtain decision vector $f^* \in \bar{\mathcal{A}}$ such that $\bar{\rho}(f^*) = d$.

The randomized stationary policy $\pi^* \sim (f^*)$ yields maximal growth rate $\rho(f)$ (or mean value of the corresponding certainty equivalent) on conditions that the growth rate $\bar{\rho}(f) \geq d$ (or mean value of the corresponding certainty equivalent is nonsmaller than d).

Illustrative Example. Consider a controlled Markov reward chain with two states and only three possible actions in state 1. Then

$$\boldsymbol{P}(f^{(1)}) = \begin{bmatrix} \frac{1}{3} & \frac{2}{3} \\ \frac{2}{3} & \frac{1}{3} \end{bmatrix}, \quad \boldsymbol{P}(f^{(2)}) = \begin{bmatrix} \frac{2}{3} & \frac{1}{3} \\ \frac{2}{3} & \frac{1}{3} \end{bmatrix}, \quad \boldsymbol{P}(f^{(3)}) = \begin{bmatrix} \frac{1}{6} & \frac{5}{6} \\ \frac{2}{3} & \frac{1}{3} \end{bmatrix},$$

are the transition probability matrices that along with transition reward matrices

$$\boldsymbol{R} = \begin{bmatrix} \frac{3}{4} & \frac{3}{10} \\ \frac{3}{4} & \frac{3}{4} \end{bmatrix}, \quad \boldsymbol{S} = \begin{bmatrix} \frac{3}{10} & \frac{3}{4} \\ \frac{3}{4} & \frac{3}{4} \end{bmatrix}$$ fully characterize the transition and reward structures of the considered Markov chain. Simple calculation yields

$$\boldsymbol{Q}(f^{(1)}) = \begin{bmatrix} \frac{1}{4} & \frac{1}{5} \\ \frac{1}{2} & \frac{1}{4} \end{bmatrix}, \quad \boldsymbol{Q}(f^{(2)}) = \begin{bmatrix} \frac{1}{2} & \frac{1}{10} \\ \frac{1}{2} & \frac{1}{4} \end{bmatrix}, \quad \boldsymbol{Q}(f^{(3)}) = \begin{bmatrix} \frac{1}{8} & \frac{1}{4} \\ \frac{1}{2} & \frac{1}{4} \end{bmatrix}$$

$$\rho(f^{(1)}) = 0.566, \qquad \rho(f^{(2)}) = 0.631, \qquad \rho(f^{(3)}) = 0.546,$$

$$\bar{\boldsymbol{Q}}(f^{(1)}) = \begin{bmatrix} \frac{1}{10} & \frac{1}{2} \\ \frac{1}{2} & \frac{1}{4} \end{bmatrix}, \quad \bar{\boldsymbol{Q}}(f^{(2)}) = \begin{bmatrix} \frac{1}{5} & \frac{1}{4} \\ \frac{1}{2} & \frac{1}{4} \end{bmatrix}, \quad \bar{\boldsymbol{Q}}(f^{(3)}) = \begin{bmatrix} \frac{1}{20} & \frac{5}{8} \\ \frac{1}{2} & \frac{1}{4} \end{bmatrix}$$

$$\bar{\rho}(f^{(1)}) = 0.681, \qquad \bar{\rho}(f^{(2)}) = 0.579, \qquad \bar{\rho}(f^{(3)}) = 0.718.$$

Obviously, stationary policy $\pi^{(2)} \sim (f^{(2)})$, resp. $\pi^{(3)} \sim (f^{(3)})$, maximizes growth rate of $\boldsymbol{U}^{\pi}(\gamma, n)$, resp. $\bar{\boldsymbol{U}}^{\pi}(\gamma, n)$. On the other hand to maximize the growth rate of $\boldsymbol{U}^{\pi}(\gamma, n)$ under condition that the growth rate of $\bar{\boldsymbol{U}}^{\pi}(\gamma, n)$ is nonsmaller than a given value d, it is necessary to randomize decisions $f^{(1)}$, $f^{(2)}$, $f^{(3)}$. For example, to maximize the growth rate of $\boldsymbol{U}^{\pi}(\gamma, n)$ under condition that the growth rate of $\bar{\boldsymbol{U}}^{\pi}(\gamma, n)$ is nonsmaller than $d = 0.6$, as a simple calculation shows, it is necessary to follow randomized policy employing in state 1 decision $f_1^{(1)}$ with probability equal to 0.16 and decision $f_1^{(2)}$ with probability equal to 0.84. This randomized policy defines optimal policy $\pi^* \sim (f^*)$ that cannot be further improved. After small algebra we get

$$\boldsymbol{Q}(f^*) = \begin{bmatrix} 0.46 & 0.12 \\ 0.5 & 0.25 \end{bmatrix}, \quad \rho(f^*) = 0.621, \quad \boldsymbol{v}(f^*) = \begin{bmatrix} 0.742 \\ 1 \end{bmatrix}$$

$$\bar{\boldsymbol{Q}}(f^*) = \begin{bmatrix} 0.18 & 0.29 \\ 0.5 & 0.25 \end{bmatrix}, \quad \bar{\rho}(f^*) = 0.600, \quad \bar{\boldsymbol{v}}(f^*) = \begin{bmatrix} 0.70 \\ 1 \end{bmatrix}.$$

Acknowledgement. This research was supported by the Czech Science Foundation under Grants 402/08/0107 and 402/07/1113.

References

1. Berman A, Plemmons RJ (1979) Nonnegative Matrices in the Mathematical Sciences. Academic Press, New York
2. Cavazos-Cadena R, Montes-de-Oca R (2003) The value iteration algorithm in risk-sensitive average Markov decision chains with finite state space. Math Oper Res 28:752–756
3. Gantmakher FR (1959) The Theory of Matrices. Chelsea, London
4. Howard RA, Matheson J (1972) Risk-sensitive Markov decision processes. Manag Sci 23:356–369
5. Sladký K (1967) On optimal service policy for several facilities (in Czech). Kybernetika 4:342–355
6. Sladký K (1976) On dynamic programming recursions for multiplicative Markov decision chains. Math Programming Study 6:216–226
7. Sladký K (2008) Growth rates and average optimality in risk-sensitive Markov decision chains. Kybernetika 44:205–226

Part VIII

Business Informatics, Decision Support and Artifical Intelligence

An EM-based Algorithm for Web Mining Massive Data Sets

Maria João Cortinhal and José G. Dias

Department of Quantitative Methods,
ISCTE Business School and CIO,
Av. das Forças Armadas, Lisboa 1649–026, Portugal
{maria.cortinhal,jose.dias}@iscte.pt

Summary. This paper introduces the PEM algorithm for estimating mixtures of Markov chains. The application to website users' search patterns shows that it provides an effective way to deal with massive data sets.

1 Introduction

Web usage mining involves using data mining techniques in the discovery of web navigation patterns from web log data. Consider a sample of n web users and let i, $i = 1, ..., n$, denote a web user. Web user i is characterized by a sequence of states $\mathbf{x}_i = (x_{i0}, x_{i1}, ..., x_{iT_i})$, where x_{it}, $t = 0, 1, ..., T_i$, denotes the tth page visited by the web user i. The length of each sequence - T_i - may differ among web users i. Under the Markov property the sequence $\mathbf{x}_i$ can be viewed as a Markov chain. Its probability is given by $p(\mathbf{x}_i) = p(x_{i0}) \prod_{t=1}^{T_i} p(x_{it}|x_{i,t-1})$, where $p(x_{i0})$ is the initial probability and $p(x_{it}|x_{i,t-1})$ is the probability that web user i is in state x_{it} at time t, given that he is in $x_{i,t-1}$ at time $t-1$. Mixture models of Markov chains allow the clustering of individuals into different behavioral segments characterized by different patterns of change [2]. Let S be the number of clusters. Then the marginal distribution of $\mathbf{x}_i$ is given by $p(\mathbf{x}_i) = \sum_{s=1}^{S} \pi_s p(\mathbf{x}_i; \boldsymbol{\theta}_s)$, where π_s is the probability that web user i belongs to segment s (with $\pi_s \geq 0$ and $\sum_{i=1}^{S} \pi_s = 1$) and $\boldsymbol{\theta}_s$ is the segment specific vector parameter. Integrating the sequential dependency within each cluster, a finite mixture of Markov chains is obtained:

$$p(\mathbf{x}_i) = \sum_{s=1}^{S} \pi_s \prod_{j=1}^{K} \lambda_{sj}^{I(x_{io}=j)} \prod_{j=1}^{K} \prod_{k=1}^{K} a_{sjk}^{\eta_{ijk}}, \tag{1}$$

B. Fleischmann et al.(Eds.), *Operations Research Proceedings 2008*,
DOI: 10.1007/978-3-642-00142-0_60, © Springer-Verlag Berlin Heidelberg 2009

where K denotes the number of states, $I(.)$ is the indicator function of the user's i initial state, η_{ijk} is a counter of the number of transitions between states done by user i, $\lambda_{sj} = p(x_{i0} = j|i \in s)$ and $a_{sjk} = p(x_{it} = k|x_{i,t-1} = j, i \in s)$ are the initial and transition probabilities, respectively. The vector of parameters is $\varphi = (\pi_1, ..., \pi_{S-1}, \boldsymbol{\theta}_1, ..., \boldsymbol{\theta}_S)$, where $\boldsymbol{\theta}_s$ includes λ_{sj}, $j = 1, ..., K$ and a_{sjk}, $j, k = 1, ..., K$. The log-likelihood function for φ is $\ell_S(\varphi; \mathbf{x}) = \sum_{i=1}^n \log p(\mathbf{x}_i; \varphi)$, and the maximum likelihood estimator (MLE) is $\hat{\varphi} = \arg\max_{\varphi} \ell_S(\varphi; \mathbf{x})$.

The Expectation-Maximization (EM) algorithm [1] has become the standard approach for finding maximum likelihood estimates of parameters in model-based clustering models. Given the vector of parameters obtained at iteration h, $\varphi^{(h)}$, the E Step computes the conditional expectation that i belongs to cluster s as $\alpha_{is}^{(h+1)} = \frac{\pi_s^{(h)} f_s(\mathbf{x}_i; \boldsymbol{\theta}_s^{(h)})}{\sum_{r=1}^S \pi_r^{(h)} f_r(\mathbf{x}_i; \boldsymbol{\theta}_r^{(h)})}$, where $f_s(\mathbf{x}_i; \boldsymbol{\theta}_s^{(h)}) = \prod_{j=1}^K [\lambda_{sj}^{(h)}]^{I(x_{io}=j)} \prod_{j=1}^K \prod_{k=1}^K [a_{sjk}^{(h)}]^{\eta_{ijk}}$. At the M Step, the EM computes a new parameter approximation:

$$\pi_s{}^{(h+1)} = \frac{1}{n} \sum_{i=1}^n \alpha_{is}^{(h+1)} \tag{2}$$

$$\lambda_{sj}^{(h+1)} = \frac{\sum_{i=1}^n \alpha_{is}^{(h+1)} I(x_{i0} = j)}{\sum_{i=1}^n \alpha_{is}^{(h+1)}} \tag{3}$$

$$a_{sjk}^{(h+1)} = \frac{\sum_{i=1}^n \alpha_{is}^{(h+1)} \eta_{ijk}}{\sum_{i=1}^n \sum_{r=1}^K \alpha_{is}^{(h+1)} \eta_{ijr}}. \tag{4}$$

The EM algorithm stops as soon as the difference between two consecutive log-likelihood values becomes smaller than a given tolerance. This algorithm may present some problems, namely being extremely slow in convergence (linear convergence). When one deals with large data sets the major problem relies on the E Step which encompasses a complete pass through the data. This burden has been mostly tackled by down scaling techniques (e.g., [4]). The main idea behind the PC-aggregated EM (PEM) algorithm is to provide better initial parameter estimates than the random initialization of the EM algorithm, reducing the number of iterations needed to converge. We propose to use principal component analysis to boost the EM algorithm speed for massive data sets. The structure of this paper is as follows. Sections 2 and 3 present the algorithm and results, respectively. The paper ends with a short conclusion.

2 The PEM Algorithm

Under the Markovian process, the sequence $\mathbf{x}_i$ can be summarized by its sufficient statistics: the initial state $\vartheta_{ij} = I(x_{i0} = j)$ and the transitions between states η_{ijk}. Without loosing information, under the Markov property $\mathbf{x}_i$ is equivalent to $\tilde{\mathbf{x}}_i = \{\vartheta_{i1}, ..., \vartheta_{iK}, \eta_{i11}, ..., \eta_{iKK}\}$. Because all vectors $\tilde{\mathbf{x}}_i$ have the same length, the matrix $\tilde{\mathbf{X}} = \{\tilde{\mathbf{x}}_i\}_{i=1}^n$, with n rows and $p = K(K+1)$ columns, contains all the sufficient statistics. As it is very unlikely that two objects have the same sequence ($\tilde{\mathbf{x}}_i = \tilde{\mathbf{x}}_{i'}$), the data set cannot be aggregated by using weights in the likelihood function. However, some sequences may be very similar.
We apply principal component analysis (PCA) to reduce data dimensionality. Let y_{ir} be the score of the object i on dimension r given by principal component PC_r, $y_{ir} = a_{1r}\breve{x}_{i1} + ... + a_{pr}\breve{x}_{ip}$.The estimated matrix $\mathbf{A}$ of coefficients (loadings) is orthogonal ($\mathbf{A}^T = \mathbf{A}^{-1}$) and consequently uncorrelated PC's are extracted by linear transformations of the original variables. Principal components are based on the covariance matrix of $\tilde{\mathbf{X}}$, which means that the original data ($\tilde{\mathbf{x}}_i$) is centered ($\breve{\mathbf{x}}_i$), resulting principal components with mean zero. The PCA computes the loadings a_{vr} in such away that PC_1 has the largest variance in the original data set, PC_2 has the second largest variance and so on. Thus, the first few PCs retain most of the variance in the original data set [3].
Let q be the number of principal components to extract with $q \ll p$; and let $\mathbf{Y}$ be an $n \times q$ matrix with the new values of the n objects projected on the q dimensions. As the first component captures the largest heterogeneity in the original patterns, it defines the first level of patterns clustering or aggregation. Thus, based on PC_1 or level 1 of the tree, one obtains two groups $\{0, 1\}$ with the mean as cut-off point, i.e., the indicator function $I(y_{i1} > 0)$ yields the classification into group 0 or 1. For example, the first two components $\{I(y_{i1} > 0), I(y_{i2} > 0)\}$ yields the aggregation of the data set into four groups.
To reproduce the total variability of the original p variables, one needs all p PCs. However, if the first q PCs account for a large proportion of the variability, one achieves the objective of dimension reduction. The proportion of the first q eigenvalues of the covariance matrix of data set $\tilde{\mathbf{X}}$ gives the proportion of explained variance, and this is the criterion we set for component extraction.[1] Let G be the number of groups with at least one observation and n_g the number of observations in group

[1] For example, for $q = 10$ components objects are allocated into a maximum of $2^{10} = 1024$ groups.

g, $g = 1, ..., G$ (only groups with $n_g \neq 0$ need to be retained) with $\sum_{g=1}^{G} n_g = n$.

The PEM algorithm comprises three main phases:

1. *PCA* phase: uses PCA to compress original data into groups. The output of this first phase and input to the next is a set of groups with their corresponding weights (n_g) and the average vector of sufficient statistics within ($\tilde{\mathbf{x}}_g = \frac{1}{n_g} \sum_{i \in g} \tilde{\mathbf{x}}_i$);
2. *Aggregate* phase: estimates model parameters using the EM algorithm on the aggregate data and random starts. Even loosing part of the information, this phase can provide good estimates. Because $G \ll n$, it is extremely fast and can be repeated several times;
3. *Disaggregate* phase: estimates model parameters using the EM algorithm with original data and phase two best parameter estimates as initial values.

All procedures were implemented in MATLAB 7.0 and ran on a Personal Computer Core2 Duo 2.2 GHz with 2.0 GB RAM.

3 Results

This paper uses the well-known msnbc.com anonymous web data in kdd.ics.uci.edu/databases/msnbc/msnbc.data.html. In our study, we used 50000 sequences of page views with at least one transition. We set the number of clusters based on the Bayesian Information Criterion, resulting in $S = 2$ (see [2] for more details on the data set).

To understand the performance of the PEM algorithm our experimental study controls: the percentage of explained variance (*ExpV*), the number of runs performed at the *aggregate* phase (*Aggr*), and the convergence tolerance level at the *aggregate* phase (*Tol*). *ExpV* was set at two levels: 80% and 90%. This leads to 10 and 13 components extracted, and 467 and 1235 number of non-empty groups, respectively for 10 and 13 components. *Aggr* was set at levels 1 and 10. At level 10, one selects the best estimate based on the maximum likelihood value. The tolerance level was set at levels 10^{-6}, 1 and 10. For each of the 12 cells of the 3×2^2 full factorial design, the PEM algorithm ran 25 times. Each run starts with a randomly generated solution. The tolerance level was set at 10^{-6} in the *disaggregate* phase. For the standard EM algorithm we have random initialization and 10^{-6} as tolerance level.

We use a percentage deviation to analyse the results. Let Res_{PEM} and Res_{EM} be the results provided by 25 runs of the PEM algorithm and the EM algorithm, respectively. The percentage deviation is defined by:

$$Dev = 100 \times \frac{Res_{PEM} - Res_{EM}}{abs(Res_{EM})}, \tag{5}$$

where $abs(x)$ is the absolute value of x. To evaluate the quality of the retrieved solutions, Res_{PEM} and Res_{EM} represent the best log-likelihood (LL) values.

Table 1. Descriptive statistics for each factor

Factor	Level	LL				Iter				Time			
		Min	Aver	Max	STD	Min	Aver	Max	STD	Min	Aver	Max	STD
ExpV	80%	-0,07	-0,07	-0,06	0,00	-40,64	-31,04	-23,55	6,47	-39,24	-27,37	-19,17	6,63
	90%	-0,03	0,01	0,02	0,02	-25,60	-13,17	-0,12	11,41	-14,03	1,97	13,50	12,22
Aggr	1	-0,07	-0,04	0,02	0,04	-29,93	-17,84	-0,12	11,76	-28,92	-13,05	10,44	17,17
	10	-0,07	-0,02	0,02	0,05	-40,64	-26,37	-2,30	13,48	-39,24	-12,36	13,50	20,37
Tol	10^{-6}	-0,07	-0,04	0,02	0,04	-29,93	-26,61	-21,06	4,21	-28,92	-14,82	9,19	17,23
	1	-0,06	-0,02	0,02	0,05	-40,64	-18,28	-2,30	17,50	-25,94	-5,29	13,50	20,16
	10	-0,07	-0,03	0,02	0,04	-36,53	-21,43	-0,12	15,32	-39,24	-18,00	5,62	19,06

A negative minimum, average or maximum percentage deviation means that EM performs better than PEM algorithm. These factors have an impact on the quality of the solutions (Table 1). On average, the quality of the best solution increases as the level of *ExpV* or *Aggr* increase. What is perhaps surprising is the strong effect on the average percentage deviation produced by the *ExpV* factor: exception made to level 10^{-6} of *Tol*, deviations are positive only if *ExpV*=90% (Table 1 and Table 2). This result was expected given that more explained variance leads to more groups. For *Tol* factor (Table 1) it is difficult to establish a direction of association.

Table 2. Average percentage deviation for each cell

Aggr	Tol	ExpV					
		80%			90%		
		LL	Iter	Time	LL	Iter	Time
	10^{-6}	-0,07	-29,93	-28,92	-0,03	-21,06	-14,03
1	1	-0,06	-23,55	-25,94	0,02	-6,62	10,44
	10	-0,06	-25,75	-25,46	0,00	-0,12	5,62
	10^{-6}	-0,06	-29,84	-25,51	0,02	-25,60	9,19
10	1	-0,06	-40,64	-19,17	0,02	-2,30	13,50
	10	-0,07	-36,53	-39,24	0,02	-23,31	-12,91

Besides the performance on retrieving good solutions, we compare both algorithms on the computer effort measured by the running time

(*Time*), in seconds, and the number of iterations (*Iter*). For the PEM algorithm we only took into account the number of iterations in the *disaggregate* phase. This allows a better understanding of the starting solution effect on the algorithm performance. *Time* is the total running time. In order to compute *Time* and *Iter* percentage deviations we use average values instead of best values. Overall, factors have an impact on *Time* and *Iter* percentage deviation (Table 1). The direction of association depends on the factor.
Analysing *LL*, *Iter* and *Time* average percentage deviation simultaneously (Table 2), it turns out that an improvement in running time is at expenses of the quality of the best retrieved solution.

4 Conclusion

This paper introduces the PEM algorithm in which principal component analysis is used to set initial estimates for the EM algorithm. The results show that the speed of the EM algorithm can be boosted without sacrificing much the quality of the solutions. In future work we intent to conduct research on the performance of the PEM algorithm for extremely large data sets.

References

1. Dempster AP, Laird N, Rubin DB (1977) Maximum likelihood for imcomplete data via the EM algorithm (with discussion). Journal of the Royal Statistical Society Series B 39:1–38
2. Dias JG, Vermunt JK (2007) Latent class modeling of website users' search patterns: Implications for online market segmentation. Journal of Retailing and Consumer Services 14(4):359–368
3. Jolliffe, IT (2002). Principal Component Analysis. Springer, New York
4. McCallum A, Nigam K, Ungar LH (2000) Efficient clustering of high-dimensional data sets with application to reference matching. In Proceedings of the sixth ACM SIGKDD International Conference on Data Knowledge Discovery and Data Mining:169–178

AgileGIST - a Framework for Iterative Development and Rapid Prototyping of DSS for Combinatorial Problems

Ulrich Derigs and Jan Eickmann

Seminar fuer Wirtschaftsinformatik und Operations Research (WINFORS), Universitaet zu Koeln, Pohligstrasse 1, D-50969, Koeln, Germany
http://www.winfors.uni-koeln.de

1 Introduction

Developing model-based decision support systems for combinatorial problems you often encounter the situation that the problem owner is not able to explicate all necessary information to set up the system of constraints and objectives but that he is able and willing to criticise solutions which are obtained from some model implementation. Thus in such scenarios it is necessary for the success of the project to design a development process based on a sequence of models with associated prototype systems and to integrate the problem owner into an iterative and experimental procedure in which (additional) components can be added or dropped from the model base. Heuristics based on indirect search have shown to be suitable and rather powerful in these environments. These problems are typical scenarios in DSS-development and have motivated the development of our new framework: they require new problem-specific representational and algorithmic knowledge which has to be acquired during an initial conceptual phase, flexibility rather than absolute accuracy as in literature problems is a dominant system requirement, and, to allow for rapid system development the use of general problem solving capabilities captured in standard coding schemes and metaheuristic strategies is more or less mandatory. In earlier papers we have reported successful applications of the so-called GIST-approach to non-standard real-world decision problems which had been given to us in form of a semi-structured planning task and for which we had to develop a decision support systems (cf.[1],[2]and[3]).In this paper we describe how evolutionary system development using the GIST-approach can be supported by a framework called AgileGIST which

B. Fleischmann et al.(Eds.), *Operations Research Proceedings 2008*,
DOI: 10.1007/978-3-642-00142-0_61,

has been motivated by two concepts from different areas: mathematical programming languages and the dependency injection framework from software engineering.

2 Indirect Search and the GIST-Approach

GIST is based on the concept of indirect (evolutionary/local) search where we work in an auxiliary search space S^{aux}, a space of simple and general, not problem specific representations. Here the definition of feasible moves is more or less trivial, a broad spectrum of metaheuristics is applicable and the operations on this abstract level do not require problem-specific knowledge. The (only) problem specific task in indirect search is to re-map, i.e. decode a given string-code.
A decoder-concept which allows flexible handling of constraints is that of a greedy decoder. In general, a decoder can be seen as a construction heuristic on S where the specific encoding $s' \in S^{\text{aux}}$ generates/determines the instructions or assignments. A common construction principle is the greedy-principle. Here, starting from the empty solution, the heuristic solution is built through a sequence of partial solutions, and the steps can be interpreted as reducing the degree of freedom by "fixing some variables" in a way which is feasible as well as optimal with respect to some myopic criteria. Thus a greedy construction algorithm is based on three concepts/modules:

- an ordering/sequencing input,
- a myopic decision rule, and
- a constraint checker.

The ordering/sequencing is captured in the encoding. This encoding is input-data "generated" by some (general) heuristic strategy. The decision rule is problem-specific, it should transform the problem's objective function into local strategies which, dependent on the instance data, generate solution extensions. The separation of the decision rule from the constraint checker reduces the complexity of implementation and supports flexibility: Confronted with problem instances from the same domain, VRP for instance, but and some difference in constraints and objectives only the checker has to be modified/adapted.

3 AgileGIST - Concept of a Development Framework

AgileGIST adopts the separation of model logic and data with the abstract modeling on type level and (late) binding of structure (model file) with values (data file). Yet, instead of expecting an external solver to construct the solution as it is the case with mathematical programming languages it contains a third component: the so called constraint checker which is a set of constraint factories i.e. independent software modules which are (only) able to check wether a given (partial) solution violates a constraint or not. Thus for all (active) constraints of the model which are specified (registered) in the model file the developer has to provide a checking algorithm coded in a callable procedural programming language which accepts the model data as parametrization. Note that the logic of the checking procedure can be encapsulated and held transparent and only the interface of all checking procedures have to be standardized. This enables iterative and distributed development as well as reuse.

3.1 Decoder-Interface

Since the decoder is responsible for converting the generic solution representation of the heuristic solver into a problem specific representation, a specific decoder is always tied to a particular type of problem. In order to allow the framework to remain independent of the problem type, a standardized interface for the decoder has to defined. The main element of this interface is a standardized generic representation that all solvers for the framework have to agree on. In our prototype we have used permutations of natural numbers as the generic representation. The permutation length is determined at runtime since it depends on the size of the specific problem instance to be solved. The required length is calculated by the decoder at the start of the solving process and handed over to the framework. The framework in turn notifies the heuristic to generate codes (permutations) of appropriate length. To control the search process, each generic solution representative (permutation) has to be decoded and evaluated in terms of an objective function. Thus, a second part of the decoder-interface consists of a function-call into the decoder that accepts a generic solution representative (permutation of the required length) as input and returns the objective function's value of this candidate as a floating point number. Note that both, input and output parameters to this function are problem independent. All problem specific knowledge resides inside the implementation of the decoder.

3.2 Symboltable and Fitness Function

While the problem-specific decoder produces solutions, the process of checking for constraint-violations is delegated to independent constraint checker software modules. To enable these constraint checkers to verify wether a given (partial) solution fulfills a particular constraint, a common data model has to be established between the decoder and the constraint checkers. In contrast to the generic solution representation used in the interface between decoder and solver, this data model is problem specific (but remains independent of the particular constraint to be checked). In addition to this, the model description specifies a fitness or objective function and a set of constraints to be checked for each solution candidate.

Parsing of the data model consists of filling a symbol-table structure with symbol-objects for the sets, parameters and variables specified in the data model. This process is independent of a specific model instance and the associated parameter values. The creation of the symboltable at this stage allows to check for inconsistencies in the data model as well as in the fitness function and constraints. Constraint-factories (described later) can check for the presence of required sets and the logic of aggregate functions in the fitness function can be verified as well.

The fitness function is expressed in terms of a mathematical expression and is transformed into an operator tree that can later be evaluated during the solution process. The operator tree contains nodes representing mathematical operations as well as symbol-nodes that are linked to particular symbols in the symbol table. After the complete parsing of a model description, the model can be bound to a datafile that specifies concrete values for the parameters and sizes for the sets and thus represents a particular instance of the problem to be solved. The framework can then check wether all necessary data is available and that for example the number of values provided for an indexed parameter corresponds to the size of the set(s) determining the dimensions of the parameter.

3.3 Constraint-Factories and Constraint-Objects

The model description comprises a set of constraints to be checked for each solution candidate. AgileGIST allows the registration of so called constraint-factories which can create constraint-objects of different types. These constraint-objects are able to check wether a specific solution satisfies a particular constraint. The checking algorithm has

to be provided by the constraint's developer once for each type of constraint.
During parsing of the model description the framework checks for each defined constraint wether a constraint factory has been registered for that particular type and asks that factory for a constraint-object. In order to allow flexible constraint-types, the constraint-definition in the modelfile allows to define parameters for the constraint factory that influence how the constraint-object is created.
In traditional mathematical programming languages only constraints described as (in-)equalitities involving linear or nonlinear mathematical functions can be specified. Constraints for combinatorial problems are often hard to express in these terms, and thus more generalized concepts of constraints have to be supported by the framework. Examples for a constraint type which cannot be expressed in terms of algebraic inequalities easily or intuitively are for instance time window constraints for vehicle routing problems. The validation of such constraints requires recursive calculation of arrival times at the different customers. This makes them computationally much more complex than the evaluation of a simple inequality.

3.4 Interaction Between Decoder and Constraint-Objects

The set of constraint-objects that are generated during model parsing are stored by the framework and the entire set can be checked during the heuristic search on request from the decoder. The usual process is as follows: Based on the permutation the decoder constructs a partial solution. This solution is specified by assigning values to the decision variables from the data model which are available in the symboltable to the decoder and then a check of the constraints is requested by the decoder. Now the framework requests each constraint to perform its check against the current state of the symboltable. If all constraints are fulfilled, the decoder can continue with the next step in its decoding process. If at least one constraint is violated, the decoder has to modify the partial solution. Only after the decoder has constructed a complete feasible solution it is evaluated according to the objective function and the resulting value is returned to heuristic as specified in the decoder-interface.An example of such a greedy decoding process for the VRP is depicted in Figure 1.

4 Preliminary Experience

We have applied a prototype implementation to different VRP-problems and the Movie Shoot Scheduling Problem. Here we were able to "'rapidly"' generate a modified system which allowed parallisation of certain activities.

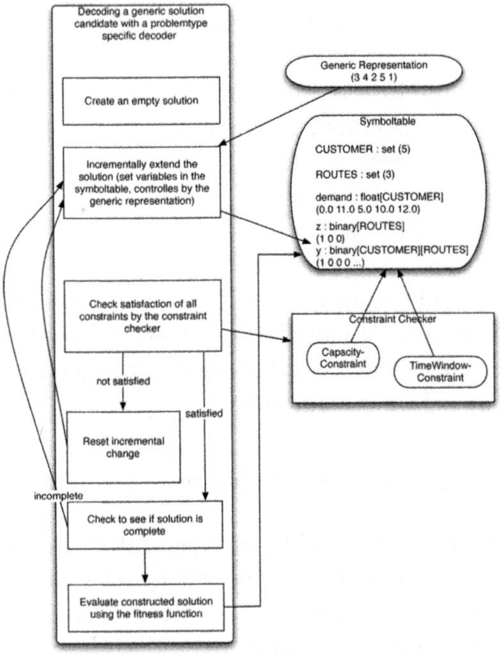

Fig. 1. Decoding process example for the VRP

References

1. Bomsdorf,F. and Derigs,U.: A model, heuristic procedure and decision support system for solving the movie shoot scheduling problem, OR Spectrum 30,751-772,(2008)
2. Derigs, U. and Döhmer, T. :ROUTER: A fast and flexible local search algorithm for a class of rich vehicle routing problems. Operations Research Proceedings 2004, 144–149.(2004)
3. Derigs, U. and Jenal, O.: A GA-based decision support system for professional course scheduling at Ford Service Organisation. OR Spectrum 27(1): 147–162, (2005).

Incorporation of Customer Value into Revenue Management

Tobias von Martens and Andreas Hilbert

Technische Universität Dresden, 01062 Dresden
{martens,hilbert}@wiid.wiwi.tu-dresden.de

1 Problem Statement

The efficient utilization of limited capacity resources, e. g. airplane seats or hotel rooms, is a prevalent success factor for service providers [1]. Hence, revenue management is applied to control the acceptance of booking requests. Though successful in the short-term, it is transaction-based so far by focusing on willingness-to-pay [8] and neglecting the establishment of relationships with long-term profitable customers. Therefore, a conceptual model of customer value-based revenue management is developed. Transaction-based optimization and booking control are enhanced by regarding customer value-related information.

2 Transaction-Based Revenue Management

Service industries are mostly characterized by limited, inflexible and perishable capacity resources as well as uncertain and heterogeneous demand. Hence, when the acceptance of lower-value booking requests that arrive early in the booking period is not limited, revenue displacement may occur. When lower-value booking requests are declined too often, instead, revenue loss can result. Therefore, revenue management is applied to utilize the capacity efficiently by controlling the acceptance of early booking requests accordingly and reserve a sufficient amount of capacity for later booking requests of higher value. Booking control can be based on the availability of different booking classes or dynamic pricing. So far, revenue management neglects relationship-focused marketing since optimization and booking control are mostly based on willingness-to-pay (i. e. prices of booking classes) and not on the customers' long-term value for the service provider [8]. Hence, prospective customers with low actual but high future contributions as

B. Fleischmann et al.(Eds.), *Operations Research Proceedings 2008*,

DOI: 10.1007/978-3-642-00142-0_62,

well as reference customers with low own but high induced contributions are mostly declined by the transaction-based approach.
The problem has been recognized but only insufficient solution approaches are provided so far: Either different revenue management strategies [9] or value classes [6] for customer segments are suggested. However, without specifying how to classify customers in the booking process and calculate the contingents assigned to the value classes.

3 Basic Idea and Tasks of Customer Value-Based Revenue Management

In order to allow for the establishment of relationships with long-term profitable customers, customer value-related information should be regarded by revenue management. Customer value as a key figure of customer relationship management represents the long-term value of a customer for a company [7] and is regarded as being closely connected to shareholder value [3]. Focusing on long-term profitable relationships requires strategical tasks, i. e. environmental analysis as well as formulation of objectives and strategies. On the tactical level, there are medium-term tasks, i. e. tactical planning, development of booking classes and pricing. On the operational level, there is a cycle of forecasting of uncertain variables in the booking period, allocation of capacity on the different booking classes and customer segments (optimization), decision on the acceptance of booking requests (booking control), adaptation of the forecast, optimization or booking control according to the actual level of bookings, and analysis, i. e. calculation of performance indicators to be used on the strategical, tactical and operational level [5]. In addition, forecasting, optimization and booking control are often based on models [12] that require model development tasks.

4 Optimization and Booking Control Approach

Optimization comprises the allocation of the available capacity on the expected demand. The results, e. g. contingents of capacity resources or bid prices (opportunity costs of capacity utilization), are used for booking control within the booking period. In transaction-based revenue management, capacity is only allocated on the different booking classes, neglecting any customer value differences between customer segments. Hence, in customer value-based revenue management, capacity has to be allocated on J booking classes and S customer segments according to value-related revenues. These represent a combination of short-term revenues (i. e. the price v_j^{BC} of the respective booking class j) and long-term value contributions (i. e. the customer value v_s^{CS} of the respective customer segment s). A weighting factor α^{IC} ($\alpha^{IC} \in [0;1]$) allows for balancing the importance of either short- or long-term value contributions in certain application areas. Formulas 1 and 2 show the value-related revenues as an additive (customer value is a monetary value) and a multiplicative combination (customer value is a score):

$$v_{sj}^{CN} = \alpha^{IC} \cdot v_j^{BC} + (1 - \alpha^{IC}) \cdot v_s^{CS} \tag{1}$$

$$v_{sj}^{CN} = \alpha^{IC} \cdot v_j^{BC} + (1 - \alpha^{IC}) \cdot \frac{v_s^{CS}}{\sum_s v_s^{CS}} \cdot v_j^{BC} \tag{2}$$

Based on the value-related revenues, the allocation of the capacity can be formulated as an LP where y_{sj}^A represent the contingents assigned to a combination of customer segment s and booking class j, b_{sj} (as an element of $\mathbf{B}$) is the amount of present bookings regarding a certain combination, m_i^{CP} is the capacity on the ith of I resources, and d_{sjt} (as an element of $\mathbf{D_t}$) is the expected remaining demand from booking interval t ($t = T, \ldots, 0$). The elements a_{ij} represent the resource utilization and have a value of 1 whenever resource i is used by booking class j, and 0 otherwise.

$$v^{CP}(\mathbf{B}, \mathbf{D_t}) = \max \sum_{s=1}^{S} \sum_{j=1}^{J} v_{sj}^{CN} \cdot y_{sj}^A$$

$$s.t. \sum_{s=1}^{S} \sum_{j=1}^{J} a_{ij} \cdot (b_{sj} + y_{sj}^A) \leq m_i^{CP} \; \forall i = 1, \ldots, I \tag{3}$$

$$0 \leq y_{sj}^A \leq d_{sjt} \; \forall s = 1, \ldots, S; \forall j = 1, \ldots, J$$

The contingents, i. e. available capacity for a customer segment requesting a certain booking class, can be derived from y_{sj}^A. These can be used by contingent control where booking requests are accepted as long as the respective contingent is positive. When a bid-price control is applied, opportunity costs have to be calculated by comparing the values v^{CP} of the remaining capacity for the rest of the booking period both in case of declining and accepting the request [4]. The matrix $\mathbf{R}$ represents the booking request from customer segment s for booking class j. A booking request is accepted as long as its value-related revenues v_{sj}^{CN} outweigh the opportunity costs:

$$v_{sj}^{CN} \geq v^{CP}(\mathbf{B}, \mathbf{D_{t-1}}) - v^{CP}(\mathbf{B} + \mathbf{R}, \mathbf{D_{t-1}}) \tag{4}$$

In order to regard the distribution of demand, Monte-Carlo simulation of different demand scenarios is suggested at which the resulting capacity values are averaged, weighted by the scenario probability [4]. Moreover, heuristic solution methods, e. g. evolutionary algorithms [11], as well as limiting the calculation to the most-utilized resources provide means to cope with the complex optimization problems.

5 Simulation Results and Evaluation

The bid-price control described above has been implemented to allow for a comparison of transaction- and customer value-based booking control and for analyzing the sensitivity of performance indicators in different environments. For the analysis, a network with two legs, a high- and low-value booking class on each possible connection as well as a high- and low-value customer segment have been simulated. Limited by the efficiency of the applied prototype, 120 runs have been calculated for each scenario. The average correlation of willingness-to-pay (i. e. requested booking class) and customer value (i. e. requesting customer segment) is represented by r ($r \in [-1;1]$) whereas $r = 1$ represents the basic assumption of transaction-based revenue management, i. e. high-value customers always request high-value booking classes, and $r = -1$, however, may occur in the presence of prospective or reference customers. At $r = 0$, the expected demand for a booking class is generated equally by both customer segments. Three methods of control have been applied according to the weighting of short- and long-term revenues: strongly transaction-based ($\alpha^{IC} = 1$), hybrid ($\alpha^{IC} = 0,5$) and strongly customer value-based ($\alpha^{IC} = 0$). For measuring the performance of the different methods, the value-related revenues with according weighting factors were used: With $\alpha^{PI} = 1$, only the short-term revenues (success, i. e. prices of all accepted booking requests) are taken into account, while $\alpha^{PI} = 0$ calculates only the long-term revenues (success potential, i. e. the customer values of those customers whose booking requests have been accepted). $\alpha^{PI} = 0,5$ represents a combined measure of short- and long-term revenues and considers the uncertainty of long-term customer values since customers whose booking requests have been accepted may not necessarily be loyal and generate revenues as predicted. Moreover, in order to evaluate the performance of the booking control in isolation, revenues can be measured in relation to the ex-post optimal solution (i. e. the allocation of the capacity on the observed booking requests), to a first-come-first-serve (fcfs) control or based on a combined indicator of both [10].

The simulation results confirm the expectation that customer value-based booking control leads to lower short-term revenues but higher long-term revenues as the correlation between willingness-to-pay and customer value becomes non-positive. In case of a positive relationship, they lead to comparable results. Furthermore, a sensitivity analysis regarding the non-homogeneous poisson arrival of booking requests shows that the gain over the fcfs control according to the combined measure ($\alpha^{PI} = 0,5$) is higher when requests of high-value customer segments arrive late in the booking period. Another sensitivity analysis confirms findings that the gain over the fcfs control is positively correlated with the amount of demand in relation to the capacity [2].
The simulations imply that customer value-based booking control is recommended when there is a non-positive correlation between willingness-to-pay and customer value [14], provided that indicators in the booking process are sufficient to classify customers into customer value segments accordingly. The appliance of (customer value-based) booking control is, furthermore, reasonable in high-demand periods and when booking requests of high value arrive late in the booking period.

6 Conclusion and Outlook on Remaining Research

Customer value-based revenue management provides means to allow for both efficient capacity utilization and profitable customer relationships. Therefore, optimization and booking control have been enhanced for taking into account customer value-related information. However, the approach requires complex optimization problems to be solved and is bound to booking processes providing suitable indicators for a classification of customers into value-based segments.
Research remains in the identification of suitable indicators and models for forecasting the customer value-related information required for optimization and booking control [13] as well as efficient optimization techniques, e. g. based on evolutionary algorithms [11]. In the presence of the importance of both revenue management and customer relationship management, customer value-based revenue management remains a research area providing meaningful insights into the competitive capability of service providers.

References

1. Anderson CK, Wilson JG (2003) Wait or buy? The strategic consumer – pricing and profit implications. Journal of the Operational Research

Society 54:299–306
2. Baker TK, Collier DA (1999) A comparative revenue analysis of hotel yield management heuristics. Decision Sciences 30:239–263
3. Berger PD, Eechambadi N, George M, Lehmann DR, Rizley R, Venkatesan R (2006) From customer lifetime value to shareholder value – theory, empirical evidence, and issues for future research. Journal of Service Research 9:156–167
4. Bertsimas D, Popescu I (2003) Revenue management in a dynamic network environment. Transportation Science 37:257–277
5. Desinano P, Minuti MS, Schiaffella E (2006) Controlling the yield management process in the hospitality business. In: Sfodera F (ed) The spread of yield management practices – the need for systematic approaches. Physica, Heidelberg
6. Esse T (2003) Securing the value of customer value management. Journal of Revenue & Pricing Management 2:166–171
7. Gupta S, Hanssens D, Hardie B, Kahn W, Kumar V, Lin N, Ravishanker N, Sriram S (2006) Modeling customer lifetime value. Journal of Service Research 9:139–155
8. Kuhlmann R (2004) Future of revenue management – Why is revenue management not working? Journal of Revenue & Pricing Management 2:378–387
9. Noone BM, Kimes SE, Renaghan LM (2003) Integrating customer relationship management and revenue management – a hotel perspective. Journal of Revenue & Pricing Management 2:7–21
10. Phillips RL (2005) Pricing and revenue optimization. Stanford University Press, Stanford
11. Pulugurtha SS, Nambisan SS (2003) A decision-support tool for airline yield management using genetic algorithms. Computer-Aided Civil & Infrastructure Engineering 18:214–223
12. Raeside R, Windle D (2005) Quantitative aspects of yield management. In: Ingold A, McMahon-Beattie U, Yeoman I (eds) Yield management – strategies for the service industries. 2nd ed. Thomson Learning, London
13. Tirenni G, Kaiser C, Herrmann A (2006) Applying decision trees for value-based customer relations management – predicting airline customers' future values. Database Marketing & Customer Strategy Management 14:130–142
14. Wirtz J, Kimes SE, Pheng JH, Theng SE, Patterson P (2003) Revenue management – resolving potential customer conflicts. Journal of Revenue & Pricing Management 2:216–226

Einfluss der Adoptoreinstellung auf die Diffusion Komplexer Produkte und Systeme

Sabine Schmidt und Magdalena Mißler-Behr

Lehrstuhl für ABWL und Besondere der Planung und des Innovationsmanagements, Brandenburgische Technische Universität Cottbus, Konrad-Wachsmann-Allee 1, 03046 Cottbus, Deutschland

Summary. Die Diffusion von Komplexen Produkten und Systemen (KoPS) auf dem Markt der Konsumenten ist mit spezifischen Risiken verbunden. Die Gründe dafür liegen in der einzigartigen Struktur der KoPS. Sie bestehen aus einer hohen Komponentenanzahl, die in Interaktionsbeziehungen zueinander stehen, projektbezogen und kundenindividuell gefertigt werden [3]. Eine hohe Produktkomplexität kann zu einem verlangsamten Diffusionsprozess führen [7]. Deshalb ist es notwendig, den Diffusionsprozess von KoPS und seine Einflussfaktoren zu untersuchen. Der Einsatz des Systemdynamischen Ansatzes ermöglicht die Entwicklung von komplexen und dynamischen Modellen. Dieses Paper analysiert den Diffusionsprozess von KoPS unter Integration des Einflussfaktors Einstellung zur Technik der privaten Haushalte. Zusätzlich unterstützt eine Primärerhebung die Phase der Modellvalidierung.

1 Motivation

In den 1990iger Jahren führten innovative Entwicklungen in der Informations- und Kommunikationstechnik verstärkt zur Herausbildung günstiger Bedingungen für den Absatz von KoPS an den privaten Nachfrager. Beispiele für KoPS sind Transportsysteme, flexible Fertigungszellen, die insbesondere im Business-to-Business Bereich bekannt sind sowie Intelligente Häuser, die den Business-to-Business-to-Consumer Bereich betreffen. Aus einer britischen Studie geht hervor, dass 21% der Bruttowertschöpfung allein durch die Produktion von KoPS im sekundären Bereich geschaffen werden, das einem monetären Wert von ungefähr 133 Milliarden £ entspricht [3].
Neue Herausforderungen ergeben sich für Unternehmen in der Produktentwicklung und Vermarktung von KoPS. Sie sind hochtechnologische, kostenintensive Güter, die speziell auf die Bedürfnisse des Nachfragers

B. Fleischmann et al.(Eds.), *Operations Research Proceedings 2008*,
DOI: 10.1007/978-3-642-00142-0_63, © Springer-Verlag Berlin Heidelberg 2009

angepasst sind und haben einen höheren Mehrwert, der sich aus der Struktur zahlreicher interagierender, hierarchisch vernetzter Komponenten ergibt, d.h. der Nutzen der Produktkomponenten im Verbund ist höher als der Nutzen einzelner Komponenten. Bei der Vermarktung ist mit verlangsamten Adoptionsprozessen zu rechnen [3, 9].
Die erfolgreiche Ausbreitung eines neuen Produktes über den Zeitverlauf wird durch den Begriff der Diffusion beschrieben. Tatsache ist, dass Merkmale wie hohe Produktkomplexität, hoher Innovationsgrad und ein für den Nachfrager unverständlicher Produktnutzen zur Verlangsamung des Diffusionsprozesses führen können. Außerdem ist der Einfluss der Adoptoreinstellung bei der Kaufentscheidung neuer Produkte mit erhöhter Komplexität von Bedeutung.
Im Folgenden wird ein systemdynamisches Modell vorgestellt, das den Einfluss der Einstellung zur Technik der privaten Haushalte auf den Diffusionsprozess von KoPS untersucht.

2 Diffusion Komplexer Produkte und Systeme

Dieser Abschnitt beantwortet die zentrale Frage, wie die Diffusion von Produktkomponenten mit Interaktionsbeziehungen (KoPS) abgebildet wird. Die Diffusionsforschung ist an der Schnittstelle von Produktinnovations- und Konsumentenforschung zu sehen [11]. Ausgangspunkt der Betrachtungen sind die Diffusionsmodelle von Peterson und Mahajan [6], die Interaktionsbeziehungen berücksichtigen ("Multi-Innovation Diffusion Models"). Basierend auf dem Diffusionsmodell von Bass [1] modellieren sie Interaktionseffekte (Komplementär, Kontingent). Der **komplementäre Effekt** besagt, dass erhöhte Absätze eines Produktes zu einer Steigerung der Absätze eines anderen Produktes führen und vice versa. Die Gleichungen zur Diffusionsdarstellung von zwei Produkten mit diesem Effekt lauten [6]:

$$n_1(t) = \frac{dN_1(t)}{dt} = (\alpha_1 + \beta_1 N_1(t) + c_1 N_2(t))(\bar{N}_1 - N_1(t)) \quad [2.1]$$

$$n_2(t) = \frac{dN_2(t)}{dt} = (\alpha_2 + \beta_2 N_2(t) + c_2 N_1(t))(\bar{N}_2 - N_2(t)) \quad [2.2]$$

Die Lösung der Differentialgleichung $n_i(t) = \frac{dN_i(t)}{dt}$ beschreibt den Anteil der Nachfrager, die das Produkt i zum Zeitpunkt t gekauft haben. $N_i(t)$ steht für die kumulierte Anzahl der Adoptoren, die das Produkt i bis zum Zeitpunkt t gekauft haben und $\bar{N}$ beschreibt das Marktpotential des Produktes i. Angelehnt an die Koeffizienten des Bass-Modells

entspricht α dem Innovationskoeffizienten und β dem Imitationskoeffizienten. Zusätzlich steht der Koeffizient c_i für die positive Interaktionsbeziehung ($c_i > 0$) [6].
Der **kontingente Effekt** betrachtet die bedingte Beziehung zwischen einem Produkt und einem anderen, d.h. Käufer des ersten Produktes kaufen ein zweites Produkt, das die Existenz des ersten voraussetzt, z.B. Personal Computer (PC) und Internetzugang. Diese Beziehung wird mit den Diffusionsgleichungen ausgedrückt:

$$n_1(t) = \frac{dN_1(t)}{dt} = (\alpha_1 + \beta_1 N_1(t))(\bar{N}_1 - N_1(t)) \quad [2.3]$$
$$n_2(t) = \frac{dN_2(t)}{dt} = (\alpha_2 + \beta_2 N_2(t))(N_1(t) - N_2(t)) \quad [2.4]$$

3 Die Einstellung zur Technik der privaten Haushalte

Bei der Einbindung der Einstellung zur Technik der privaten Haushalte in die Diffusionsforschung gilt es abzuschätzen, ob ein Zusammenhang zwischen der Einstellung eines Nachfragers und seinem Kaufverhalten besteht. Die Theorie des Konsumentenverhaltens findet bis auf Ausnahmen in der Innovationsmanagementliteratur insbesondere der Neuproduktinnovation weniger Berücksichtigung [11]. Die Nachfrage nach einem Produkt resultiert aus einer unterschiedlichen Gewichtung ökonomischer, psychischer und sozialer Faktoren des Nachfragers [2]. Einstellungen gehören zu den wesentlichen Einflussparametern des Konsumentenverhaltens [4]. Die Einstellung wird von Trommsdorff [10] als "Zustand einer gelernten und relativ dauerhaften Bereitschaft, in einer entsprechenden Situation gegenüber dem betreffenden Objekt regelmäßig mehr oder weniger stark positiv bzw. negativ zu reagieren" definiert. Neuere Forschungen verweisen darauf, Einstellungen und Verhalten in einer wechselseitigen Beeinflussung (Rückkopplung) zu analysieren [5], die das systemdynamische Modell aufgreift (vgl. Abb. 1). Das Ergebnis einer Primärerhebung, die bei 1.002 privaten Haushalten (Rücklaufquote 15,4%) in drei Städten von Oktober bis November 2007 durchgeführt wurde, bekräftigt die Annahme, dass von einem positiven Einfluss der Einstellung auf den Diffusionsprozess auszugehen ist. Die privaten Haushalte wählten mit insgesamt 70,8% den Bereich des starken bis äußerst starken Einflusses. Jedoch hängt die Intensität von der Person und dem technischen Anwendungsbereich ab.
Zur Modellierung der Diffusion eines KoPS (bestehend aus drei Produktkomponenten) und der Nachfragereinstellung ist eine passende Untersuchungsmethode auszuwählen, die berücksichtigt, dass es sich bei

der Diffusion um einen äußerst dynamischen Prozess handelt, der aufgrund von Wechselbeziehungen (Feedback) seiner Systemelemente in Gang kommt und aus sich selbst heraus das System wieder stimuliert. Der Systemdynamische Ansatz ist ein experimentelles Instrument, berücksichtigt Feedbackbeziehungen und Zeitverzögerungen [8]. Das analytische Instrument des "Stock und Flow Diagramms" ist das Simulationsmodell des Systemdynamischen Ansatzes. Die Gesamtheit der Regelkreise bilden ein Differenzialgleichungssystem. Beispielhaft für die Basisdynamik stehen die Gleichungen der Bestandsgröße (Gl. 3.1) und der Flussgröße (Gl. 3.2), bezogen auf die Periode t.

Einstellung zur Technik PHH(t) [Einstellung zur Technik Einheiten]
= Einstellung zur Technik PHH(t-1)
+ Veränderung der Einstellung zur Technik(t)-Risiko(t) [3.1]

Risiko(t) [Einstellung zur Technik Einheiten/Jahr]
*= Einstellung zur Technik PHH(t)*Risikofaktor(t)* [3.2]

Der Wert der Einstellung ist nicht absolut anzusehen, sondern ändert sich im Verhältnis zum Ausgangswert. Veränderungen des Wertes bewirken Erfahrungen mit dem KoPS und das wahrgenommene Kaufrisiko des KoPS (Gl. 3.2). Der Erfahrungswert hängt von der Anzahl der bisher gekauften KoPS ab. Das Risiko wird mit Hilfe eines Faktors zwischen 0% und 100% festgelegt (Gl. 3.2). Der Risikofaktor wurde aus der empirischen Datenanalyse ermittelt. Zur Bestimmung des Risikofaktors wurde das Risiko in drei Gruppen unterteilt: risikoscheu (Skalenwerte 1-2), risikoneutral (3-5) und risikofreudig (6-7). Der Anteil der risikoscheuen Haushalte beträgt damit 34,4%.

4 Modellverhalten

Entscheidende Analysegrößen des systemdynamischen Modells sind die Bestands- und Flussgrößen. Der Simulationszeitraum des Diffusionsprozesses umfasst 12 Jahre (1998-2009), damit jede der drei Produktkomponenten zumindest einmal den Zyklus Erstkauf und Wiederholungskauf durchlaufen hat. Als äquivalente Beispiele wurden für die drei Produktkomponenten PC und Internetzugang (kontingente Beziehung) sowie PC und Digitalkamera (komplementäre Beziehung) gewählt, da reale Datenwerte (Statistisches Bundesamt Wiesbaden) als Ausstattungsgrad (je 100 Haushalte) der deutschen Privathaushalte zur Verfügung standen. Die Produktkomponenten befinden sich im Jahr 1998 in unterschiedlichen Lebenszyklusphasen. Da in t=1998 der PC in der Sättigungsphase und das Internet in der Wachstumsphase waren, gibt es bereits Adoptoren. Abbildung 1 veranschaulicht den Diffusionsprozess des KoPS mit einem Basisszenario und zwei Zusatzszenarien, die Veränderungen im Kaufrisiko (um 20% erhöhtes oder verringertes Risiko gegenüber dem Basisszenario) aufzeigen. Beispielhaft zeigt die Produktkomponente A den Verlauf, der der Kurvenform von Produktkomponente B und C ähnlich ist. Kommt es zu einer Risikominderung beschleunigt sich der Diffusionsverlauf der Produktkomponente A im Vergleich zum Basisszenario.

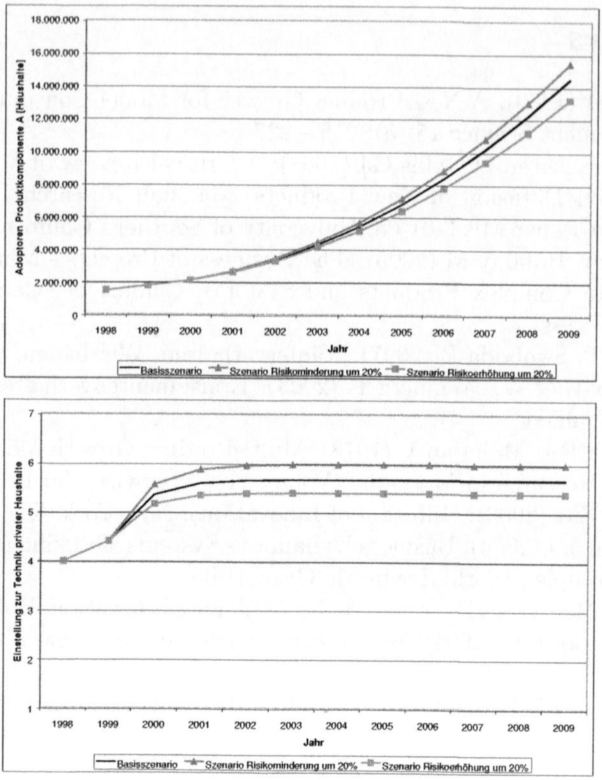

Fig. 1. Diffusion der Produktkomponente A des KoPS und Einstellung zur Technik

5 Zusammenfassung

Hochgradige Innovationen, zu denen die KoPS gehören, treten verstärkt auf den Markt der Konsumenten. Erhöhte Produktkomplexität und Nachfragerunsicherheiten in Bezug auf den Produktnutzen und die schrittweise Beschaffung können zu einem verzögerten Diffusionsprozess führen. Mit Hilfe des Systemdynamischen Ansatzes wurde in einer ersten Analyse ein Modell entwickelt, das den Diffusionsprozess von KoPS unter dem Einfluss der Adoptoreinstellung betrachtet.

References

1. Bass FM (1969): A New Product Growth for Model Consumer Durables. Management Science 15 (5): 215 - 227
2. Chandrasekaran D, Tellis GJ (2008): A Critical Review of Marketing Research on Diffusion of New Products. Marshall Research Paper Series, Working Paper MKT 01-08, University of Southern California, 38-80
3. Davies A, Hobday M (2005): The Business of Projects - Managing Innovation in Complex Products and Systems. Cambridge, Cambridge University Press
4. Foscht T, Swoboda B (2007): Käuferverhalten. Wiesbaden, Gabler
5. Kroeber-Riel W, Weinberg P (2003): Konsumentenverhalten. München, Verlag Vahlen
6. Peterson RA, Mahajan V (1978): Multi-Product Growth Models, In: Seth J (eds) Research in Marketing-Volume 1. Greenwich, Jai Press, 201-232
7. Rogers EM (2003): Diffusion of Innovations. New York, The Free Press
8. Sterman JD (2000): Business Dynamics - Systems Thinking and Modeling for a Complex World. Irwin Mc Graw Hill
9. Tidd J, Bessant J, Pavitt K (2005): Managing Innovation. Hoboken, Wiley
10. Trommsdorff V (2004): Konsumentenverhalten. Stuttgart, Kohlhammer GmbH
11. Trommsdorff V, Steinhoff F (2007): Innovationsmarketing. München, Vahlen

Data Similarity in Classification and Fictitious Training Data Generation

Ralf Stecking[1] and Klaus B. Schebesch[2]

[1] Faculty of Economics, University of Oldenburg, D-26111 Oldenburg, Germany
ralf.w.stecking@uni-oldenburg.de

[2] Faculty of Economics, University "Vasile Goldiş", Arad, Romania
kbsbase@gmx.de

Summary. Classification methods work best when all training cases are uniquely labeled and when the number of cases grossly exceeds the number of features available. In real world applications concerning classification of the expected behavior of clients, of the expected performance of firms or products, to name just a few, data sparseness may occur for many reasons. This clearly reduces the potential performance of very powerful classifiers like distance-kernel based Support Vector Machines (SVM). Assuming that the data observed so far are not erroneous, i.e. falsely labeled, etc., one might wonder whether it is possible to replicate to some extend what humans seem to accomplish with ease in some contexts, namely generalizing from a few (possibly complicated) examples into many more valid training examples. However, such human skills may heavily depend on geometric or other intuition about a particular application domain, for instance image classification, where humans have at least observed very big numbers of image examples during their everyday lives enabling them the envisage image templates. In our application of credit client scoring such intuition about possible neighborhoods in client data is improbable and the high dimensional feature vectors also do not map into meaningful low dimensional "images". In this paper we use a procedure which is generating fictitious training data from the existing empirical data by using a simple notion of similarity and also by using information about "feasible changes" in credit client data, i.e. by producing a large number of labeled examples which are plausible unseen credit clients. We then show by means of extensive computational experiments that under certain conditions such fictitious training examples are improving expected out-of-sample classification performance of our credit client scoring models.

B. Fleischmann et al.(Eds.), *Operations Research Proceedings 2008*,
DOI: 10.1007/978-3-642-00142-0_64,

1 Introduction

The search for improved out-of-sample performance of empirical classification tasks includes various procedures which attempt to guide the use of powerful standard tools like Support Vector Machines (SVM). In a broad sense such guidance entails emphasizing certain features of training examples to a context neutral learning machine. Given a set of $N > 0$ training examples $\{x_i, y_i\}$, $i = 1, ..., N$, with $x_i \in \mathbb{R}^m$ (m input features) and labels $y_i \in \{-1, 1\}$ (for simplicity, say) one would ideally request joint and conditional probabilities of the simultaneous occurrence of features in a training example. Such a probability for the occurrences of the ith training example p_i, is in fact needed in order to evaluate the classification errors, for instance by $\sum_{i=1}^{N}(y_i - s(x_i))^2 p_i$ with an estimated data model $s(x)$. Such probabilities p_i are quite obviously not available in many real world applications or extremely difficult to evaluate. While in a given classification problem we typically do not have but a very small fraction of all feasible (possible) input-output combinations, in certain approaches to enhance the performance of classifiers (in manifold learning) we would like to know if certain sets of input combination are forbidden, in order to concentrate on the feasible ones. This is in close recognition to the fact that not all domains are usefully modeled by "vector space" models (models which implicitly assume that additional feasible training examples can be generated by linearly combining existing examples, see the discussion in [3]). Directly avoiding non-feasible input or input-output combinations may be a very difficult task as well, hence one may ask the simpler question of which further feasible examples follow from the available training examples. By emphasizing the available empirical examples by means of their feasible neighbors via generating fictitious training examples one may (hope to) draw the classification model away from covering potentially infeasible regions at the expense of classification performance on the feasible regions. In various forms such attempts have been made in the past, especially in image classification. In these domains we find early work concerning "hints" to help neural networks to learn symmetries in data [1] and "invariant" deformations of digital images subjected to SVM classification [6]. In [6] so called *virtual support vectors* are employed to enhance classification performance. Support vectors are those training examples which are near the class boundaries of a SVM solution and which permit training of a classifier with the same out-of-sample performance as a classifier trained on all the other (redundant, etc.) training examples as well.

Many real word classification problems like credit client scoring already suffer from relative sparseness (i.e. low N or few total number of training examples as would be required for "safe" classification of m-dimensional feature vectors). Although SVM are known to behave comparatively well (due to the margin maximization principle) in situations of relative data sparseness (cf. [5] as a technical reference and [4] for credit scoring), increasing the number of points which are effectively contributing to the shape of the separating function may be expected to still improve out-of-sample classification performance, which is in fact confirmed by [6] and also in a later general article [2]. In the present contribution our goal is to study the role of such newly generated training points in the context of empirical data where we have no or much less intuition about what constitutes feasible neighbors of the given data. Building on some preliminary results by using most simple replicas of the original empirical data [10] we now introduce somewhat more involved neighbors which are in part generated by verifying feasibility constraints of the empirical application. Finally we should state what our data model $s(x)$ looks like in the event of using SVM (for details we refer to [5] and SVM and credit scoring [4]). In order to follow the exposition in the sequel it suffices to know that the SVM finally produces a decision rule (a separating function) of the type $y^{pred} = \mathbf{sign}(s(x)) = \mathbf{sign}\Big(\sum_{i=1}^{N} y_i \alpha_i^* k(x_i, x) + b^*\Big)$, with $0 \leq \alpha_i^* \leq C$ and b^* the result of the SVM optimization. Important is to note that $\alpha_i^* > 0$ (a support vector i referred to as "SV" if $\alpha_i^* < C$ and as "BSV" if $\alpha_i^* = C$ in the sequel) actively contribute to the decision function by invoking the ith training example via a user defined kernel $k(,)$ which in most cases is selected to be a semi-positive, symmetric and distance dependent function.

2 Fictitious Training Data Generation and Evaluation

The initial empirical data set for our credit scoring models is a sample of 658 clients for a building and loan credit with seven metric, two ordinal and seven nominal input variables which are coded into a 40 dimensional input vector. It contains 323 defaulting and 335 non defaulting credit clients [7]. Fictitious data points are generated by changing categories in nominal or ordinal input variables. A new fictitious *nominal* data point is generated by switching the outcome of the original data point to any of the other possible categories. For *ordinal* variables only switches to the neighboring categories are performed. By allowing only *one* switch in *one* nominal or ordinal input variable (while preserving the outcomes of all metric variables as well as the target label) one gets

feasible fictitious data points *close* to the original data. Furthermore, a switched input pattern always is exclusive, i.e. there are no double fictitious credit client combinations. Altogether, 10000 fictitious data points are generated in this way. In a first step 3290 (= 5×658) points were sampled randomly and added to the initial data set. The *full fictitious* data set consists of 3948 (658 original plus 3290 fictitious) data points. Then, we follow a two step procedure for SVM model building, proposed by [6]. In a first step, a SVM is trained using the initial data set of 658 clients. The resulting support vectors are extracted and fictitious data points are generated by switching categories, as outlined before. Finally, SVM with five different kernel functions are used for classifying good and bad credit clients. Detailed information about kernels, hyperparameters and tuning can be found in [9].
In Table 1 for each kernel the number of support vectors (SV), the number of bounded support vectors (BSV) and the tenfold cross validation classification error is shown for five different models trained on *(i)* the *Full Fictitious* data set with additional data points around *each* initial point, *(ii)* the *SV Fictitious* data set with additional points generated around the extracted support vectors and *(iii)* the *Real* initial credit data set with 658 clients. Three classification measures are reported: The total error is the percentage of those clients classified incorrectly relative to all credit clients. The alpha error is the percentage of *accepted bad* relative to all bad clients and the beta error is the percentage of *rejected good* relative to all good clients. Using *Full Fictitious* data, the classification error of the linear model decreases clearly, when compared to the real data performance (total error turns down from 29.5 to 25.7 %). For models with sigmoid and polynomial (2nd deg.) kernel the total error almost remains the same, indeed with changing alpha and beta error. The highly non linear models (Polynomial 3nd deg. and RBF) finally do not profit at all by using this type of fictitious data. With *SV Fictitious* data we detect improvement in classification performance in all but the sigmoid model when compared to the *Real* data performance. Especially the RBF kernel shows very good classification results. The number of training cases for this type of fictitious data decreases, which reduces computation time. The linear kernel has no bounded support vectors at all. In previous work [8] it was shown, that in this case the SVM can be replaced by a simple linear discriminant analysis, that is estimated just on the support vectors.

Table 1. Evaluation and comparison of five SVM with different kernel functions. Each model is trained and evaluated on two fictitious data sets (*Full Fict.* and *SV Fict.*). *Real data* models are trained and evaluated without using any additional fictitious data points. Tenfold cross validation error for five SVM models trained on real and fictitious data sets is evaluated for real data points only.

SVM-Kernel	No. of Cases	No. of SV	No. of BSV	Alpha Error	Beta Error	Total Error
Linear						
Full Fict.	3948	43	2115	25.4 %	26.0 %	25.7 %
SV Fict.	768	43	0	32.2 %	22.4 %	27.2 %
Real data	658	41	316	29.1 %	29.9 %	29.5 %
Sigmoid						
Full Fict.	3948	23	2637	33.1 %	26.0 %	29.5 %
SV Fict.	310	11	103	13.3 %	51.0 %	32.5 %
Real data	658	17	544	28.8 %	29.6 %	29.2 %
Polyn. 2nd deg.						
Full Fict.	3948	187	1967	28.2 %	26.6 %	27.4 %
SV Fict.	1060	94	96	19.5 %	31.6 %	25.7 %
Real data	658	63	392	27.2 %	28.1 %	27.7 %
Polyn. 3rd deg.						
Full Fict.	3948	636	1030	31.9 %	30.2 %	31.0 %
SV Fict.	3715	532	119	23.8 %	30.5 %	27.2 %
Real data	658	216	211	27.9 %	29.0 %	28.4 %
RBF						
Full Fict.	3948	574	1255	31.3 %	26.6 %	28.9 %
SV Fict.	3113	419	97	25.7 %	23.3 %	24.5 %
Real data	658	179	252	28.2 %	23.9 %	26.0 %

3 Conclusions

After investigating the effect of the simplest placement of fictitious training points, namely seeding the vicinity of each training point with randomly drawn points, having the same label as the original data point (see also [10]) we now turn to the problem of whether *newly generated* training points do actually express feasible domain data. In the context of credit scoring and in many other classification problems it is quite obvious that *not every* combination of input features can be a feasible case description (e.g. a client) for any of the classes. Consequently we generated *feasible fictitious* data points by switching categories of nominal or ordinal variables. As a result we found that in most cases the generalization error decreases when using additional fictious data points for SVM model training. This is especially true, when placing fictitious data points around support vectors of the respective SVM models only.

References

1. ABU-MOSTAFA, Y.S. (1995): Hints. *Neural Computation* 7, 639-671.
2. DECOSTE, D. and SCHÖLKOPF, B. (2002): Training Invariant Support Vector Machines. *Machine Learning* 46, 161-190.
3. DUIN, R.P.W. and PEKALSKA, E. (2007): The Science of Pattern Recognition. Achievements and Perspectives. In: W. Duch, J. Mandziuk (eds.), Challenges for Computational Intelligence, *Studies in Computational Intelligence*, Springer.
4. SCHEBESCH, K.B. and STECKING, R. (2005): Support vector machines for credit applicants: detecting typical and critical regions. *Journal of the Operational Research Society, 56(9), 1082-1088.*
5. SCHÖLKOPF, B. and SMOLA, A. (2002): *Learning with Kernels.* The MIT Press, Cambridge.
6. SCHÖLKOPF, B., BURGES, C. and VAPNIK, V. (1996): Incorporating Invariances in Support Vector Learning. In: von der Malsburg, C., von Seelen, W., Vorbrüggen, J.C., Sendhoff, B. (Eds.): Artificial Neural Networks – ICANN'96. Springer Lecture Notes in Computer Science, Vol. 1112, Berlin, 47-52.
7. STECKING, R. and SCHEBESCH, K.B. (2003): Support Vector Machines for Credit Scoring: Comparing to and Combining with some Traditional Classification Methods. In: Schader, M., Gaul, W., Vichi, M. (Eds.): *Between Data Science and Applied Data Analysis.* Springer, Berlin, 604-612.
8. STECKING, R. and SCHEBESCH, K.B. (2005): Informative Patterns for Credit Scoring Using Linear SVM. In: Weihs, C. and Gaul, W. (Eds.): *Classification - The Ubiquitous Challenge.* Springer, Berlin, 450-457.
9. STECKING, R. and SCHEBESCH, K.B. (2006): Comparing and Selecting SVM-Kernels for Credit Scoring. In: Spiliopoulou, M., Kruse, R., Borgelt, C., Nürnberger, A., Gaul, W. (Eds.): *From Data and Information Analysis to Knowledge Engineering.* Springer, Berlin, 542-549.
10. STECKING, R. and SCHEBESCH, K.B. (2008): Improving Classifier Performance by Using Fictitious Training Data? A Case Study. In: Kalcsics, J., Nickel, S. (Eds.): *Operations Research Proceedings 2007.* Springer, Berlin 89-94.

Journal Ratings and Their Consensus Ranking

Stefan Theussl and Kurt Hornik

Dep. of Statistics and Mathematics, Wirtschaftsuniversität Wien, Austria
{Stefan.Theussl,Kurt.Hornik}@wu-wien.ac.at

Summary. In this paper we explore the possibility of deriving consensus rankings by solving consensus optimization problems, characterizing consensus rankings as suitable complete order relations minimizing the average Kemeny-Snell distance to the individual rankings. This optimization problem can be expressed as a binary programming (BP) problem which can typically be solved reasonably efficiently. The underlying theory is discussed in Sect. 1. Applications of the proposed method given in Sect. 2 include a comparison to other mathematical programming (MP) approaches using the data set of Tse [9] and establishing a consensus ranking of marketing journals identified by domain experts from a subset of the Harzing journal quality list [2]. In Sect. 3 we discuss computational details and present the results of a benchmark experiment comparing the performance of the commercial solver CPLEX to three open source mixed integer linear programming (MILP) solvers.

1 Consensus Journal Ranking

Journal rankings are increasingly being employed in order to decide upon incentives for researchers, promotion, tenure or even library budgets. These rankings are based on the judgement of peers or are employed in scientometric approaches (e.g., citation frequencies, acceptance rate, etc.). Considering n journals $\mathcal{J} = \{J_1, \ldots, J_n\}$ ranked in M different journal rankings $\mathcal{D} = \{D_1, \ldots, D_M\}$ one might be interested in deriving a consensus ranking by comparing the performance of each journal in the set $\mathcal{J}$ on each ranking D_m.

Paired comparisons for a journal ranking D_m induce a *relation* (more precisely, endorelation) R_m on the set of journals $\mathcal{J}$. A consensus ranking is a suitable aggregation of the relation *profile*—the collection of relations $\mathcal{R} = \{R_1, \ldots, R_M\}$—into a single relation R, i.e., an endorelation on $\mathcal{J}$ which is at least complete, reflexive and transitive.

B. Fleischmann et al.(Eds.), *Operations Research Proceedings 2008*,
DOI: 10.1007/978-3-642-00142-0_65,

Regnier [8] suggested to determine R in a suitable set $\mathcal{C}$ of possible consensus relations by solving (a non-weighted variant of) the problem

$$\sum_{m=1}^{M} w_m d(R, R_m) \Rightarrow \min_{R \in \mathcal{C}}$$

where d is a suitable dissimilarity (distance) measure. Accordingly, in this *optimization* approach consensus relations are described as the ones that "optimally represent the profile" where the average distance d to the individual rankings is to be minimized. As a natural way to measure distance between preference relations we use the Kemeny-Snell distance [5] which for preference relations coincides with the *symmetric difference distance* d_Δ.
The resulting optimization problem can be expressed as a BP problem which can typically be solved reasonably efficiently although this combinatorial optimization problem is known to be $\mathcal{NP}$-complete [10]. Let $r_{ij}(m)$ and r_{ij} be the incidences of relations R_m and R, respectively, and $c_{ij} = \sum_m (2 w_m r_{ij}(m) - 1)$ then the order relation R can be obtained by solving

$$\sum_{i \neq j} c_{ij} r_{ij} \Rightarrow \max$$

subject to the constraints that the r_{ij} be the incidences of a preference order, i.e., r_{ij} meet the binarity, reflexivity and transitivity conditions (see [3] for more details).
Finally, we note that given a relation profile there is not necessarily a unique solution to the above optimization problem. For this reason we use a rather brute-force branch and cut approach to find all consensus solutions. This allows for a more detailed interpretation of the well founded preference structure found.

2 Application

As an initial illustration we apply our method to the data set originally employed by Tse [9] to obtain linear attribute weights for seven marketing journals. Tse compared his results to rankings established in an earlier paper by Ganesh et al. [1]. Horowitz [4] proposed another MP approach and applied it to Tse's data in its original (raw) form as well as in an adapted (normalized) form.
Table 1 shows the ranking obtained by Ganesh et al., the ratings obtained by Tse and Horowitz as well as the scores obtained by deriving a consensus ranking—a symmetric difference preference (SD/P)

ordering—from Tse's data (SD/P (Tse)). Note that, whereas in the studies of Ganesh et al., Horowitz and Tse higher scores correspond to higher ranked journals in our approach the most preferred journals get the lowest scores (ranks).

Table 1. Comparison of Ranking Results

	Ganesh et. al.	Tse	Horowitz (raw)	Horowitz (norm.)	SD/P (Tse)	SD/P (JQL)
Journal of Marketing Research (JoMR)	7	4.324	0.528	0.731	2.5	1.5
Marketing Science (MrkS)	6	0.928	0.442	0.547	2.5	4.5
Journal of Marketing (JroM)	5	3.941	0.594	0.032	2.5	3.0
Journal of Consumer Research (JoCR)	4	3.919	0.869	1.075	2.5	1.5
Journal of Advertising Research (JoAR)	3	0.891	0.175	−0.461	6.5	4.5
Journal of Retailing (JroR)	2	0.584	0.035	−0.933	5.0	6.5
Industrial Marketing Management (InMM)	1	0.785	0.068	−0.990	6.5	6.5

The results of our consensus method show that four journals (JoMR, MrkS, JroM, JoCR) are most preferred and two are least preferred (JoAR, InMM). Comparing our findings to the results obtained in the earlier studies we see that there is high agreement on the top journals but lower agreement further down. E.g., JroR is ranked higher using our approach.

In a second step we extended our previous analysis using data from the Harzing journal quality list (JQL) [2]. The JQL contains data on 852 business and management journals rated in 17 different rankings from which we transcribed ratings for 82 marketing journals in 13 rankings (subsequently referred as JQL data).

A consensus ranking based on data from the JQL (SD/P (JQL)) for the seven marketing journals are presented in Table 1. Interestingly, although we used a different data set there is still high agreement with the results obtained by methods using Tse's data. In comparison to SD/P (Tse) we see that there is a finer grained preference ordering, e.g., MrkS and JroM are not perceived as the most preferred journals anymore and constitute a new group.

In a second illustration we derive a consensus ranking of marketing journals from the JQL. As not every journal is rated in each of the 13 rankings we were facing journals having a lot of missing values. After examination of the data set we chose to remove journals appearing

in less than 25% of the ratings. Hereafter, we replaced the remaining missing values in the incidence matrices with a value of zero reasonably indicating that this journal is not rated above the others. This lead to a final data set containing 64 journals.
Table 2 shows the scores of the top 20 ranked journals obtained by our consensus method (SD/P) compared to the ranking EJIS07 [2] obtained by a method proposed by Mingers and Harzing [6]. All of the most preferred journals in SD/P have also a high rating in EJIS07. The lower the preference for a particular journal the higher the probability that it has a lower ranking in EJIS as well. Although in general there is a high rank correlation (0.89 in absolute value), considerable differences can be observed for particular journals (e.g., AoTR in Table 2).

Table 2. Relation Scores SD/P on JQL

	SD/P	EJIS07
Journal of Applied Psychology (JoAP)	3.0	4
Journal of Consumer Research (JoCR)	3.0	4
Journal of Marketing (JroM)	3.0	4
Journal of Marketing Research (JoMR)	3.0	4
Marketing Science (MrkS)	3.0	4
International Journal of Research in Marketing (IJoRiM)	8.0	4
Journal of Product Innovation Management (JoPIM)	8.0	3
Journal of Retailing (JroR)	8.0	3
Journal of the Academy of Marketing Science (JotAoMS)	8.0	4
Marketing Letters (MrkL)	8.0	3
Annals of Tourism Research (AoTR)	11.0	4
Journal of Consumer Psychology (JoCP)	12.0	3
European Journal of Marketing (EJoM)	16.5	2
Industrial Marketing Management (InMM)	16.5	2
Journal of Advertising (JroA)	16.5	2
Journal of Advertising Research (JoAR)	16.5	2
Journal of Business Research (JoBR)	16.5	3
Journal of International Marketing (JoIM)	16.5	3
Journal of Public Policy & Marketing (JoPP&M)	16.5	2
Psychology and Marketing (PsaM)	16.5	2

Fig. 1 shows the Hasse Diagram of the consensus ranking of the top 20 marketing journals using the JQL data. Instead of incomparability in standard Hasse diagrams, elements in the same layer indicate equivalent preference.
We are also developing a methodology for determining the "best" k journals, i.e., finding a relation R which minimizes $\sum_m d(R, R_m)$ over all relations for which "winners" are always strictly preferred to "losers", without any further constraints on the relations between pairs of winners or pairs of losers. Applying this "social choice function" on the set $\mathcal{J}$ of journals using the data from the JQL we would choose

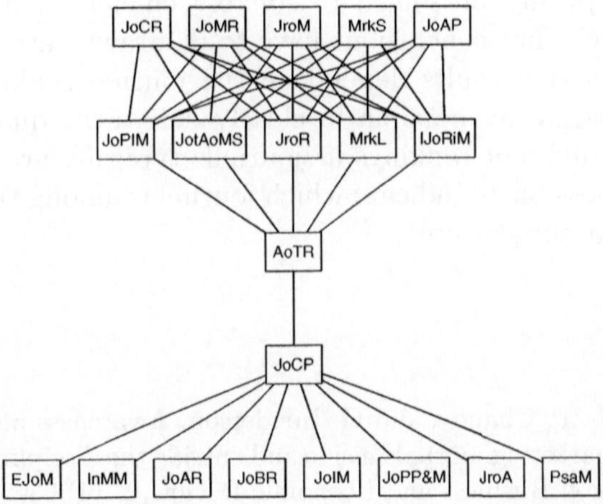

Fig. 1. Consensus Journal Ranking of top 20 Marketing Journals

the following journals in increasing order (the first would be choosen if $k = 1$, if $k = 2$ the second is additionally chosen, etc.): JoMR, JroM, JoCR, MrkS and JoAP.

3 Computational Details and Benchmark Results

For our analysis we used R, a language for statistical computing and graphics [7]. Currently, interfaces to the MILP solvers CPLEX, GNU Linear Programming Kit (GLPK), lp_solve, and COIN-OR SYMPHONY are available in R.
To rank 82 journals in 13 rankings a BP problem containing 6642 objective variables and $547,965$ constraints is to be solved. On a machine with an Intel Xeon processor with 2.33 GHz and 16 GB of memory run times differ considerably between the solvers. Whereas CPLEX was capable to solve this problem in around 11 seconds, SYMPHONY needed 13 minutes and lp_solve 48 minutes. GLPK could not solve this problem within a reasonable amount of time.

4 Conclusion

We presented a method for deriving consensus rankings by solving consensus optimization problems minimizing average distance between the individual ratings. This problem can be formulated as a BP problem which can typically be solved reasonably efficiently. Nevertheless, it has to be noted that conclusions have to be drawn carefully from such an analysis as the results depend on the journal rankings employed. Although it is hardly possible to clearly derive the quality of a journal from the different rankings as journals typically are not uniformly ranked, it is possible to indicate which journal is among the top journals in a particular subject area.

References

1. G. Ganesh, P. Chandy, and G. Henderson. Awareness and evaluation of selected marketing journals inside and outside the discipline: An empirical study. Akron Business and Economic Review, 21(4):93-106, 1990.
2. A.-W. Harzing. Journal quality list, December 2007. `http://www.harzing.com/jql.htm`.
3. K. Hornik and D. Meyer. Deriving consensus rankings from benchmarking experiments. In R. Decker and H.-J. Lenz, editors, Advances in Data Analysis (Proceedings of the 30th Annual Conference of the Gesellschaft für Klassifikation e.V., Freie Universität Berlin, March 8-10, 2006, Studies in Classification, Data Analysis, and Knowledge Organization, pages 163-170. Springer-Verlag, 2007.
4. I. Horowitz. Preference-neutral attribute weights in the journal-ranking problem. The Journal of the Operational Research Society, 54(5):452-457, 2003.
5. J. G. Kemeny and J. L. Snell. Mathematical Models in the Social Sciences, chapter II. MIT Press, Cambridge, 1962.
6. J. Mingers and A.-W. Harzing. Ranking journals in business and management: A statistical analysis of the Harzing data set. European Journal of Information Systems, 16(4):303-316, 2007.
7. R Development Core Team. R: A Language and Environment for Statistical Computing. R Foundation for Statistical Computing, Vienna, Austria, 2008. ISBN 3-900051-07-0.
8. S. Régnier. Sur quelques aspects mathématiques des problèmes de classification automatique. ICC Bulletin, 4:175-191, 1965.
9. A. C. B. Tse. Using mathematical programming to solve large ranking problems. The Journal of the Operational Research Society, 52(10):1144-1150, 2001.
10. Y. Wakabayashi. The complexity of computing medians of relations. Resenhas, 3(3):323-349, 1998.

The Effect of Framing and Power Imbalance on Negotiation Behaviors and Outcomes

Ali Fehmi Ünal and Gül Gökay Emel

Department of Business Administration. Uludag University, Görükle Campus, 16059 Bursa, Turkey. {afunal,ggokay}@uludag.edu.tr

1 Introduction

Negotiation can be defined as a joint decision making process where two or more parties are trying to influence the other about the allocation of resources or division of gains for the purpose of achieving own or mutual interests [2]; [16]. Negotiators often fail to reach Pareto optimal solutions when there is integrative potential that expand the pie and yield higher joint outcomes [2]; [16]; [19]. Literature showed that framing of conflicts and power relations are two widely acknowledged factors that affect the negotiation process and outcomes. According to Emerson "the power of A over B is equal to and based upon the dependence of B upon A" [7]. The power imbalance empirically manifests when high-power and low-power parties initialize a supply-demand relationship in which these demands are contradictory to supplier's desires [7]. Nevertheless, decision makers -therefore negotiators- systematically violate the requirements of rational choice and deviate from rationality because of imperfections of human perception and decisions [9]; [14]; [18]. One of the hidden traps in decision making is reference framing which was first introduced by Prospect theory [9]; [18]. It is proposed that people normally perceive outcomes as gains and losses rather than final states of outcomes and valuation of any outcome, defined as gains or losses, depends on its location relative to the reference point which is assigned a value of zero. Further, they stated that people are more risk averse for positive but risk-seeking for negative outcomes. Thus, the way a problem is framed can dramatically influence our choices.

1.1 Framing when Parties Have Unequal Power

The literature has shown that equal power dyads achieve higher joint outcomes by focusing on the integrative side of the negotiation and

B. Fleischmann et al.(Eds.), *Operations Research Proceedings 2008*,
DOI: 10.1007/978-3-642-00142-0_66, © Springer-Verlag Berlin Heidelberg 2009

under unequal power condition the lower power player was found to be responsible for driving a solution of higher joint gains [12] . Another factor that facilitates integrative solutions is powerful parties' positive emotion which shapes the quality of negotiation process and outcomes [1]. While low power negotiators are not willing to accept the agreements that reflects the power difference, high power parties strive to reach agrements resulting a larger payoff for the high power negotiator [11]. Previous frame research includes diversified findings. Some studies have shown that loss frame produces less cooperation, more demands, fewer concessions, more impasse [2]; [5]; [14], some reported contrary findings when payoff from settling is risky [3], and some other studies indeed reported no relation between frame and cooperation [4]. Literature also provides evidence that the concessions made by the loss framed opponent loom larger due to loss aversion [5]; [6]. Another finding is that when the counterpart has a loss frame, he/she is perceived as more cooperative but more importantly negotiators generally demand more and concede less when their opponent has a loss rather than a gain frame [4]; [5]. In sum, following hypotheses are derived from prior findings in the literature and current theoretical considerations that were mentioned.

Hypothesis1a. High power parties with loss frames will obtain higher individual outcomes when low power parties have gain frames compared to when have loss frames.

Hypothesis1b. High power parties with loss frames will obtain higher individual outcomes than high power parties with gain frames when low power parties are gain framed.

Hypothesis2a. The mutual outcomes will be lowest when high power negotiators have loss frames and low-power negotiators have gain frames.

Hypothesis2b. The mutual outcomes will be highest when high-power negotiators have gain frames and low power parties have loss frames.

Hypothesis3a. Low power parties with loss frames will reach better individual outcomes than low power parties with gain frames when high-power parties have loss frame.

Hypothesis3b. Low power parties with gain frames will get higher individual outcomes than low power parties with loss frames when high power parties are gain framed.

Hypothesis4. The difference between individual outcomes will be highest in favor of high power parties when they have loss frames and low power parties have gain frames.

2 Method

2.1 Participants and Design

Ninety two business administration undergraduates at the University of Uludag participated in the experiment for extra course credit. Participants were randomly assigned to dyads and to a roll of either a recruiter (high power) or a candidate (low power) (adapted from [21]). The experimental design manipulated recruiters' frame [gain (RG) vs. loss (RL)] and candidates' frame [gain (CG) vs. loss (CL)] as between-subjects variable and power (high and low) as a within subject variable. This led to two cells in which parties are in the same frame condition (RG / CG and RL/CL) and two cells in which parties are in different frame condition (RG/CL and RL/CG).

2.2 Procedure

Participants involved in a job interview between a recruiter and a candidate. They were given 25 minutes to read the instructions regarding their roles, the issues, their payoff schedules and 35 minutes to negotiate. To settle, participants had to reach agreement on each of five issues. The issues were differently valued by the recruiter and candidate, thus negotiators were able to make trade-offs (logrolling) between issues. In the gain frame condition every alternative was represented with a positive number (gains) and in the loss frame condition, alternatives were represented with negative numbers that express the losses.

2.3 Experimental Manipulations

Manipulation of Power. The actual power was manipulated by the number of alternatives [12] ; [13] and perceived relative power was manipulated by giving information about own/counterparts' alternatives and values attributed to outcomes of the relationship [21] and by assigning higher legitimate and reward power [8].

Manipulation of Frame. The frame condition was manipulated as in previous research [14]; [5]; [6]. The participants were informed that: *Any concession by the candidate/recruiter will result in serious gains/losses for your company/you. Please do not forget that your primary objective is to maximize/minimize the monetary gains/losses for the company/yourself. What is being expected from you is to provide the counterpart to concess as much as possible so that you can ascend/descend your monetary gains/losses to a top/the lowest level that is 1800 YTL/0 YTL.*

2.4 Results

The hypotheses concerned the anticipation of two dependent variables. The individual outcomes were based on the terms of agreement and the level of integrativeness was measured by summing the individual payoffs. *Power manipulation* was adequate. Sense of power in high power condition was higher (M= 81,26 / M= 42.93); $F(3, 88) = 42.169$, $p < 0.001$. *Manipulation of frame* was also successful. Loss frame produced a higher sensitiveness for losing money (M = 5.67 / M =.327); $F(3, 88) = 16.650$, $p < 0.01$. Within 46 dyads, 43 reached agreement. An initial ANOVA (analysis of variance) showed that effect of frame conditions on integrativeness did not remain significant. Thus, H2a and H2b were not supported, $F(3, 39) < 1$. An additional ANOVA and further post hoc Dunnet test provided evidence that high power parties with loss frames were able to achieve a higher proportion of the rewards when low power parties were in gain frame condition (M= 1275 YTL) compared to when in loss frame condition(M= 1002 YTL). High power parties when have loss frames also obtained higher individual outcomes (M= 1275) compared to high power parties with gain frames (M= 996.15) when low power parties have gain frames. Thus, H1a and H1b were supported, $F(2, 30) = 4.303$, $p < 0.05$. H3a and H3b were about the individual outcomes of the low power parties. Results were as predicted. Low power parties with loss frames obtained higher individual outcomes than low power parties with gain frames when high power parties were in loss frame condition (M = 924 vs. M = 626) and low power parties with gain frames obtained higher individual outcomes when high power parties were in gain frame condition (M = 906.92 vs. M = 626). Thus, H3a and H3b were supported, $F(2, 30)=5.248$, $p < 0.05$. Hypotheses 4 predicted that the difference between individual outcomes. Although the results were significant, post hoc Dunnet test showed that the difference between individual outcomes was not significant between RL/CG and RG/CL conditions but was significant in all other conditions, $F(3, 39) = 2.963$, $p < 0.05$. These findings provide partial support for H4.

3 Discussion and Limitations

This purpose of this study was to investigate the effects of gain-loss frames in power-asymmetric negotiations. Predictions regarding the integrativeness of the agreements were not supported. High difference in perceived relative power [12] [11], lack of prior experience and related knowledge to raise mutual outcomes [17]; [20] and individualistic instructions in frame manipulation [6] could have directed the parties to maximize own rather than mutual outcomes. An important finding in our experimental study is the interaction of power with frame conditions to determine the level of individual outcomes. As aspirations are related to individual outcomes [21], we interpret that the frame conditions are more likely to affect individual rather than mutual outcomes. The finding that low power parties reach higher individual outcomes when they were in the same frame condition with the high power parties is in consistence with the study of Olekalns and Smith(2005) [15] which showed that perceived similarity is more important than individual's goals when it comes to cooperation. A last finding is that high power parties when have loss frames seem to get what they want from the negotiation unless low power parties also have loss frames. We would like to mention the limitations to this study. Although experimental studies have high internal validity thus, are powerful to demonstrate a causal effect between independent and dependent variables, generalizability of findings to other settings and populations has to be considered carefully. Further, it was found that Pareto improvements are not always preferred by negotiators [10]. Therefore, more satisfying rewards, like monetary incentives, could be used to retain more realism. To conclude, this study underscores the importance of knowledge about power asymmetric relations and awareness about the imperfections of human perception and decisions traps in managing organizations rationally and effectively.

References

1. Anderson, C. and Thompson, L. L., (2004). Affect From The Top Down: How Powerful Individuals' Positive Affect Shapes Negotiations. Organizational Behavior and Human Decision Processes, 95: 125-139.
2. Bazerman, M. H., and Neale, M. A. (1992). Negotiating Rationally. New York: Free Press.
3. Bottom, W. P. (1998). Negotiator Risk: Sources of Uncertainty and the Impact of Reference Points on Negotiated Agreements. Organizational Behavior And Human Decision Processes, Vol. 76, No. 2, November, pp. 89-112.
4. De Dreu, C. K. W., Emans, B. and Van de Vliert, E. (1992). Frames Of Reference And Cooperative Social Decision Making. European Journal of Social Psychology, 22, 297-302.
5. De Dreu, C. K.W., Carnevale, P.J., Emans, B., and Van de Vliert, E. (1994). Effects Of Gain-Loss Frames İn Negotiation: Loss Aversion, Mismatching And Frame Adoption. Organizational Behavior and Human Decision Processes, 60, 90-107.
6. De Dreu C. K. W and McKusker, c. (1997). Gain-Loss Frames And Cooperation İn Two-Person Social Dilemmas: A Transformational Analysis. Journal of Personality and Social Psychology, 72 (5), 1093-1106.

7. Emerson, R. M. (1962). Power-Dependence Relations. American Sociological Review, 27, 31-40.
8. French, J. R. P., Jr., and Raven, B. (1959). The Bases Of Social Power. In D. Cartwright (Ed.), Studies in social power, 150-167. Ann Arbor: University of Michigan Press.
9. Kahneman, D., and Tversky, A. (1979). Prospect Theory: An Analysis Of Decision Making Under Risk. Econometrica, 67, 263-291.
10. Korhonen, P., Phillips, J., Teich, J. and Wallenius, J. (1998). Are Pareto İmprovements Always Preffered By Negotiators? Journal of Multi-Criteria Decision Analysis, 7, 1-2.
11. Mannix, E. A. (1993). The İnfluence Of Power, Distribution Norms And Task Meeting Structure On Resource Allocation In Small Group Negotiation. The International Journal of Conflict Management, 4 (1), 5-23.
12. Mannix, E. A. and M. A. Neale. (1993). "Power Imbalance And The Pattern Of Exchange İn Dyadic Negotiation," Group Decision and Negotiation, 2, 119-133.
13. McAlister, L., Bazerman, M. H., and Fader, P. (1986). Power And Goal Setting In Channel Negotiations. Journal of Marketing Research, 23, 228-236.
14. Neale, M. A. and Bazerman, M. H. (1985). The Effects Of Framing And Negotiator Overconfidence On Bargaining Behavior And Outcomes. Academy of Management Journal, 28, 34-49.
15. Olekalns, M. and Smith P. L. (2005). Cognitive Representations Of Negotiation. Australian Journal of Management, 30 (1), 57-76.
16. Raiffa, H. (1982). The Art and Science of Negotiation, Cambridge, MA: Belknap/Harvard University Press.
17. Thompson, L. (1990). An Examination Of Naive And Experienced Negotiators. Journal of Personality and Social Psychology, 59, 82-90.
18. Tversky A., and Kahneman, D. (1981). The framing of decisions and the psychology of choice. Science, 211(4481), 453-458.
19. Walton, R. E. and McKersie R. B. (1965). A Behavioral Theory of Labor Negotiations. ILR Press Ithaca, New York.
20. Weingart, L. R., Hyder, E. B., and Prietula, M. J. (1996). Knowledge Matters: The Effect Of Tactical Descriptions On Negotiation Behavior And Outcome. Journal of Personality and Social Psychology, 70 (6), 1205-1217.
21. Wolfe, J. Rebecca and McGinn K. L., (2005). Perceived Relative Power and Its Influence On Negotiations. Group Decision and Negotiation, 14, 3-20.

Response Mode Bias Revisited – The "Tailwhip" Effect

Rudolf Vetschera[1], Christopher Schwand[2], and Lea Wakolbinger[1]

[1] University of Vienna, Vienna, Austria
{rudolf.vetschera,lea.wakolbinger}@univie.ac.at
[2] IMC University of Applied Sciences, Krems, Austria
christopher.schwand@fh-krems.ac.at

1 Introduction

Expected utility theory is an important instrument for decision making under risk. To apply it in a prescriptive context, the decision maker's utility function needs to be elicited. Several methods for utility elicitation have been proposed in the literature which, from a theoretical point of view, should lead to identical results.

However, beginning in the late 1970s, researchers have discovered that theoretically equivalent methods of utility elicitation lead to different results [4, 6, 7]. One important bias phenomenon which was identified in this context is the "Response mode bias", which describes an inconsistency in elicitation results of the Certainty Equivalence (CE) and the Probability Equivalence (PE) methods.

Both methods are based on the comparison between a certain payoff x_s and a lottery, which yields a better outcome x_h with probability p and a worse outcome x_l with probability $1-p$. In the CE method, the decision maker is informed about x_h, x_l and p and has to provide a value x_s which makes him or her indifferent between the lottery and the certain payment. In the PE method, the decision maker provides a value for p in response to the three other parameters.

Using a set of 10 questions, Hershey et al. [6] found that in problems involving only losses, the PE method led to a significantly higher number of subjects exhibiting risk seeking behavior than the CE method. This phenomenon, which was later on labeled the "Response mode bias" was particularly noticeable in lotteries involving a comparatively high probability of loss.

B. Fleischmann et al.(Eds.), *Operations Research Proceedings 2008*,
DOI: 10.1007/978-3-642-00142-0_67, © Springer-Verlag Berlin Heidelberg 2009

Subsequent studies mainly focussed on testing possible explanations for this phenomenon like scale compatibility [3, 5] or probability weighting [10], or developed alternative procedures for utility elicitation [2].
Our current study returns to the original research on the response mode bias and seeks to improve our understanding of this phenomenon in two ways:

1. Rather than just classifying responses from subjects as risk-averse, risk-neutral or risk-seeking, we use a cardinal measure of risk attitude based on an approximation of the Arrow-Pratt coefficient of risk aversion. This enables us to study not only the occurrence, but also the strength of the bias phenomenon and to analyze cases in which it is present, but not strong enough to bring about a full swing in risk attitude.
2. We use this more sensitive instrument to study lotteries involving a wide range of probability values to examine the response mode bias under different settings.

2 Hypotheses

Based on the previous literature, we formulate the following hypotheses concerning the response mode bias:

Hypothesis 1. For decision problems involving losses, utility functions elicited using the CE method will exhibit a significantly higher risk seeking attitude compared to utility functions elicited using the PE method.

This hypothesis directly builds upon the results of Hershey et al. [6]. Since they found this phenomenon mainly in problems involving a high probability of loss, and probability levels have been shown to influence bias [8], we also expect:

Hypothesis 2. The shift towards risk seeking in the CE method will be weaker in decision problems involving low probabilities of losses.

Furthermore, empirical evidence [1, 9] also suggests a reversal of the phenomenon between loss and gain domains:

Hypothesis 3. The shift in risk attitudes between CE and PE elicitation methods will be reversed in lotteries involving gains compared to lotteries involving losses.

3 Experimental Setup and Measurement

In total, we performed a series of three experiments. The first experiment, which we do not report here because of space limitations, was based on a replication of the experiments by Hershey et al. [6] and involved ten lotteries in the loss domain only. To test the hypotheses formulated above, we then conducted a series of two experiments using the five sets of values shown in Table 1, which were used both for gains and losses. All questions involved a certain outcome and a lottery, which led to an outcome of zero or the uncertain outcome with the probability indicated in Table 1. The order of all questions was randomized and experimental conditions were set up so that subjects were not able to record their answers between questions.

Table 1. Questions used in the experiments

Question	Certain outcome	Probability	Uncertain outcome
Q1	€100	5%	€2,000
Q2	€10	5%	€200
Q3	€10	50%	€20
Q4	€90	90%	€100
Q5	€1,900	95%	€2,000

For the two experiments, the samples consisted of undergraduate student subjects at two different institutions in Austria. All experiments took place in class sessions to maximize participation and minimize potential contact between subjects across treatments. The experiments were conducted in introductory courses, so subjects were not familiar with the methods studied. Participants were randomly assigned to one of two treatment groups; each group answered questions according to one of the two elicitation methods considered. The composition of the two samples is shown in Table 2.

Each question posed in the experiments is analyzed separately. The answer provided by the subject provides one point of the utility function, the other outcomes (zero and the uncertain payoff of the lottery) form the two endpoints of the utility scale, which can be fixed at zero and one. Based on these values, we approximate the first derivative of the utility function below and above the certain outcome. The average of these two values was used as an approximation of the first derivative and their difference as an approximation of the second derivative. The

Table 2. Composition of Samples

Experimental Groups	Elicitation Method	Sample Composition Male	Female
Experiment 2	CE	11 (35%)	20 (65%)
	PE	13 (34%)	25 (66%)
Experiment 2	total	24 (35%)	45 (65%)
Experiment 3	CE	11 (46%)	13 (54%)
	PE	15 (68%)	7 (32%)
Experiment 3	total	26 (57%)	20 (43%)

ratio of these two values therefore can be considered as an approximation of the Arrow-Pratt coefficient of risk aversion.

4 Results

When data from all experiments is analyzed as one data set, the two methods do not lead to significantly different results, thus hypothesis 1 must be rejected. However, when lotteries involving low and high probabilities are analyzed separately, a clear picture begins to emerge. For the following analysis, questions Q1 and Q2 are considered as low probability lotteries, questions Q4 and Q5 as high probability lotteries, and question Q3, which involves a probability of 50%, is not used. Table 3 shows the average values of the approximate Arrow-Pratt indices for lotteries in the loss domain in the two experiments.

Table 3. Arrow-Pratt indices for low and high probabilities – loss domain

		Experiment 2		Experiment 3	
Method		Low	High	Low	High
CE	Mean	-0.2445	-1.3136	0.1413	-1.2200
	Std	0.9569	0.7731	0.9216	0.7386
PE	Mean	-1.3547	-0.3091	-1.5126	0.1101
	Std	0.9027	1.2750	0.5211	1.2038
	t-test	***6.2045	***- 5.5207	***8.7070	*** -5.7906

***: $p < 0.0001$

These results clearly show the expected relationship: for lotteries involving high probabilities, the Arrow-Pratt index is considerably lower

for the CE method, indicating a more risk seeking behavior, while for lotteries involving low probabilities, the effect is reversed. It should also be noted that in experiment 2, subjects under all experimental conditions reacted on average risk seeking, thus the differences indicated in our experiment could not have been found using just a classification into risk averse and risk seeking behavior. This clearly shows the advantage of using a cardinal measure of risk attitude.
The effect can also be graphically illustrated by plotting the distribution of Arrow-Pratt indices across subjects for the four conditions. Figure 1 clearly shows the resulting "Tailwhip" effect.

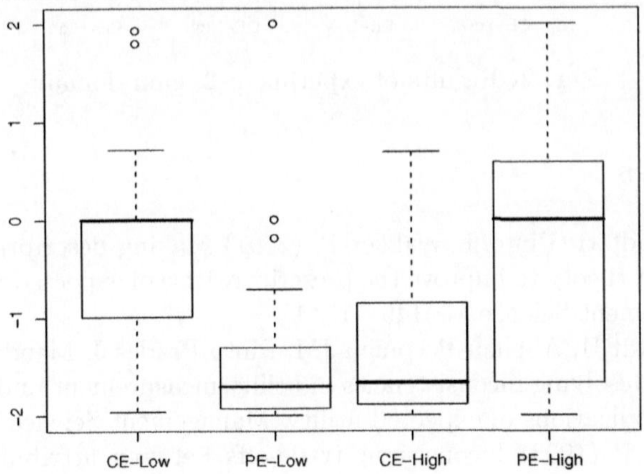

Fig. 1. Results of experiment 2, loss domain

When considering gains (Figure 2), the effect is still very strongly present for lotteries involving low probabilities, where also the predicted reversal between losses and gains can be observed. These differences were significant in both experiments ($t = 17.25, p < 0.0001$ for experiment 2 and $t = 10.66, p < 0.0001$ for experiment 3). However, for lotteries with high probabilities, the expected reversal did not take place, as Figure 2 shows for experiment 2. In experiment 2, this effect (which contradicted our expectations) was significant ($t = -3.75, p = 0.0003$); but in experiment 3, it was not.
Our results therefore clearly show that the effect of elicitation methods on the estimated utility function can be measured quite precisely by using an approximation of the Arrow-Pratt index as a cardinal indicator of risk attitude. Furthermore, we have observed a strong impact of both the probability levels involved and the outcome domain on the direction and strength of the response mode bias.

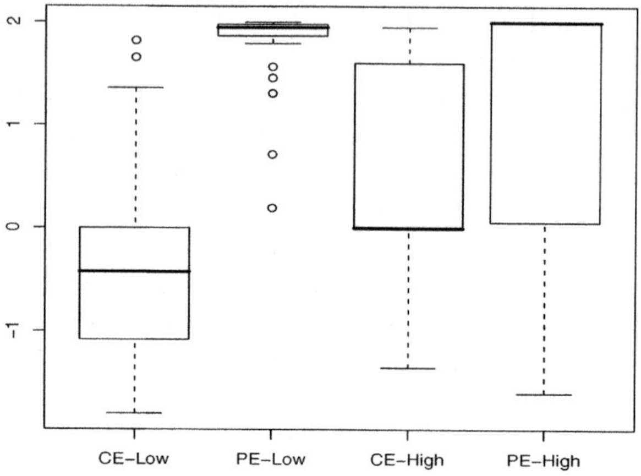

Fig. 2. Results of experiment 2, gain domain

References

1. Bleichrodt H, Pinto J, Wakker P, (2001) Making descriptive use of the prospect theory to improve the prescriptive use of expected utility theory. Management Science 47: 1498–1514.
2. Bleichrodt H, Abellan-Perpiñan JM, Pinto-Prades J, Mendez-Martinez I (2007) Resolving inconsistencies in utility measurement under risk: Tests of generalizations of expected utility. Management Science 53: 469-482.
3. Delquié P (1993) Inconsistent trade-offs between attributes: New evidence in preference assessment biases. Management Science 39: 1382–1395.
4. Fishburn P, Kochenberger G (1979) Two-piece von Neumann-Morgenstern utility functions. Decision Sciences 10:503–518.
5. Fischer, G. W. and Hawkins, S. A.: (1993), Strategy compatibility, scale compatibility, and the prominence effect, Journal of Experimental Psychology: Human Perception and Performance 19(3): 580-597.
6. Hershey J, Kunreuther H, Shoemaker P (1982) Sources of bias in assessment procedures for utility functions. Management Science 8:936–954.
7. Hershey J, Schoemaker P (1985) Probability versus certainty equivalence methods in utility measurement: Are they equivalent? Management Science 31:1213–1231.
8. Johnson E, Schkade D, (1989), Bias in utility assessments: Further evidence and explanations. Management Science 35: 406–424.
9. Kahneman D, Tversky A (1979) Prospect Theory: An analysis of decision under risk. Econometrica 47: 263–291.
10. Wu G, Gonzalez R (1996) Curvature of the probability weighting function. Management Science 42: 1676–1690.

Enhancing Target Group Selection Using Belief Functions

Ralf Wagner and Jörg Schwerdtfeger

University of Kassel, Mönchebergstrasse 17, D-34125 Kassel, Germany
rwagner@wirtschaft.uni-kassel.de,
schwerdtfeger.joerg@googlemail.com

Summary. One of the most critical decisions in customer relationship management, interactive marketing, and in direct marketing is the assignment of customers to groups with similar response characteristics. This decision usually is based on previous purchase behavior and socio-demographic variables. Unfortunately, information from different sources is likely to provide the managers with conflicting indications of group memberships. In this study we utilize Multi Sensor Clustering and a Rough Sets approach for assigning individuals to groups. We outline the combination of different assignment recommendations by means of the Evidence Theory and the Transferable Belief Model. For an empirical application we rely on a set transaction data recorded from a customer loyalty program.

1 Introduction

The assignment of customers to groups with similar response characteristics is an awkward task [6]. Usually different data describing customers is available, but to make reliable decisions these information need to be combined. We classify customers by means of their purchase history and additional data. For combining these different sources of information in decision making the *Dempster-Shafer Theory* (DST) [1] provides us with a suited and flexible framework. Firstly we calculate the belief functions and classify the customers applying the Rough Set Theory (RST) [2]. This results in a *Multi Sensor Clustering* (MST) [4]. Secondly we utilize the *Transferable Belief Model* (TBM) [3] to combine conflicting recommendations to draw a final conclusion for each customer. This paper is organized that in section 2 the formalisms of DST and RST are introduced. In section 3 the MST and the TBM are outlined. Subsequently, in section 4 the empirical results are discussed. Final conclusions are drawn in section 5.

B. Fleischmann et al.(Eds.), *Operations Research Proceedings 2008*,

DOI: 10.1007/978-3-642-00142-0_68,

2 Linking Dempster-Shafer Theory with Rough Sets

Let Θ be the *frame of discernemet*, which includes mutually exclusive and exhaustive atomic elements $(a, b, c, \dots)$. These represent hypotheses, labels, or statements of a problem domain. 2^Θ is the set of subsets $\mathcal{A}, \mathcal{B}, \dots$ of Θ. If the function $m : 2^\Theta \to [0,1]$ satisfies $m(\emptyset) = 0$ and $\sum_{X \in 2^\Theta} m(X) = 1$, then m is a *basic probability assignment* (*bpa*) over Θ. The quantity $m(X)$ for any $X \in 2^\Theta$ is a *mass of belief* that is committed to X. From the *bpa* a *belief function* $bel : 2^\Theta \to [0,1]$ is derived by: $bel(X) = \sum_{\mathcal{A} \subseteq X} m(\mathcal{A}) \forall X \in 2^\Theta$. This captures the *total degree of belief* given to X. Noticeably, $bel(X)$ represents the amount of support given to X. The *potential* amount of support given to X is calculated by the *plausibility function* $pl : 2^\Theta \to [0,1]$ defined as $pl(X) = 1 - bel(\bar{X})$, with $\bar{X}$ describing the complement of X. *Dempster's rule of Combination* enables the a combination of two distinct sources of evidence by allocating the two *bpa*: $m'(X) = \sum_{\mathcal{A} \cap \mathcal{B} = X} m_1(\mathcal{A}) \cdot m_2(\mathcal{B}) \forall X \in 2^\Theta$. This conjunctive sum may produce subnormal beliefs (i.e. $m'(\emptyset) > 0$), which have to be normalized by dividing $m'(X)$ by $1 - m'(\emptyset)$ for $X \neq \emptyset$. The assignment of the belief function represents the degree of belief to the elements of 2^Θ with various ways of assessing statistical evidence [1]. Let $\mathbf{AT} = (\mathcal{U}, \mathcal{AT})$ be an information system with U depicting a set of customers and $\mathcal{AT}$ the non-empty set of attributes. If a target information (e.g., buying a product) is available, the information system is a decision system $\mathbf{AT} = (\mathcal{U}, \mathcal{AT} \cup \{d\})$, with the decision attribute $d \neq \mathcal{AT}$ [2]. For any information system with $\mathbf{AT} = (\mathcal{U}, \mathcal{AT})$ we determine the subsets $B \subseteq \mathcal{AT}$ and $Z \subseteq U$. We asses Z by a B-Lower approximation $\underline{B}Z = \{z \in \mathcal{U} | [z]_B \cap Z \neq \emptyset\}$ and a B-Upper approximation $\overline{B}Z = \{z \in \mathcal{U} | [z]_B \subseteq Z\}$. All customers in $\underline{B}Z$ are clearly classified as an element from Z and all customers in $\overline{B}Z$ are possible elements from Z with respect to $\mathcal{AT}$. Noticeably, RST enables an assessment of the relevance of each attribute with respect to d. The *reduct* is the set of relevant attributes, which is a subset of $\mathcal{AT}$. We assess the quality γ for the B-Lower and B-Upper approximation by $\underline{\gamma}_B(Z) = \frac{\sum_{i=1}^{r(d)} |\underline{B}Z_i|}{|\mathcal{U}|}$ and $\overline{\gamma}_B(Z) = \frac{\sum_{i=1}^{r(d)} |\overline{B}Z_i|}{|\mathcal{U}|}$. $\underline{\gamma}_B(Z)$ is the relation of correct classified customers to the number of all customers in the decision system and $\overline{\gamma}_B(Z)$ is the relation of the possibly classified customers to the number of all customers (*quantitative* approach [5]). The *qualitative* approach implies a refinement:

$$bel(X) = \frac{Dis[\underline{B} \bigcup_{i \in X} Z_i]}{Dis[\mathcal{U}]} \quad \text{and} \quad pl(X) = \frac{Dis[\overline{B} \bigcup_{i \in X} Z_i]}{Dis[\mathcal{U}]},$$

with $Dis[set]$ denoting the number of distinct customers according to a set of attributes.

3 Transferable Beliefs and Multi Sensor Clustering

The TBM allows us to allocate a belief higher than null to the empty set: $M(\emptyset) > 0$ [3]. Due to this, the definition for the belief and plausibility changes: $BEL(X) = \sum_{\mathcal{A}\subseteq X} M(\mathcal{A}) \forall X \in 2^{\Theta}, X \neq \emptyset$ and $PL(X) = \sum_{\mathcal{A}\cap X\neq\emptyset} M(\mathcal{A}) = BEL(\Theta) - BEL(\bar{X}) \forall X \in 2^{\Theta}$. In contrast to the DST both the belief and the plausibility include a proportion allocated to the empty set. Moreover, in the TBM two rules–the *conjunctive* and *disjunctive* rule–are considered: $M(X) = \sum_{\mathcal{A}\cap\mathcal{B}=X} M_1(\mathcal{A}) \cdot M_2(\mathcal{B}) \forall X \in 2^{\Theta}$ and $M(X) = \sum_{\mathcal{A}\cup\mathcal{B}=X} M_1(\mathcal{A}) \cdot M_2(\mathcal{B}) \forall X \in 2^{\Theta}$. The conjunctive rule is equivalent to Dempster's rule without the normalization term. The disjunctive rule is allocated from the product of elements of 2^{Θ} to the union of sets and not to the intersection. If two different sources of evidence are combined, the intersection frequently is a empty set. Therefore, the mass allocated to $M(\emptyset)$ expresses the conflict value CV for the fusion of the two sources of evidence.
In MSC the information recorded from each sensor is a source of evidence [4]. Thus, the main problem of decision making is the combination of different sources of evidence when the CV is high. The CV provides us with an indicator of reliability for the different sources. With an increasing CV the reliability of sensors analyzing the same object decreases. The MSC is implemented as follows: Let S be the number of sensors. We compute the CV of all $s = 2, \ldots, S$ combinations $\binom{S}{s}$ and use the set $\mathcal{S} \leftarrow \min_{s=2,\ldots,S} CV$.

4 Empirical Results

4.1 Data Set

In the empirical application we rely on a set transaction data recorded over three years from a customer loyalty program of a German grocery chain. The data cover a description of the card holders including demographic variables and Sinus Milieu assignments. Additionally, the data cover weekly advertising information as well as products featured in flyers. For this study the top 706 customers (buying in average at least once a week products of a category in a particular outlet) were selected from a total of 15,542 loyalty card holders.

4.2 Advertising Response Types

To answer the question whether or not the flyer triggers additional purchases, the buying history of each customer is split in two equal periods. By analyzing the first period we identify the products frequently bought by the customer (bought at least every second week). If these articles are bought by the same customer in a promotion period of the second period, we do not know whether the flyer triggered the purchase or not. But, if a customer buys an advertised article in the second period, which he never bought before, he clearly reacts to the flyer. As a result two advertising reaction types are defined. Type I ($a \in \Theta$): Customers did not buy a featured article. Thus, there is no reaction to the flyer. Type II ($b \in \Theta$): Customers bought a featured article and there is a reaction to the flyer. This decision variable and 15 attributes make up our decision system used to calculate the belief and plausibility functions with the formalism of the RST. Results from both the qualitative and the quantitative approach are depicted in Table 1. In contrast to conventional analysis (revealing only two attributes having statistical significant impact) it becomes obvious from the table, that indeed five attributes are needed to distinguish between the response types. Proceeding stepwise we identify those customers with a belief $m(\{b\}) > .35$ for type II. The highest belief value for the type II is obtained with the qualitative approach. Table 2 depicts the decision rules of the two qualitative solutions. By interpreting the

Table 1. Advertising response types

Reduct of attributes	Approach	$m(\{a\})$	$m(\{b\})$	$m(\{a,b\})$	$\lvert\mathcal{U}\rvert$
Age$^{(1)}$, Time$^{(2)}$, Day$^{(3)}$, SinusM$^{(4)}$, Volume$^{(5)}$	quan.	.248	.431	.321	706
	qual.	.342	.532	.126	461
Age$^{(1)}$, Time$^{(2)}$, SinusM$^{(4)}$, SVolume$^{(6)}$, BL$^{(7)}$	quan.	.218	.368	.414	706
	qual.	.329	.467	.204	370

$^{(1)}$age $^{(2)}$daytime $^{(3)}$weekday, $^{(4)}$Sinus Milieu, $^{(5)}$customers' annual purchases, $^{(6)}$annual category purchases, $^{(7)}$brand loyalty

Table 2. Decision Rules' Antecedents for Rules with Consequence: Type II

(1): (> 60) $\wedge$ (2): [8;12) $\wedge$ (3): [WE;TH] $\wedge$ (4): (TRAD$^{\alpha}$) $\wedge$ (5): (2500;3000]
(1): (40;50] $\wedge$ (2): [12;16) $\wedge$ (3): [WE;TH] $\wedge$ (4): (LEAD$^{\beta}$) $\wedge$ (5):(3000;3500]
(1): (> 60) $\wedge$ (2): [8;12) $\wedge$ (4): (TRAD$^{\alpha}$) $\wedge$ (6): (> 30) $\wedge$ (7): true
(1): (40;50] $\wedge$ (2): [12;16) $\wedge$ (4): (LEAD$^{\beta}$) $\wedge$ (6): (> 30) $\wedge$ (7): true

$^{(1)}$age $^{(2)}$daytime $^{(3)}$weekday, $^{(4)}$Sinus Milieu, $^{(5)}$customers' annual purchases, $^{(6)}$annual category purchases, $^{(7)}$brand loyalty
$^{\alpha}$ Traditional milieu, $^{\beta}$ Leadership milieu

rules solving these two decision systems, the characteristics of type II customers become tangible for marketing managers. For instance, the rule in the first line of Table 2 says that a customer aged more than 60 years, shopping in the time span between 8 and 12 a.m., shopping on Wednesdays or Thursdays, steaming from the traditional Sinus milieu, spending between 2.500 and 3000 Euros in the grocery outlet, and will react to featuring products in the flyer. Noticeably, the marketing managers need not consider customers individually, but are provided with a decision minimizing the CV of sensors for each customer.

4.3 Premium Customers

Mostly, the customers buy premium products from very few categories only or switch from low-priced brands when premium products are featured or are on sale. Thus, we distinguish three customer types. Type I: "Customers, who only buy products from the premium assortment." Type II: "...middle price segment." and Type III:"...lower..." We consider $\kappa = 9$ product categories (ketchup, pasta, bread, coffee, etc.) as sensors. The percentages of premium, middle-class, and low-price brands bought by the customers define the degree of beliefs according to the product categories. For each category we compute the *bpa* and combine them with the conjunctive rule. categories, the number of customers grasped by the analysis varies with the number of sensors and the grouping of sensors.

Table 3. Matching with varying sensors

Sensors	# Categories	Customers	$\overline{CV}$	std (CV)
2 Groups	4-1	171	.76	.26
	2-3	53	.68	.32
3 Groups	3-1-1	166	.43	.27
	2-2-1	58	.46	.29

The first line in Table 3 is read as follows: If only two groups of sensors are considered, with four categories assigned to the first group and one category assigned to the second group, a total of 171 from our 707 customers under consideration match this scheme. The arithmetic mean of conflict $\overline{CV} = .76$ is high and substantive with respect to the standard deviation of .26. Consequently, a conventional classification (e.g., by means of discriminant analysis using these $\kappa = 5$ categories) is likely to be unreliable. The main result of this analysis is the reduction of the conflict sum by incorporating additional groups of sensors and providing the sensors with a lower amount of information.

5 Conclusions

In this paper we introduce an innovative methodology to assign individual customers to groups of managerial relevance, i.e. premium buyers. This methodology roots in the Rough Set Theory, but adds the decision formalism from the Dempster-Shafer Theory and the Transferable Belief Model. This enables us to overcome the the problem of conflicting recommendations generated by evaluation different data describing the one and same customer. We consider both the qualitative and the quantitative approach of Multi-Sensor Clustering, but only the qualitative approach turned out to perform well in our empirical application. Moreover, using a data instance commonly used in retailing and direct marketing we found the conflict sum to decrease with the incorporation of additional sensors and providing the sensors lower amount of information. This result clearly needs to be validated, but might guide a venue to better decisions in customer relationship management and targeting of direct marketers.

References

1. Shafer G (1976) A Mathematical Theory of Evidence. Princeton University Press
2. Pawlak Z (1982) Rough Sets. Int. J. Computer Sci. 11:341–356
3. Smets P, Kennes R (1994) The Transferable Belief Model. Art. Intelligence. 66:191–234
4. Ayoun A, Smets P (2001) Data Association in Multi-Target Detection Using the Transferable Belief Model. Int. J. Int. Sys. 16:1167–11182
5. Klopotek MA, Wierzchon ST (2002) Empirical Models for the Dempster-Shafer Theory. In: Srivastava RP, Mock TJ(eds) Belief Functions in Business Decisions. Springer, Berlin Heidelberg New York 62–112
6. Wagner R, Scholz SW, Decker R (2005) The Number of Clusters in Market Segmentation. In: Baier D, Decker R, Schmidt-Thieme L (eds) Data Analysis and Decision Support. Springer, Berlin Heidelberg New York 157–176
7. Skowron A, Grzymala-Busse J (1994) From Rough Set Theory to Evidence Theory. In: Yager RR, Kacprzyk J, Fedrizzi M (eds) Advantages in the Dempster-Shafer Theory of Evidence. Wiley, New York 193–236

Part IX

Discrete and Combinatorial Optimization

A Constraint-Based Approach for the Two-Dimensional Rectangular Packing Problem with Orthogonal Orientations

Martin Berger, Michael Schröder, and Karl-Heinz Küfer

Fraunhofer Institute for Industrial Mathematics, Fraunhofer-Platz 1, 67663 Kaiserslautern, Germany
martin.berger@itwm.fraunhofer.de

Summary. We propose a constraint-based approach for the two-dimensional rectangular packing problem with orthogonal orientations. This problem is to arrange a set of rectangles that can be rotated by 90 degrees into a rectangle of minimal size such that no two rectangles overlap. It arises in the placement of electronic devices during the layout of 2.5D System-in-Package integrated electronic systems. Moffitt et al. [2] solve the packing without orientations with a branch and bound approach and use constraint propagation. We generalize their propagation techniques to allow orientations. Our approach is compared to a mixed-integer program and we provide results that outperform it.

1 Introduction

Rectangular packing problems occur in many real world applications and challenge researchers from operations research, constraint programming, artificial intelligence and many more. We address the two-dimensional rectangular packing problem with orthogonal orientations (RPWO) which is to arrange rectangles that can be rotated by 90 degrees into a rectangular container such that no two overlap. Minimizing the container size is an $\mathcal{NP}$-hard combinatorial optimization problem. Such a problem arises in 2.5D System-in-Package (SiP) layout design [3]. 2.5D SiP is a modern integration approach of heterogeneous electronic components on modules which are stacked or folded vertically [4]. The component placement on each module involves a RPWO. A multitude of approaches from metaheuristics, linear, mixed integer, nonlinear and constraint programming have been developed for rectangular packing. However, the orientation is often disregarded and most of the approaches can not integrate important constraints arising in SiP design. Therefore, we extend the meta-constraint satisfaction problem (CSP) approach of Moffitt et al. [2]. Their branch and bound solves

DOI: 10.1007/978-3-642-00142-0_69,

minimal area packing and uses constraint propagation. We generalize the propagation algorithms for orientations and show the relation of the meta-CSP approach to mixed-integer programming (MIP). We compare our extended meta-CSP to an MIP and provide numerical results that outperform it. In conclusion, our approach solves RPWO and can be extended to address the wiring of SiP components in order to serve as algorithm of an SiP design automation tool.

2 Problem Formulation

We denote $\mathcal{R} := \{r_1, \ldots, r_n\}$ as rectangle set with index set $\mathcal{I} := \{1, \ldots, n\}$; $w_i, h_i \in \mathbb{N}$ represent the width and height, $x_i, y_i \in \mathbb{N}_0$ the coordinates of the lower left corner and $o_i \in \{0, 1\}$ models the orientation of rectangle r_i; $W, H \in \mathbb{N}$ represent the width and height of the container with upper bounds $W_{\max}, H_{\max}$. We formulate RPWO as minimal half perimeter packing problem with linear objective $f := W + H$:

$$\begin{aligned} &\min f && \text{subject to} && \text{(RPWO)} \\ &x_i + s_i^x \leq W, && W \leq W_{\max}, && (1) \\ &y_i + s_i^y \leq H, && H \leq H_{\max}, && (2) \\ &(1 - o_i)w_i + o_i h_i = s_i^x, && o_i w_i + (1 - o_i)h_i = s_i^y, \quad \forall i \in \mathcal{I}, && (3) \\ &(x_i + s_i^x \leq x_j) \vee (x_j + s_j^x \leq x_i) && && (4) \\ &\vee\, (y_i + s_i^y \leq y_j) \vee (y_j + s_j^y \leq y_i), && \forall i, j \in \mathcal{I}, i < j. && \end{aligned}$$

(1-2) ensure the rectangle containment, (3) define the size of the oriented rectangles and (4) make sure that no two overlap by arranging them left (d_{iLj}), right (d_{iRj}), below (d_{iBj}) or above (d_{iAj}) of each other.

3 Meta-CSP Model

In [2] minimal area rectangular packing with fixed orientations is approached. Instead of searching x_i, y_i, meta-variables C_{ij} are introduced for each non-overlapping constraint. C_{ij} ranges in domain $D(C_{ij}) := \{d_{iLj}, d_{iRj}, d_{iBj}, d_{iAj}\}$ and represents the geometric relation between r_i and r_j. The search tree is branched over the C_{ij} and pruned with constraint propagation. Propagation uses incrementally maintained graphs that describe a partial solution. Figure 1 shows a complete solution and the corresponding packing of the rectangles. A partial packing for a subset $\mathcal{S}$ of $\mathcal{R}$ is described by two sets of inequalities $u + s \leq v$, $\mathcal{C}_h$ for the horizontal and $\mathcal{C}_v$ for the vertical geometric relations. $\mathcal{C}_h$ and $\mathcal{C}_v$ are encoded in two weighted directed graphs G_h and G_v that represent the left and below precedences of r_i and r_j. The weights of the edges from

r_j to r_i are given by the size $-s$ of r_i.[1]. Figure 2 shows G_h and G_v for the example in figure 1.

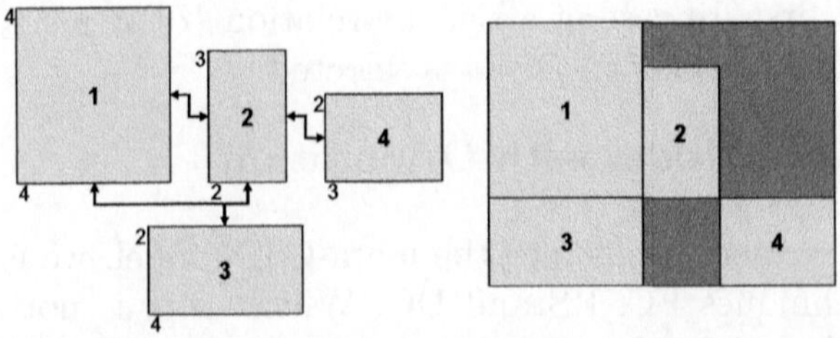

Fig. 1. Example for a complete meta-CSP solution and its packing.

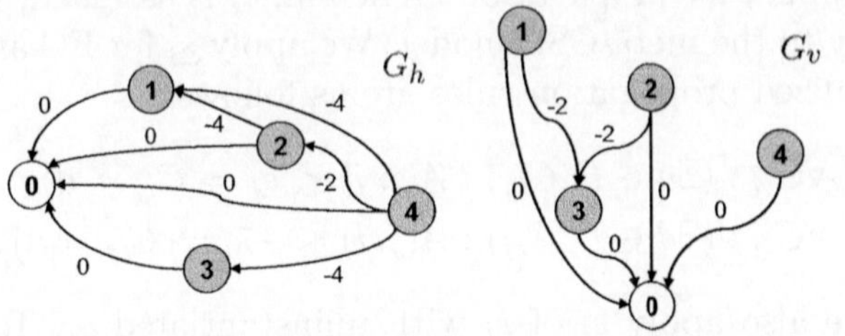

Fig. 2. Precedence graphs for the example in figure 1.

Then, $\mathcal{C}_h$ ($\mathcal{C}_v$) is consistent iff G_h (G_v) has no negative cycle. Negative cycles can be detected with the all-pairs shortest path matrix A_h (A_v) of G_h (G_v). A_h and A_v are updated with Floyd-Warshall in $O(n^3)$. With G_h, G_v, A_h, A_v propagation in $O(1)$ is applied during search:

- Forward checking (FC) removes inconsistent values of C_{ij} with respect to a partial assignment. To check if value $\{d : u + s \leq v\} \in D(C_{ij})$ is consistent, $A(i, j)$ must not be less than s.
- Removal of subsumed variables (RS) assigns C_{ij} transitively implied by C_{ik} and C_{kj}. A value $\{d : u + s \leq v\} \in D(C_{ij})$ of C_{ij} is satisfied iff the shortest path from r_j to r_i is smaller than $-s$.

Detecting cliques of displacement (DC) is another pruning technique. Partial solutions with cliques consisting of pairwise, horizontally or vertically, aligned rectangles whose alignment exceeds the container, lead to a dead-end. A greedy heuristic is applied to a horizontal and vertical displacement graph to detect such cliques early. Furthermore, to avoid symmetric solutions, symmetry breaking before and during the search is applied in the branch and bound of [2]. A dynamic most

[1] W.l.o.g., the edges are weighted with the negative sizes of the rectangles in order to illustrate the graph algorithms canonically.

constrained variable first heuristic orders C_{ij} of large rectangle pairs first. Those relations of C_{ij} that require the minimal increase in area are selected first. In case of a tie, the relation $\{d : u + s \leq v\}$ with the least amount of slack $A(i,j) - s$ is selected.

4 Meta-CSP Model with Orientation

To introduce orientations into the meta-CSP model, we generalize the pruning techniques FC, RS and DC. When an o_i is not assigned, we either use the minimal $\underline{s}_i = \min(w_i, h_i)$ or maximal side lengths $\overline{s}_i = \max(w_i, h_i)$ of r_i for propagation. When an o_i is assigned, we use s_i^x, s_i^y analogously to the meta-CSP model. We apply $\underline{s}_i$ for FC and $\overline{s}_i$ for RS. The generalized propagation rules are as follows:

$$\forall C_{ij} : (\exists d \in D(C_{ij}) : A(i,j) < \underline{s}_i \Rightarrow C_{ij} \neq d)\,, \tag{5}$$

$$\forall C_{ij} : (\exists d \in D(C_{ij}) : A(j,i) \leq -\overline{s}_i \Rightarrow C_{ij} = d)\,. \tag{6}$$

For DC, we also apply $\underline{s}_i$ of r_i with uninstantiated o_i. To strengthen propagation we propose new pruning techniques that incorporate the orientations. We assign o_i of r_i which only fits into the container with a certain o_i. The rules for exceeding $W_{\max}$ are as follows, analogous rules follow for the lower bound $\underline{y}_i$ and $H_{\max}$:

$$\forall i \in \mathcal{I} : (\underline{x}_i + h_i > W_{\max} \Rightarrow o_i = 0)\,, \tag{7}$$

$$(\underline{x}_i + w_i > W_{\max} \Rightarrow o_i = 1)\,. \tag{8}$$

In a similar way we assign o_i of r_i which can only precede r_j with a certain o_i. The rule for the left precedence (d_{iLj}) is as follows, analogous rules follow for the other geometric relations:

$$\forall i,j \in \mathcal{I}, i < j : (C_{ij} = d_{iLj} \wedge \underline{x}_i + h_i > \overline{x}_j \Rightarrow o_i = 0)\,, \tag{9}$$

$$(C_{ij} = d_{iLj} \wedge \underline{x}_i + w_i > \overline{x}_j \Rightarrow o_i = 1)\,. \tag{10}$$

Conversely, an instantiated o_i is propagated on $\overline{x}_i, \overline{y}_i$:

$$o_i = 0 \Rightarrow \overline{x}_i \leftarrow \min\left(\overline{x}_i, W_{\max} - w_i\right), \tag{11}$$

$$\overline{y}_i \leftarrow \min\left(\overline{y}_i, H_{\max} - h_i\right).$$

Again, an analogous rule is applied for the instantiation $o_i = 1$. Furthermore, we propose a symmetry breaking technique which imposes a lex-leader constraint on equal sized rectangles by removing one horizontal and one vertical geometric relation of the corresponding C_{ij}.

In addition to the meta-variables C_{ij} we have to search the orientations o_i. With our generalization we can instantiate C_{ij} and o_i order-independently. We instantiate o_i and o_j right after the meta-variable C_{ij} has been assigned. This ordering only marginally weakens propagation and leads to smaller search trees than ordering the o_i first. To strengthen our exhaustive search method we initialize it with an upper bound on f. We obtain the bound from a feasible solution constructed by a greedy best-fit packing heuristic.

5 Mixed-Integer Formulation

We formulate RWPO as a mixed-integer program. Therefore, we have to resolve the disjunctive non-overlapping constraints (4) through a big-M relaxation and auxiliary variables $z_{ij}^k \in \{0,1\}, k = 1, \ldots, 4$:

$$
\begin{aligned}
x_i + s_i^x &\leq x_j + (1 - z_{ij}^1) W_{\max}, & y_i + s_i^y &\leq y_j + (1 - z_{ij}^3) H_{\max}, \\
x_j + s_j^x &\leq x_i + (1 - z_{ij}^2) W_{\max}, & y_j + s_j^y &\leq y_i + (1 - z_{ij}^4) H_{\max}, \\
1 &\geq z_{ij}^1 + z_{ij}^2, & 1 &\geq z_{ij}^3 + z_{ij}^4, \\
1 &\leq z_{ij}^1 + z_{ij}^2 + z_{ij}^3 + z_{ij}^4. & &
\end{aligned}
$$

By definition, this MIP is similar to the meta-CSP: $C_{ij} = d_{iLj}$ corresponds to $z_{ij}^1 = 1$, $C_{ij} = d_{iRj}$ to $z_{ij}^2 = 1$, $C_{ij} = d_{iBj}$ to $z_{ij}^3 = 1$ and $C_{ij} = d_{iAj}$ corresponds to $z_{ij}^4 = 1$.

Theorem 1 (Proved in [1]). *For RWPO with fixed orientations is the value of f for the linear relaxation of the MIP model equal to the lower bound of f in the meta-CSP model.*

6 Numerical Results

We implemented our extended meta-CSP approach with the generalized techniques in Ilog Solver 6.5 and the MIP model in Ilog Cplex 11.1. For a fair comparison we applied the same symmetry breaking in both models. We tested the models on problem instances of size $n = 5, \ldots, 27$ with rectangles whose sizes are inspired by electronic devices typically found in SiPs. The test instances can be found in [1]. We impose a runtime limit of 600 CPU seconds.

Table 1 shows that our approach outperforms the MIP. This is due to good initial upper bounds for f from the greedy heuristic and constraint propagation which shrinks the search tree. Our approach is significantly faster than the MIP which already fails for $n \geq 10$ to find optimal

Table 1. Runtimes (less than 600 CPU sec.) to prove optimality.

	Runtime CPU sec.	
n	meta-CSP	MIP
5	0,07	0,03
6	0,11	0,44
7	0,34	9,42
8	5,34	175,98
9	0,31	19,69
10	7,23	–
11	–	–
12	–	–
13	125,69	–
14	–	–
15	349,98	–

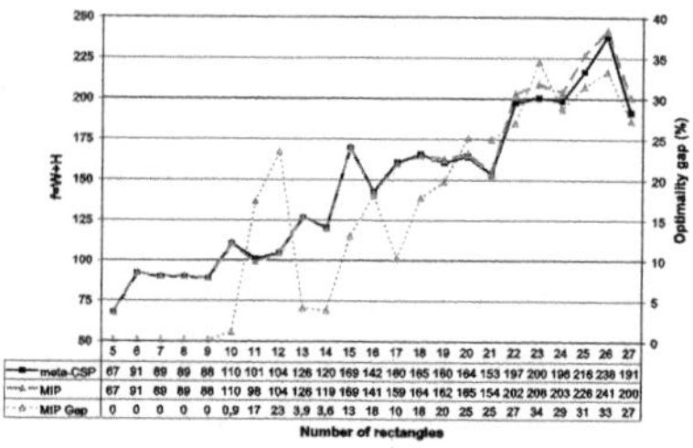

Fig. 3. Values for f and optimality gap for the MIP.

solutions. Our approach also solves to optimality for $n = 10, 13, 15$. The solution quality is comparable for $n \leq 21$ but solutions of our approach are considerably better for $n \geq 22$ (see figure 3).

7 Conclusion

We extend the meta-CSP approach to RWPO where orientations are allowed. Therefore, we are flexible enough to instantiate the C_{ij} and o_i in any order and propagate between C_{ij}, o_i, x_i and y_i whenever possible. We show that the meta-CSP approach is similar to an MIP formulation but produces a smaller search tree due to constraint propagation.
In future work we will introduce a second objective for the wiring of the SiP components in order to use it for SiP design automation.

Acknowledgement. This work originates from the PhD activities of Martin Berger and is funded by the Fraunhofer ITWM.

References

1. M. Berger, M. Schröder, and K.-H. Küfer. A constraint programming approach for the two-dimensional rectangular packing problem with orthogonal orientations. Technical Report 147, Fraunhofer Institute for Industrial Mathematics, Oktober 2008.
2. M. D. Moffitt and M. E. Pollack. Optimal rectangle packing: A meta-CSP approach. In Proceedings of the 16th International Conference on Automated Planning and Scheduling, 2006.
3. D. D. Polityko. Physikalischer Entwurf für die vertikale SiP Integration. PhD thesis, Berlin Institute of Technology, June 2008.
4. C. Richter, D. D. Polityko, J. Hefer, S. Guttowski, H. Reichl, M. Berger, U. Nowak, and M. Schröder. Technology aware modeling of 2.5D-SiP for automation in physical design. In Proceedings of the 9th Electronics Packaging Technology Conference, pages 623-630, December 2007.

Detecting Orbitopal Symmetries

Timo Berthold and Marc E. Pfetsch

Zuse Institute Berlin, Germany
{berthold,pfetsch}@zib.de

Symmetries are usually not desirable in integer programming (IP) models, because they derogate the performance of state-of-the-art IP-solvers like SCIP [1]. The reason for this is twofold: Solutions that are equivalent to ones already discovered are found again and again, which makes the search space "unnecessarily large". Furthermore, the bounds obtained from the linear programming (LP) relaxations are very poor, and the LP-solution is almost meaningless for the decision steps of the IP-solving algorithm. Overall, IP-models mostly suffer much more from inherent symmetries than they benefit from the additional structure.
Margot [4, 5] and Ostrowski et al. [7, 8] handle symmetries in general IPs, without knowing the model giving rise to the instance.
Kaibel et al. [2, 3] took a polyhedral approach to deal with special IP-symmetries. They introduced orbitopes [3], which are the convex hull of 0/1-matrices of size $p \times q$, lexicographically sorted with respect to the columns. For the cases with at most or exactly one 1-entry per row, they give a complete and irredundant linear description of the corresponding orbitopes. These orbitopes can be used to handle symmetries in IP-formulations in which assignment structures appear, such as graph coloring problems; see the next section for an example.
All of the above approaches assume that the symmetry has been detected in advance or is known. Therefore, automatic detection of symmetries in a given IP-formulation is an important task.
In this paper, we deal with the detection of orbitopal symmetries that arise in the orbitopal approach. While this problem is polynomially equivalent to the graph automorphism problem, whose complexity is an open problem, orbitopal symmetries can be found in linear time, if at least the assignment structure is given. Otherwise we show that the problem is as hard as the graph automorphism problem.

B. Fleischmann et al.(Eds.), *Operations Research Proceedings 2008*,
DOI: 10.1007/978-3-642-00142-0_70, © Springer-Verlag Berlin Heidelberg 2009

1 Symmetries in Binary Programs

In this section, we introduce symmetries in binary programs and their detection by color-preserving graph automorphisms.
For any $k \in \mathbb{N}$, let $[k]$ denote the set $\{1, \ldots, k\}$. Let $m, n \in \mathbb{N}$, $A \in \mathbb{R}^{m \times n}$, $b \in \mathbb{R}^m$, and $c \in \mathbb{R}^n$. We deal with *binary programs (BPs)* in the following form:

$$\begin{aligned} \min \quad & c^T x \\ \text{s.t.} \quad & Ax \le b \\ & x_j \in \{0,1\} \qquad \text{for all } j \in [n]. \end{aligned} \tag{1}$$

Without loss of generality, we assume that there is no zero row in A and that no two rows in A are positive multiples of each other. Let N denote the number of nonzero entries of A. For $k \in \mathbb{N}$, let $\mathfrak{S}(k)$ denote the full symmetric group of order k.
For $x \in \mathbb{R}^n, \sigma \in \mathfrak{S}(n)$, we write $\sigma(x)$ for the vector which is obtained by permuting the components of x according to σ, i.e., $\sigma(x)_i = x_{\sigma(i)}$. For $\sigma \in \mathfrak{S}(m)$ and $\pi \in \mathfrak{S}(n)$, we write $A(\sigma, \pi)$ for the matrix which is obtained by simultaneously permuting the rows of A according to σ and the columns of A according to π.
Let $\mathcal{F} \subseteq \{0,1\}^n$ be the set of feasible solutions of BP (1). If there is a permutation $\sigma \in \mathfrak{S}(n)$ such that $x \in \mathcal{F}$ if and only if $\sigma(x) \in \mathcal{F}$, then σ is called a *symmetry* of $\mathcal{F}$. Obviously, the set of all symmetries of $\mathcal{F}$ is a subgroup of $\mathfrak{S}(n)$. Clearly, it is $\mathcal{NP}$-complete to determine whether a binary program has a non-trivial symmetry group.
To avoid this complexity, one concentrates on finding symmetries of the BP-formulation. Focusing on the BP, however, implies that the symmetry depends on the problem formulation. We give a formal definition of symmetry groups, which is similar to the one of Margot [4, 5].

Definition 1. *A subgroup $\mathfrak{G}$ of the full symmetric group $\mathfrak{S}(n)$ is a symmetry group of BP (1), if and only if there is a subgroup $\mathfrak{H}$ of $\mathfrak{S}(m)$, s.t. the following conditions hold for all elements $\pi \in \mathfrak{G}$:*

(i) $\pi(c) = c$,
(ii) there exists $\sigma \in \mathfrak{H}$ s.t. $\sigma(b) = b$ and $A(\sigma, \pi) = A$.

We reduce the problem of finding symmetries in an BP to a graph automorphism problem. Let $V_{\text{row}} := \{u_1, \ldots, u_m\}$, $V_{\text{col}} := \{v_1, \ldots, v_n\}$, $V := V_{\text{col}} \cup V_{\text{row}}$, $E := \{\{u_i, v_k\} \in V_{\text{row}} \times V_{\text{col}} \mid a_{ik} \neq 0\}$, and $G = (V, E)$. Note that G is bipartite and that the size of G is in $O(N)$.
The coefficients of the BP are treated as follows. We introduce a color $\gamma(r) \in \mathbb{N}$ for each value in the set $\{b_1, \ldots, b_m\}$. We assign the color $\gamma(r)$

to all vertices $u_i \in V_{\text{row}}$ with $b_i = r$. The same is done for the objective c and the "variable vertices" V_{col}.
We proceed analogously for the matrix coefficients. If at least one $(i,j) \in [m] \times [n]$ with $a_{ij} = r$ exists, color $\hat{\gamma}(r)$ is assigned to all edges $\{u_i, v_j\}$ with $a_{ij} = r$. We call the edge- and vertex-colored graph G the *coefficient graph* of BP (1).

Definition 2. *Given a graph $G = (V, E)$, a mapping $\zeta \colon V \mapsto V$ is called an automorphism of G, if it preserves adjacency: $\{u,v\} \in E \Leftrightarrow \{\zeta(u), \zeta(v)\} \in E$. If for a given vertex and edge coloring, the colors of all vertices and edges stay the same under ζ, we call the automorphism* color-preserving.

For two different graphs G and $\hat{G}$, a *color-preserving isomorphism* is defined analogously.
Note that it is neither known whether the decision problem "Is there a non-trivial automorphism of G?" can be solved in polynomial time nor whether it is $\mathcal{NP}$-complete. Nevertheless, there are codes, such as nauty [6], which solve practically relevant graph automorphism and isomorphism instances within reasonable time. Note that computing color-preserving graph automorphisms is polynomially equivalent to computing graph automorphisms. Symmetry detection can be reduced to finding color-preserving graph automorphisms:

Proposition 1. *Every symmetry of a binary program induces a color-preserving automorphism of its coefficient graph and vice versa.*

In the following, we want to concentrate on symmetries, which arise from permuting blocks of variables as in the orbitope approach. As an example consider the *maximal k-colorable subgraph problem.* Given a graph G and a number $k \in \mathbb{N}$, the task is to find a subset of vertices $\hat{V} \subseteq V$ such that the subgraph induced by $\hat{V}$ is k-colorable. The standard BP-model for the maximal k-colorable subgraph problem uses binary variables x_{vc} that determine whether color c is assigned to vertex v:

$$\begin{array}{lrl}
\max & \sum_{v \in V} \sum_{c \in [k]} x_{vc} & \\
\text{s.t.} & \sum_{c \in [k]} x_{vc} \leq 1 & \text{for all } v \in V \\
& x_{uc} + x_{vc} \leq 1 & \text{for all } \{u,v\} \in E \text{ and } c \in [k] \\
& x_{vc} \in \{0,1\} & \text{for all } v \in V \text{ and } c \in [k].
\end{array}$$

For a given k-colorable subgraph, permuting the colors in $[k]$ yields an orbit of $k!$ structurally identical solutions. If we consider x as a 0/1-matrix of size $|V| \times k$, permuting color classes corresponds to permuting

columns—"blocks of variables"—of the matrix x. Each column/block consists of $|V|$ variables x_{vc} belonging to a particular color c.
We generalize this structure as follows: Let $q \in \mathbb{N}$ divide n and $\mathcal{C} := \{C_1, \dots, C_q\}$ be a partition of the column index set of A into q distinct subsets of the same cardinality. We call C_j a *variable block.*
Groups acting on variable blocks, like $\mathfrak{S}(k)$ in the k-colorable subgraph example, are called orbitopal symmetries. More precisely, the group $\mathfrak{S}(q)$ is called an *orbitopal symmetry* of BP (1), if for all $j, \hat{\jmath} \in [q]$ there exists a bijection $\pi_{j\hat{\jmath}}\colon C_j \to C_{\hat{\jmath}}$ such that $c_k = c_{\pi_{j\hat{\jmath}}(k)}$ for all $k \in C_j$, and for every row $i \in [m]$

$$\sum_{k \in C_j} a_{ik} x_k \qquad + \sum_{k \in C_{\hat{\jmath}}} a_{ik} x_k \qquad + \sum_{\ell \neq j, \hat{\jmath}} \sum_{k \in C_\ell} a_{ik} x_k \quad \leq b_i \tag{2}$$

there exists a row $\hat{\imath} \in [m]$ that has the form

$$\sum_{k \in C_j} a_{i\pi_{j\hat{\jmath}}(k)} x_k \quad + \sum_{k \in C_{\hat{\jmath}}} a_{i\pi_{j\hat{\jmath}}^{-1}(k)} x_k \quad + \sum_{\ell \neq j, \hat{\jmath}} \sum_{k \in C_\ell} a_{ik} x_k \quad \leq b_{\hat{\imath}}. \tag{3}$$

Let $j, \hat{\jmath} \in [q]$ and $\sigma_{j\hat{\jmath}}\colon [m] \to [m]$, $i \mapsto \hat{\imath}$ be the mapping which links the rows of A. Note that, since there are no identical rows, $\sigma_{j\hat{\jmath}} = \sigma_{\hat{\jmath}j}^{-1}$, in particular $\sigma_{jj} = \mathrm{id}$. For the k-colorable subgraph problem, there is an orbitopal symmetry acting on the blocks of variables associated to a common color.
The set of maps $\pi : [n] \to [n]$ defined by $\pi_{j\hat{\jmath}}$ on C_j, $\pi_{j\hat{\jmath}}^{-1}$ on $C_{\hat{\jmath}}$, and the identity on the remaining elements forms a symmetry of BP (1). Indeed, condition (i) of Definition 1 is fulfilled and the requirements (2) and (3) show that condition (ii) holds as well.

2 Complexity of Detecting Orbitopal Symmetries

In the following, we want to describe a polynomial time algorithm, which is able to verify whether a partition $\mathcal{C}$ induces an orbitopal symmetry of an BP without having knowledge of the mappings π, σ.

Definition 3. *Let $S \subseteq [n]$, $\ell \in \mathbb{N}$, $\bowtie \in \{\leq, =, \geq\}$.*

(i) A linear constraint of the form $\sum_{k \in S} x_k \bowtie \ell$ is called a leading constraint *of $\mathcal{C}$, if it stays invariant under the mappings $\pi_{j\hat{\jmath}}$ and contains exactly one variable from each variable block C_j.*
(ii) We call a set $\mathcal{S} := \{S_1, \dots, S_p\}$ of leading constraints a leading system *of $\mathcal{C}$ if every variable is contained in exactly one S_i.*

In the above definition, we identify a leading constraint with its set of variable indices. Note that for a leading system $n = pq$ holds.
For the maximal k-colorable subgraph problem, the set packing constraints $\sum_{c \in [k]} x_{vc} \leq 1$ form a leading system. The leading constraints determine the orbit of a variable under the symmetry $\mathfrak{S}(q)$.

Proposition 2. *Given a set of variable blocks $\mathcal{C}$ and a leading system $\mathcal{S}$ of $\mathcal{C}$, verifying that they describe an orbitopal symmetry of the binary program (1) is possible in $O(qmN)$ time.*

Proof. Let $j, \hat{j} \in [q]$ and $i \in [p]$. Let $\{k\} = S_i \cap C_j$ and $\{\hat{k}\} = S_i \cap C_{\hat{j}}$. Recall that S_i is invariant under $\pi_{j\hat{j}}$, which means that $\pi_{j\hat{j}}(k) = \hat{k}$. Hence, constructing the maps $\pi_{j\hat{j}}$ elementwise is possible in $O(N)$ time. For every constraint (2) of BP (1), row (3) describes its image under $\pi_{j\hat{j}}$. If this image constraint does not exist in BP (1), $\mathcal{C}$ and $\mathcal{S}$ do not yield a symmetry of the BP.
Searching the image row is possible in $O(N)$ time. This search has to be performed for all rows. Hence, checking a variable block pair can be achieved in $O(mN)$ time. It suffices to only check block pairs $\{1, j\}$, since $\pi_{j\hat{j}} = \pi_{1\hat{j}} \circ \pi_{1j}^{-1}$. We get an overall running time of $O(qmN)$. □

The next lemma is tailored towards the typical case, in which removing the leading constraints decomposes the BP into blocks. Let $G(\mathcal{S})$ be the coefficient graph of BP (1) without the leading constraints $\mathcal{S}$.

Lemma 1. *Let be $\mathfrak{S}(q)$ be an orbitopal symmetry of BP (1). If for all rows, which are not leading constraints, all non-zeros are within one variable block, then $G(\mathcal{S})$ is partitioned into q components which are pairwise color-preserving isomorphic.*

Note that these components do not have to be connected.

Theorem 1. *For a given leading system $\mathcal{S}$, detecting the corresponding orbitopal symmetry is possible in $O(N)$ time.*
If there are no leading constraints, detecting orbitopal symmetries is as hard as the graph isomorphism problem.

Proof. Following Lemma 1, we detect the orbitopal symmetry by determining the connected components of the graph $G(\mathcal{S})$. Using a breadth-first-search, this takes $O(|V| + |E|) = O(N)$ time. For checking that all components are pairwise isomorphic, it is sufficient to show that all components are isomorphic to the first component. Let $j \in [q] \setminus \{1\}$. The mappings π_{1j} can be constructed in $O(n)$ time as in the proof of Proposition 2 by evaluating $\mathcal{S}$.

For testing whether the π_{1j} describe color-preserving isomorphisms, each edge of the graph has to be regarded. This takes $O(N)$ time.
The reduction to graph isomorphism is achieved as follows. Let two graphs G and $\hat{G}$ with the same number of vertices and edges be given. For each vertex v in one of the graphs, we introduce a distinct binary variable x_v and for each edge $\{u, v\}$, we introduce a set packing constraint $x_u + x_v \leq 1$. The variables are partitioned into two disjoint blocks C_U, C_V. The BP has an orbitopal symmetry arising from the blocks C_U and C_V, if and only if G and $\hat{G}$ are isomorphic. □

One can show that the detecting leading constraints within a BP is also polynomially equivalent to a graph isomorphism problem.
Finally, we want to investigate the case in which the leading constraints do not yield a complete system. As an example, think of a graph coloring model, which uses variables x_{vc} to assign color c to vertex v, connected by leading constraints $\sum_c x_{vc} = 1$, which ensure that each vertex is colored exactly once. It uses one additional variable y_c per block, which indicates whether color c is used or not. This case can be handled by the following result, which we state without proof:

Corollary 1. *If there is a constant number of variables per block, which are not contained in any leading constraint, detecting orbitopal symmetries is possible in $O(N)$ time.*

References

1. T. Achterberg. Constraint Integer Programming. PhD thesis, Technische Universität Berlin, 2007.
2. V. Kaibel, M. Peinhardt, and M. E. Pfetsch. Orbitopal fixing. In M. Fischetti and D. Williamson, editors, Proc. of the 12th IPCO, volume 4513 of LNCS, pages 74-88. Springer-Verlag, 2007.
3. V. Kaibel and M. E. Pfetsch. Packing and partitioning orbitopes. Mathematical Programming, 114(1):1-36, 2008.
4. F. Margot. Pruning by isomorphism in branch-and-cut. Mathematical Programming, Series A, 94:71-90, 2002.
5. F. Margot. Symmetric ILP: Coloring and small integers. Discrete Optimization, 4:40-62, 2007.
6. B. McKay. nauty User's Guide (version 2.4), 2007.
7. J. Ostrowski, J. Linderoth, F. Rossi, and S. Smirglio. Orbital branching. In M. Fischetti and D. Williamson, editors, Proc. of the 12th IPCO, volume 4513 of LNCS, pages 104-118. Springer-Verlag, 2007.
8. J. Ostrowski, J. Linderoth, F. Rossi, and S. Smirglio. Constraint orbital branching. In A. Lodi, A. Panconesi, and G. Rinaldi, editors, Proc. of the 13th IPCO, volume 5035 of LNCS, pages 225-239. Springer-Verlag, 2008.

Column Generation Approaches to a Robust Airline Crew Pairing Model For Managing Extra Flights*

Elvin Çoban[1], İbrahim Muter[1], Duygu Taş[1], Ş. İlker Birbil[1], Kerem Bülbül[1], Güvenç Şahin[1], Y. İlker Topçu[2], Dilek Tüzün[3], and Hüsnü Yenigün[1]

[1] Sabancı University, Orhanlı-Tuzla, 34956 Istanbul, Turkey
{elvinc,imuter,duygutas}@su.sabanciuniv.edu,
{sibirbil,bulbul,guvencs,yenigun}@sabanciuniv.edu
[2] Istanbul Technical University, Maçka-Beşiktaş, 34367 Istanbul, Turkey
ilker.topcu@itu.edu.tr
[3] Yeditepe University, Kayışdağı-Kadıköy, 34755 Istanbul, Turkey
dtuzun@yeditepe.edu.tr

1 Introduction

The airline crew pairing problem (CPP) is one of the classical problems in airline operations research due to its crucial impact on the cost structure of an airline. Moreover, the complex crew regulations and the large scale of the resulting mathematical programming models have rendered it an academically interesting problem over decades. The CPP is a tactical problem, typically solved over a monthly planning horizon, with the objective of creating a set of crew pairings so that every flight in the schedule is covered, where a crew pairing refers to a sequence of flights operated by a single crew starting and ending at the same crew base.

This paper discusses how an airline may hedge against a certain type of operational disruption by incorporating robustness into the pairings generated at the planning level. In particular, we address how a set of extra flights may be added into the flight schedule at the time of operation by modifying the pairings at hand and without delaying or canceling the existing flights in the schedule. We assume that the set of potential extra flights and their associated departure time windows are

* This research has been supported by The Scientific and Technological Research Council of Turkey under grant 106M472.

DOI: 10.1007/978-3-642-00142-0_71, © Springer-Verlag Berlin Heidelberg 2009

known at the planning stage. We note that this study was partially motivated during our interactions with the smaller local airlines in Turkey which sometimes have to add extra flights to their schedule at short notice, e.g., charter flights. These airlines can typically estimate the potential time windows of the extra flights based on their past experiences, but prefer to ignore this information during planning since these flights may not need to be actually operated. Typically, these extra flights are then handled by recovery procedures at the time of operation which may lead to substantial deviations from the planned crew pairings and costs. The reader is referred to [3] for an in-depth discussion of the conceptual framework of this problem which we refer to as the Robust Crew Pairing for Managing Extra Flights (RCPEF). In [3], the authors introduce how an extra flight may be accommodated by modifying the existing pairings and introduce a set of integer programming models that provide natural recovery options without disrupting the existing flights. These recovery options are available at the planning stage and render operational recovery procedures that pertain to crew pairing unnecessary.

The main contribution of this work is introducing a column generation algorithm that can handle the robust model proposed in the next section. This model poses an interesting theoretical challenge and is not amenable to a traditional column generation algorithm designed for the conventional CPP. We point out that in [3] the authors explicitly generate all possible crew pairings and solve the proposed integer programs by a commercial solver. This approach is clearly not computationally feasible for large crew pairing instances, and in the current work we present our preliminary algorithms and results for large instances of RCPEF. We demonstrate the proposed solution approaches on a set of actual data acquired from a local airline [2].

2 Robust Airline Crew Pairing Problem

In this section, we first introduce the proposed robust model and then discuss the difficulties that arise while solving this model by conventional methods. This leads us to the two solution approaches presented in this paper.

In [3], the authors examine several recovery options for managing the extra flights at the planning level. They classify the possible solutions into two types:

- **Type A.** Two pairings are selected and (partially) swapped to cover an extra flight.

- **Type B.** One pairing with sufficient connection time between two consecutive legs is modified to cover an extra flight.

In this work, we incorporate one Type A and two Type B solutions as illustrated in Figure 1 where the estimated time window of the extra flight k is depicted by the shaded rectangles. In Figure 1(a), the original pairings p and q, covering the flight legs i_1, i_2 and j_1, j_2, respectively, are partially swapped so that the extra flight k is inserted into the flight schedule (Type A). The resulting pairings after swapping are illustrated in the figure where the term deadhead refers to a repositioning of crew members to another airport as passengers on a flight, train, etc. In Figures 1(b) and 1(c), the original pairing p is modified to accommodate the extra flight k (Type B). The feasibility rules that define both Type A and B solutions are explained in detail in [3].

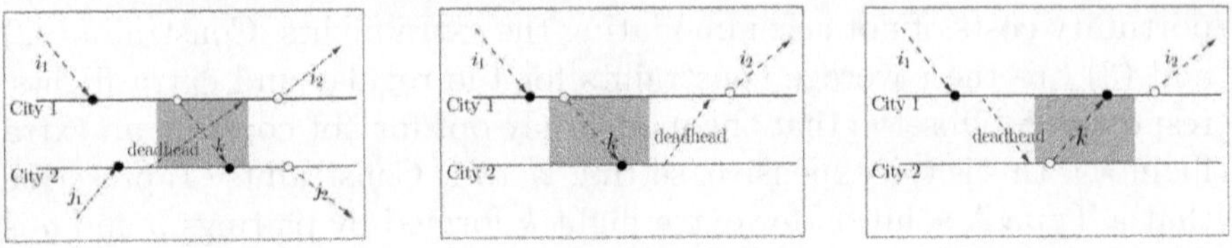

(a) Pairings p,q are swapped. (b) Pairing p is modified. (c) Pairing p is modified.

Fig. 1. Recovery options for covering the extra flight k.

The proposed robust mathematical model is given below:

$$\min \sum_{p\in\mathcal{P}} c_p y_p + \sum_{k\in\mathcal{K}} d_k z_k +$$

$$\sum_{k\in\mathcal{K}} d_k \left[\sum_{p\in\mathcal{P}} (1-y_p)\bar{a}_{kp} + \sum_{p,q\in\mathcal{P}} (1-x^k_{(p,q)})\bar{a}_{pqk} \right] \tag{1}$$

s.t

$$\sum_{p\in\mathcal{P}} a_{ip} y_p \geq 1, \qquad \forall i \in \mathcal{F}, \tag{2}$$

$$\sum_{p\in\mathcal{P}} \bar{a}_{kp} y_p + \sum_{p,q\in\mathcal{P}} \bar{a}_{pqk} x^k_{(p,q)} \geq 1 - z_k, \qquad \forall k \in \mathcal{K}, \tag{3}$$

$$2\bar{a}_{pqk} x^k_{(p,q)} \leq y_p + y_q, \qquad \forall p,q \in \mathcal{P}, \forall k \in \mathcal{K} \tag{4}$$

$$y_p \in \{0,1\}, \qquad p \in \mathcal{P}, \tag{5}$$

$$z_k \in \{0,1\}, \qquad k \in \mathcal{K}, \tag{6}$$

$$x^k_{(p,q)} \in \{0,1\}, \qquad p,q \in \mathcal{P}, k \in \mathcal{K}, \tag{7}$$

where $\mathcal{F}$ is the set of all flights, $\mathcal{K}$ is the set of all extra flights, and $\mathcal{P}$ is the set of all feasible pairings. Here, c_p is the cost of pairing p, and d_k is the opportunity cost of failing to cover extra flight k. Furthermore, we define the parameters $a_{ip} = 1$ if flight i is included in pairing p, and 0 otherwise; $\bar{a}_{kp} = 1$ if extra flight k can be inserted into pairing p as a Type B solution, and 0 otherwise; and $\bar{a}_{pqk} = 1$ if pairings p and q can form a Type A solution to cover extra flight k, and 0 otherwise. The decision variable y_p is set to 1 if pairing p is selected, and 0 otherwise. Also, let $x^k_{(p,q)}$ be an auxiliary binary variable that takes the value 1 if two pairings p and q forming a Type A solution for extra flight k are both included in the solution, and 0 otherwise. Finally, we define the binary variable z_k equal to 1 if no Type A or B solution is present in the solution for extra flight k, and 0 otherwise.
The objective (1) minimizes the sum of the pairing costs and the opportunity costs of not accommodating the extra flights. Constraints (2) and (3) are the coverage constraints for the regular and extra flights, respectively. Observe that the model may opt for not covering an extra flight k if this is too expensive, setting z_k to 1. Constraints (4) prescribe that a Type A solution for extra flight k formed by pairings p and q is only possible if both of these pairings are selected.
The formulation (1)-(7) has both exponentially many variables, one for each pairing, and exponentially many constraints of type (4) which makes it both practically and theoretically challenging. Typical crew pairing models incorporate exponentially many variables, but have a fixed number of constraints and are solved by traditional column generation approaches where the pricing subproblem is a multi-label shortest path problem solved over an appropriate flight/duty network. (See [1] for a review of these concepts.) In our proposed robust model, the number of constraints (4) is not known a priori and depends on the pairings present in the model. Thus, ideally this formulation requires simultaneous row and column generation. In the next section, we present our preliminary algorithms developed for the problem RCPEF.

3 Solution Approaches

In both approaches presented here, the primary goal is to fix the number constraints in the model before applying column generation.

The Static Approach

In the "static" approach, all pairings that construct Type A solutions are generated a priori before column generation is applied to the linear

programming (LP) relaxation of (1)-(7). To this end, we identify all possible flights and connections that may appear in a pairing before covering an extra flight or its associated deadhead by a breadth-first-search and then construct pairings over this reduced network. We next run pairwise feasibility checks on these generated pairings to determine Type A solutions. Thus, all constraints (4) are identified and added to the model along with the associated auxiliary variables $x^k_{(p,q)}$ before the column generation procedure is invoked to identify Type B solutions and new pairings that may lower the objective function value. Upon termination of the column generation procedure, a primal heuristic is applied if the LP optimal solution is not integral.
We point out that the static approach is an exact method for solving the LP relaxation of (1)-(7) because all constraints (4) are explicitly included in the model. Clearly, the computational effort for this algorithm will be excessive for large instances of RCPEF.

The Dynamic Approach

In the static approach, all constraints (4) are incorporated into the formulation prior to column generation. In the "dynamic" approach, we opt for the complete opposite for speed. We exclude all variables $x^k_{(p,q)}$ and constraints (4) from the formulation and dynamically generate pairings that reduce the objective and yield Type B solutions. After the column generation terminates, we check whether the available pairings yield any Type A solutions and add the associated constraints and variables to the model. Next, we solve the LP relaxation of (1)-(7) with the available constraints and variables and invoke a primal heuristic, if necessary, in order to obtain an integer feasible solution to RCPEF.
The proposed dynamic approach does not necessarily provide an optimal solution to the LP relaxation of (1)-(7) because pairings leading to Type A solutions may be missed during the column generation. In order to reach a compromise between speed and solution quality, we promote that at least N (partial) pairings that may potentially form Type A solutions are kept on each node during the pricing subproblem. At the end of the pricing subproblem, such pairings are added to a special pool. This pool is examined for Type A solutions after the column generation terminates.

4 Computational Results

In this section, we present our preliminary results on the proposed static and dynamic approaches. Our primary goal is to illustrate the trade-off between robustness (as indicated by the number of Type A and B solutions obtained) and computational effort. We conducted a numerical study on two sets of actual data. The results are presented in Table 1 where in each cell the number of Type A solutions identified is followed by the solution time in parentheses. No more than 2 Type B solutions per extra flight were obtained in all cases.

Table 1. Comparison of the number of Type A solutions and CPU times for the dynamic and static approaches.

$\|\mathcal{F}\|$	$\|\mathcal{K}\|$	**Dynamic**					**Static**
		$N = 0$	$N = 10$	$N = 50$	$N = 100$	$N = 500$	
	1	0(0.17)	10(0.31)	14(0.62)	18(0.71)	18(0.96)	18(0.50)
42	2	0(0.17)	10(0.32)	86(0.67)	98(0.75)	118(1.03)	118(0.70)
	3	0(0.20)	10(0.42)	93(0.71)	106(0.81)	128(1.31)	128(1.34)
	1	10(0.71)	40(1.07)	64(1.39)	64(1.70)	64(2.57)	64(1.83)
96	2	20(0.86)	80(1.15)	128(1.64)	128(1.98)	128(2.87)	128(5.86)
	3	38(0.86)	60(1.25)	136(1.76)	141(2.06)	141(3.42)	141(9.70)

Two trends are clear from Table 1. First, the performance of the dynamic approach depends critically on the value of N. There is a threshold value for N above which extra solution time is spent with no additional benefit. Second, the dynamic approach outperforms the static approach for large problem instances.

5 Future Research

The results in Section 4 point to a clear need for simultaneous row and column generation for solving the proposed model. We are going to pursue this interesting direction in the future.

References

1. Klabjan D (2005) Large-scale models in the airline industry. In: Desaulniers G, Desrosiers J, Solomon MM (eds) Column Generation. Springer
2. Çoban E (2008) Column generation approaches to a robust airline crew pairing model for managing extra flights. MS Thesis, Sabancı U, Istanbul
3. Tekiner H, Birbil Şİ, Bülbül K (2008) Robust crew pairing for managing extra flights. To appear in Computers and Operations Research, DOI: 10.1016/j.cor.2008.07.005

Approximation Polynomial Algorithms for Some Modifications of TSP

Edward Gimadi

Sobolev Institute of Mathematics SB RAS, pr. Koptyga 4, Novosibirsk, Russia
gimadi@math.nsc.ru

Summary. In the report polynomial approximation algorithms with performance guarantees are presented for some modifications of TSP: for the minimum-weigt 2-PSP on metric distances and for the maximum-weight m-PSP in Euclidean space $\mathbf{R}^k$. [1]

1 Introduction

The Traveling Salesman Problem (TSP) is one of the popular combinatorial problems [14]. The problem is MAX SNP-hard: existence of a polynomial approximation scheme for it yields P = NP. A natural generalization of TSP is a problem of finding several edge-disjoint Hamiltonian circuits with extreme total edge weight. The problem is known as m-Peripatetic Salesman Problem (m-PSP). It was introduced by Krarup [10] and has network design and scheduling applications. More detailed information on motivation and application see in [2]. De Kort [8] proved that the 2-PSP is NP-hard by constructing a polynomial-time reduction from the Hamiltonian Path Problem. By similar arguments one can show that m-PSP is NP-hard for each $m > 2$. The problem does not admit any constant-factor approximation in the general case. However, like that of TSP, the minimization version of 2-PSP (2-PSP$_{\min}$) admits constant factor approximations in metric case. Note that maximization version of the problem (2-PSP$_{\max}$) admits constant factor approximations even in general case. The currently best result for this problem is a 3/4-approximation algorithm with time complexity $O(n^3)$ [1].

[1] Research was supported by Russian Foundation for Basic Research (projects 08-01-00516 and 07-07-00222).

B. Fleischmann et al.(Eds.), *Operations Research Proceedings 2008*,
DOI: 10.1007/978-3-642-00142-0_72,

In this paper considered modifications of TSP are Metric 2-$\mathrm{PSP}_{\min}$ and Euclidean m-$\mathrm{PSP}_{\max}$ focusing attention on results attained recently. Further let $G = (V, E)$ be a complete undirected graph with n vertices. The edges of the graph are weighted with functions $w : E \to R$.

2 Approximation Algorithms for the Metric 2-$\mathrm{PSP}_{\min}$

It is supposed that the triangle inequality holds.

2.1. In [3] for the minimum weight metric 2-PSP (Metric 2-$\mathrm{PSP}_{\min}$) NP-hardness was established and $(9/4 + \varepsilon)$-approximation algorithms with running time $O(n^3)$ was presented. The algorithm based on designing first Hamiltonian circuit H_1 by the 3/2-approximation Christofides-Serdyukov's algorithm [6, 13]. After that a second circuit H_2 is designed that is edge-disjoint with H_1 and whose weight is at most twice the weight of H_1. Later the similar performance ratio for the Metric 2-$\mathrm{PSP}_{\min}$ was also announced in [7].

2.2. Recently for the Metric 2-$\mathrm{PSP}_{\min}$ an improved approximation algorithm was presented by Ageev and Pyatkin [2].

Theorem 1. *The algorithm of Ageev and Pyatkin finds a feasible solution of the metric* 2-$PSP_{\min}$ *whose weight is at most twice the weight of the optimum in time* $O(n^2 \log n)$.

While above mentioned Christofides-Serdyukov's algorithm exploits transforming a minimum weight spanning tree to TSP tour, in [2] two edge-disjoint spanning trees of minimum total weight transform to pair of edge-disjoint Hamiltonian circuits. At that strikingly that several edge-disjoint spanning trees of minimum total weight can be found in running time $O(n^2 \log n + m^2 n^2)$, where m is the number of trees [12].

2.3. An interest subclass of considered problem is 2-$\mathrm{PSP}_{\min}(1, 2)$ when edge weights equal to 1 and 2. This problem is metric also.

In [7] performance ratio of about 197/144 was announced for this problem in assumption that performance ratio 7/6 holds for solution of $\mathrm{TSP}_{\min}$(1,2) on input graph, found by algorithm presented in [11].

The following statement gives the better performance ratio using the reduction the 2-$\mathrm{PSP}_{\min}$(1,2) to the 2-$\mathrm{PSP}_{\max}$(0,1), whose 3/4-approximation solution can be found by the algorithm from [1].

Theorem 2. *Let we have a* ρ_1*-approximation algorithm A solving the* $TSP_{\max}(1, 2)$. *Then the* 2-$PSP_{\min}(1, 2)$ *can be solved with a performance ratio at most* $(1 + \rho_1)$ *in time determined by the algorithm A.*

Thus for the 2-$\mathrm{PSP}_{\min}$(1,2) an approximate solution with a total weight of at most 5/4 times the optimum can be found in $O(n^3)$ running-time. An improved result was obtained recently in [9]. Let's formulate improved result for more general problem 2-PSP$_{\min}$, where edge weights possess arbitrary values in the interval $[1, q]$, $q \geq 1$.

Theorem 3. *[9] The problem 2-PSP$_{\min}$ with edge weights in the interval $[1, q]$ can be solved in the running time $O(n^3)$ with performance ratio that is at most $\frac{4+q}{5}$.*

A crucial point for establishing this theorem is the following

Statement 1 *[9] In n-vertex 4-regular graph a pair of edge-disjoint partial tours with total number of edges at least $8n/5$ can be found in running time $O(n^2)$.*

Corollary 1 *The problem 2-PSP$_{\min}$(1,2) can be solved in $O(n^2)$ time with performancc ratio that is at most $6/5$.*

2.4. More complicated Metric 2-PSP$_{\min}$ we have in the case of two independent weight functions $w_1 : E \to R$, $w_2 : E \to R$. In this case it is required to find two edge-disjoint Hamiltonian circuits $H_1 \subset E$ and $H_2 \subset E$ minimizing $W_1(H_1) + W_2(H_2)$.
For this problem 12/5-approximation algorithm with the time complexity $O(n^3)$ was presented in [3]. From the beginning of the algorithm in [3] two approximate solutions H_1 and H_2 of TSP$_{\min}$ with the weight functions w_1 and w_2 respectively are found by the 3/2-approximation Christofides-Serdyukov's algorithm [6, 13]. After that a second circuit H_2 is transformed in H_2' such that H_2' is edge-disjoint with H_1 and whose weight is at most twice the weight of H_1. Then roles of graphs H_1 and H_2 are exchanged and the pair (H_1, H_2') or (H_1', H_2) of minimum total weight is chosen as an approximate solution.
Let's give the following statement for the problem 2-PSP(1,2) with two independent weight functions.

Theorem 4. *Let A_{TSP} be ρ_2-approximation algorithm with running time $O(p(n))$ for solving $TSP_{\min}(1, 2)$. Then for solving 2-$PSP_{\min}(1, 2)$ with two independent weight functions, $(1+0.5\rho_2)$-approximation algorithm can be constructed with the same time complexity.*

Corollary 2 *Let Algorithm A_{TSP} from [5] be used. Then the problem 2-$PSP_{\min}(1, 2)$ with two weight functions can be solved in polynomial time with performance ratio that is at most $11/7$ of the optimum weight.*

The proof relies on the performance ratio $\rho_2 = 8/7$ of the polynomial algorithm from [5]. However the degree of the polynomial is very high: $O(n^{K+4})$, where the constant K in is equal to 21.
Thus, use of Algorithm A_{TSP} from [11] (its performance ratio equals 7/6) according to Theorem 4 implies slightly greater value 19/12 of performance ratio, but in the running time $O(n^3)$, that is much smaller.

3 Euclidean m-PSPmax

3.1. Preliminaries.
PSP is called Euclidean, if vertexes in graph G correspond to points in Euclidean space $\mathbf{R^k}$, and edge weights equal to lenghts of relative edges. Below we describe the polynomial approximation algorithm $\mathcal{A}$ for solving Euclidean m-PSPmax and present conditions of its asymptotic optimality.
Let $\mathcal{M}^* = \{I_1, \ldots, I_\mu\}$ be a family of edges (intervals) of the maximum-weight matching in the complete undirected graph G; $\mu = \lfloor n/2 \rfloor$. Let t be a natural parameter such that $2 \le t \le n/4$ and $\mathcal{M}^*$ consists of $(t-2)$ "light" and the rest "heavy" edges. A family of adjacent edges is called *interval chain (I-chain)*. One of two extreme edges of I-chain is called *leading*, and another extreme edge is called *driven*. A pair of I-chains is adjacent if their leading edges are connected.

3.2. Algorithm $\mathcal{A}$ for Euclidean m-PSPmax.
The algorithm $\mathcal{A}$ consists of Preliminary Stage and Common Stage for $i = 1, \ldots, m$.
On the **Preliminary Stage** the maximum weight matching $\mathcal{M}^*$ and its partition on $(\mu - t + 2)$ heavy and $(t-2)$ light are constructed. From the beginning all of edges in $\mathcal{M}^*$ are nonadjacent.
On **Common Stage** i, using the same $\mathcal{M}^*$ and its partition on light and heavy edges, the Hamiltonian cycle H_i is designed, that is edge-disjoint with $H_1, \ldots, H_{i-1}$. Designing H_i consists of three Steps:

Step 1. Among of first t I-chains connect a pair of nonadjacent leading edges with minimal angle between them. After that these edges become adjacent and one of two driven edges is assigned as the leading edge of the new I-chain. Repeat Step 1 while the number of I-chains becomes of $(t-1)$. Then relocate I-chains such that the left edge of first I-chain and the right edge of $(t-1)$-th I-chain would be unmarked.

Step 2. Each of light edges locate between pairs of I-chains such that it would be nonadjacent with extreme edges of these I-chains. After that such light edge becomes adjacent to these intervals and we obtain a common sequence $I_1, I_2, \ldots, I_\mu$ of adjacent edges of $\mathcal{M}^*$. Using endpoints of the interval, denote $I_j = (x_j, y_j)$, $j = 1, \ldots, \mu$.

Step 3. Construct an approximate solution as a set E_i of edges, that form a Hamilton cycle H_i bypassed the given n points in $\mathbf{R^k}$.
3.1. $j := 1$ and $E_i := \{I_1, I_\mu\}$. Mark the edges I_1, I_μ and go to 3.2.
3.2. If $w(x_j, x_{j+1}) + w(y_j, y_{j+1}) \ge w(x_j, y_{j+1}) + w(y_j, x_{j+1})$ then $E_i := E_i \cup (x_j, x_{j+1}) \cup (y_j, y_{j+1})$, else $E_i := E_i \cup (x_j, y_{j+1}) \cup (y_j, x_{j+1})$.
3.3. $j := j + 1$. If $j < \mu$ then go on to 3.2, else proceed to 3.4.

3.4. If n is odd, then the n-th point x_0, not represented as an endpoint of an interval from $\mathcal{M}^*$, is inserted in the forming cycle as follows: the edge $e = (x_1, y_1) \in E_i$ is replaced with the edges $(x_1, x_0), (x_0, y_1)$.

Produce some statements useful for an analysis of the algorithm $\mathcal{A}$.

Lemma 1. *[14] The total weight of the heavy edges from $\mathcal{M}^*$ is at least $W(\mathcal{M}^*)\left(1 - \frac{t-2}{\mu}\right)$.*

Lemma 2. *[14] Let $\alpha \leq \frac{\pi}{2}$ be an angle between two edges $I = (x, y)$ and $I' = (x', y')$ in $\mathcal{M}^*$. Then the following inequalities hold:*
$$w(I) + w(I') \geq \max\begin{cases} w(x, x') + w(y, y') \\ w(x, y') + w(y, x') \end{cases} \geq \Big(w(I) + w(I')\Big)\cos\frac{\alpha}{2}.$$

Lemma 3. *[14] Let the constant γ_k be depend on dimension k of the space $\mathbf{R}^k$ only. Then the minimal angle between pairs of t line intervals in $\mathbf{R}^k$ is at least $\alpha_k(t)$ such that*
$$\sin^2\frac{\alpha_k(t)}{2} \leq \gamma_k t^{-2/(k-1)}.$$

The following statement follows from Brooks theorem:

Lemma 4. *The number of independent (mutually nonadjacent) leading intervals during Step 1 of designing graph H_i is at least*
$$t_i = \begin{cases} t, & \text{if } i = 1; \\ \left\lfloor \frac{t}{2i-2} \right\rfloor, & \text{if } 1 < i \leq m. \end{cases}$$

Lemma 5. *For a performance ratio of the algorithm $\mathcal{A}$ we have*
$$\frac{W(\bigcup_{i=1}^{m} H_i)}{W(\bigcup_{i=1}^{m} H_i^*)} \geq \frac{1}{m}\sum_{i=1}^{m}\left\{1 - \frac{2t-1}{n} - \gamma_k t_i^{-2/(k-1)}\right\}$$
$$\geq 1 - \frac{2t-1}{n} - \gamma_k\left(\frac{t}{2m-2}\right)^{-2/(k-1)}.$$

Using Lemmas 1– 5, we can conclude the following main statement for the Euclidean m-PSPmax:

Theorem 5. *Let the number m of salespersons be at most $n^{\frac{1}{k+1}}$. Then m-PSP$_{\max}$ in Euclidean space $\mathbf{R}^k$ can be solved asymptotically optimal in time $O(n^3)$ by the algorithm $\mathcal{A}$ with a parameter $t = \lceil n^{(k-1)/(k+1)} \rceil$.*

Note that the running time $O(n^3)$ of the algorithm $\mathcal{A}$ is determined by finding of the maximum weight matching on the Preliminary Stage.

Acknowledgement. The author appreciates their colleagues A. Ageev, A. Baburin, Y. Glazkov, A. Glebov and A. Pyatkin for effective cooperation within common grants RFBR.

References

1. Ageev A., Baburin A, Gimadi E. (2007)A 3/4–Approximation Algorithm for Finding Two Disjoint Hamiltonian Cycles of Maximum Weight. *Journal of Applied and Industrial Mathematics.* 1(2): 142–147.
2. Alexander A. Ageev and Artem V. Pyatkin (2008) A 2-Approximation Algorithm for the Metric 2-PSP. *WAOA 2007 (Eds.: C. Kaklamanis and M. Skutella), LNCS, Springer-Verlag Berlin Heidelberg,* 4927: 103–115.
3. Baburin A., Gimadi E., Korkishko N.(2004) Approximation algorithms for finding two disjoint Hamiltonian circuits of minimal total weight. *Discrete Analysis and Operations Research. Ser. 2.* 11(1): 11–25 (in Russian).
4. Baburin E., Gimadi E.(2005) Approximation algorithms for finding a maximum-weight spanning connected subgraph with given vertex degrees. *Oper. Res. Proc. 2004, Intern. Conf. OR 2004, Tilburg. Springer, Berlin,* 343–351.
5. Berman P., Karpinski M. (2006) 8/7–approximation algorithm for (1,2)-TSP. *Proc. 17th ACM-SIAM SODA*: 641–648.
6. Christofides N. (1976) Worst-case analysis of a new heuristic for the traveling salesman problem. *Tech. Rep. CS-93-13, Carnegie Mellon University.*
7. Croce F. D., Pashos V. Th., Calvo R. W. (2005) Approximating the 2-peripatetic salesman problem. *7th Workshop on MAPS 2005. (Siena, Italy, June 6-10)*: 114–116.
8. De Kort, J.B.J.M. (1993) A branch and bound algorithm for symmetric 2-peripatetic salesman problems. *European J. of Oper. Res.* 70: 229–243.
9. Gimadi E.Kh., Glazkov Y.V., Glebov A.N. (2007) Approximation algorithms for two salespersons in complete graph with edge weights 1 and 2. *Discrete Analysis and Operations Research. Ser. 2, Novosibirsk.* 14(2): 41–61 (in Russian).
10. Krarup J. (1975) The peripatetic salesman and some related unsolved problems. *Combinatorial programming: methods and applications (Proc. NATO Advanced Study Inst., Versailles, 1974)*: 173–178.
11. Papadimitriu C. H., Yannakakis M. (1993) The traveling salesman problem with distance One and Two. *Math. Oper. Res..* 18(1): 1–11.
12. Roskind J., Tarjan R.E. (1985) A note on finding minimum-cost edge-disjoint spanning trees. *Math. Oper. Res.* 10(4): 701–708.
13. Serdjukov A.I. (1978) Some extremal bypasses in graphs. *Upravlyaemye Sistemy, Novosibirsk* 17(89): 76–79 (in Russian).
14. The Traveling Salesman Problem and its Variations. Gutin G., Punnen A. P. (eds.) (2002) *Kluwer Acad. Publishers, Dordrecht/Boston/London.*

A Robust Optimization Approach to R&D Portfolio Selection

Farhad Hassanzadeh[1], Mohammad Modarres[2], and Mohammad Saffari[2]

[1] Department of Industrial Engineering, Sharif University of Technology, Tehran, Iran fhzadeh@gmail.com
[2] Department of Industrial Engineering, Sharif University of Technology, Tehran, Iran
{modarres,m_saffari}@mehr.sharif.ir

Summary. Intense competition in the current business environment leads firms to focus on selecting the best R&D project portfolio. Achieving this goal is tied down by uncertainty which is inherent in all R&D projects and therefore, investment decisions must be made within an optimization framework accounting for unavailability of data. In this paper, such a model is developed to hedge against uncertainty. The robust optimization approach is adopted and the problem is formulated as a robust zero-one integer programming model to determine the optimal project portfolio. An example is used to illustrate the benefits of the proposed approach.

1 Introduction

R&D activities are becoming more and more essential to gain long-term survival and growth for a majority of firms. The purpose of project portfolio decision is to allocate the limited set of resources to various projects in a way that balances risk, reward, and alignment with corporate strategy [2]. But unfortunately in the R&D project portfolio decision, much of the information required to make decisions is uncertain.

2 Literature Review

Studies on R&D project portfolio selection can be divided into three major categories: strategic management tools, benefits measurement

B. Fleischmann et al.(Eds.), *Operations Research Proceedings 2008*,
DOI: 10.1007/978-3-642-00142-0_73,

methods, and mathematical programming approaches [4]. One of the mathematical formulations which is based on uncertain data is Fuzzy set theory. As an example, Wang and Hwang (2007) used options approach instead of traditional discounted cash flow to evaluate the value of each R&D project and determine the optimal project portfolio. Robust optimization is a new approach which is a novel one for R&D project portfolio selection. The objective of this paper is to develop such a model to optimize the R&D portfolio for the risk-averse decision maker in an uncertain R&D environment.

3 Robust Optimization Approach for Modeling Uncertainty

We rely extensively on the robust optimization tools developed by Bertsimas and Sim (2004) to handle uncertain parameters.
Let

$$\Lambda = \{\mathbf{A} \in \mathbf{R}^{m \times n} | a_{ij} \in [\bar{a}_{ij} - \hat{a}_{ij}, \bar{a}_{ij} + \hat{a}_{ij}] \quad \forall i,\ j,\ \sum_{(i,j)\in J} \frac{|a_{ij} - \bar{a}_{ij}|}{\hat{a}_{ij}} \leq \Gamma\} \tag{1}$$

The robust problem is then formulated as:

$$\begin{aligned} &\mathit{Minimize} && \acute{\mathbf{c}}\mathbf{x} \\ &\mathit{Subject\ to} && \mathbf{Ax} \leq \mathbf{b} \quad \forall \mathbf{A} \in \Lambda \\ & && \mathbf{1} \leq \mathbf{x} \leq \mathbf{u} \end{aligned} \tag{2}$$

Theorem 1. (Bertsimas and Sim (2004)) *The uncertain linear programming problem has the following robust, linear counterpart:*

$$\begin{aligned} &\mathit{Minimize} && \acute{\mathbf{c}}\mathbf{x} \\ &\mathit{Subject\ to} && \sum_j \bar{a}_{ij}x_{ij} + q_i\Gamma + \sum_{j:(i,j)\in J} r_{ij} \leq b_i \quad \forall i \\ & && q_i + r_{ij} \geq \hat{a}_{ij}y_j \quad \forall (i,j) \in J \\ & && -\mathbf{y} \leq \mathbf{x} \leq \mathbf{y}, \mathbf{1} \leq \mathbf{x} \leq \mathbf{u} \\ & && \mathbf{q} \geq \mathbf{0}, \mathbf{r} \geq \mathbf{0}, \mathbf{y} \geq \mathbf{0}. \end{aligned} \tag{3}$$

Proof. : See Bertsimas and Sim (2004).

4 Robust Optimization Framework for R&D Project Selection

4.1 R&D Project Valuation: a Compound Options Valuation Model

In addition to selecting an appropriate set of R&D projects, the valuation of R&D projects is important. We options valuation approach to evaluate each R&D project and mainly rely on work of Geske [3] and Perlitz [5].

4.2 Model Formulation of the R&D Portfolio Selection

For the proposed R&D selection model, let's adopt the following notations:

n The total number of candidate projects
v_i The uncertain compound option value of candidate project i
B_t The budget available for stage t
c_{it} The uncertain investment cost of candidate project i during stage t
l_{it} labor (in working months) required to implement project i at stage t
L_t labor (in working months) available to staff projects at stage t

$$x_i = \begin{cases} 1 & if\ project\ i\ is\ selected\ for\ funding, \\ 0 & otherwise, \end{cases}$$

The robust Model is as follows:

$$Max \quad \sum_{j=1}^{n} (v_i - c_{i1})x_i \tag{4}$$

$$s.t. \quad \sum_{i=1}^{n} c_{it}x_i \le B_t \quad \forall t \tag{5}$$

$$\sum_{i=1}^{n} l_{it}x_i \le L_t \quad \forall t \tag{6}$$

$$x_i \in \{0,\ 1\} \quad \forall i \tag{7}$$

The objective 4 of this model is to maximize the total benefit of the R&D investment portfolio. Other constraints are about project spending categories, required personnel, and decision variables.

5 Illustrative Example

In this section, an example of R&D project portfolio selection problem in the pharmaceutical industry is presented to illustrate the developed approach ([6]; [7]). A pharmaceutical company has 20 candidate R&D projects, where each project has three stages. We assume that the times to maturities of the first and second options for all projects are all set to 3 and 10 years, respectively. The preferred development budgets for stages 1, 2, and 3 are 270, 985, and 1975, respectively. Similarly, the preferred capacities of R&D staff for three stages are 375, 1965, and 1320, respectively. Table 1 lists estimated volatility as well as estimates of R&D staffs required for project stages. Table 2 presents the uncertain development costs and estimated present value of cash inflows of each project as interval numbers.

Table 1. Estimated R&D staffs required and projects volatilities

Projects	Required Staff (in working months) Stage 1	Stage 2	Stage 3	Volatility
P1	6	72	50	80%
P2	12	80	48	70%
⋮	⋮	⋮	⋮	⋮
P20	48	230	160	20%

Table 2. Estimated R&D staffs required and projects volatilities

Projects	Investment Costs Initial (c_{i1})	Stage 2(c_{i2})	Stage 3(c_{i3})	NPV of Inflows at t=0	Project Option Value
P1	(1.8,2.2)	(27.0,33.0)	(27.0,33.0)	(45.0,55.0)	(24.0,33.9)
P2	(2.7,3.3)	(45.0,55.0)	(40.5,49.5)	(90.0,110.0)	(46.5,68.2)
⋮	⋮	⋮	⋮	⋮	⋮
P20	(45.0,55.0)	(117.0,143.0)	(315.0,385.0)	(1035,1265)	(643.7,940.8)

We first employ Geske's valuation approach (section 4.1) for each of 20 projects to analytically determine the compound option value of each project. Following option valuation of each project, we formulate the R&D portfolio selection problem (6) while taking Bertsimas and Sim's budget of uncertainty approach (section 3) in the following sense:

$$\sum_{i=1}^{n}\sum_{t=1}^{T}\frac{|c_{it}-\bar{c}_{it}|}{\hat{c}_{it}}+\frac{|v_{it}-\bar{v}_{it}|}{\hat{v}_{it}}\leq \Gamma \tag{8}$$

$$\Gamma \in [0, n(T+1)]$$

In the following, we solve the problem and analyze the results for values of Γ in [0, n(T+1)]. Figure 1 shows the realized value of selected portfolios via different values of Γ in [0,40]. The non-increasing shape of the objective function goes back to the fact that as uncertainty of the environment grows, the uncertain problem parameters gain more volatility. It's also obvious that for $\Gamma \geq 11.80$, the objective function fairly remains constant while $\Gamma \geq 20.00$ can impose no further decline on the realized objective function (1797.9).
Table 3 shows the portfolio of selected projects along with corresponding values and sizes for interval values of Γ with 0.01 approximations. It is observed that when uncertainty is low ($\Gamma \leq 5$), the projects combination of the optimal portfolio changes erratically and the portfolio value drops rapidly, while medium uncertainty ($5 \leq \Gamma \leq 10$) gives rise to more "robust" portfolios. Furthermore, there is an inclination to form smaller portfolios as the uncertainty grows. This is inevitable, because when parameters uncertainty increases, the look-for-feasible nature of robust optimization confronts tighter budget constraints to satisfy and therefore, fewer projects qualify to enroll the optimal portfolio.

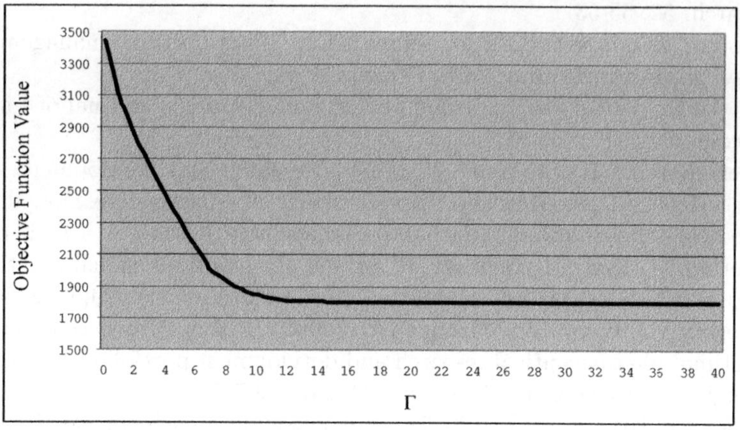

Fig. 1. Objective function value versus Γ

6 Conclusions

In this paper, a robust optimization approach was developed to select a set of uncertain R&D projects from a pool of candidate projects. We

Table 3. Optimal project portfolios for diverse uncertain environments

Γ	Portfolio of selected projects	Portfolio Size	Portfolio value
[0.01 , 0.14]	1, 5, 10, 11, 12, 14, 15, 16, 17, 18, 19, 20	12	[3430.2 , 3390.7]
[0.15 , 0.38]	2, 3, 5, 10, 11, 12, 14, 15, 17, 18, 19, 20	12	[3373.4 , 3300.3]
[0.39 , 0.76]	1, 2, 3, 5, 10, 12, 14, 15, 17, 18, 19, 20	12	[3290.5 , 3173.0]
[10.43 , 11.79]	1, 2, 5, 7, 10, 11, 12, 14, 15, 18, 19, 20	12	[1831.9 , 1812.2]
[11.80 , 40.00]	2, 3, 5, 12, 14, 15, 16, 18, 19, 20	10	[1812.1 , 1797.9]

adopted real option valuation approach and applied the proposed robust optimization approach on a real-world example and showed how detailed projects data can sum up to a simple project selection recipe supported by advanced mathematical formulas which account for uncertainty. This in essence provides a very useful decision making instrument for managers who are typically not interested in detailed data, and rather decide based on some qualitative/graphic tools.

References

1. Bertsimas D. & Sim M. (2004) The price of robustness, Operations Research, 52, 35-53.
2. Cooper RG, Edgett SJ, Kleinschmidt EJ. (1998) Portfolio management for new products. Reading, MA: Perseus Books.
3. Geske R. (1979) The valuation of compound options. Journal of Financial Economics; 7(1): 63-81.
4. Heidenberger K, Stummer C. (1999) Research and development project selection and resource allocation-a review of quantitative modelling approaches. International Journal of Management Review; 1: 197-224.
5. Perlitz M, Peske T, Schrank R. (1999) Real options valuation: the new frontier in R&D project evaluation? R&D Management; 29(3): 255-69.
6. Rogers MJ, Gupta A, Maranas CD. (2002) Real options based analysis of optimal pharmaceutical research and development portfolios. Industry and Engineering Chemistry Research; 41(25): 6607-20.
7. Wang J, Hwang W.-L. (2007) A fuzzy set approach for R&D portfolio selection using a real options valuation model. Omega; 35: 247-257.

A New Branch and Bound Algorithm for the Clique Partitioning Problem

Florian Jaehn and Erwin Pesch

Institute of Information Systems, University Siegen, 57068 Siegen, Germany
{florian.jaehn,erwin.pesch}@uni-siegen.de

Abstract

This paper considers the problem of clustering the vertices of a complete, edge weighted graph. The objective is to maximize the edge weights within the clusters (also called cliques). This so called Clique Partitioning Problem (CPP) is NP-complete, but it has several real life applications such as groupings in flexible manufacturing systems, in biology, in flight gate assignment, etc.. Numerous heuristic and exact approaches as well as benchmark tests have been presented in the literature. Most exact methods use branch and bound with branching over edges. We present tighter upper bounds for each search tree node than those known from literature, improve constraint propagation techniques for fixing edges in each node, and present a new branching scheme.

1 Introduction

The Clique Partitioning Problem (CPP) is to find a partition of a complete, weighted graph into non-overlapping subsets of arbitrary size. The CPP is NP-complete unless all edge weights are positive or all weights are negative (see e. g. [3]). Theoretical aspects of this problem are discussed by [5] and [4] present a cutting plane algorithm as well as benchmark tests. Different publications refer to these tests. E. g. [1] apply simulated annealing and [2] present an ejection chain heuristic as well as a branch and bound method.

Let us introduce a formal description of the CPP. Consider a complete, weighted graph $G = (V, E, W)$ consisting of a set of vertices $V = \{1, 2, \ldots, a\}$, a set of edges $E = (e_{ij}) \subset V \times V$, and a set of weights

B. Fleischmann et al.(Eds.), *Operations Research Proceedings 2008*,
DOI: 10.1007/978-3-642-00142-0_74, © Springer-Verlag Berlin Heidelberg 2009

$W = (w_{ij}), \quad i, j \in V, \ w_{ij} \in \mathbb{R}$. The clique partitioning problem is to find an equivalence relation on V, so that the sum of the edge weights of all vertex pairs in relation is maximized. This is equivalent to finding a partition of V into cliques, i. e. vertex subsets, so that the sum of the edge weights within the cliques is maximized. With binary variables

$$x_{ij} = \begin{cases} 1 & \text{if vertices } i \text{ and } j \text{ are in relation (are in the same clique),} \\ 0 & \text{otherwise} \end{cases}$$

for all edges (i, j) the CPP can be described by the following model (see [2]):

$$\begin{aligned} \max \quad & \sum_{1 \le i < j \le a} w_{ij} \cdot x_{ij} \\ \text{s.t.} \quad & x_{ij} + x_{jk} - x_{ik} \le 1 \quad \text{for } 1 \le i < j < k \le a \\ & x_{ij} - x_{jk} + x_{ik} \le 1 \quad \text{for } 1 \le i < j < k \le a \\ & -x_{ij} + x_{jk} + x_{ik} \le 1 \quad \text{for } 1 \le i < j < k \le a \\ & x_{ij} \in \{0, 1\} \quad \text{for } 1 \le i < j \le a \end{aligned} \tag{1}$$

The constraints guarantee the transitivity of the relation: If vertex i and j belong to the same clique and vertices j and k do, then vertices i, j, k belong to the same clique.

We use a branch and bound algorithm for solving this problem. The binary branching procedure decides for every edge of the graph, whether it is selected and included in a potential solution or not. After each branching, constraint propagation is applied in order to find other edges that now must or must not be selected. Afterwards an upper bound is determined for each search tree node.

The presented algorithm has been implemented and tested using benchmark tests found in various publications. The results are very promising and lead to significant runtime savings. However, tests are not yet finished and therefore not yet ready for publication.

2 The Search Tree

At first, a lower bound $\underline{g}$ with a corresponding feasible solution for the clique partitioning problem will be determined, using an arbitrary heuristic algorithm. We have used the ejection chain algorithm presented by [2].

The search tree structures as follows: At the root, there is the initial graph $G(V, E, W)$. Then branching with two child nodes and a

subsequent constraint propagation follows. In the first node, a specific variable x_{ij} is explicitly set to one. In other words, edge $(i,j) \in E$ is selected. Through constraint propagation further variables may be fixed implicitly. The second node explicitly fixes the same edge $x_{ij} = 0$ ((i,j) is deselected) and again constraint propagation may fix further variables implicitly.
For each node an upper bound will be evaluated (see below). A node is fathomed if its upper bound is not greater than the overall lower bound, or if all variables are fixed. If a node is not fathomed, two new child nodes derive from the current node through branching.
The outline of the search tree still misses details on how the upper bounds are obtained, how to determine which edges are to be chosen for branching, and how constraint propagation works. This will be delineated in the next sections.

3 Upper Bounds

The objective of the CPP is the sum of the edge weights within all cliques. Relaxing the triangle restrictions (see (1)) leads to a very simple initial upper bound $\overline{g}_0^*$, which is the sum of all positive edge weights:

$$\overline{g}_0^* := \sum_{1 \le i < j \le a} \max\{w_{ij}, 0\}$$

This upper bound can similarly be applied to each node λ of the search. If a positive edge weight is deselected or if a negative edge weight is selected, the upper bound is reduced accordingly:

$$\overline{g}_\lambda^* = \overline{g}_0^* - \sum_{\substack{1 \le i < j \le a \\ x_{ij} \text{ fixed}}} \left((1 - x_{ij}) \cdot \max\{w_{ij}, 0\} - x_{ij} \cdot \min\{w_{ij}, 0\} \right)$$

However, this upper bound cannot efficiently limit the search space. Thus, we lower this upper bound by taking into account the triangular restrictions. Consider a triple of vertices of the complete graph in which two edge weights are positive and one is negative, e. g. as shown in Figure 1. Both positive edges are included in $\overline{g}_0^*$, although both can only be selected in a feasible solution if the negative edge is selected, too. Thus, a triple of vertices i, j and k in which x_{ij}, x_{ik} and x_{jk} are not fixed, can reduce the initial upper bound by

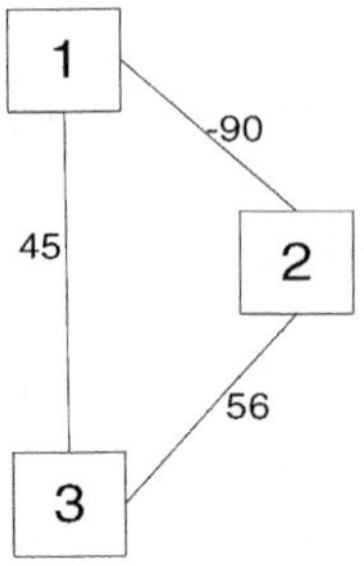

Fig. 1. Complete subgraph

$$\nabla(i,j,k) := \begin{cases} \min\{|w_{ij}|, |w_{ik}|, |w_{jk}|\} & \text{if } (w_{ij} < 0 \wedge w_{ik}, w_{jk} > 0) \\ & \vee (w_{ik} < 0 \wedge w_{ij}, w_{jk} > 0) \\ & \vee (w_{jk} < 0 \wedge w_{ij}, w_{ik} > 0) \\ 0 & \text{else} \end{cases} .$$

Thus, in Figure 1 the upper bound could be reduced by 45.
An edge disjunctive set of such triples of vertices (determined greedily) leads to a tighter upper bound. In fact, compared to $\overline{g}_0^*$, this upper bound diminished the difference between initial upper bound and optimal solution by more than 60 % in each of the elven benchmark tests used.

4 Branching

Which edge should be chosen next for branching? In order to answer this question, let us investigate how the search space is split.
A child node in which $x_{ij} = 1$ has less feasible solutions than the child node in which $x_{ij} = 0$ (i. e. the assumption that two vertices must be in the same clique is more restrictive than the assumption that these vertices must not be in the same clique). However, in order to let the branch and bound algorithm terminate quickly, it is important to reduce the size of the search space of both nodes. This can be achieved if there is a low upper bound in the node defined by $x_{ij} = 0$. A low upper bound could lead to the elimination of nodes or enables constraint propagation (to be described below) to furthermore reduce the search space. Thus, we will always branch over the edge (i, j) which in case of $x_{ij} = 0$ reduces the upper bound most. For the sake of shortness we omit the procedure how to determine the edge that reduces the upper bound most.

5 Constraint Propagation

After each branching, two consistency tests are performed. One examines the triangular conditions, i. e. the fact that for three vertices $i, j, k \in V$ it is not possible that $x_{ij} = x_{ik} = 1$ and $x_{jk} = 0$. Thus, constraints are propagated in order to determine whether branching leads to further edges to be fixed that must or must not be included in a feasible solution. The other test examines whether some edges must or must not be selected in order to keep the upper bound above the lower bound.

5.1 First Consistency Test

Let us assume that x_{ij}, $i, j \in V$ has been fixed during last branching. If $x_{ij} = 0$, we may conclude for every $k \in V \backslash \{i, j\}$ that $x_{ik} = 0$ and/or $x_{jk} = 0$. If this condition is not fulfilled, then the node leads to an infeasible solution and need not be examined. Otherwise it might appear, that only one edge is selected (say $x_{ik} = 1$), and the other one is not yet selected or deselected. In this case x_{jk} is implicitly set to zero. An analogous test is performed if $x_{ij} = 1$. Note that all implicit selections or deselections will be propagated until no propagation is possible any longer.

5.2 Second Consistency Test

A node λ is fathomed if the (local) upper bound is not greater than the (global) lower bound. If the selection (deselection) of an unfixed edge leads to such a case, we can implicitly deselect (select) this edge. So, for each edge we either have to determine how much the upper bound is lowered if this edge is deselected (this has already been done while determining the next edge for branching), or how much the upper bound is lowered if this edge is selected. If the absolute value of any of these numbers is greater than the difference between upper and lower bound, the according edges can be selected or deselected.

6 Summary

An exact solution method for the clique partitioning problem has been presented. The branch-and-bound algorithm varies from former approaches concerning the search tree, the upper bounds, and makes use of a stronger constraint propagation.

References

1. de Amorim, S., Barthélemy, J.-P., and Ribiero, C. C. (1992). Clustering and clique partitioning: simulated annealing and tabu search approaches. Journal of Classification, 9:17-41.
2. Dorndorf, U. and Pesch, E. (1994). Fast clustering algorithms. ORSA Journal on Computing, 6: 141-153.
3. Dyer, M. and Frieze, A. (1985). On the complexity of partitioning graphs into connected subgraphs. Discrete Applied Mathematics, 10: 139-153.
4. Grötschel, M. and Wakabayashi, Y. (1989). A cutting plane algorithm for a clustering problem. Mathematical Programming B, 45: 52-96.
5. Grötschel, M. and Wakabayashi, Y. (1990). Facets of the clique partitioning polytope. Mathematical Programming A, 47: 367-387.

On the Benefits of Using NP-hard Problems in Branch & Bound

Jörg Rambau and Cornelius Schwarz

University of Bayreuth, Bayreuth, Germany
{joerg.rambau,cornelius.schwarz}@uni-bayreuth.de

Summary. We present a Branch-and-Bound (B&B) method using combinatorial bounds for solving makespan minimization problems with sequence dependent setup costs. As an application we present a laser source sharing problem arising in car manufacturing.

1 Introduction

Some car manufactures use laser welding technology for the assembly of car bodys. The equipment in a welding cell consists of a number of welding robots and one or more laser sources, each of which can supply more than one robot, but only one at a time. In usual settings only a small fraction of the process time is spent with welding. This motivates the idea of sharing laser sources between robots. Because production cycle times must not be exceeded, the question is: "How many laser sources are needed to process a given set of welding tasks with a given set of robots in a given time?" To answer this question, we propose the *Laser Source Sharing Problem (LSP)*: Given a set of robots, a set of welding tasks and a set of laser sources, find a *scheduled tour* (i.e., an order of job start and end points together with start and end times) for each welding robot and an assignment of robots to laser sources so that

- all jobs are served,
- robots assigned to identical laser sources never weld simultaneously,
- the makespan is minimized.

This problem was introduced in [6], where a mixed-integer model for the special case of fixed robot tours was developed. An extension to integrate tour optimization was proposed in [8], but could not be solved on real-world scales (3–6 robots, 1–3 sources, ≈ 30 jobs).

B. Fleischmann et al.(Eds.), *Operations Research Proceedings 2008*,
DOI: 10.1007/978-3-642-00142-0_75,

When we drop the resource sharing constraint, we obtain a *vehicle routing problem (VRP)* with makespan minimization. Classical exact approaches to solve large VRPs use column generation in mixed-integer models, see [4] or [7]. However: The makespan objective yields large integrality gaps in the Master Problem, and – because of few servers for many jobs – the columns are dense.
We propose a combinatorial B&B algorithm based on partial schedules. Such algorithms are common in project scheduling, and a key problem is to find good lower bounds. Most lower bound constructions in project scheduling are based on precedence constraints, for instance critical paths calculations, see [2] or [3]. Since our problem does not contain precedence constraints, we follow a different method.
Our contribution is a new B&B algorithm that solves NP-hard sub-TSPs, which provide much better bounds than LP relaxations of common mixed-integer models. This is the first algorithm that can solve industrial-scale LSP-instances to proven optimality. Since estimating real robot driving times is a non trivial practical problem, all computations had to be done with artificial data, generated from real-world welding plans, though. We are currently working on providing more realistic data using KuKa SimPro. Moreover, collision avoidance is not yet part of the algorithm but can and will be integrated later.
We believe that bounds from the solutions of NP-hard subproblems may also be helpful for other makespan minimization problems.

2 Problem Definition

Let R be a set of robots, J a set of jobs and L a set of laser sources. Each robot $r \in R$ has a nullposition o_r, where the tour has to start and to end. Each job $j \in J$ has two end positions j_a, j_b. If the service of a job starts at j_a it has to finish at j_b, and vice versa. Let p_j be the processing time of Job $j \in J$. We denote the driving time of Robot r from Positions q_i to q_j by $\delta_r(q_i, q_j)$. We also introduce a latency δ_l for laser sources. When l switches robots then there is a delay of δ_l.
The task is to assign each $j \in J$ to a robot $r \in R$, each robot $r \in R$ to a laser source $l \in L$, and to create a scheduled tour for every robot through all assigned jobs so that

- each job is assigned to exactly one robot,
- each robot is assigned to exactly one laser source,
- jobs assigned to robots sharing a laser source do not overlap in time.

The cost of a scheduled tour is the time length, i.e., the time when the robot finishes its tour at o_r. The goal is to minimize the makespan, which is the maximum over the tour costs. If we restrict to one robot (the *1-server problem*) and set for all jobs $j_a = j_b, p_j = 0$ we get a TSP, which is NP-hard. Thus, the laser sharing problem is also NP-hard.
The LSP can be interpreted as a *vehicle routing problem*, where vehicles correspond to robots. The task is to find a route for every vehicle with minimum makespan subject to the resource constraints. From a scheduling point of view we can interpret the robots as machines, resulting in a parallel-machine scheduling problem with sequence dependent setup costs. The laser sources are resources with the condition that each machine can only use a unique resource.

3 The Algorithm

We already showed that the TSP is a special case of our problem. Since TSPs of the usual scale of the LSP (around 30 jobs) are relatively easy to solve nowadays, we can use TSPs as relaxations. In the next section we will see that this yields better and faster bounds than LP relaxations of mixed-integer-models of the LSP.
Assume that an assignment $\mathcal{J} : R \to 2^J$ of robots to jobs and an assignment $\mathcal{L} : R \to L$ of laser sources to robots are fixed. The resulting problem is called *LSP($\mathcal{J}$,$\mathcal{L}$)*. If resource constraints are neglected, then we can solve the 1-server problems separately by auxiliary TSPs, see [5]. The duration (in the LSP) of any tour t will be denoted by $\ell(t)$. The set of jobs served in Tour t is denoted by $J(t)$.
LSP($\mathcal{J}$,$\mathcal{L}$) can now be solved as follows: Assume that for each robot r we are given a partial scheduled tour t_r ending in q_r with duration $\ell(t_r)$. Then no scheduled tour starting with t_r visiting all jobs in $\mathcal{J}(r)$ can finish earlier than the concatenation of t_r and an optimal TSP tour $t_r^{\text{TSP}}(K_r, q_r)$ starting at q_r, visiting all jobs in $K_{t_r} := \mathcal{J}(r) \setminus J(t_r)$, and ending at o_r. Thus, a lower bound of LSP($\mathcal{J}$,$\mathcal{L}$) with given scheduled prefix tours $(t_r)_{r\in R}$ is given by $\max_{r\in R}\big(\ell(t_r) + \ell(t_r^{\text{TSP}}(K_{t_r}, q_r))\big)$. Now we can solve LSP($\mathcal{J}$,$\mathcal{L}$) using B&B with a node for each $(t_r)_{r\in R}$ and child nodes corresponding to all single-job extensions of a single t_r.
We summarize the algorithm for LSP($\mathcal{J}$,$\mathcal{L}$): For $r \in R$, a set of jobs K, and a start position q we denote by $\text{TSP}_r(K, q)$ a call to an exact TSP oracle solving the 1-server problem of Robot r starting at Position q, processing all jobs in K, and ending at o_r. The set $T^{\text{TSP}} := \{t_r^{\text{TSP}} \mid r \in R\}$ stores the solutions of the 1-server problems. Similarly, $T := (t_r)_{r\in R}$

keeps a partial scheduled tour t_r for every robot. We write μ for the best upper bound and λ for the lower bound of the current node:

Algorithm 1 (Combinatorial Branch-and-Bound for LSP($\mathcal{J}$,$\mathcal{L}$))

```
INPUT:  Data of LSP(𝓡,𝓛)
OUTPUT: A set T_OPT with optimal scheduled tours {t_r}_{r∈R}.
```

1. initialize: $t_r := ()$ for all $r \in R$
2. set $\mu := \infty$, $T_{\mathrm{OPT}} := T := \big(()\big)_{r\in R}$ *// empty tours*
3. CBB(T)
4. return T_{OPT}

Procedure CBB(T) for $T = (t_r)_{r\in R}$:

1. for all $r \in R$,
   set q_r to the last position of t_r or o_r if $t_r = ()$ and set
   $$t_r^{\mathrm{TSP}} := \mathrm{TSP}_r(\mathcal{J}(r) \setminus J(t_r), q_r)$$
2. set $\lambda := \max_{r\in R}\big(\ell(t_r) + \ell(t_r^{\mathrm{TSP}})\big)$
   // length of partial scheduled tour t_r completed with t_r^{TSP}
3. if $\lambda > \mu$ return *// pruning*
4. if $J(t_r) = \mathcal{J}(r)$ *// all jobs are scheduled*
   a) complete the tours, i.e., append o_r to t_r $\forall\ r \in R$
   b) set $\mu_{\mathrm{new}} := \max_{r\in R} \ell(t_r)$ *// makespan of current solution*
   c) if $\mu_{\mathrm{new}} < \mu$, set $T_{\mathrm{OPT}} := T$, $\mu := \mu_{\mathrm{new}}$ *// new best solution*
   d) return
5. for all $r \in R$, $j \in \mathcal{J}(r) \setminus J(t_r)$, $(q_{\mathrm{start}}, q_{\mathrm{end}}) \in \{(j_a, j_b), (j_b, j_a)\}$
   // run through all not yet scheduled jobs
   a) append $(q_{\mathrm{start}}, q_{\mathrm{end}})$ to t_r
      // Job j will be welded next from q_{start} to q_{end}
   b) set the start time of j to
   $$\max\{\ell(t_r) + \delta_r(q_r, q_{\mathrm{start}}),\ \max_{\substack{s\in \mathrm{R}\setminus r\\ \mathcal{L}(s)=\mathcal{L}(r)}} \ell(t_s) + \delta_{\mathcal{L}(s)}\}$$
      // earliest time so that r can reach the job
      // and so that the laser source is available again
   c) CBB(T)
   d) remove j from t_r

Remark 1 *At any time, we can pipe the tour information given by the TSP call into a scheduling heuristic for fixed tours, e.g., [6], which gives us a feasible solution for the LSP. This primal information is not available from LP-relaxations.*

If the assignments $\mathcal{J}$ and $\mathcal{L}$ are prescribed, then we end up in a cluster-first-schedule-second approach, which yields an optimal solution whenever we guess the "right" assignments. Since $|R|$ and $|L|$ are small, we can enumerate all $\mathcal{L}$s. The number of potential job-robot assignments, however, is too big for a naive enumeration. However, we can reduce this big set to a small set of candidates without missing an optimal assignment. To this end, we branch over partial job-server assignments. Whenever we reach a leaf, we run an heuristic scheduling algorithm to generate a feasible solution using only one laser source. The value of this solution provides an upper bound for LSP. The lower bounds in the nodes are obtained by solving the 1-server TSPs for the partially assigned jobs. Every leaf with a lower bound not worse than the best global upper bound is a candidate. Finally, the LSP can now be solved exactly by solving the LSP($\mathcal{J}$,$\mathcal{L}$) for all candidate assignments.

4 Computational Results

We now compare the lower bounds to the LP relaxations of two mixed-integer-models. The first one is an improved version of [8] which uses linear ordering variables and many big-M constraints. Since good LP-bounds in scheduling often come from time indexed variables we also compare to a model based on a time expanded networks.
Our test instances consist of randomly selected jobs of a real welding plan from a car manufacturer with three robots. Unfortunately, it is a non trivial practical problem to get real robot driving times. We used Euclidean distances on the 2D projection of the welding points as an approximation. The comparison was done on a Intel Core 2 Duo processor with 3 Ghz and 4 GB memory running Ubuntu Linux 8.04 in 64 bit mode. For the LP relaxations we used Ilog Cplex 11.1 (barrier for time discrete networks and dual simplex for linear orderings). The TSP relaxations were solved by `concorde` [1]. In the following table the lower bounds from various root relaxations for typical instances of $LSP(\mathcal{L},\mathcal{J})$ with given optimal assignments are listed. The respective optima were calculated by our method. For non-optimal assignments, LP-relaxations are no better.

problem	lin. ordering		time-exp. netw.		TSP		optimum
	cpu/s	value	cpu/s	value	cpu/s	value	
10jobs	0.08	24.7	24.5	24.7	0.02	24.7	24.7
16jobs	0.02	18.0	200.0	17.1	0.02	20.3	20.3
18jobs	0.09	20.0	282.8	19.1	0.02	22.7	22.8
20jobs	0.03	20.0	463.3	19.1	0.02	23.0	23.0
34jobs	0.11	24.7	2605.9	31.4	0.07	31.4	31.4

We see that, for a use in our B&B, linear ordering relaxations are too weak for the large instance and time expanded network relaxations are way too slow. The TSP relaxation is the strongest and the fastest throughout and was the only one with which the full B&B for the 34jobs-LSP could finish within a couple of hours.

5 Conclusions

We showed that NP-hard subproblems have the power to provide much better bounds in B&B algorithm than classical LP based approaches. The key lies inside the problem scale: A large-scale for the original problem (here: LSP) may be small for the subproblem (here: TSP). It remains to verify the method on real-world welding data and to integrate collision avoidance in the B&B.

References

1. D.L. Applegate, R.E. Bixby, V. Chvátal, and W.J. Cook. The Traveling Salesman Problem - A Computational Study. Princeton University Press, 2006.
2. P. Brucker and S. Knust. Complex Scheduling. Springer-Verlag, Berlin, Heidelberg, New York, 2006.
3. E. Demeulemeester. Optimal Algorithms for various classes of multiple resource-CPCPs. PhD thesis, Katholik Universiteit Leuven, 1996.
4. J. Desroisers, Y. Dumas, M.M. Solomon, and F. Soumis. Time constraint routing and scheduling. In M. Ball, T.L. Magnanti, C.L. Monma, and G. Nemhauser, editors, Network Routing, volume 8 of Handbooks in Operations Research and Management Science, pages 35-140. Elsevier, Amsterdam, 1995.
5. M. Dror, (edt). Arc Routing. Theory, Solutions and Applications. Springer-Verlag, 2001.
6. M. Grötschel, H. Hinrichs, K. Schröer, and A. Tuchscherer. Ein gemischt-ganzzahliges lineares Optimierungsproblem für ein Laserschweißproblem im Karosseriebau. Zeitschrift für wissenschaftlichen Fabrikbetrieb, 5:260-264, 2006.
7. S.O. Krumke, J. Rambau, and L.M. Torres. Realtime dispatching of guided and unguided automobile service units with soft time windows. In R. Möhring et al., editor, Algorithms - ESA 2002, volume 2461 of LNCS, pages 637-648. Springer, 2002.
8. T. Schneider. Ressourcenbeschränktes Projektscheduling zur optimierten Auslastung von Laserquellen im Automobilkarosseriebau. Diplomarbeit, University of Bayreuth, 2006.

Automatic Layouting of Personalized Newspaper Pages

Thomas Strecker and Leonhard Hennig

DAI-Labor TU Berlin, Ernst-Reuter-Platz 7, 10587 Berlin
{thomas.strecker,leonhard.hennig}@dai-labor.de

Summary. Layouting items in a 2D-constrained container for maximizing container value and minimizing wasted space is a 2D Cutting and Packing (C&P) problem. We consider this task in the context of layouting news articles on fixed-size pages in a system for delivering personalized newspapers. We propose a grid-based page structure where articles can be laid out in different variants for increased flexibility. In addition, we have developed a fitness function integrating aesthetic and relevance criteria for computing the value of a solution. We evaluate our approach using well-known layouting heuristics. Our results show that with the more complex fitness function only advanced C&P algorithms obtain nearly-optimal solutions, while the basic algorithms underperform.

1 Introduction

Automatic layouting is the task of selecting elements from a given set and placing them on a two-dimensional plane such that the resulting layout optimally fills the available area, e.g. minimizes wasted space. The general class of this kind of problems is known as Cutting and Packing [9], which is an NP-hard problem [3]. Therefore, the task is to constrain the search space and still find nearly-optimal layouts. If elements have a value, this is known as an output maximization problem. Hence, the algorithm does not necessarily need to minimize wasted space, but rather maximizes the overall value of the packed elements. Solutions to this problem include variants of greedy search and Genetic Algorithms [7], a greedy randomized adaptive search procedure (GRASP) [1] and tabu search [2]. Many algorithms focus on ranking elements which are then laid out according to a placement strategy, e.g. proceeding from the top-left corner to the bottom right, greedily filling the next available space.

B. Fleischmann et al.(Eds.), *Operations Research Proceedings 2008*,
DOI: 10.1007/978-3-642-00142-0_76,

Laying out articles on a newspaper page is a special type of C&P. To produce a visually pleasing layout which conforms as much as possible to a typical newspaper page layout, aesthetic criteria, such as page whitespace and article whitespace, must be considered as contributing to the value of the solution [5]. Gonzáles et al. [4] employ a greedy Simulated Annealing algorithm for creating a column-based layout for on-line articles. However, their approach only minimizes inter-article whitespace without restricting the page height, thus rendering it unusable for fixed page sizes. Jacobs et al. [6] use a dynamic programming approach based on pre-defined page templates. The drawback of this approach is the need for manually specifying the templates. Goldenberg employs a Genetic Algorithm to layout variable-sized documents with the goal of minimizing the required area and hence the amount of whitespace [3]. He considers the aspect ratio of the resulting area as an aesthetic constraint and reports that items tend to be laid out in rows and columns if their respective heights or widths are similar. The approach cannot discard any elements and therefore cannot be applied for pages with fixed size.

Our Contribution

In our approach, we use a fixed-size page and subdivide it into columns and rows, similar to a typical newspaper page. To increase the flexibility of the page's layout, we allow an article to be laid out in different forms, e.g. spanning a single or multiple columns, or including an optional media element. We also extend the basic greedy top-left placement strategy by allowing it to skip areas which are not beneficial, thus avoiding to discard elements that do not fit the next free space.
In addition, our method computes an element's value as a combination of its inherent value, i.e. its relevance for a user, and its contribution towards an aesthetic overall page layout. We compare the resulting layouts to those using a baseline function optimizing wasted space only. For the evaluation of our approach we use a set of algorithms that have been shown to result in nearly-optimal values for typical packing tasks: various greedy algorithms and a Simulated Annealing approach [4, 7, 8]. To compute the optimal values achievable for a dataset, we use an exhaustive search.

2 Laying Out Newspaper Pages

We subdivide a page into a fixed-size 4-by-16 grid, where each cell corresponds to 6 lines of text (see Figure 1 (a)). An article may be laid out in several variants depending on article properties such as text length and/or the availability of media elements. Each article is transformed into a set of candidate items for the layouting process, discarding variants which do not fulfill basic aesthetic constraints, e.g. aspect ratio or headline/text ratio. There are at most 8 items per article, corresponding to an article laid out across 1, 2, 3 or 4 columns, combined with using a media element or not. The layout algorithm is responsible for ensuring that at most one item per set is used.

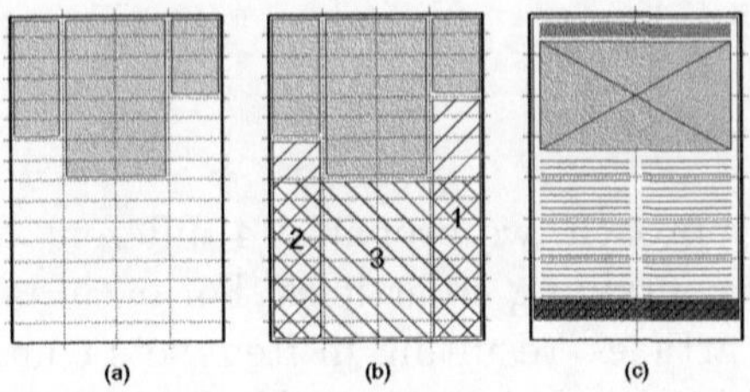

Fig. 1. Layout Grid and Article Coverage

Placement Strategy

Items are placed on the page in a left-to-right and top-to-bottom fashion, in the order computed by the layouting algorithm. The placement algorithm computes the location and extent of the next rectangular empty area (see Figure 1 (b), area 1). Since this may result in areas that are too small for any of the remaining articles, we extend this greedy approach by allowing the algorithm to skip areas of the page, giving new spaces (see Figure 1 (b), areas 2 and 3).

Computing the Page Score

We use two fitness functions for computing the value of an item. Our baseline function f_1 simply measures the page coverage of an item, i.e. the ratio of its area to the page area. To address personalization and aesthetic issues, our second fitness function f_2 considers the user relevance r_i of and the amount of whitespace c_i (empty text lines in the last row of cells due to variable text length, see 1 (c)) within an item i as additional factors. The value of an article is computed as the sum of its user relevance, the item coverage and its fractional contribution to the page coverage:

$$f_2(i) = r_i + c_i + \frac{w_i * h_i}{WH - \sum_{\substack{j \in \mathcal{B} \\ j \neq i}} w_j h_j}, \tag{1}$$

where w_i and h_i are the width and height of item i, W and H are the width and height of the entire page, and $\mathcal{B}$ is the set of items already placed on the page. The total page score is then computed as the sum over all item values.

3 Experiments

3.1 Experimental Setup

To evaluate our approach, we prepared 4 different test data sets from article collections containing 30 articles. For each data set we used the first 10, 20 or all articles, resulting in item sets containing between 31 and 129 items. Most sets do include at least one full-page article, and a number of item combinations that can be laid out to completely fill the page. For computing the optimal solution value we use an exhaustive search. Like Julstrom [7], we present two slightly different realizations of greedy search, namely an absolute (AG) and a relative greedy (RG) variant, as well as a third variant (EG) based on Sahni's ϵ-approximate algorithm [8] for $k = 1$. We also implemented a Simulated Annealing (SA) approach [4], using 50-fold cross-validation to optimize the swap rate and maximum number of iterations. Maximal page scores were achieved for $T_{max} = 1000 * |items|$ and $swap = 50\%$.

3.2 Results

We evaluate each algorithm in combination with each fitness function, repeating this procedure for each data set and data set size, giving a total of 24 evaluation runs for each algorithm. For the greedy algorithms, we compute the best solution for each run. The SA algorithm was run 50 times for each evaluation run to compute mean and standard deviation of its solutions, as well as the average error rate of the mean solution. When using f_1, all algorithms consistently find optimal solutions as computed by an exhaustive search, with only a few exceptions. This is to be expected, as the function simply minimizes page whitespace. Therefore, we do not show these results. The picture is somewhat different for f_2. Here we find that the basic greedy heuristics consistently underperform (see Table 1). Since the value of an item is spread across its area, the greedy heuristics have a strong tendency

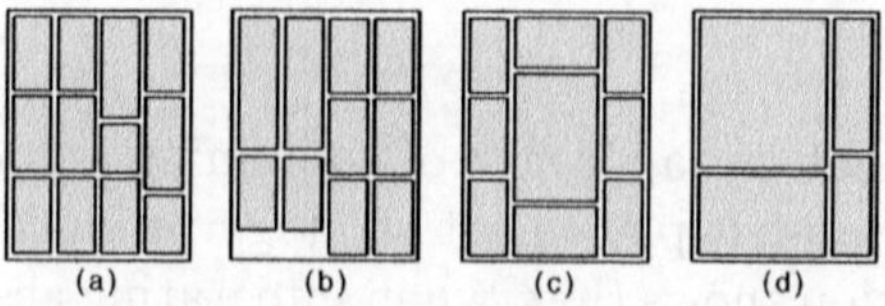

Fig. 2. Generated Layout Examples

to prefer smaller articles. In other words, the algorithms select many relevant articles, which leads to rather unpleasing, fragmented layouts, as shown in Figure 2 (a) and (b).

The EG heuristic consistently outperforms the AG and RG algorithms, since it can place items other than the most dense first. The SA algorithm in turn easily outperforms all other algorithms, almost always finding nearly-optimal solution values (see Table 1), but has a higher runtime complexity than the greedy heuristics. The pages (c) and (d) in Figure 2 illustrate two sample pages generated using f_1 and f_2. Page (c) has been found using f_2 and corresponds to a page with a quality close to the optimum. Page (d) has been found using f_1. In both cases larger articles have been put on the page which results in pages looking aesthetically more pleasing.

We note that for some data sets the optimal solution value is actually lower when using more articles. This can be explained by the way the placement strategy works: The larger data set includes an item with a higher density than the smaller set, which is then placed first. The sequence and size of rectangles computed for placing the next item may then be worse than layouting sequence of the smaller data set. This emphasizes the importance of the placement strategy.

Table 1. Performance with Fitness Function f_2 (Opt. unknown for Runs 6,12)

No.	Opt.	AG		RG		EG		SA		
		Value	%E	Value	%E	Value	%E	Mean	StdDev	%E
1	2.2068	1.2681	42.54	1.2681	42.54	2.1139	4.21	2.2035	0.0165	0.15
2	3.3767	2.3017	31.84	2.3017	31.84	3.1762	5.94	3.3767	0.0000	0.00
3	4.3688	3.0223	30.82	3.0223	30.82	3.5354	19.08	4.3658	0.0053	0.07
4	4.6967	3.0111	35.89	3.0111	35.89	3.4244	27.09	4.6842	0.0878	0.27
5	8.0428	5.4261	32.53	5.4261	32.53	6.9553	13.52	7.7069	0.2059	4.18
6	(8.7666)	6.6678	?	6.6678	?	9.0800	?	9.2685	0.2756	?
7	5.6004	4.1745	25.46	4.1745	25.46	5.1356	8.3	5.5743	0.0402	0.47
8	7.2056	5.3049	26.38	5.3049	26.38	5.7886	19.67	6.6959	0.3336	7.07
9	7.5765	4.8854	35.52	4.8854	35.52	6.8242	9.93	6.9890	0.2464	7.75
10	6.7334	4.4884	33.34	4.4884	33.34	6.3152	6.21	6.7210	0.0577	0.18
11	7.8890	6.5950	16.4	6.5950	16.4	7.8198	0.88	7.6860	0.1370	2.57
12	(8.5275)	5.8718	?	5.8718	?	8.7147	?	8.1817	0.2216	?

4 Conclusion

We have shown that our approach of combining a grid-based structural template with a top-left placement strategy produces high-quality layouts. Our evaluation shows that, when integrating aestethic criteria and user relevance into the fitness function, more attention must be paid to the layouting algorithm and placement strategy. Further investigation into the choice of the fitness function and its weight parameters is an interesting challenge for future research, e.g. for unique branding and advanced personalization of the newspaper page layouts.

References

1. R. Alvarez-Valdes, F. Parreñno, and J. Tamarit. A grasp algorithm for constrained two-dimensional non-guillotine cutting problems. Technical report, University of Valencia, 2004.
2. R. Alvarez-Valdes, F. Parreñno, and J. Tamarit. A tabu search algorithm for two-dimensional non-guillotine cutting problems. European Journal of Operational Research, 183(3):1167-1182, 2007.
3. E. Goldenberg. Automatic layout of variable-content print data. Master's thesis, Information Infrastructure Laboratory, HP Laboratories, 2002.
4. J. González, I. Rojas, H. Pomares, M. Salmerón, and J. Merelo. Web newspaper layout optimization using simulated annealing. IEEE Transactions on Systems, Man and Cybernetics, Part B, 2002.
5. S. J. Harrington, J. F. Naveda, R. P. Jones, P. Roetling, and N. Thakkar. Aesthetic measures for automated document layout. In Proc. of DocEng '04, pages 109-111. ACM Press, 2004.
6. C. Jacobs, W. Li, E. Schrier, D. Bargeron, and D. Salesin. Adaptive document layout. Commun. ACM, 47(8):60-66, 2004.
7. B. A. Julstrom. Greedy, genetic, and greedy genetic algorithms for the quadratic knapsack problem. In Proc. of GECCO '05, pages 607-614. ACM Press, 2005.
8. S. Sahni. Approximate algorithms for the 0/1 knapsack problem. J. of the ACM, 1975.
9. G. Wäscher, H. Haußner, and H. Schumann. An improved typology of cutting and packing problems. European J. of Operational Research, 127(3):1109-1130, December 2007.

A Tabu Search Approach to Clustering

Marcel Turkensteen and Kim A. Andersen

CORAL - Center of OR Applications in Logistics, Aarhus School of Business, Denmark

Summary. In Clustering Problems, groups of similar subjects are to be retrieved from large data sets. Meta-heuristics are often used to obtain high quality solutions within reasonable time limits. Tabu search has proved to be a successful methodology for solving optimization problems, but applications to clustering problems are rare. In this paper, we construct a tabu search approach and compare it to the existing k-means and simulated annealing approaches. We find that tabu search returns solutions of very high quality for various types of cluster instances.

1 Introduction

Clustering problems are the problems of constructing groups of subjects such that each group shares relevant characteristics [4]. Clustering problems are encountered in many fields of research, such as pattern recognition, marketing, and computer science. The relevant characteristics of subjects, or the *attributes*, take many forms, such as binary (yes or no) or ratio scores. When the attributes are measured on a ratio scale, the Minimum Sum of Squares Criterion (MSSC) is often taken to measure the dissimilarity between subjects [4]. This measure takes the sum of the squared distances between each pair of subjects in the same cluster.

Because of the complexity of MSSC Clustering Problems, *heuristics* are often applied to solve them. A commonly used cluster heuristic is k-means (KM), introduced in [5]. It is extended in [7] with the so-called Ward's heuristic from [9] to generate starting solutions. However, KM is a local search method, meaning that it is prone to being trapped in a local optimum [10]. Methods that use and guide other heuristics in order to produce solutions beyond the local optima that are normally

B. Fleischmann et al.(Eds.), *Operations Research Proceedings 2008*,

DOI: 10.1007/978-3-642-00142-0_77, © Springer-Verlag Berlin Heidelberg 2009

produced are *meta-heuristics* [3]. A popular meta-heuristic for MSSC Clustering problems is simulated annealing (SA); see [2]. It produces good cluster solutions and is relatively easy to implement.
A promising meta-heuristic is *tabu search (TS)* [3]. It has been applied to clustering [1, 8], but we address the following potential improvements that have not yet been considered in the existing literature.

- When methods are compared, in particular to KM, starting solutions are often randomly generated or unspecified; see e.g. [1]. We also use Ward's algorithm to generate high quality initial solutions, for TS as well as its competitors.
- Current tabu search comparisons consider clustering problems of at most 100 subjects. We address large databases of up to 5,000 subjects, which are likely to occur in practice.
- We include a more effective *diversification* stage; see Section 2.

2 Tabu Search for Clustering

The clustering problem can be formally stated as follows. Let $\mathcal{X}$ denote the set of all cluster solutions on n subjects. Given a particular cluster solution $x \in \mathcal{X}$ the function $f(x)$ measures the quality of that particular cluster. The problem can be formally stated as follows:

$$\begin{aligned} &\min f(x) \\ &s.t. \;\; x \in \mathcal{X}. \end{aligned} \tag{1}$$

Tabu search requires an initial solution x, which belongs to the solution space $\mathcal{X}$, i.e. $x \in \mathcal{X}$. It then tries to improve the solution through an iterative process, called *intensification.* The purpose of this phase is to determine the best solution in some part of the solution space containing the starting solution. During this process, deteriorating moves are allowed as well. A *tabu list* is being kept in order to prevent cycling through the same solutions. After an intensification phase, the tabu search can move to a new solution through *diversification.* In the diversification phase, the search is redirected from the current solution to a different part of the solution space. A new intensification phase then takes place. The tabu search procedure alternates between these two phases a predetermined number of times, saving the best solution $x^* \in \mathcal{X}$ found so far.
In the intensification phase, solutions are searched through iteratively. The iterative procedure is performed as follows: Given a solution x,

there is a predefined set of solutions $\mathcal{N}(x) \subset \mathcal{X}$ that can be reached from x, called the *neighborhood* of x. The solution with the lowest costs in the neighborhood is chosen as the next solution; this may even be a solution with higher costs than the current one. However, some solutions are placed in a so-called *tabu list* and are not available. After a move from solution x to $\hat{x}$ ($x \mapsto \hat{x}$) has been made, the inverse move ($\hat{x} \mapsto x$) is added to the tabu list in order to prevent the tabu search from being caught in the same set of solutions. The *length of the tabu list* determines how long a certain move is forbidden. It may be useful to make a move even though it is in the tabu list. A move in the tabu list is allowed when the *aspiration criterion* is met, for example, when the forbidden move leads to a better solution than the current best one. Finally, the *number of iterations* determines how many iterations are made in the intensification phase. Important decisions in the diversification phase are the *number of diversifications* and the *diversification strategy*, i.e., the procedure with which the solution of a new intensification phase is constructed.

The tabu search algorithm from [1] is quite basic as no diversification is done. Moreover, there are no reports on the effects of changing intensification decision rules on the performance of tabu search. A more extensive tabu search algorithm is presented in [8]. This method forces pairs of similar subjects into the same cluster. It then diversifies the search by allowing the subjects of one of these pairs to be in different clusters. For larger instances, it means that either a large number of diversifications needs to be performed, or the search is restricted to a small part of the solution space.

In our TS procedure, we assume that the initial solution is given. The intensification stage is as follows. The neighborhood of a solution is the set of all solutions in which the cluster membership of one subject is changed. The number of iterations is not fixed, but the intensification is terminated if no new best solutions are obtained in 100 iterations. We maintain a relatively long tabu list of length 180 to prevent cycling between the same solutions.

We apply the following diversification strategy. Take, in each cluster, the subject with the longest distance to its respective cluster center; use these k subjects as new cluster seeds, i.e, assign each of the $n - k$ other subjects to the closest seed point. This strategy brings the search to a better than random strategies. We also find that about 8 diversifications is sufficient.

3 Computational Experiments

Comparative studies between methods on meta-heuristics for clustering have been sparse. Meta-heuristics have often only been tested on specific case studies without any mentioning of the used parameter values.

In a typical clustering study, there are many issues that need to be addressed, but we limit ourselves to minimization of the MSSC score, given a fixed number of k clusters. For a discussion of the entire cluster process, we refer to [4].

We choose the commonly used KM version from [5]. Our SA algorithm is experimentally determined and has the following cooling schedule (initial temperature 2500, $\alpha = 0.95$, freezing temperature 0.00001, the number of tries $\frac{1}{2}kn$ before stability is achieved at a given temperature). Starting solutions are both randomly generated ("random") and generated using Ward's algorithm ("Ward"). All procedures have been implemented in C++ and compiled with the Visual Studio 5.0 compiler. All tests were performed on an HP 2.00 GHz computer with 0.97 GB RAM using a Windows NT operating system.

Firstly, we consider the so-called Milligan instances, generated with the cluster instance generator from [6]. We use 120 instances of 200 subjects each of seven different cluster problem types. In Table 1, the results are presented as follows. There are no optimal MSSC scores available. Therefore, we choose to divide the MSSC score of a method for instance by the best score found for that instance. The averages of these scores are then computed and reported for each instance type and each method. A score of 1 of algorithm A for instance type i means that A always finds the best solution for type i among our tested algorithms. This relative measure is taken, because comparing average MSSC scores would bias the results towards the instances with large MSSC scores.

Table 1. Relative average MSSC scores

	Random start				Ward start		
Type	SA	TS	KM	Ward	SA	TS	KM
1 Normal	1.4164	1.0404	7.1704	1.0096	1.0048	1	1.0075
2 20% outliers	1.1288	1	2.9523	1.1250	1.0273	1	1.0592
3 20% outliers	1.0591	1.0055	2.1914	1.1271	1.0138	1	1.0686
4 Error (low)	1.2437	1.0693	5.2270	1.0113	1.0027	1	1.0037
5 Error (high)	1.1221	1.0054	3.0897	1.0442	1.0070	1	1.0339
6 Noise dim.	1.0015	1.0005	1.1600	1.0943	1.0014	1.0002	1.0586
9 Standardized	1.2562	1.0051	1.3272	1.1250	1.0420	1	1.0031
11 Random data	1.2043	1.0173	3.2637	1.0833	1.0135	1.0002	1.0306

Table 1 shows that all methods, but in particular KM, perform clearly better when departing from an initial solution generated with Ward's algorithm, in particular for the instance types 1 and 9, which contain clear cluster structures. Overall, TS returns the best results. TS and SA take much more time than KM. However, we find that performing a large number of KM runs from different starting points does not lead to high quality solutions.
Meta-heuristics can be used to find high quality solutions for large practical instances instances as well. In order to reduce search times, we randomly select a subset of 5,000 potential candidate moves in each TS and SA iteration. Moreover, the procedures are terminated after 1,000 seconds of CPU time; this includes the time for performing Ward's algorithm. We solve eight randomly generated instances of size 3,000 and 5,000 each which have 4 attributes with normally distributed attribute scores in 4 clusters. The instances contain both badly and well separated clusters.

Table 2. Average MSSC scores and solution times for large randomly generated cluster instances

		SA		TS		K-means	
Initial sol.	n	MSSC	Time	MSSC	Time	MSSC	Time
Random	3000	11187	1000	11627	945	9727	6
start	5000	19796	1000	20089	949	41292	18
Ward	3000	11525	1000	11188	550	11966	37
start	5000	20090	1000	18527	508	19135	145

The results indicate that TS, departing from Ward's solution, obtains the best results. Remarkably, SA performs very well from a random start, possibly because it explores a large variety of solutions initially, but it obtains poor solutions from Ward's starting solutions. For $n = 5000$, an adjustment of the initial temperature to 25 is necessary. The TS parameter choice appears to be more robust. However, for further research, a balanced set of large cluster instances needs to be developed.[1]

4 Conclusions and Future Research

In this paper, we present a tabu search approach that is applicable to large cluster instances. We found that this approach produces very good

[1] The instances from Milligan et al. are only up to 200 subjects

solutions in comparison with the commonly used methods simulated annealing and k-means. Meta-heuristics are most effective when the clusters are not well-separated.
An interesting direction of future research is to compare cluster methods for other similarity measures than the commonly studied MSSC; see [10]. Another interesting direction of future research is the application of meta-heuristics to very large databases.

References

1. K.S. Al-Sultan and M. Maroof Khan. Computational Experience on Four Algorithms for the Hard Clustering Problem. Pattern Recognition Letters, 17:295-308, 1996.
2. M.J. Brusco, J.D. Cradit, and S. Stahl. A Simulated Annealing Heuristic for a Bicriterion Partitioning Problem in Market Segmentation. Journal of Marketing Research, 39:99-109, 2002.
3. F. Glover and M. Laguna. Tabu Search. Kluwer Academic Publishers, 1999.
4. A.K Jain, M.N. Murty, and P.J. Flynn. Data Clustering: A Review. ACM Computing Surveys, 31(3):264-323, 1999.
5. J. MacQueen. Some Methods for Classification and Analysis of Multivariate Observations. In Proceedings of Fifth Berkeley Symposium on Mathematical Statistics and Probability, pages 281-297. University of California Press, 1967.
6. G.W. Milligan and M.C. Cooper. Methodology Review: Clustering Methods. Applied Psychological Measurement, 11(4):329-354, December 1987.
7. G. Punj and D.W. Stewart. Cluster Analysis in Marketing Research: Review and Suggestions for Application. Journal of Marketing Research, 20:134-148, 1983.
8. C.S. Sung and H.W. Jin. A Tabu-Search-Based Heuristic for Clustering. Pattern Recognition, 33:849-858, 2000.
9. J.H. Ward. Hierarchical Grouping to Optimize an Objective Function. Journal of the American Statistical Assciation, 58(301):236-244, March 1963.
10. M.Wedel and W.A. Kamakura. Market Segmentation: Conceptual and Methodological Foundations. International Series in Quantitative Marketing. Kluwer Academic Publishers, 2nd edition, 2000.

Parallel Computation for the Bandwidth Minimization Problem

Khoa T. Vo and Gerhard Reinelt

Institute of Computer Science, University of Heidelberg, Germany
{khoa.vo,gerhard.reinelt}@informatik.uni-heidelberg.de

1 Introduction

The bandwidth minimization problem is a classical combinatorial optimization problem that has been studied since around 1960. It is formulated as follows. Given a connected graph $G = (V, E)$ with n nodes and m edges, find a labeling π, i.e. a bijection between V and $\{1, 2, \ldots, n\}$, such that the maximum difference $|\pi(u) - \pi(v)|$, $uv \in E$, is minimized. This problem is NP-hard even for binary trees [3]. Though much work on the bandwidth problem has been done, the gaps between best known lower and upper bounds for benchmark problems is still large.
Applications of the bandwidth problem can be found in many areas: solving systems of linear equations, data storing, electronic circuit design, and recently in topology compression of road networks [7].
Due to the hardness of bandwidth minimization, much research has dealt with heuristic methods. These range from structured methods to metaheuristic methods, genetic algorithm and scatter search. The reader can refer to [1] for detailed references in this areas. For solving the problem to optimality, dynamic programming and branch-and-bound have been used but with little success for sparse graphs. Caprara and Salazar [2] have proposed new lower bounds and strong integer linear programming (ILP) formulations to compute the bounds more effectively. Their method is therefore used in our parallel implementation documented in this paper. We will report about computational results for problem instances with up to 200 nodes.
In the following let $d(u, v)$ denote the distance between two nodes $u, v \in V$, i.e. the smallest number of edges of a path connecting them. The maximum distance from u to nodes $s \in S$ is denoted by $d(u, S)$ and $N_k(u)$ is the set of nodes whose distance from u is at most k.

B. Fleischmann et al.(Eds.), *Operations Research Proceedings 2008*,
DOI: 10.1007/978-3-642-00142-0_78, © Springer-Verlag Berlin Heidelberg 2009

2 The Algorithm

The algorithms basically checks whether a specified bandwidth ϕ can be realized. To this end it successively assigns labels to nodes and checks bounds on the overall bandwidth implied by these *partial labelings*. If all nodes can be labeled, then bandwidth ϕ is feasible.
We use the strongest ILP relaxation from Caprara and Salazar [2]. The partial labeling is performed from two sides. $L = \{u_h : \pi_{u_h} = h, h = 1, ..., k\}$ is the set of nodes labeled from left to right and $R = \{v_i : \pi_{v_i} = n - i + 1, i = 1, ..., q\}$ the set of nodes labeled from right to left. F is the set of remaining free nodes.
For a node $v \in F$, $f_v = \max\{(n - i + 1) - h\phi : u_i \in R \cap N_h(v)\}$ is its smallest feasible label and $l_v = \min\{h\phi + i : u_i \in L \cap N_h(v)\}$ is its largest feasible label, where $v \in F$ is at distance h from $u_i \in L$ and $u_i \in R$. A layout is only feasible if $f_v \leq \pi_v \leq l_v$. In addition, $N_1^L(v) = \{u : d(u, L) = d(v, L) - 1\}$ is the set of nodes whose distance to L is shorter than that of v by one unit, $N_1^R(v)$ is defined analogously. The ILP relaxation used by Caprara and Salazar is the following.

$$
\begin{array}{llr}
\min & \phi & \\
& f_v \leq \pi_v \leq l_v, & v \in F, \\
& \pi_{u_h} = h, & h = 1, ..., k, \\
& \pi_{v_i} = n - i + 1, & i = 1, ..., q, \\
& \pi \in \Pi, & \\
& l_{u_h} = h, & h = 1, ..., k, \\
& \phi \geq i - h, & (u_i, u_h) \in E
\end{array}
$$

$$
l_v = \left\{ \begin{array}{ll} \max \tau_v^v & \\ \phi \geq \tau_v^v - \tau_u^v, & u \in N_1^L(v) \\ \tau_u^v \leq l_u, & u \in N_1^L(v) \\ \tau^v \in \Pi_{|N_1^L(v) \cup \{v\}|} & \end{array} \right. , v \in F \qquad (1)
$$

$$
\begin{array}{lr}
f_{v_i} = n - i + 1, & i = 1, ..., q, \\
\phi \geq i - h, & (v_i, v_h) \in E
\end{array}
$$

$$
f_v = \left\{ \begin{array}{ll} \min \rho_v^v & \\ \phi \geq \rho_u^v - \rho_v^v, & u \in N_1^R(v) \\ \rho_u^v \geq f_u, & u \in N_1^R(v) \\ \rho^v \in \Pi_{|N_1^R(v) \cup \{v\}|} & \end{array} \right. , v \in F
$$

We now describe how feasibility of a partial labeling is tested.
The function *ExpandLayout* expands the partial layout while maintaining feasibility, i.e., $f_v \leq \pi_v \leq l_v$ for all remaining free nodes. First new

bounds for f_v and l_v are set according to their definitions and this constraint is tested. If a violation is found, the algorithm stops. If not, the second step proceeds to tighten the bound as described in the inner ILPs for l_v and f_v as in ILP (1), and the bounds are tested again. If no violation is detected, then *Labeling* is recursively called by *ExpandLayout* to label the next position. The procedures stop if all nodes have been labeled (meaning that a solution has been found) or if all possibilites have been tried without success.
The algorithm is outlined in Figure 1.

```
Init: F = V; l_v = n; f_v = 1; k = 0; q = 0
bool Labeling (φ, direction, F, k, q, f, l)
begin
  if F==∅ then
    return true
  end if
  Left = {v ∈ F : f_v = k + 1}; {nodes can be labeled from the left side}
  Right = {v ∈ F : l_v = n − q}; { nodes can be labeled from the right side}
  if direction==LeftToRight OR |Left| ≤ |Right| then
    for v ∈ Left do
      F = F\{v}; k = k + 1; π_v = k;
      ExplandLayout(LEFT);
    end for
  else
    for v ∈ Right do
      F = F\{v}; q = q + 1; π_v = n − q + 1;
      ExpandLayout(RIGHT);
    end for
  end if
end
```

Fig. 1. Checking Bandwidth Feasibility

As shown in [2] the execution time of *Labeling* to test one position is $O(n^2 logn + nm)$. Notice that the worst-case execution time of feasibility testing for a given bandwidth value is still exponential. If one is only interested in finding good lower and upper bounds one could set a CPU time limit to the algorithm. But since we want to compute exact optimum solutions we do not use this option. Depending on the indicated direction, the algorithm performs labeling from either both sides or only from left to right.

3 Parallelization

Obviously, the above algorithm is suited for parallelization. We have chosen the open-source software framework ALPS [8], a COIN-OR project, for implementing a parallel version. ALPS supports branch-and-bound (and also branch-and-cut which might be useful for subsequent research on the bandwidth problem) and is designed for Linux systems running MPICH [6]. It uses a master-hub-worker scheme and supports load balancing which is very important in our case.
Information in an ALPS-based program is denoted as *knowledge*, derived from class `AlpsKnowledge`. There are four types of knowledge:

- *Model*
  contains the data describing the graph and the bandwidth problem. It is implemented in the class `BwModel`.
- *Solution*
  description of feasible solution (implemented in class `BwSolution`).
- *Tree node*
  describes the data and methods in a node of the tree search (implemented in class `BwTreeNode`).
  The member function *process()* determines how a node will be processed (fathomed, solution found, or continuation of search) and the function *branch()* decides how child nodes are generated.
- *Subtree*
  contains the hierarchy of tree nodes (in class `AlpsSubTree`). ALPS communicates between processes at this level rather than the node level to avoid transfering many times.

Knowledge is stored and managed in so-called knowledge pools. Communication is carried out through knowledge brokers. These brokers are available for all processes and are responsible for sending, receiving, and routing all of the knowledge mentioned above. ALPS currently supports two communication protocols: single-process and MPICH on Linux clusters. Searching is driven by the functions *process()* and *branch()* of the class `BwTreeNode`. Each tree node keeps its partial labeling state with an object of class `PartialLabeling` whose member function *Labeling()* realizes the algorithm of Figure 1.
The search is initialized with a single root node and is divided into three phases: *ramp-up* for initialization, *steady* for searching, and *ramp-down* at the end. For balancing load effectively ALPS defines the concepts of static load balancing in the ramp-up phase and dynamic load balancing in the steady phase. We use the so-called *spiral* scheme for static load balancing, where the master generates child nodes and distributes them

to hubs where they do the same to their workers. The reason is that, initially nodes are created evenly, as can be seen in the algorithm in Figure 1.

In the steady phase, the nodes in one branch of the search tree may be pruned so this branch has little work left, while the other branches are running heavily. Therefore dynamic load balancing plays an important role. ALPS manages this at two levels: *intra-cluster* dynamic load balancing performed by a hub to balance its cluster, and *inter-cluster* dynamic load balancing by the master to reallocate jobs between clusters. We use dynamic load balacing at both levels in our implementation.

Best-first search is chosen as the overall search strategy. Notice that no quality value or lower bounds for subproblems have been used in [2]. It is possible to introduce bounds by solving the relaxation of ILP (1) and using its objective value for guiding the tree search. However, the running time gets larger with the addition of using an LP solver. This feature is not yet included in the current system, but we plan to have it in the next version.

It turned out to be helpful to run a fast heuristic in the ramp-up phase. We used the enhanced GPS heuristic [7]. While the running time for finding lower bounds did not get smaller, upper bounds can be found more quickly for most instances in the benchmark suite. For example, upper bound for instance `impcol_b` (59 nodes) can be found in 37.9 seconds while the program without this heuristic takes longer than 2 hours. In the case of instance `west0156` (156 nodes), the numbers are 4.5 seconds compared to more than 1.5 hours.

4 Computational Results

In this section we report the computational results of our program. We compare with the best known results in the literature [1] and use the same benchmark suite as in their work. Computations were carried out on the cluster of the Interdisciplinary Center for Scientific Computing (IWR) in Heidelberg ([4]) This cluster consists of 127 nodes, each equipped with two Dual Core AMD 2.8 GHz processors with 8 GB RAM memory running Debian 4.0. The cluster uses Myricom 10G for the network layer with MPICH-MX and the program is compiled with an Intel compiler. The running time was limited to 30 minutes. Only new results are reported in Table 1, we obtained equally good results for other instances in [1].

Table 1. Computational Results

Instance	Nodes	Edges	Best in [1]			Our result		
			LB	UB	% Gap	LB	UB	% Gap
bcspwr03	118	179	10	11	10.00	10	10	0.00
bcsstk22	110	254	9	10	11.11	10	10	0.00
can__144	144	576	13	14	7.69	13	13	0.00
gre__115	115	267	20	24	20.00	21	23	9.52
gre__185	185	650	17	22	29.41	18	22	22.22
impcol_b	59	281	19	21	10.53	20	20	0.00
west0156	156	371	34	37	8.82	34	36	5.88
will199	199	660	57	67	17.54	59	69	16.95
Average					8.93			6.71

5 Conclusion

The bandwidth minimization problem is a difficult NP-hard problem, and solving it to optimality remains challenging. With a parallel implementation we could find better lower and upper bounds for benchmark problems with up to 200 nodes. However, this is only a first step. In the future, we plan like to strengthen our parallel solver with valid inequalities and move to a branch-and-cut scheme instead of branch-and-bound.

References

1. Campos V, Pinana E, Marti R (2006) Adaptive Memory Programming for Matrix Bandwidth Minimization. Technical Report, University of Valencia
2. Caprara A, Salazar JJ (2005) Laying Out Sparse Graphs with Provably Minimum Bandwidth. INFORMS Journal on Computing 17:356–373
3. Garey M, Graham R, Johnson D, Knuth D. (1978) Complexity results for bandwidth minimization. SIAM J. Applied Mathematics 34:477–495
4. IWR Heidelberg Linux Cluster System. *http://helics.uni-hd.de*
5. Ralphs TK (2003) Parallel branch and cut for capacitated vehicle routing. Parallel Computing 29: 607–629
6. MPICH2 *http://www-unix.mcs.anl.gov/mpi/mpich*
7. Suh J, Jung S, Pfeifle M, Vo KT, Oswald M, Reinelt G (2007) Compression of Digital Road Networks. Advances in Spatial and Temporal Databases, 10th International Symposium, SSTD 2007:423–440
8. Xu Y, Ralphs TK, Ladanyi L, Saltzman M (2005) ALPS: A Framework for Implementing Parallel Search Algorithms. The Proceedings of the Ninth INFORMS Computing Society Conference, 319

Strengthening Gomory Mixed-Integer Cuts

Franz Wesselmann[1], Achim Koberstein[1], and Uwe H. Suhl[2]

[1] Decision Support & Operations Research Lab, Universität Paderborn, 33098 Paderborn, Germany
{wesselmann,koberstein}@dsor.de

[2] Institut für Produktion, Wirtschaftsinformatik und Operations Research, Freie Universität Berlin, 14195 Berlin
uwe.suhl@fu-berlin.de

Summary. Gomory mixed-integer cuts play a central role in solving mixed-integer linear programs and are an integral part of state-of-the-art optimizers. In this paper we present some of the the existing approaches for improving the performance of the Gomory mixed-integer cut. We briefly discuss the ideas of the different techniques and compare them based on computational results.

1 Introduction

Mixed-integer programming has become a widely used tool for modeling and solving real-world management problems. Applications originate from different domains such as telecommunication, transportation or production planning. Recently there has been an enormous performance improvement of standard software for solving mixed-integer programs (MIPs). These improvements are due to several reasons. Besides faster computers and improved implementations of the simplex method [11], enhanced cutting plane methods have brought about a major reduction of the time needed to solve many MIPs to proven optimality.
One of the most well-known cutting planes is the Gomory mixed-integer (GMI) cut. These cutting planes were proposed by Gomory [10] in the 1960s and had the reputation of being useless in practice for a long time. This changed in the 1990s [5] when GMI cuts were integrated into state-of-the-art optimizers like MOPS [13] and proved their practical value. Due to their computational importance [8], improvements in the performance of the GMI cuts are likely to cause further progress in solving hard real-world MIPs.
In the remainder of this paper we will briefly recapitulate the separation of GMI cuts and give a short introduction into three known approaches

B. Fleischmann et al.(Eds.), *Operations Research Proceedings 2008*,
DOI: 10.1007/978-3-642-00142-0_79,

to strengthen them: k-cuts, reduce-and-split cuts and lift-and-project cuts. Finally, we will present some computational results.

2 Gomory Mixed-Integer Cuts

Consider a mixed-integer linear program in the form

$$\min\left\{c^T x : Ax \geq b, x \geq 0, x_j \in \mathbb{Z} \text{ for } j \in N_I\right\} \qquad \text{(MIP)}$$

where c, $x \in \mathbb{R}^n$, $A \in \mathbb{R}^{m \times n}$ and $N_I \subseteq N = \{1, \ldots, n\}$. The linear programming relaxation (LP) of (MIP) is obtained by omitting the integrality conditions on x_j for all $j \in N_I$.
Given a basis of (LP), let B index the basic and J index the nonbasic variables. Furthermore, let x^* denote the optimal solution to (LP). A Gomory mixed-integer cut is generated from a simplex tableau row associated with a basic integer-constrained variable having a fractional value. Suppose that such a tableau row is given in the form

$$x_i = \bar{a}_{i0} - \sum_{j \in J} \bar{a}_{ij} x_j \tag{1}$$

for some $i \in B \cap N_I$. The GMI cut generated from (1) is

$$\sum_{j \in J \cap N_I : f_{ij} \leq f_{i0}} f_{ij} x_j + \sum_{j \in J \cap N_I : f_{ij} > f_{i0}} \frac{f_{i0}(1 - f_{ij})}{1 - f_{i0}} x_j + \sum_{j \in J \setminus N_I : \bar{a}_{ij} \geq 0} \bar{a}_{ij} x_j + \sum_{j \in J \setminus N_I : \bar{a}_{ij} < 0} \frac{f_{i0}(-\bar{a}_{ij})}{1 - f_{i0}} x_j \geq f_{i0}, \tag{2}$$

where $f_{ij} = \bar{a}_{ij} - \lfloor \bar{a}_{ij} \rfloor$ and $f_{i0} = \bar{a}_{i0} - \lfloor \bar{a}_{i0} \rfloor > 0$. The validity of the inequality (2) can easily be checked by applying mixed-integer rounding to the tableau row (1).
The quality of a GMI cut $\alpha^T x \geq \beta$ can, for instance, be measured by computing the Euclidean distance between the hyperplane defined by the cut and the solution x^*:

$$d(\alpha, \beta) = \frac{\beta - \alpha^T x^*}{\|\alpha\|} = \frac{\beta}{\|\alpha\|}. \tag{3}$$

3 Strengthening

Developing algorithms to strengthen Gomory mixed-integer cuts is a quite vital research topic. In the following we will present some of the latest progress.

K-Cuts

A simple idea to obtain different variations of GMI cuts from a single simplex tableau row was studied by Cornuéjols et al. [9]. They propose to multiply a given tableau row (1) by an integer $k \neq 0$ before generating a GMI cut. This multiplication affects the size of the fractional part of the right-hand side and consequently produces a set of different GMI cuts. In [9] it is proved that in the pure integer case k-cuts perform variable-wise better than the GMI cut with exactly fifty percent probability. Unfortunately, coefficients of the continuous variables tend to deteriorate with increasing values of k in the mixed-integer case.

Reduce-and-Split Cuts

Andersen et al. [1] developed an alternative idea to improve GMI cuts. Their approach is based on the observation that the size of the coefficients on the continuous variables in a GMI cut depends on the size of the corresponding entries in the simplex tableau row and that the size of these coefficients affects the quality of the cut (3). Their algorithm is designed to reduce the size of the coefficients on the continuous variables by forming linear combinations of simplex tableau rows. Consider an additional simplex tableau row:

$$x_k = \bar{a}_{k0} - \sum_{j \in J} \bar{a}_{kj} x_j \tag{4}$$

To improve the GMI cut generated from (1), the latter row is combined with (4) by adding $\delta \in \mathbb{Z}$ times (4) to (1):

$$x_i + \delta x_k = \bar{a}_{i0} + \delta \bar{a}_{k0} - \sum_{j \in J} \left(\bar{a}_{ij} + \delta \bar{a}_{kj}\right) x_j$$

As the procedure aims at reducing the coefficients of the variables $j \in J \backslash N_I$, the value of δ is chosen such that it minimizes the function

$$h(\delta) = \sum_{j \in J \backslash N_I} \left(\bar{a}_{ij} + \delta \bar{a}_{kj}\right)^2 .$$

Lift-and-Project Cuts

Lift-and-project cuts were proposed by Balas et al. [3] and are a special class of disjunctive cuts [2]. These cutting planes are based on the argument that every 0-1 (or binary) variable x_i has to satisfy the disjunction

$$(-x_i \geq 0) \vee (x_i \geq 1) \tag{5}$$

in a feasible solution to a (mixed) 0-1 program. Balas et al. showed that it is possible to convexify the set of these feasible points by sequentially generating cutting planes which are valid for disjunctions of the form (5) on the 0-1 variables. The most-violated (deepest) lift-and-project cut $\alpha x \geq \beta$ from the above disjunction can be obtained by solving a so-called cut generating linear program (CGLP)

$$\begin{array}{llllll} \min \alpha x^* - \beta & & & & & \\ s.t. & & & & & \\ \alpha & -uA & & +u_0 e_i & & \geq 0 \\ \alpha & & -vA & & -v_0 e_i & \geq 0 \\ -\beta & +ub & & & & = 0 \\ -\beta & & +vb & & +v_0 & = 0 \\ & ue & +ve & +u_0 & +v_0 & = 1 \end{array} \tag{CGLP$_i$}$$

where $e = (1, ..., 1)^T$ and e_i is the i-th unit vector. The last equation is called normalization constraint and truncates the polyhedral cone defined by the remaining inequalities. Computational experience within a branch-and-cut framework is reported in [4].

Balas and Perregaard [6] discovered a precise correspondence between bases of (LP) and (CGLP) which allows for a more efficient generation of lift-and-project cuts. In particular, they presented an algorithm which optimizes (CGLP) by performing pivots on the (LP) simplex tableau. This algorithm can alternatively be seen as a procedure for systematically improving GMI cuts.

4 Computational Results

In this section we report on computational experiments obtained with MOPS 9.13 running on a PC with a 3.4 GHz Intel Xeon processor and 4 GB of RAM. We chose a test set of 126 instances taken from the MIPLIB 3.0 [7] and the Mittelmann MIP collection [12].

The four variants we consider are plain GMI cuts, k-cuts [9], a variant of the reduce-and-split cuts [1] and the Balas-Perregaard procedure [6] for generating lift-and-project cuts from the simplex tableau.

We use these cutting planes to tighten the LP relaxation at the root node. Then, we assess their computational effectiveness by solving the instances in the test set with the branch-and-bound code of MOPS (cut-and-branch). The time limit for solving each instance is one hour.

First, we evaluate the performance of the different cut generators by looking at the total number of instances each of them is able to solve to optimality. Table 1 shows that lift-and-project cuts and reduce-and-

Table 1. Number of instances solved by variants

variant	# instances solved in t seconds				
	$t \le 10$	$10 < t \le 600$	$600 < t \le 1800$	$1800 < t \le 3600$	$3600 < t$
GMI cuts	31	33	12	7	43
K-cuts	31	35	10	8	42
L&P cuts	29	37	14	5	41
R&S cuts	29	37	9	10	41

split cuts perform worse on some easy instances due to the amount of additional computational work. On the other hand, they allow for solving two additional instances to optimality within the imposed time limit.

In a second experiment, we compare lift-and-project (and reduce-and-split) cuts to GMI cuts based on the partitioning of the solution times defined in Table 1 to more precisely analyze the benefit of strengthened GMI cuts. The underlying question is how the strengthened cuts per-

Table 2. Comparing lift-and-project, reduce-and-split and GMI cuts

variant	geometric mean of solution times on instances solved with GMI cuts in t seconds				
	$t \le 10$	$10 < t \le 600$	$600 < t \le 1800$	$1800 < t \le 3600$	$3600 < t$
GMI cuts	0.89	113.86	912.87	2462.63	3600.00
L&P cuts	1.45	123.60	988.31	1013.05	3321.34
R&S cuts	1.19	111.09	885.19	2417.88	3058.88
# instances	31	33	12	7	43

form on instances which are easy or relatively hard to solve with GMI cuts. Table 2 shows that lift-and-project cuts are performing particularly well on instances that are hard to solve using only GMI cuts. For example, consider the instances which can be solved with GMI cuts in between half an hour and an hour of running time (i.e. the fifth column in Table 2). The geometric mean of the solution times needed to solve the 7 instances in this group decreases by about 50% using lift-and-project cuts.

References

1. Kent Andersen, Gérard Cornuéjols, and Yanjun Li. Reduce-and-split cuts: Improving the performance of mixed-integer Gomory cuts. Management Science, 51(11):1720-1732, 2005.
2. Egon Balas. Disjunctive programming. Annals of Discrete Mathematics, 5:3-51, 1979.
3. Egon Balas, Sebastián Ceria, and Gérard Cornuéjols. A lift-and-project cutting plane algorithm for mixed 01 programs. Mathematical Programming, 58(1-3):295-324, 1993.
4. Egon Balas, Sebastián Ceria, and Gérard Cornuéjols. Mixed 0-1 programming by lift-and-project in a branch-and-cut framework. Management Science, 42(9):1229-1246, 1996.
5. Egon Balas, Sebastian Ceria, Gérard Cornuéjols, and N. Natraj. Gomory cuts revisited. Operations Research Letters, 19:1-9, 1996.
6. Egon Balas and Michael Perregaard. A precise correspondence between lift-and-project cuts, simple disjunctive cuts, and mixed integer Gomory cuts for 0-1 programming. Mathematical Programming Series B, 94(2-3):221-245, 2003.
7. Robert E. Bixby, Sebastián Ceria, Cassandra M. McZeal, and Martin W. P. Savelsbergh. An updated mixed integer programming library: MIPLIB 3.0. Optima, 58:12-15, 1998.
8. Robert E. Bixby and Edward Rothberg. Progress in computational mixed integer programming - a look back from the other side of the tipping point. Annals of Operations Research, 149(1):37-41, 2007.
9. Gérard Cornuéjols, Yanjun Li, and Dieter Vandenbussche. K-cuts: A variation of Gomory mixed integer cuts from the LP tableau. INFORMS Journal on Computing, 15(4):385-396, 2003.
10. Ralph E. Gomory. An algorithm for the mixed integer problem. Technical Report RM-2597, The RAND Cooperation, 1960.
11. Achim Koberstein. The Dual Simplex Method, Techniques for a fast and stable implementation. PhD thesis, Universität Paderborn, 2005.
12. Hans Mittelmann. Decision tree for optimization software: Benchmarks for optimization software. http://plato.asu.edu/bench.html.
13. Uwe H. Suhl. MOPS - Mathematical OPtimization System. European Journal of Operations Research, 72:312-322, 1994.

Part X

Forecasting, Econometrics and Game Theory

Applied Flexible Correlation Modeling

Frederik Bauer and Martin Missong

Bremen University
{frederik.bauer,missong}@uni-bremen.de

Estimation of correlations of asset returns lies at the core of modern portfolio management. For the seminal Dynamic Conditional Correlation model [1], several generalizations have been proposed recently. In this contribution, we focus on the Flexible Dynamic Conditional Correlation model proposed in [2]. Using both simulation exercises and applications to observed return data, we show that the flexible specification performs well only in very restrictive cases, contradicting the "flexibility" of the approach. However, our results indicate that model performance can be improved substantially by a particular adjustment of the variance specification.

1 Dynamic Conditional Correlation Models

The Dynamic Conditional Correlation (DCC) model [1] separates variance modelling from correlation modelling. Let $\mathbf{r}_t$ denote the $N \times 1$ time-series vector collecting N series of returns at time t. In the following $\mathbf{R}_t$ will denote the $N \times N$ time varying correlation matrix and $\mathbf{D}_t$ the $N \times N$ time varying diagonal matrix with the conditional standard deviations, i.e. with elements $\mathrm{diag}(\sqrt{h_{1t}}, \dots, \sqrt{h_{Nt}})$. Assuming normally distributed returns, the seminal dynamic conditional correlation (DCC) introduced in [1] is then given by:

$$\mathbf{r}_t|_{I_{t-1}} \sim N(\mathbf{0}, \mathbf{D}_t\mathbf{R}_t\mathbf{D}_t), \tag{1}$$

$$\mathbf{D}_t{}^2 = \mathrm{diag}\{\boldsymbol{\omega}\} + \mathrm{diag}\{\boldsymbol{\kappa}\} \circ \mathbf{r}_{t-1}\mathbf{r}'_{t-1} + \mathrm{diag}\{\boldsymbol{\lambda}\} \circ \mathbf{D}_{t-1}{}^2, \tag{2}$$

$$\epsilon_t = \mathbf{D}_t^{-1}\mathbf{r}_t, \tag{3}$$

$$\mathbf{Q}_t = \mathbf{S}(1 - \alpha - \beta) + \alpha\epsilon_{t-1}\epsilon'_{t-1} + \beta\mathbf{Q}_{t-1}, \tag{4}$$

$$\mathbf{R}_t = \mathrm{diag}\{\mathbf{Q}_t\}^{-1}\mathbf{Q}_t\{\mathbf{Q}_t\}^{-1}. \tag{5}$$

B. Fleischmann et al.(Eds.), *Operations Research Proceedings 2008*,
DOI: 10.1007/978-3-642-00142-0_80, © Springer-Verlag Berlin Heidelberg 2009

According to (2), conditional variances follow univariate GARCH processes and $\boldsymbol{\epsilon_t}$ in (3) denotes the GARCH standardized returns. Other specifications in the first step conditional variance estimation are also allowed, like e.g. the GJR. With (4) and (5), a GARCH structure is also assumed for the dynamics of the conditional correlation matrix of these returns. The restriction $\alpha + \beta < 1$ on the scalar parameters α and β ensures that the correlation process is mean reverting. In (4), the matrix of constant terms is given by $\mathbf{S}(1 - \alpha - \beta)$, where $\mathbf{S}$ can be estimated consistently by the sample correlation matrix of normalized returns, $\boldsymbol{\epsilon_t}$. The DCC will thus have $N(N-1)/2 + 2$ parameters to be estimated in the second step, i.e. the conditional correlation estimation, where only two parameters will be estimated by maximum likelihood. Note that in the DCC model the same dynamic correlation structure applies to all assets. This may be a sensible restriction only for a small collection of assets. The Flexible Dynamic Correlation (FDCC) model presented by [2] addresses this feature of the DCC model. They propose to replace (4) by

$$\mathbf{Q}_t = \boldsymbol{\gamma}\boldsymbol{\gamma}' + \boldsymbol{\alpha}\boldsymbol{\alpha}' \circ \boldsymbol{\epsilon}_{t-1}\boldsymbol{\epsilon}_{t-1}' + \boldsymbol{\beta}\boldsymbol{\beta}' \circ \mathbf{Q}_{t-1}, \tag{6}$$

with $\boldsymbol{\alpha} = [\alpha_1 * \mathbf{i_1}' \quad \alpha_2 * \mathbf{i_2}' \dots \alpha_w * \mathbf{i_w}']'$ and $\mathbf{i_w}$ being a vector of ones with size equal to the number of elements in the sector w. $\boldsymbol{\beta}$ and $\boldsymbol{\gamma}$ are defined accordingly. As $\boldsymbol{\alpha}$, $\boldsymbol{\beta}$ and $\boldsymbol{\gamma}$ each consist of W different sector-parameters, the total number of parameters in the FDCC reduces to $3W$. According to [2] the correlation process is stable as long as $\alpha_i\alpha_j + \beta_i\beta_j < 1$ for all $i, j = 1, 2, \dots W$.

2 Correlation Dynamics in the FDCC Model

The increased flexibility of the FDCC when compared to the DCC stems from the fact that in the FDCC correlation dynamics are only equal for return subsets. However, the FDCC turns out to be by far more restrictive than the FDCC with respect to the long run correlation dynamics: The long run correlation matrix in the FDCC is given by $\boldsymbol{\gamma}\boldsymbol{\gamma}' \div (1 - \boldsymbol{\alpha}\boldsymbol{\alpha}' - \boldsymbol{\beta}\boldsymbol{\beta}')$, where $\div$ denotes element-wise division. It follows that this matrix consists of W^2 blocks of equal parameters. Hence assets of the same sectors are assumed to be perfectly correlated in the long-run, as the long-run correlation matrix has blocks of ones on the diagonal after applying normalization (5). Therefore, it is an empirical question to determine whether the unrealistic assumption of perfect correlation in the long run leads to an unrealistic short-run behavior of the conditional intra-group correlations, too.

We illustrate this point by referring to the data used and the estimation results reported in [2]. They used 20 assets divided into three sectors with data of the Italian MIBTEL index. The three sectors are the industrial sector (10), the finance sector (4) and the service sector (6) with number of elements in brackets. This means that each of the sectors consists of further subsectors (20 subsectors in total), which are the returns that are modelled by the FDCC with $N = 20$ and $W = 3$. Reference [2] reports rather high $\hat{\gamma}$-values as compared to the $\hat{\boldsymbol{\alpha}}$ and $\hat{\boldsymbol{\beta}}$ estimates, see Table 1 below (*column 'Billio' subcolumn 'True'*). With these values, "news" introduced through $\boldsymbol{\epsilon}_t\boldsymbol{\epsilon}_t{}'$ in (6) have almost no impact. Furthermore the high $\hat{\gamma}$-values are accumulated heavily because of high $\hat{\boldsymbol{\beta}}$-values, so that the correlation process is governed by the constant terms given by $\boldsymbol{\gamma}\boldsymbol{\gamma}'$. It follows that the conditional correlation process will quickly approach a correlation matrix with blocks of equal elements which stay almost constant. By re-estimating conditional correlations with the parameters in [2], this turned out to be the case after no more than 5 periods: Then the block-structure in $\mathbf{R_t}$ gets evident, and the conditional correlation of returns for, say, the subsectors "Cars" and "Chemicals" approach unity, as these subsectors belong to the same sector ("Industrial sector"). We therefore conjecture that the elements in $\boldsymbol{\gamma}$ should be small. A graphical illustration is given in the next section.
Hence, the particular correlation processes in the FDCC may turn out to be not at all "flexible" in empirical applications. However, it is obvious that the matrix of constant terms, $\boldsymbol{\gamma}\boldsymbol{\gamma}'$ in (6) may be replaced by the variance targeting constraint, i.e. $\boldsymbol{\gamma}\boldsymbol{\gamma}' = (1 - \boldsymbol{\alpha}\boldsymbol{\alpha}' - \boldsymbol{\beta}\boldsymbol{\beta}') \circ \mathbf{S}$. This specification is even mentioned, but not investigated further in [2]. Note that with variance targeting the FDCC proves to be more flexible than the DCC in any case, as in the former α- and β- parameters are allowed to vary across sectors (groups of assets). Hence, we will investigate this specification in detail in the next section.

3 Simulations Results

At first we re-estimated the FDCC model with the same model specification and the MIBTEL-dataset used in [2]. We came to different results especially for the $\boldsymbol{\gamma}$-parameters, see Table 1 (*column 'Estimated' subcolumn 'True'*). A possible explanation for the divergence of results will be given later. Furthermore our estimates show very similar values between sectors (for detailed tests of the DCC vs. the FDCC we refer to [5]).

Next, we simulated data sets according to several specifications of the FDCC model and checked the accuracy of the estimation process using the root mean squared error (RMSE). All simulations generate multivariate normal data governed by an GJR(1,1)-FDCC data generating process (DGP). The underlying parameters of the GJR(1,1)-DGP are those estimated from the MIBTEL dataset. For the FDCC-DGP we use our estimated parameters, parameters reported in [2] and some made up parameters sets. For each simulated dataset we use a sample size of 3100 observations. We discard the first 100 observations to let the process "run in", i.e. we only use the last 3000 simulated observations for the estimation. The simulation was rerun 1000 times. Of course we use 20 assets divided in 3 sectors with 10, 4 and 6 elements respectively as in [2].

Table 1 summarizes the simulation results. The column '*True*' reports the underlying parameters of the data generating process (DGP). The column '*RMSE*' reports the root mean squared error of the estimated parameters. 'Billio' uses the parameters of the standard FDCC reported in [2] to simulate the data. 'Estimated' uses our parameter estimates of the same dataset. The next three columns report the results for the FDCC with variance targeting. 'Different' uses parameters values which are quite dissimilar between sectors. On the other side 'Similar' uses quite similar parameter values between sectors in the DGP. 'Estimated' uses our parameter estimates again, but this time for simulating the FDCC with variance targeting. Only in this last column are the root mean squared errors very small, indicating a reasonable estimation accuracy.

Table 1. Results of FDCC Simulations

	Billio		Estimated		Different		Similar		Estimated	
	True	*RMSE*	*True*	*RMSE*	*True*	*RMSE*	*True*	*RMSE*	*True*	*RMSE*
α_1	0.0946	0.1275	0.0700	0.0266	0.0950	0.0399	0.0225	0.0276	0.0925	0.0025
α_2	0.0645	0.1248	0.0623	0.0419	0.0650	0.0264	0.0300	0.0380	0.0907	0.0034
α_3	0.0193	0.1954	0.0744	0.0378	0.0200	0.0132	0.0500	0.0445	0.1035	0.0030
β_1	0.6306	0.3792	0.9975	0.0011	0.6300	0.2764	0.8900	0.6796	0.9919	0.0011
β_2	0.9845	0.1834	0.9981	0.0091	0.9850	0.3013	0.8600	0.9555	0.9920	0.0013
β_3	0.9246	0.2398	0.9970	0.0061	0.9200	0.5690	0.8300	0.6443	0.9904	0.0011
γ_1	0.7069	11.2870	-0.0025	0.0089						
γ_2	0.9993	1.1483	0.0059	0.0233						
γ_3	0.0391	2.6046	0.0029	0.0232						

Before discussing the simulation results, we use the parameters reported above to illustrate the potential problems with the correlation dynamics in the FDCC as discussed in the preceding section. We readdress the example of conditional return correlations of the subsectors "Cars" and "Chemicals", both belonging to the industrial sector. We will show the temporal sequence of conditional correlations resulting from parameters estimated with the dataset of [2]. They are displayed in Figure 1 for several model specifications: For the FDCC with parameters given in [2], these conditional correlations broadly stick to their long run value 1, which is a questionable feature. On the other hand, for the FDCC model with variance targeting, the conditional correlations show a much more volatile pattern, similar to the dynamics typically reported in empirical applications of the DCC. FDCC parameters from our re-estimation lead to a similar pattern in this particular example.

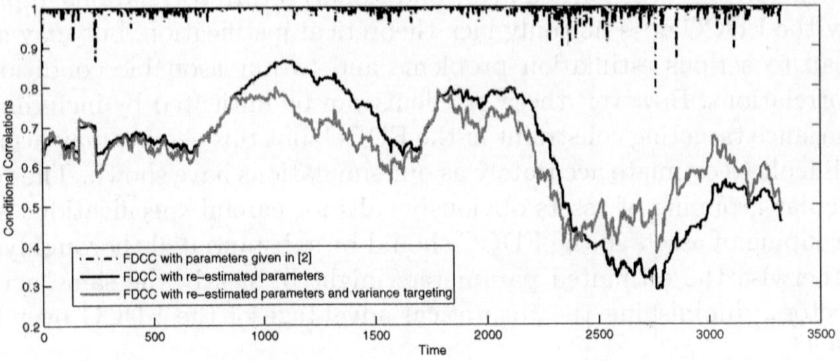

Fig. 1. Correlations of Sectors 'cars' and 'chemicals'

We encountered massive problems while estimating the standard FDCC (without variance targeting) during our simulations. Hence, the "irregular" shape of the likelihood function is one possible explanation why estimates can differ as is the case for our parameters when compared to those those reported in [2].

The parameters of a DCC model can be quite accurately estimated on average. There are some biases in the DCC estimation as discussed in [6], but in no way are they comparable in magnitude to those in the estimation of the FDCC model. This holds especially for the standard FDCC, but also for the FDCC with variance targeting. In some cases can we estimate the FDCC parameters reasonably accurate, especially if the parameters are very similar between sectors and the model is nearly integrated, i.e. $\alpha_i\alpha_j + \beta_i\beta_j$ is nearly equal to one. In this case occurrence of extreme parameter observations seem to vanish, but here the DCC model may outperform the FDCC, because of additional parameter uncertainty in the latter.

In two cases are the parameters of the FDCC estimated with especially great inaccuracy: If parameters are quite different between sectors, or the process is not nearly integrated. The former is exactly the case for which the FDCC is designed. The latter is not frequently observed in practice, as DCC-type models, similarly as GARCH models, are often nearly integrated empirically.

4 Conclusion

In this study, we critically analyzed the FDCC model. We showed that the behavior of the long run (unconditional) return correlations implied by the FDCC does not only lack theoretical justification, but may also lead to serious estimation problems and to unreasonable conditional correlations. However, these problems can be mitigated by including a variance targeting constraint in the FDCC, but this version may also be difficult to estimate accurately as our simulations have shown. The particular grouping of assets obviously calls for careful specification: Any grouping of assets in the FDCC should be substantiated thoroughly, as otherwise the estimated parameters might be nearly the same across sectors, diminishing the theoretical advantage of the FDCC over the DCC.

References

1. Engle R (2002) Dynamic conditional correlation: A simple class of multivariate generalized autoregressive conditional heteroskedasticity models. Journal of Business & Economic Statistics 20:339–350
2. Billio M, Caporin M, Gobbo M (2006) Flexible dynamic conditional correlation multivariate GARCH models for asset allocation. Applied Financial Economics Letters 2:123–130
3. Engle R, Colacito R (2006) Testing and valuing dynamic correlations for asset allocation. Journal of Business & Economic Statistics 24:238–253
4. Engle R, Mezrich J (1996) GARCH for groups. Risk 9:36-40
5. Bauer F, Missong M (2008) Dynamic conditional correlation modelling - Complexity vs. feasibility. Working Paper in Empirical Economics, Bremen University
6. Engle R (2007) Modeling correlations with factor DCC. Festschrift for David Hendry

Forecasting Behavior in Instable Environments

Otwin Becker[1], Johannes Leitner[2], and Ulrike Leopold-Wildburger[2]

[1] Department of Economic Theory I, Alfred Weber Institute, University of Heidelberg

[2] Department of Statistics and Operations Research, Karl Franzens University of Graz, Universitaetsstrasse 15/E3, 8010 Graz, Austria
ulrike.leopold@uni-graz.at

1 Introduction

In many economic situations periodically occurring changes in behavior of the involved agents can be observed. These changes have the characteristics of abrupt structural breaks. The behavior often seems to switch between regimes as if there were constant relationships between economic variables between these breaks. Examples are the alternating price determination of sellers and buyers on a market out of equilibrium or the periodical development of price cartels and the resulting switches in prices. In this study, we want to analyze the expectation formation of participants of a laboratory experiment subject to regime switches.

Despite the practical relevance there is hardly any experimental contribution regarding the reaction of economic decision-makers to structural breaks. The only systematic experiment in this context was performed by [2]. The authors found inconsistent evidence on the performance of judgmental forecasts versus statistical procedures in the literature and therefore wanted to experimentally test the individual performance of subjects. The noise levels of the time series, the type of break (abrupt or creeping) and the direction of the break was systematically varied between 10 time series. The participants were told that the time series may contain structural changes. The judgmental forecasts were found to perform significantly worse than statistical procedures. The subjects were trying to read too much signal into the series and their forecasts contained excessive noise.

In our experiment the forecasting performance will be analyzed but the main interest is the explanation of the average forecasts in order to understand how the subjects react on the break. The most simple case

B. Fleischmann et al.(Eds.), *Operations Research Proceedings 2008*,

DOI: 10.1007/978-3-642-00142-0_81,

for a structural break in our data generating process is a constant shift. Three time series are applied in the experiment subject to one break and two breaks respectively. Between the breaks the data generating process remains constant. By these means the behavior of the subjects in several regimes and their reactions on the breaks can be observed. Recently [1] presented a simple heuristic for the modeling of average forecasts of the subjects. The authors showed that the model forecasts the behavior of the subjects better than the Rational Expectations Hypothesis (REH) when indicators are in the information set of the participants. This heuristic is restricted to stable and stationary time series. We will apply a modified version of the model to the forecasts in the setting with the structural breaks.
We find that the behavior of the subjects after the break is best described by a transition phase. When the new level of the series has established several periods after the break the information before the break (i.e. especially earlier turning points) is gradually ignored. We also find that the human performance compared to statistical procedures is rather poor.

2 The Experiment

The task of the participants is the judgmental prediction of a time series, i.e. forecasting the next period by eyeballing the past observations without any help from statistical or econometrical models. The time series s_t^1, s_t^2 and s_t^3 are used for three different versions of the experiment. These three series are derived from the stationary time series x_t. The base series x_t is a realization of the stochastic difference equation

$$x_t = x_{t-1} - int(\frac{1}{2} \cdot x_{t-2}) + u_t \tag{1}$$

with $x_1 = 7, x_2 = 12$, the endogenous variable x_t and the white noise u_t. The uniformly distributed variable u_t represents the realizations drawn from a six-sided dice, i.e. $u_t \in \{1, 2, 3, 4, 5, 6\}$. The function *int* ensures that all values of the base series and the indicators are integer. The time series has 60 realizations. In Figure 1 x_t and s_t^1 are graphed. For s_t^1 a constant shift of five units was added to x_t in period 22:

$$s_t^1 = \begin{cases} x_t & \text{for } 1 \leq t \leq 21 \\ x_t + 5 & \text{for } 22 \leq t \leq 60 \end{cases} \tag{2}$$

The break was in an upswing period of x_t so the time series changed from $s^1_{21} = 6$ to $s^1_{22} = 14$ whereas the maximum value of s^1_t observed until period 21 is 12 and the maximum change of the time series is only 5 (see Figure 1). For this reason the break was obvious to the participants. The two remaining time series were generated in order to analyze cases of unapparent breaks. For s^2_t a constant positive shift of 7 units was added to x_t in period 26. Despite its dimension this break is hidden in a downswing. The second break retracts s^2 to the initial level of x_t. The downward break occurs in an upswing period and therefore it is also unapparent.

$$s^2_t = \begin{cases} x_t & \text{for } 1 \leq t \leq 25 \\ x_t + 7 & \text{for } 26 \leq t \leq 43 \\ x_t & \text{for } 44 \leq t \leq 60 \end{cases} \tag{3}$$

The third time series s^3_t was generated with two positive shifts:

$$s^3_t = \begin{cases} x_t & \text{for } 1 \leq t \leq 25 \\ x_t + 7 & \text{for } 26 \leq t \leq 41 \\ x_t + 14 & \text{for } 42 \leq t \leq 60 \end{cases} \tag{4}$$

The first structural break in s^3 is identical to s^2 and the second shift is placed in periods in which their occurrence was not obvious.

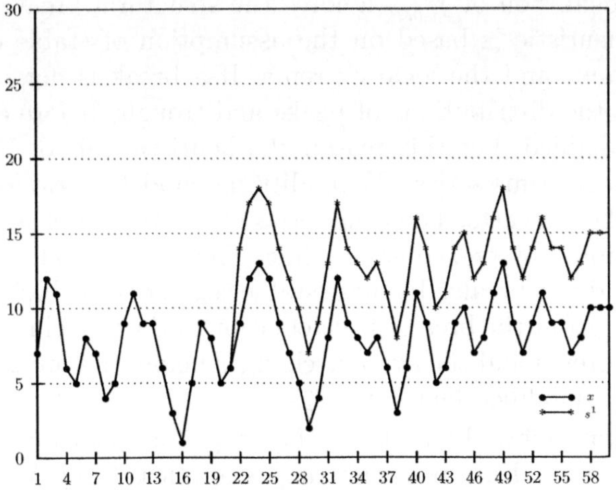

Fig. 1. The time series x_t and s^1_t

The time series were unlabelled and no context was given to the subjects. They were not made aware of the fact that structural breaks could occur. The chart graphing all past observations in each period was the only source of information for their decision making.
In the experiment 120 subjects participated altogether, 40 in each version. Each subject participated only once and predicted one of the three time series. All of these subjects were undergraduate students of economics or business administration without special knowledge of time series analysis.
The experiment was computer-based. The participants did not see a history of past values at the beginning of the experiment. Only the first value of each series ($s_1^1 = s_1^2 = s_1^3 = 7$) was graphed on the screen when the subjects made their first forecasts f_2 for the second period. Each subject made 59 forecasts. The first six periods are considered as practice phase and the analysis will be limited to periods 7-60.
Each participant i was paid for his forecasts with the function $p_t^i = 20 \cdot max\{3 - |s_t - f_t^i|; 0\}$, i.e. in each period he received 60 cents for an exact prediction and 40 (20) cents for a deviation of one (two) unit(s). The average payment was about 7€ for a duration of about 30 minutes.

3 Experimental Results

We are interested in the ability of the b&l model and REH[3] in explaining the average forecasts in the three versions of the experiment. The standard definition of $f_{t,b\&l}^s$ ignores the structural break and the fact that the heuristic is based on the assumption of stable distributions of the bounds and the local extrema. If a break is not considered in the model the distributions of peaks and troughs before and after the break are unified. For this reason the heuristic cannot be applied to non-stationary time series. We modify the model to test for level adaptation of the forecasts. The time series s_t^1 has two stable sections which are interrupted by one constant shift. The series s_t^2 and s_t^3 have three sections and two breaks. In these stable segments the b&l heuristic can be applied. After the break the calculation of the model is restarted, i.e. all local extrema and the average changes before the break are ignored. We call this modified heuristic $f_{t,b\&l}^{s,m}$.
The (naturally) few observations of average forecasts and the two models after the break do not allow any statistical tests. We look at the

[3] The calculation of the values of the bounds & likelihood heuristic and the Rational Expectations Hypothesis are presented in detail in [1]

absolute deviations of the bounds & likelihood heuristic $f^s_{t,b\&l}$ and its modified version $f^{s,m}_{t,b\&l}$ from the average forecasts. This comparison favors the standard model immediately after the break. This means that, in the first periods after the break, the modified model forecasts the subjects' average opinion worse.

It can be hypothesized that the subjects do not adapt immediately to the new level of the time series but that they use the model after a transition phase. We apply a mixed model for the explanation of the average forecasts of the subjects. For version 1 of the experiment we define the model $f^{s^1,mix}_{t,b\&l}$:

$$f^{s^1,mix}_{t,b\&l} = \begin{cases} f^{s^1}_{t,b\&l} & \text{for } t = 7,\dots,30 \\ f^{s^1,m}_{t,b\&l} & \text{for } t = 31,\dots,60 \end{cases} \tag{5}$$

Hence after the break we still use the standard model until period 30. The modified versions of the b&l heuristic in versions 2 include two restarts of the calculations of bounds and likelihoods in periods 26 and 44 which are followed by 10 periods transition phase in each case. Considering the transition phases, the mixed model for version 2 is defined as follows:

$$f^{s^2,mix}_{t,b\&l} = \begin{cases} f^{s^2}_{t,b\&l} & \text{for } t = 7,\dots,35 \\ f^{s^2,m_1}_{t,b\&l} & \text{for } t = 36,\dots,53 \\ f^{s^2,m_2}_{t,b\&l} & \text{for } t = 54,\dots,60 \end{cases} \tag{6}$$

$$f^{s^3,mix}_{t,b\&l} = \begin{cases} f^{s^3}_{t,b\&l} & \text{for } t = 7,\dots,35 \\ f^{s^3,m_1}_{t,b\&l} & \text{for } t = 36,\dots,51 \\ f^{s^3,m_2}_{t,b\&l} & \text{for } t = 52,\dots,60 \end{cases} \tag{7}$$

We test for this ability by estimating a linear time series regression. In this model, the $f^{s,mix}_{t,b\&l}$/REH are the independent variables z_t and the average forecasts are the dependent variables:

$$f_{t,avg} = \beta_1 + \beta_2 \cdot z_t + \beta_3 \cdot z_{t-1} + \beta_4 \cdot f_{t-1,avg} + \epsilon_t \tag{8}$$

The results of these regressions are presented in Table 1:

Model (8) accounts for remanence effects in the forecasts and both models. The results of both models are very comparable. The β_1 (intercept) coefficients are not significantly different from 0 in all cases.

Table 1. Results of the Multiple Linear Regression

Version	Model	β_1	β_2	β_3	β_4	R^2	DW
1	mixed b&l	0.711 (0.360)	1.041 (0.039)	-0.522 (0.122)	0.430 (0.119)	0.962	1.957
	REH	-0.991 (0.544)	1.008 (0.062)	-0.257 (0.168)	0.342 (0.139)	0.923	1.747
2	mixed b&l	0.694 (0.384)	0.996 (0.451)	-0.281 (0.138)	0.238 (0.136)	0.945	1.977
	REH	-0.991 (0.544)	1.008 (0.062)	-0.257 (0.168)	0.342 (0.139)	0.923	1.747
3	mixed b&l	0.553 (0.593)	0.693 (0.071)	0.029 (0.122)	0.279 (0.135)	0.930	1.910
	REH	-0.991 (0.544)	1.008 (0.062)	-0.257 (0.168)	0.342 (0.139)	0.923	1.747

The important β_2 values are significantly different from 0 for all models. Only for the mixed b&l in version 3 the value differs significantly from 1. The lagged variables and the average forecasts can be neglected as the β_3 and β_4 coefficients show.
The pairwise comparison of the models between all experiments shows that the mixed b&l model performs at least equivalent to the REH. It explains 93% of the variance of average forecasts in version 3 and slightly more in the other versions. The Durbin Watson values are in all cases closer to the critical value of 2 which indicates that the regression model for the b&l heuristic does not suffer from autocorrelated residuals.
Based on these results we conclude that the heuristic explains forecasting behavior very well when the first periods after the break are considered as a transition phase in which the old rules are still valid. After these periods the subjects use the information as it is described by the modified b&l heuristic.

References

1. Becker, O., Leitner, J., Leopold-Wildburger, U., 2007. Heuristic modeling of expectation formation in a complex experimental information environment. European Journal of Operational Research, 975-985.
2. O'Connor, M., Remus, W., Griggs, K., 1993. Judgemental forecasting in times of change. International Journal of Forecasting 9, 163-172.

Der Einfluss von Kostenabweichungen auf Nash-Gleichgewichte in einem nicht-kooperativen Disponenten-Controller-Spiel

Günter Fandel und Jan Trockel

FernUniversität in Hagen, Lehrstuhl für Betriebswirtschaftslehre,
Universitätsstr. 41, 58084 Hagen,
{guenter.fandel,jan.trockel}@fernuni-hagen.de

1 Einleitung

In diesem Beitrag wird die effiziente Gestaltung von Überprüfungen der Bestellmengen, die von einem Disponenten durchgeführt werden, durch einen Controller untersucht. [2] beschreibt - basierend auf den Arbeiten von [3] und [1] - ein effizientes Design von Kontrollen für das Risikomanagement als Nahtstelle zwischen vollständigem Vertrauen und vollständigem Misstrauen auf der Basis des spieltheoretischen Modells Inspection Game. Bei [7] wird das Inspection Game hinsichtlich einer Nichtentdeckung und eines Fehlalarms untersucht. Diese Sichtweise der Nichtentdeckung wird auf das modifizierte Spiel übertragen. Es spiegelt sich im Modell von [4] an der Stelle wider, an der angenommen wird, dass ein Controller durch ein niedriges Prüfniveau das nicht-methodische Arbeiten seitens des Disponenten nicht aufdeckt. Demgegenüber wird nun ein weiterer Schritt in das im Folgenden zu analysierende Modell integriert: Durch das Einfügen der Wahrscheinlichkeit - Aufdecken der fehlerhaften Arbeitsweisen durch das Top-Management - wird die Möglichkeit bzw. Gefahr modelliert, dass eben dieses Fehlverhalten nicht nur unentdeckt bleibt, sondern auch gewährleistet wird, dass der Controller, der in einem lateralen Verhältnis zum Disponenten steht, ebenfalls bestraft wird, sofern bei niedrigem Kontrollniveau der Fehler des Disponenten verborgen bleibt. In dem nachfolgenden Modell, das auf dem Ansatz von [4] basiert, wird analysiert, inwieweit eine große Kostenabweichung aufgrund einer nicht optimal gewählten Bestellmenge gegenüber einer geringen Kostenabweichung die Entscheidungen der Spieler und folglich die gleichgewichtige Lösung in diesem Spiel beeinflusst.

B. Fleischmann et al.(Eds.), *Operations Research Proceedings 2008*,
DOI: 10.1007/978-3-642-00142-0_82,

2 Erweiterung des Ansatzes von Fandel/ Trockel

Das nachfolgende Spiel basiert auf dem extensiven Baum analog zu der Idee von Fandel und Trockel [4]. Auch die Auszahlungen, die bei [4] definiert wurden, gelten hier. Da aber die Wahrscheinlichkeit p_a nicht strategisch von den Akteuren bestimmt werden kann, lassen sich die Knoten 4 und 5 im extensiven Spielbaum zusammenfassen, und die zugehörige Normalform des Spiels wird durch die Abbildung 1 dargestellt.

Dabei gelten die folgenden Symbole:

- Z, V Grundvergütung,
- S Strafe für den Disponenten, falls die nicht-methodisch durchgeführte Bestimmung der Bestellmenge entdeckt wird; Strafe für den Controller, falls das Management aufdeckt, dass der Controller die nicht-methodische Bestimmung der Bestellmenge durch den Disponenten nicht entdeckt hat,
- B_D Bonus, falls die durchgeführte Tätigkeit des Disponenten vom Controller als korrekt eingestuft wird,
- B_C Bonus, falls der Controller die nicht-methodisch durchgeführte Arbeit des Disponenten als solche aufdeckt,
- L Mußegewinn, den der Disponent durch seine nicht-methodisch durchgeführte Arbeit erhält,
- K zusätzliche Kosten aufgrund der intensiven Prüfung der vorliegenden Daten,
- p_m Wahrscheinlichkeit, dass der Disponent seine Entscheidung methodisch und nicht nach Gutdünken (on instinct) durchführt und
- p_h Wahrscheinlichkeit, dass der Controller die vorliegenden Daten ordnungsgemäß und intensiv prüft.

Die Wahrscheinlichkeit p_a beschreibt - wie bereits angedeutet -, inwieweit das Management durch Überprüfung herausfindet, dass Disponent und Controller schlecht gearbeitet haben.
Die weitergehende Untersuchung erfolgt nun in zwei Schritten. Zuerst wird das Nash-Gleichgewicht in gemischten Strategien bestimmt, bevor analysiert wird, welchen Einfluss eine Kostenabweichung aufgrund einer falsch bestimmten optimalen Bestellmenge auf die Entscheidungsfindung besitzt. Plant der Disponent schlecht (on instinct), so weicht er von der optimalen Bestellmenge ab. Er wird eine Bestellmenge realisieren, die dazu führt, dass die Kosten für das Unternehmen ansteigen. Übersteigt die Kostendifferenz ΔK einen Wert ε eines vernachlässigbaren Intervalls, so soll der Fall vorliegen, dass der Disponent schlecht geplant hat (vgl. [4]). Es wird dabei angenommen, dass

Controller / Disponent	hohes Kontrollniveau (h) p_h	geringes Kontrollniveau (nh) $(1-p_h)$
methodisch bestimmte Bestellmenge (m) p_m	$V-K$ / $Z+B_D$	V / $Z+B_D$
nicht-methodisch bestimmte Bestellmenge "on instinct" (nm) $(1-p_m)$	$V+B_C-K$ / $Z-S+L$	$V-p_a\cdot S$ / $Z+B_D+L-p_a\cdot(S+B_D)$

Fig. 1. Bi-Matrix für den extensiven Spielbaum nach Fandel und Trockel[4]

diese Abweichung in einem linearen und direkten Zusammenhang zu den Bestrafungen S steht. Einen Mußegewinn L erhält der Disponent, wenn er seine Arbeit schnell und ohne Anwendung methodischer Modellierung durchführt. Wird dies jedoch durch eine intensive Kontrolle seitens des Controllers und im Extremfall durch das Management selbst aufgedeckt, so erfährt der Disponent eine Strafe S. Der Controller hingegen wird einen festen Bonus erhalten, falls er falsches Verhalten aufdeckt, so dass dies nicht als Funktion in Abhängigkeit der Kostenabweichung dargestellt werden kann. Erfährt er jedoch eine Bestrafung durch das Management aufgrund eines niedrigen Prüfniveaus bei gleichzeitigem Fehlverhalten des Disponenten, so kann diese Bestrafung als Funktion in Abhängigkeit von der Kostenabweichung definiert werden. Somit wird untersucht, inwieweit durch bestimmte Konstellationen der Wahrscheinlichkeitswerte $(1 - p_m)$ bzw. $(1 - p_h)$ die Gefahr gegeben ist, dass Knoten 5 (vgl. [4]) zum Nash-Gleichgewicht wird (Schritt 1). Weiterhin wird dann diskutiert, inwieweit die Kostenabweichungen vom Optimum der Planung diese Wahrscheinlichkeiten beeinflussen (Schritt 2).

Schritt 1:

Es wird das Nash-Gleichgewicht (vgl. [5]) im vorliegenden Spiel unter Berücksichtigung gemischter Strategien über die Berechnung der besten Antwort des einen Spielers auf eine gegebene beste Antwort des zweiten Spielers (vgl. [6]) bestimmt, da unter den Bedingungen $K > 0$, $B_C > K - p_a \cdot S$, $B_D > L - S$ und $L > p_a \cdot (S + B_D)$, die hier gelten sollen, kein Gleichgewicht in reinen Strategien existiert.
Anhand der obigen Matrix kann die Reaktionskorrespondenz der beiden Spieler beschrieben werden. Für den Controller und den Disponenten erhält man die folgenden Strategienübersichten:

$$s_C = \begin{cases} (h) & \text{für } p_m < \dfrac{B_C - K + p_a \cdot S}{B_C + p_a \cdot S}, \\ (h, nh) & \text{für } p_m = \dfrac{B_C - K + p_a \cdot S}{B_C + p_a \cdot S}, \\ (nh) & \text{für } p_m > \dfrac{B_C - K + p_a \cdot S}{B_C + p_a \cdot S}, \end{cases} \text{und } s_D = \begin{cases} (m) & \text{für } p_h > \dfrac{L - p_a \cdot (S + B_D)}{(1 - p_a) \cdot (S + B_D)}, \\ (m, nm) & \text{für } p_h = \dfrac{L - p_a \cdot (S + B_D)}{(1 - p_a) \cdot (S + B_D)}, \\ (nm) & \text{für } p_h < \dfrac{L - p_a \cdot (S + B_D)}{(1 - p_a) \cdot (S + B_D)}. \end{cases}$$

Wie erkennbar wird, ist in den optimalen Strategienkombinationen der beiden Spieler immer noch der Fall möglich, dass nicht-methodisches Planen des Disponenten durch ein niedriges Prüfniveau des Controllers nicht aufgedeckt wird. Dieser Ansatz zeigt auf, dass auch der Controller in einem lateralen Inspection Game nicht unbedingt im Sinne der Gewinnmaximierung des Unternehmens handeln muss.

Schritt 2:

Hier wird nun untersucht, welche Auswirkungen einzelne Parameter auf die Lösung des Spiels aufweisen. Zuerst wird die Bestrafung der beiden Spieler infolge der Kostenabweichung, die sich aufgrund einer nicht korrekt gewählten Bestellmenge ergibt, modelliert. Dabei wird im Folgenden $S = c_D \cdot \Delta K$ für den Disponenten und $S = c_C \cdot \Delta K$ für den Controller unterstellt. c_D bezeichne den Faktor der Bestrafung für den Disponenten und c_C den Faktor für den Controller. Aus Vereinfachungsgründen wird weiter davon ausgegangen, dass $c_D = c_C = 1$ gilt. Als Gleichgewichtspunkt in gemischten Strategien erhält man dann

$$(p_h^*, p_m^*) = \left(\frac{L - p_a \cdot (\Delta K + B_D)}{(1 - p_a) \cdot (\Delta K + B_D)}, \frac{B_C - K + p_a \cdot \Delta K}{B_C + p_a \cdot \Delta K} \right).$$

Anhand dieses Gleichgewichts wird im Folgenden analysiert, welchen

Einfluss der exogene Wert S bzw. die vom Disponenten verursachte Kostenabweichung ΔK auf diesen Gleichgewichtszustand nehmen. Eine höhere Kostenabweichung auf das vorliegende Nash-Gleichgewicht ergibt eine Verschiebung des Gleichgewichts. Dies bedeutet, dass mit zunehmendem ΔK die Wahrscheinlichkeit eines methodischen Vorgehens seitens des Disponenten zunimmt, jedoch die Wahrscheinlichkeit für eine intensive Kontrolle durch ein hohes Prüfniveau des Controllers abnimmt:

$$\frac{\partial p_m^*}{\partial \Delta K} > 0 \text{ und } \frac{\partial p_h^*}{\partial \Delta K} < 0.$$

Der Gleichgewichtspunkt im modifizierten Inspection Game wandert folglich nach rechts unten.

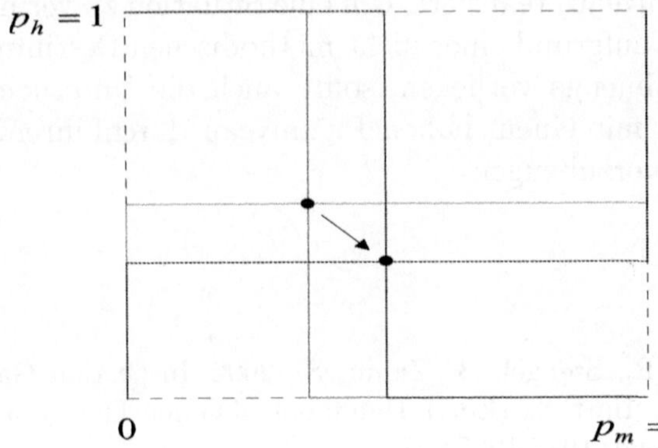

Fig. 2. Auswirkungen einer steigenden Kostenabweichung auf das Nash-Gleichgewicht

Eine höhere mögliche Kostenabweichung veranlasst den Disponenten dazu, die optimale Bestellmenge bzw. eine derart geringe Abweichung von dieser zu realisieren, so dass die Bedingung $\Delta K < \varepsilon$ erfüllt wird. Dabei bezeichnet ε weiterhin das Intervall, in dem eine Kostenabweichung vom Minimum keinerlei Auswirkungen auf die Auszahlungen der Spieler hat, da nahezu der optimale Gewinn erreicht wird. Der Controller wird daraufhin nicht mehr intensiv prüfen. Dies kann unter Umständen dazu führen, dass der Fehler des Disponenten bei der Bestimmung der optimalen Bestellmenge unentdeckt bleibt und folglich die Unternehmensführung ihre Aufgaben intensiv wahrnehmen muss, um das Fehlverhalten der Akteure zu erkennen.

3 Fazit

In diesem Beitrag wird gezeigt, welchen Einfluss hohe Kostenabweichungen auf das Verhalten eines Disponenten und eines Controllers haben. Es wird analysiert, wie das Nash-Gleichgewicht in gemischten Strategien in Richtung der Strategienkombination (methodische Bestimmung der Bestellmenge, niedriges Prüfniveau) mit höheren Kostenabweichungen und Strafzahlungen wandert. Setzt die Unternehmensleitung hohe Strafen für Fehlverhalten an, so kann zwar nicht ganz ausgeschlossen werden, dass der Disponent und der Controller keinen im Unternehmenssinne maximalen Gewinn realisieren, aber es wird zumindest die Wahrscheinlichkeit der Gefahr von nicht korrektem Verhalten erheblich reduziert und ein Nash-Gleichgewicht in der Nähe der reinen Strategienkombination (methodische Bestimmung der Bestellmenge, niedriges Prüfniveau) realisiert. Um eine Situation zu vermeiden, in der höhere Kosten aufgrund einer nicht-methodischen Bestimmung der optimalen Bestellmenge vorliegen, sollte auch die Unternehmensleitung ihre Aufgaben mit einem hohen Prüfniveau durchführen, um diesem Fehlverhalten vorzubeugen.

References

1. Avenhaus, R., Stengel, B., Zamir, S., 2002. Inspection Games. In: Aumann, R. J., Hart, S., (Eds.). Handbook of Game Theory, 3, Amsterdam: North-Holland, 1947-1987.
2. Biermann, B., 2006. Die Anwendung spieltheoretischer Methoden zur Definition eines optimalen Kontrolldesigns. Dissertation, Norderstedt: Books on Demand GmbH.
3. Borch, K., 1982. Insuring and Auditing the Auditor. In: Deistler, M., Fürst, E., Schwödiauer, G., (Eds.). Games, economic dynamics, time series analysis. Wien: Physica, 117-126.
4. Fandel, G., Trockel, J., 2008. Stockkeeping and controlling under game theoretic aspects. Discussion paper, No. 420, FernUniversität in Hagen.
5. Nash, J., 1951. Non-cooperative Games. Annals of Mathematics, 54, 286-295.
6. Osborne, M. J., 2004. An introduction to Game Theory. Oxford [u. a.]: Oxford University Press.
7. Rinderle, K., 1996. Mehrstufige sequentielle Inspektionsspiele mit statistischen Fehlern erster und zweiter Art. Dissertation, Hamburg: Kovac.

Multilayered Network Games: Algorithms for Solving Discrete Optimal Control Problems with Infinite Time Horizon via Minimal Mean Cost Cycles and a Game-Theoretical Approach

Dmitrii Lozovanu[1] and Stefan Pickl[2]

[1] Institute of Mathematics and Computer Science, Academy of Sciences, Academy str., 5, Chisinau, MD–2028, Moldova
lozovanu@math.md

[2] Institut fur Angewandte Systemwissenschaften und Wirtschaftsinformatik, Fakultät fur Informatik, Universität der Bundeswehr, München
stefan.pickl@unibw.de

Summary. We extend a classical discrete optimal control problem in such a way that the feasible sets and the costs depend now on the parameter t: Such problems occur for example in (multilayered) emission trading games which are introduced by the authors [8]. Here, we develop two possible characterizations and solution principles. One is based on a classical linear programming approach. The other one exploits a game-theoretic treatment. Via these aprroaches we can solve the problem even for the case that we handle with an arbitrary transition function τ.

1 Introduction

We study the following discrete optimal control problem with infinite time horizon and varying time of states' transitions.
Let the dynamical system L with a finite set of states $X \subseteq R^n$ be given, where at every discrete moment of time $t = 0, 1, 2, \ldots$ the state of L is $x(t) \in X$. Assume, that the control of the system L at each time-moment $t = 0, 1, 2, \ldots$ for an arbitrary state $x(t)$ is realized by using the vector of control parameters $u(t) \in R^m$ for which a feasible set $U_t(x(t))$ is given, i.e. $u(t) \in U_t(x(t))$. For arbitrary t and $x(t)$ on $U_t(x(t))$ we introduce an integer function $\tau : U_t(x(t)) \to N$ which relates to each control $u(t) \in U_t(x(t))$ an integer value $\tau(u(t))$.

B. Fleischmann et al.(Eds.), *Operations Research Proceedings 2008*,
DOI: 10.1007/978-3-642-00142-0_83,

This value expresses the time of system's passage from the state $x(t)$ to the state $x\big(t+\tau\left(u(t)\right)\big)$ if the control $u(t) \in U_t\left(x(t)\right)$ has been applied at the moment t for a given state $x(t)$. The dynamics of the system L is described by the following system of difference equations

$$\begin{cases} t_{j+1} &= t_j + \tau\left(u(t_j)\right); \\ x(t_{j+1}) &= g_{t_j}\left(x(t_j), u(t_j)\right); \\ & u(t_j) \in U_{t_j}\left(x(t_j)\right); \\ & j = 0, 1, 2, \ldots, \end{cases} \tag{1}$$

where

$$x(t_0) = 0, \;\; t_0 = 0 \tag{2}$$

is a given starting representation of the dynamical system L. We suppose that the functions g_t and τ are known and t_{j+1} and $x(t_{j+1})$ are determined uniquely by $x(t_j)$ and $u(t_j)$ at every step j.

Let $u(t_j),\;\; j = 0, 1, 2, \ldots,$ be a control, which generates the trajectory $x(0),\ x(t_1),\ x(t_2),\ \ldots,\ x(t_k),\ \ldots$. For this control we define the mean integral-time cost by a trajectory

$$F_{x_0}\left(u(t)\right) = \lim_{k \to \infty} \frac{\sum\limits_{j=0}^{k-1} c_{t_j}\left(x(t_j), g_{t_j}\left(x(t_j), u(t_j)\right)\right)}{\sum\limits_{j=0}^{k-1} \tau(u(t_j))} \tag{3}$$

where $c_{t_j}\big(x(t_j), g_{t_j}\left(x(t_j), u(t_j)\right)\big) = c_{t_j}\left(x(t_j), x(t_{j+1})\right)$ represents the cost of the system L to pass from the state $x(t_j)$ to the state $x(t_{j+1})$ at the stage $[j, j+1]$.

We consider the problem of determining the time-moments $t = 0,\ t_1,\ t_2,\ \ldots,\ t_{k-1},\ \ldots$ and the vectors of control parameters $u(0),\ u(t_1),\ u(t_2),\ \ldots,\ u(t_{k-1}),\ \ldots$ which satisfy conditions (1), (2) and minimize the function (3).

In the case $\tau \equiv 1$ this problem is similar to the control problem with unit time of states transitions from [1, 2, 6]. The problem of determining the stationary control with unit time of states transitions has been studied in [4, 6, 9]. In the cited papers it is assumed that $U_t(x(t))$, g_t and c_t do not depend on t, i.e. $g_t = g$, $c_t = c$ and $U_t(x) = U(x)$ for $t = 0, 1, 2, \ldots$.

R. Bellman [1] proved that for this stationary case of the problem with unit time of states transistions there exists an optimal stationary control $u^*(0)$, $u^*(1)$, $u^*(2)$, $\ldots$, $u^*(t)$, $\ldots$, such that

$$\lim_{k\to\infty} \frac{\sum_{t=0}^{k-1} c\big(x(t), g\,(x(t), u^*(t))\big)}{k} =$$

$$= \inf_{u(t)} \lim_{k\to\infty} \frac{\sum_{t=0}^{k-1} c\big(x(t), g\,(x(t), u(t))\big)}{k} = \lambda < \infty.$$

Furthermore in [6, 9] it is shown that the stationary case of the problem can be reduced to the problem of finding the optimal mean cost cycle in a graph. Based on these results in [4, 6, 9] polynomial-time algorithms for finding an optimal stationary control are proposed.
Here, we extend the results mentioned above for the general case of the problem with an arbitrary transition-time function τ. We show that this problem can by formulated as the problem of determining optimal mean cost cycles in a graph.

2 The Main Results

The main results we propose here are concerned with determining the stationary control parameter in the general case for the problem from Section 1. We show that this problem can be reduced to the following optimization problem on certain graph.
Let a strongly connected directed graph $G = (X, E)$ which represents the states transitions of the dynamical system L be given. An arbitrary vertex x of G corresponds to a state $x \in X$ and an arbitrary directed edge $e = (x, y) \in E$ expresses the possibility of system L to pass from the state $x(t)$ to the state $x(t + \tau_e)$, where τ_e is the time of system's passage from the state x to the state y through the edge $e = (x, y)$. So, on the edge set E it is defined the function $\tau : E \to R^+$ which relates to each edge a positive number τ_e which means that if the system L at the moment of time t is in the state $x = x(t)$ then the system can reach the state y at the time moment $t + \tau_e$ if it passes through the edge $e = (x, y)$, i.e. $y = x(t + \tau_e)$. Additionally, on the edge set E it is defined the cost function $c : E \to R$, which relates to each edge the cost c_e of the system's passage from the state $x = x(t)$ to the state $y = x(t + \tau_e)$ for an arbitrary discrete moment of time t.

So, finally to each edge two numbers c_e and τ_e are associated. On G we consider the following problem: Find a directed cycle C^* such that

$$\frac{\sum\limits_{e\in E(C^*)} c_e}{\sum\limits_{e\in E(C^*)} \tau_e} = \min_C \frac{\sum\limits_{e\in E(C)} c_e}{\sum\limits_{e\in E(C)} \tau_e}.$$

Algorithms based on the parametrical method for solving this problem in the case of positive costs c_e and positive transition time τ_e on the edges have been proposed in [3, 5].
Here we shall use the following linear programming approach:
Minimize

$$z = \sum_{e\in E} c_e \alpha_e \tag{4}$$

subject to

$$\begin{cases} \sum\limits_{e\in E^+(x)} \alpha_e - \sum\limits_{e\in E^-(x)} \alpha_e = 0, \quad \forall x \in X; \\ \sum\limits_{e\in E} \tau_e \alpha_e = 1, \\ \qquad \alpha_e \geq 0, \quad \forall e \in E, \end{cases} \tag{5}$$

where $E^-(x) = \{e = (y,x) \in E \,|\, y \in X\}$; $E^+(x) = \{e = (x,y) \in E \,|\, y \in X\}$.

The following lemma holds.

Lemma 1. *Let $\alpha = (\alpha_{e_1}, \alpha_{e_2}, \dots, \alpha_{e_m})$ be the set of feasible solutions of the system (5) and $G_\alpha = (X_\alpha, E_\alpha)$ be the subgraph of G, induced by the set of edges $E_\alpha = \{e_i \in E \,|\, \alpha_{e_i} > 0\}$. Then an arbitrary extreme point $\alpha^\circ = (\alpha^\circ_{e_1}, \alpha^\circ_{e_2}, \dots, \alpha^\circ_{e_m})$ of the polyhedron set determined by (5) corresponds to a subgraph $G_{\alpha^\circ} = (X_{\alpha^\circ}, E_{\alpha^\circ})$ which has a structure of simple directed cycle and vice versa: If $G_{\alpha^\circ} = (X_{\alpha^\circ}, E_{\alpha^\circ})$ is a simple directed cycle in G then the solution $\alpha^\circ = (\alpha^\circ_{e_1}, \alpha^\circ_{e_2}, \dots, \alpha^\circ_{e_m})$ with*

$$\alpha_{e_i} = \begin{cases} \dfrac{1}{\sum\limits_{e\in E_{\alpha^\circ}} \tau_e} & \text{if } \; e_i \in E_{\alpha^\circ}, \\ \quad 0 & \text{if } \; e_i \notin E_{\alpha^\circ} \end{cases}$$

represent an extreme point of set of the solution (5).

Based on Lemma 1 the following theorem can be proved.

Theorem 1. *The optimal solution* $\alpha^* = (\alpha^*_{e_1}, \alpha^*_{e_2}, \dots, \alpha^*_{e_m})$ *of problem (4), (5) corresponds to a minimal mean cycle* $C^* = G_{\alpha^*}$ *in* G, *i.e.*

$$\alpha^*_{e_i} = \begin{cases} \dfrac{1}{\sum\limits_{e \in E(C^*)} \tau_e}, & \text{if } e \in E(C^*); \\ 0, & \text{if } e \notin E(C^*), \end{cases}$$

where $E(C^*)$ *is the set of edges of the directed cycle* C^*. *So, for the optimal value* z^* *of the objective function* z *it holds*

$$z^* = \frac{\sum\limits_{e \in E(C^*)} c_e}{\sum\limits_{e \in E(C^*)} \tau_e}.$$

Using these results new algorithms for solving the minimum mean cost cycle problem can be derived.
The linear programming model (4), (5) can be used for solving our problem in the case when the transition times τ_e are positive. In the general case when these parameters may be negative the minimal mean cost cycle problem can be reduced to the following fractional linear programming problem:

Minimize

$$z = \frac{\sum\limits_{e \in E} c_e \alpha_e}{\sum\limits_{e \in E} \tau_e \alpha_e}$$

subject to

$$\sum_{e \in E^+(x)} \alpha_e - \sum_{e \in E^-(x)} \alpha_e = 0, \ x \in X;$$

$$\sum_{e \in E} \alpha_e = 1;$$

$$\alpha_e \geq 0, \ e \in E,$$

where $E^-(x) = \{e = (y, x) \in E \,|\, y \in X\}$; $E^+(x) = \{e = (x, y) \in E \,|\, y \in X\}$.
We compare this approach with a game-theoretical treatment which offers the possibility to deal with cooperative and non-cooperative behavior in such decision processes.

3 Conclusion-Outlook

We extend a classical discrete optimal control problem: Algorithms for solving such an optimal control problem are proposed. Additionally, a game theoretical characterization for the considered problem is indicated. Hereby some results from [7] can be extended. Within this game theoretic extension the non-cooperative and the cooperative case can be compared. Via these approaches we can solve the problem even for the case that we handle with an arbitrary transition function τ.
The obtained results can be used in general decision making systems., esp. in the so called multilayered decision problems of emission trading markets.

References

1. Bellman R. (1959) Functional equations in the theory of dynamic programming, XI-Limit theorems, Rand. Circolo Math. Palermo 8(3):343–345
2. Bellman R., Kalaba R. (1965) Dynamic programming and modern control theory. Academic Press, New York and London
3. Christofides N., Graph Theory: An Algorithmic Approach, Academic Press, New York, London, San Francisco (1975).
4. Karp R. (1978) A characterization of the minimum cycle mean in a digraph. Discrete Mathematics 23(3):309–311
5. Lawler E., Optimal cycles in doubly weighted directed linear graphs. Int. Symp. on Theory of Graphs, Dunod, Paris, p. 209, (1996).
6. Lozovanu D. (1991) Extremal-combinatorial problems and algorithms for its solving. Kishinev, Stiinta (in Russian).
7. Lozovanu D., Pickl S. (2007) Algorithms and the calculation of Nash equilibria for multi-objective control of time-discrete systems and polynomial-time algorithms for dynamic c-games on networks. European Journal of Operational Research 181:1214-1232
8. Lozovanu D., Pickl S. (2008)Optimization and Multi-Objective Control of Time-Discrete Systems. Springer Verlag.
9. Romanovski I. (1967) Optimization of stationary control of discrete deterministic processes. Cybernetics 2:66–78

Competitive Facility Placement for Supply Chain Optimization

Andrew Sun and Hugh Liu

UTIAS, 4925 Dufferin Street, Toronto, Ontario, Canada, M3H-5T6
{andrew.sun,liu}@utias.utoronto.ca

1 Introduction

The decision of where to place a company's distribution and outlet facilities when entering a new market is a crucial one faced by today's retail company supply chain planners. This problem alone is complicated enough and is currently an active area of research [1], [2]. The problem is further complicated when there exists a rival company intent on servicing the same market by implementing a parallel distribution network. Similar competitive problems have been studied by [3] and [4].
In this problem, two competitor companies are both interested in an answer to the same two questions:

1. Where should the distribution centre be located? and,
2. Where should the retail outlets be located (what areas of the market are to be serviced)?

Both companies answer these questions with the understanding that its rival will attempt to maximize its own profits. There is one leader company which moves first. A follower company moves second and makes its decision with the knowledge of the leader company's move. In this paper, a bilevel optimization method is presented for solving the dynamic and continuous competitive facility placement problem.

2 Problem Formulation

In this paper, the market is limited to a 1-D space centred at the origin on the closed interval of $[-xlim, xlim]$. On this interval, there exists

DOI: 10.1007/978-3-642-00142-0_84,

a known fixed market demand represented by the *Demand Function*, $D(x)$. The profits *extractable* from the demand landscape, however will only be a fraction of the total possible $D(x)$ and will generally decrease with distance from the distribution center. This attenuation of profits with distance is motivated by transportation costs associated with moving goods from the distribution center to the retail outlet and is denoted by the function $A_i(x)$ for the *ith* company, where $i = 1, 2$.
Along with the problem of selecting the distribution facility location, the companies also choose what areas of the market to service. The *ith* company decides on a *Service Coverage Function* which represents how much *effort* to invest in servicing a particular region of the market. This function, denoted by $S_i(x)$, can alternatively be thought of as any utility measurement of the sales activity in that region (eg. advertising cost, salaries for sales people, size of the retail outlets). The cost associated with $S_i(x)$ is the *Cost-to-Service Function*, $C_i(x)$.
The conflict between the two companies is a consequence of the fixed market demand. At any given point, the demand awarded to any one company is a function of the service coverage function value at that particular region compared to the service coverage function value of its rival company. The company that invests more service coverage receives a greater portion of the market demand for that region.
A summary of the game is given as follows:

- **Players:** Company A and Company B.
- **Player Action Sets:** When it is Company A's turn, it must first decide on a location to place its distribution centre (x_A) within the 1-D interval of $[-xlim,\ xlim]$. A service coverage function ($S_A(x)$)must also be selected. $S_A(x)$ is the linear interpolant spline that runs through the points with values $[s_0,\ s_1,\ \dots\ ,\ s_n]$ at the respective x-coordinates of $\left(\frac{2\cdot xlim}{n}\right) \cdot j \quad where \quad j \in [0, 1, \cdots, n]$. Therefore, each player has $n + 2$ parameters to choose (1 parameter for the facility location and $n + 1$ parameters that define the service coverage function $S_i(x)$). An identical action set exists for Company B.
- **Order of Play:** Company A moves first followed by Company B.
- **Payoffs:** Payoff to Company A is defined as:

$$\int_{-xlim}^{xlim} A(x)D(x)\exp\left[-\frac{S_B(x)}{S_A(x)}\right] - C_A(x) \cdot S_A(x)dx \tag{1}$$

The payoff to Company B is the same as equation 1, but replacing subscripts B with subscripts A and vice versa.

- **Constraints:** The sole constraint on the problem is that neither company can apply a negative service in the game space. A negative service area is meaningless for this problem since we are only concerned with selling goods and not purchasing. Thus

$$S_A(x) \geq 0, \quad S_B(x) \geq 0, \quad -xlim \leq x \leq xlim \tag{2}$$

3 Method

A bi-level optimization routine adapted from [5] is used to find Stackelberg equilibrium solutions (see figure 1). The outer optimizer is a genetic optimizer which optimizes the spline points and distribution facility location for the leader firm. The inner loop holds the leader firm parameters constant and optimizes profit for the follower firm. A gradient based quasi-newton DFP method is used for the inner loop. Calculation of sensitivities is done by a central finite difference method with a fixed step size.

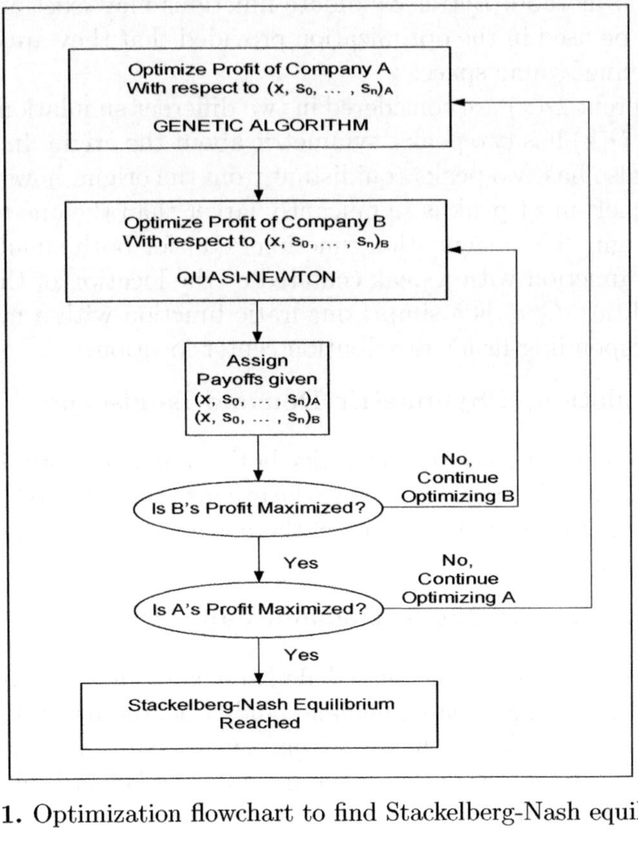

Fig. 1. Optimization flowchart to find Stackelberg-Nash equilibria.

4 Simulation Results

The hypothetical scenario dealt with in the following simulations is that of a new consumer product being introduced into a market that has a known fixed demand. Two companies (called A and B) manufacture, distribute, and sell the product and are both interested in selling in the targeted one-dimensional market. Company A moves first (the leader) followed by company B (the follower) who has the luxury of observing Company A's move. Specification of the demand, attenuation and cost-to-service functions are provided below. Two simulations are presented. The first consists of a demand landscape with 2 symmetric peaks both equidistant from the origin. The second also consists of 2 peaks with pinnacles equidistant from the origin, but in this case one of the two peaks is made to be much larger than the other.

4.1 Function Definitions

Selection of the functions for simulation are done to illustrate the optimization method. Better candidate functions may exist and can just as easily be used in the optimization provided that they are continuous in the defined game space.

Two different $D(x)$ are considered in two different simulations. In simulation 1, $D(x)$ has two peaks, symmetric about the origin. In simulation 2, $D(x)$ also has two peaks equidistant from the origin; however, in this case, the left most peak is significantly larger than the one to the right of the origin. The attenuation function $A(x)$ for both simulations is a gaussian function with a peak centered at the location of the distribution facilities. $C(x)$ is a simple quadratic function with a minimum at the corresponding firm's distribution center location.

4.2 Simulation 1: Symmetric Demand Landscape

When encountering symmetric peaks, both companies optimize to dominate only one of the peaks, leaving the other to be exploited by its rival, see figure 2. In the symmetric case, there is a slight advantage in being a follower firm.

4.3 Simulation 2: Biased Demand Landscape

Simulation results for the biased demand landscape are provided in figure 3. In a strongly biased demand landscape, there is a strong leader firm advantage. The implication from this scenario is that if there is a clear discrepancy in peaks in the demand market, a leader firm will typically do better.

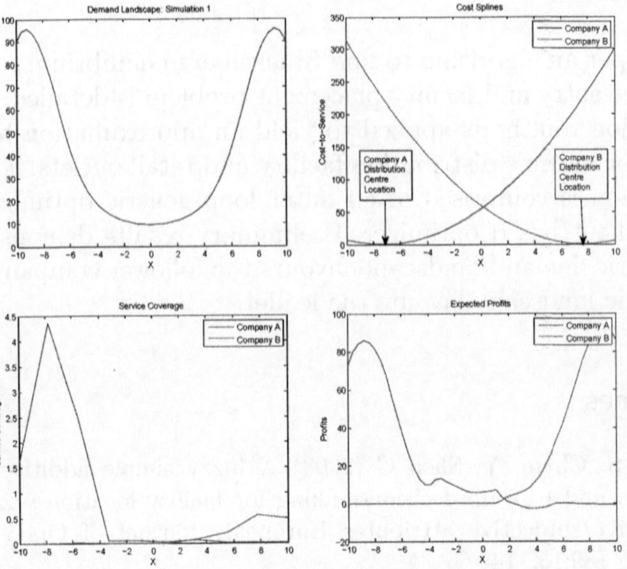

Fig. 2. Simulation 1: Symmetric Demand Landscape

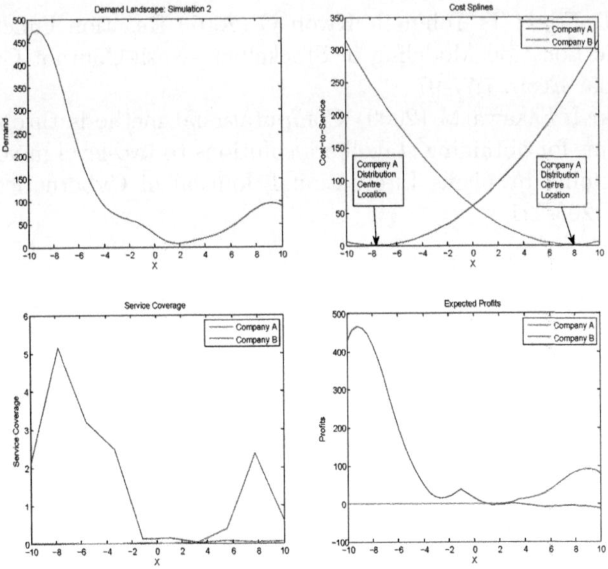

Fig. 3. Simulation 2: Biased Demand Landscape

5 Conclusion

In this paper an algorithm to find Stackelberg equilibrium solutions to the market entry and facility placement problem is detailed. A bi-level optimization routine is applied to yield an approximation to the best location for a firm's distribution facility and retail outlets. The bi-level optimization is comprised of an outer loop genetic optimizer and an inner gradient based optimizer. Preliminary results demonstrate that a symmetric demand landscape favours the follower company while an asymmetric landscape favours the leader.

References

1. Chou S, Chang Y, Shen C (2008) A fuzzy simple additive weighting systme under group decision-making for facility location selection with objective/subjective attributes European Journal of Operational Research, 189:132-145
2. Dupont L (2008) Branch and bound algorithm for a facility location problem with concave site dependent costs International Journal of Production Economics, 112:245-254
3. Aboolian R, Berman O, Krass D (2007) Competitive Facility Location and Design Problem European Journal of Operational Research, 182:40-62
4. Miller T, Friesz T, Tobin R, Kwon C (2007) Reaction Function Based Dynamic Location Modeling in Stackelberg-Nash-Cournot Competition Netw Spat Econ, 7:77-97
5. Nishizaki I, Sakawa M (2000) Computational methods through genetic algorithms for obtaining stakelberg solutions to two-level mixed zero-one programming problems International Journal of Cybernetics and Systems, 31:203-221

Part XI

Linear, Nonlinear and Vector Optimization

New Optimization Methods in Data Mining

Süreyya Özöğür-Akyüz[1], Başak Akteke-Öztürk[1], Tatiana Tchemisova[2], and Gerhard-Wilhelm Weber[1]

[1] Institute of Applied Mathematics, METU, Turkey,
{sozogur,boztur,gweber}@metu.edu.tr
[2] Department of Mathematics, University of Aveiro, Portugal
tatiana@ua.pt

1 Introduction

Generally speaking, an optimization problem consists in maximization or minimization of some function (objective function) $f : S \to \mathbf{R}$. The *feasible* set $S \subseteq \mathbf{R}^n$ can be either finite or infinite, and can be described with the help of a finite or infinite number of equalities and inequalities or in the form of some topological structure in $\mathbf{R}^n$. The methods for solution of a certain optimization problem depend mainly on the properties of the objective function and the feasible set. In this paper, we discuss how specific optimization methods of optimization can be used in some specific areas of data mining, namely, in *classification* and *clustering* that are considered interrelated [11].

2 Clustering

Clustering is an unsupervised learning in which data are separated into clusters according to their similarity. It has many applications, including decision-making, machine-learning, pattern classification, etc. Alternatively, it may support preprocessing steps for other algorithms, such as classification and characterization, operating the detecting clusters [6].

2.1 Optimization Models for Clustering Problems

Assume that we have a finite set X of points (patterns) in the n−dimensional space $\mathbf{R}^n : X = \{x^1, x^2, \ldots, x^M\}$, where $x^k \in \mathbf{R}^n (k = 1, 2, \ldots, M)$. Given a number $q \in \mathbf{N}$, we are looking for q subsets $C^i, i = 1, 2, \ldots, q$, such that the medium distance between the elements in each subset is minimal and the following conditions are satisfied: $1. C^i \neq \emptyset, (i = 1, 2, \ldots, q), 2. X = \bigcup_{i=1}^{q} C^i$. As a measure of similarity we use any distance function. Here for the sake of simplicity we consider Euclidean distance $\|\cdot\|_2$. The sets $C^i (i = 1, 2, \ldots, q)$, introduced

B. Fleischmann et al.(Eds.), *Operations Research Proceedings 2008*,
DOI: 10.1007/978-3-642-00142-0_85,

above are called *clusters* and the problem of determination of clusters is the *clustering problem.* When the clusters can overlap, the clustering problem is *fuzzy.* If we request additionally: $3. C^i \bigcap C^j = \emptyset$ if $i \neq j$, then we obtain a *hard* clustering problem. Let us assume that each cluster C^i, can be identified by its *center* or *centroid*, defined as (see [3]) $c^i := \frac{1}{|C^i|} \sum_{x \in C^i} x$, where $|C^i|$ denotes a cardinality of the cluster C^i. Then the clustering problem can be reduced to the following optimization problem, which is known as a *minimum sum of squares clustering* [4]:

$$\begin{array}{l} \min \ \frac{1}{M} \sum_{i=1}^{q} \sum_{x \in C^i} ||c^i - x||_2^2 \\ \text{such that} \ \ C = \{C^1, C^2, \ldots, C^q\} \in \bar{C}, \end{array} \tag{1}$$

where $\bar{C}$ is a set of all possible $q-$partitions of the set X. The clustering problem (1) can be rewritten as single *mixed-integer* minimization problem as follows:

$$\begin{array}{l} \min \ \ \frac{1}{M} \sum_{j=1}^{M} \sum_{i=1}^{q} w_{ij} ||x^j - c^i||_2^2, \\ \text{such that} \ \ w_{ij} \in \{0, 1\}, \ \sum_{i=1}^{q} w_{ij} = 1 \\ (i = 1, 2, \ldots, q) \ \ (j = 1, 2, \ldots, M). \end{array} \tag{2}$$

Here, centroids are rewritten as $c^i := (\sum_{j=1}^{M} w_{ij} x^j)/(\sum_{j=1}^{M} w_{ij})$, w_{ij} is the association weight of the pattern x^j with cluster i given by

$$w_{ij} = \begin{cases} 1, & \text{if pattern } j \text{ is allocated to cluster } i, \\ 0, & \text{otherwise.} \end{cases}$$

It can be shown that (2) is a global optimization problem with possibly many local minima [3]. In general, solving the global optimization problem is a difficult task. This makes it necessary to develop clustering algorithms which compute the local minimizers of problem (2) separately. In [3], the optimization techniques are suggested that are based on nonsmooth optimization approach. Finally, note that the clustering problems (1) and (2) can be reformulated as an unconstrained non smooth and non convex problem

$$\min f(c^1, c^2, \ldots, c^q), \tag{3}$$

where $f(c^1, c^2, \ldots, c^q) = \frac{1}{M} \sum_{i=1}^{M} \min_j ||c^j - x^j||_2^2$. Since the function $\psi(y) = ||y - c||_2^2$ $(y \in \mathbf{R}^n)$, is separable (as a sum of squares), the function $\varphi(x^i) = \min_j ||c^j - x^j||_2^2$ is *piece-wise separable.* It is proved in [2] that the function $f(c^1, c^2, \ldots, c^q)$ is piecewise separable as well. The special separable structure of this problem together with its non smoothness allows a corresponding analysis and specific numerical methods related with *derivative free optimization.*

2.2 Cluster Stability Using Minimal Spanning Trees

Estimation of the appropriate number q of clusters is a fundamental problem in cluster analysis. Many approaches to this problem exploit the *within-cluster dispersion matrix* (defined according to the pattern of a covariance matrix). The span of this matrix (column space) usually decreases as the number of groups rises and may have a point in which it "falls". Such an "elbow" on the graph locates in several known methods, a "true" number of clusters. Stability based approaches, for the cluster validation problem, evaluate the partition's variability under repeated applications of a clustering algorithm on samples. Low variability is understood as high consistency of the results obtained and the number of clusters that minimizes cluster stability is accepted as an estimate for the "true" number of clusters. In [10], a statistical method for the study of cluster stability is proposed. This method suggests a geometrical stability of a partition drawing samples from the partition and estimating the clusters by means of each one of the drawn samples. A pair of partitions is considered to be consistent if the obtained divisions match. The matching is measured by a *minimal spanning tree (MST)* constructed for each one of the clusters and the number of edges connecting points from different samples is calculated. MSTs are important for several reasons: they can be quickly and easily computed with the help known methods of *discrete* optimization, they create a sparse subgraph which reflects some essence of the given graph, and they provide a way to identify clusters in point sets.

3 Classification in Statistical Learning

Classification is a supervised learning in which the classification function is determined from the set of examples so called training set.

3.1 Classification by SVM

In this paper, we concentrate on *support vector machines (SVMs)* as one important classification tool that uses continuous optimization [7]. A SVM is a classification method based on finding a discriminative function which maximizes the distance between two class of points. More formally, let (x, y) be an (input,output) pair, where $x \in \mathbf{R}^n$ and $y \in \{-1, 1\}$ and x comes from some input domain X and similarly y comes from some output domain Y. A training set is defined by l input-output pairs by $S = \{(x_i, y_i)\}_{i=1}^{l}$. Given S and a set of functions $\mathcal{F}$ we search for a candidate function $f \in \mathcal{F}$ such that $f : x \mapsto y$. We refer to this candidate function as a *hypothesis* [5]. The classes are separated by an affine function, hyperplane $\langle w, x\rangle + b = 0$, where $w \in \mathbf{R}^n$ is a normal vector (weight vector) helping to define the hyperplane, $b \in \mathbf{R}$ is the bias term [5], and $\langle \cdot, \cdot \rangle$ denotes the scalar product. Hence, given a set of examples S, the SVM separates it into two groups by a hyperplane. In linearly inseparable cases, one can define a *non-linear mapping* ϕ which transforms the input space into a higher dimensional *feature space* that we will refer to as the SVM. The original points are separable in feature space. But the mapping can be of very high-dimension or even infinite. Hence, it is hard to interpret decision (classification) functions which are expressed as $f(x) = \langle w, \phi(x)\rangle + b$. Following the notation of [5], the *kernel function* is defined as an inner product of two points under the mapping ϕ, i.e., $\kappa(x_i, x_j) = \langle \phi(x_i), \phi(x_j)\rangle$ which can also be explained as the similarity between two points and the optimization problem for separating two classes is expressed as follows:

$$\begin{array}{rl} & \min_{\xi,\mathbf{w},b} \|\mathbf{w}\|_2^2 + \mathcal{C}\sum_i \xi_i \\ \text{such that} & y_i \cdot (\langle \mathbf{w}, \phi(\mathbf{x_i})\rangle + b) \geq 1 - \xi_i \quad (i = 1, 2, \ldots, m), \end{array} \tag{4}$$

where $\mathcal{C}$ is an error constant to penalize tolerance variable, slack variable, ξ. The dual problem in the soft margin case looks as follows:

$$\begin{array}{rl} \max_\alpha & \sum_{i=1}^m \alpha_i - \frac{1}{2}\sum_{i=1}^m y_i y_j \alpha_i \alpha_j \kappa(x_i, x_j), \\ \text{such that} & \sum_{i=1}^m y_i \alpha_i = 0, \quad 0 \leq \alpha_i \leq \mathcal{C} \quad (i = 1, 2, \ldots, m), \end{array} \tag{5}$$

where the vector α is dual variable so called *support vectors.* The solution of the optimization problem (5) yields a maximal margin hyperplane that defines our SVM. In [8], we propose a combination of infinitely many kernels in Riemann Stieltjes integral form for binary classification to allow all possible choices of kernels into the kernel space which makes the problem infinite in both dimensions and number of constraints, a so called *infinite programming (IP).* Based on motivation in [9], we can define our *infinite* learning problem as follows:

$$\begin{array}{l} \max\limits_{\theta\in\mathbf{R},\beta} \ \theta \ (\beta : [a, b] \to \mathrm{R} \text{ monotonically increases}), \\ \text{such that } \int_\Omega \left(\frac{1}{2}S(\omega, \alpha) - \sum_{i=1}^l \alpha_i\right) d\beta(\omega) \geq \theta \ \forall \alpha \in A, \\ \int_\Omega d\beta(\omega) = 1. \end{array} \tag{6}$$

Here, $S(\omega, \alpha) := \frac{1}{2}\sum_{i,j=1}^l \alpha_i \alpha_j y_i y_j \kappa(x_i, x_j, \omega)$, $A := \{\alpha \in \mathbf{R}^l \mid 0 \leq \alpha_i \leq \mathcal{C} \ \ (i = 1, 2, \ldots, l)$, and $\sum_{i=1}^l \alpha_i y_i = 0\}$. Let $T(\omega, \alpha) := S(\omega, \alpha) - \sum_{i=1}^l \alpha_i$, and $\Omega := [0, 1]$. Having introduced Riemann-Stieltjes integrals via functions β, we can now reinterpret the latter ones by (probability) measures. Herewith, (6) turns into the following form:

$$\begin{array}{ll} \max\limits_{\theta\in\mathbf{R},\beta} \ \theta \ (\beta : \text{ a positive measure on } \Omega), & \\ \text{such that} & \theta - \int_\Omega T(\omega, \alpha) d\beta(\omega) \leq 0 \ \forall \alpha \in A, \ \int_\Omega d\beta = 1. \end{array} \tag{7}$$

It is evident [1] that problem (7) is an *infinite programming (IP) problem.* The dual to (7) is:

$$\begin{array}{ll} \min\limits_{\sigma\in\mathbf{R},\rho} \ \sigma & (\rho : \text{ a positive measure on} A), \\ \text{such that} & \sigma - \int_A T(\omega, \alpha) d\rho(\alpha) \leq 0, \ \forall \omega \in \Omega, \ \int_A d\rho(\alpha) = 1. \end{array} \tag{8}$$

Assume that there exist pairs $(\beta,\ \theta)$ and $(\rho,\ \sigma)$ of feasible solutions of problems (7) and (8) which are complementary slack, i.e., $\sigma^* = \int_A T(\omega,\alpha)d\rho^*(\alpha)$ and $\theta^* = \int_A T(\omega,\alpha)d\beta^*(\omega)$. Then, β has measure only where $\sigma = \int_A T(\omega,\alpha)d\rho(\alpha)$ and ρ has measure only where $\theta = \int_\Omega T(\omega,\alpha)d\beta(\omega)$ which implies that both solutions are optimal for their respective problems. The regularity condition of problem (8) is analyzed in [8]. The so-called *reduction ansatz* enables the Implicit Function Theorem for reducing an infinite number of constraints to a finite number [14]. Of course, this can also be achieved by a smart *discretization.* Note that we can also focus on parametric classes of probability measures; then our IP problems turn to *SIP* (semi-infinite programming) problems; eventually, when applying any of the those three approaches, we arrive at a finitely constrained program.

3.2 Max-min Separability

According to [2], the problem of supervised data classification can be reduced to a number of set separation problems. For each class, the training points belonging to it have to be separated from the other training points using a certain, not necessarily linear, function. This problem is formulated in [2] as a nonsmooth optimization problem with max-min objective function. Let A and B be given disjoint sets containing m and p vectors from $\mathbf{R}^n$, respectively: $A = \{a^1,\ldots,a^m\}$, $B = \{b^1,\ldots,b^p\}$. Let $H = \{h_1,\ldots,h_l\}$ be a finite set of hyperplanes, where h_j is given by $\langle x_j, z\rangle - y_j = 0$ $j = (1,2,\ldots,l)$ with $x_j \in \mathbf{R}^n$, $y_j \in \mathbf{R}$. Let $J = \{1,2,\ldots,l\}$. Consider any partition of J in the form $J^r = \{J_1,\ldots,J_r\}$, where $J_k \neq \emptyset, k = 1,\ldots,r$; $J_k \bigcap J_s = \emptyset$, if $k \neq s$; $\bigcup_{k=1}^r J_k = J$. Let $I = \{1,\ldots,r\}$. A particular partition $J^r = \{J_1,\ldots,J_r\}$ of the set J defines the following max-min type function:

$$\varphi(z) = \max_{i\in I}\min_{j\in J_i}(\langle x_j, z\rangle - y_j) \quad (z \in \mathbf{R}^n). \tag{9}$$

We say that the sets A and B are *max-min separable* if there exist a finite number of hyperplanes, H, and a partition J^r of the set J such that for all $i \in I$ and $a \in A$ we have $(\langle x_j, a\rangle - y_j) < 0$ and for any $b \in B$ there exists at least one j such that $(\langle x_j, b\rangle - y_j) > 0$. It follows from the definition above that if the sets A and B are max-min separable then $\varphi(a) < 0$ for any $a \in A$ and $\varphi(b) > 0$ for any $b \in B$, where the function φ is defined by (9). Thus the sets A and B can be separated by a function represented as a max-min of linear functions. The problem of the max-min separability is reduced to the following optimization problem:

$$\min f(x,y) \quad \text{such that } (x,y) \in \mathbf{R}^{l\times n} \times \mathbf{R}^l, \tag{10}$$

where the objective function f is $f(x,y) = f_1(x,y) + f_2(x,y)$. Here, $f_1(x,y) = \frac{1}{m}\sum_{k=1}^m \max[0, \max_{i\in I}\min_{j\in J_i}(\langle x_j, a^k\rangle - y_j + 1)]$, $f_2(x,y) =$

$\frac{1}{p}\sum_{s=1}^{p} \max[0, \min_{i\in I} \max_{j\in J_i}(-\langle x_j, b^s\rangle + y_j + 1)]$. f_1 and f_2 are piece-wise linear, hence f is piece-wise linear and piece-wise separable. In [2] it is shown that the calculation of subgradients for f may become difficult. The authors propose a derivative free algorithm for minimizing max-min-type functions, which solved large scale problems with up to 2000 variables efficiently.

4 Conclusion

This paper introduces some recent optimization methods developed in data mining by some modern areas of clustering and classification. There is a great potential of important OR applications, and of future research waiting.

References

1. E.J. Anderson and P. Nash, John Wiley and Sons Ltd, *Linear Programming in Infinite-Dimensional Spaces*, 1987.
2. Bagirov, A.M., and Ugon, J., *Piecewise partially separable functions and a derivative-free algorithm for large scale nonsmooth optimization*, Journal of Global Optimization 35 (2006) 163-195.
3. Bagirov, A.M., and Yearwood, J., *A new nonsmooth optimization algorithm for minimum sum-of-squares clustering problems*, EJOR 170, 2 (2006) 578-596.
4. Bock, H.H., *Automatische Klassifikation*, Vandenhoeck, Göttingen (1974).
5. Cristianini, N., and Shawe-Taylor, J., *An introduction to Support Vector Machines and other Kernel-Based Learning Methods*, Cambridge University Press (2000).
6. Han, J., and Kamber, M., *Data Mining: Concepts and Techniques, The Morgan Kaufmann Series in Data Management Systems*, Jim Gray, Series Editor Morgan Kaufmann Publishers (2000).
7. Hastie, T., Tibshirani, R., and Freedman, J., *The Elements of Statistical Learning - Data Mining, Inference and Prediction*, Springer Series in Statistics, 2001.
8. Özöğür-Akyüz, S. and Weber, G.-W., *Learning with Infinitely Many Kernels via Semi-Infinite Programming*, in ISI Proceedings of 20th Mini-EURO Conference *Continuous Optimization and Knowledge-Based Technologies*, Neringa, Lithuania, May 20-23, 2007.
9. Sonnenburg, S., Raetsch, G., Schafer, C. and Schoelkopf, B. (2006), Large scale multiple kernel learning, J. Machine Learning Research 7, (2006) 1531–1565.
10. Volkovich, Z.V., Barzily, Z., Akteke-Öztürk, B., and Weber, G.-W., *Cluster stability using minimal spanning trees*, submitted to TOP.
11. Weber, G.-W., Taylan, P., Özögür, S., and Akteke-Öztürk, B., Statistical learning and optimization methods in data mining, in: *Recent Advances in Statistics*, Turkish Statistical Institute Press, Ankara (2007) 181-195.

The Features of Solving of the Set Partitioning Problems with Moving Boundaries Between Subsets

Tetyana Shevchenko[1], Elena Kiseleva[1], and Larysa Koriashkina[1,2,3]

[1] Department of Calculating Mathematics and Mathematical Cybernetics
Dnipropetrovsk National University
49010 Dnipropetrovsk, 72 Gagarin avenue, Ukraine
{sheva2985,kiseleva47,koryashkinals}@mail.ru

[2] Department of Systems Analysis and Management
National mining university
49027 Dnipropetrovsk, 19 Charles Marx avenue, Ukraine

[3] Department of Higher mathematics
Odessa National Academy of Telecommunications named after O.S.Popov
65029 Odessa, Kuznechnaya street 1, Ukraine

Summary. Problems and methods presented in this paper synthesize foundations of theory of continuous set partitioning problems (SPP) and optimal control of systems described by ordinary differential equations. In order to mathematically formulate SPP quite often one should take into account the temporal and spatial changes of object or process state. Some of such models concerned with problems of preservation of the environment were learning by our scientists. Mathematical models of problems mentioned above are new from the viewpoint of problem statement and interesting to further generalization and developing of theoretical results which could be used in practice widely. A common SPP could be formulated as follows: it is necessary to partition a given area (set) into a finite number of disjoint subsets that satisfies certain restrictions so that the objective function reaches an extreme value. We propose a new problem statement, which differs from known ones in the following way: the desired set partition is dynamic in consequence of 1) a function which describes the certain object or process state varies with time; 2) a function, choice of which has an influence on state of this object or process, is defined by partition of considered set each moment of time. This problem amounts to optimal control one for which one should write out the necessary conditions of optimality in the form of Pontrjagin's maximum principle. The constructed algorithm for solving such problems bases on combining both the methods of solving continuous SPP and methods of optimal control theory. With a view to investigate the properties of solutions of new set partition-

B. Fleischmann et al.(Eds.), *Operations Research Proceedings 2008*,
DOI: 10.1007/978-3-642-00142-0_86,

ing problem we realized the series of computational experiments and made qualitative analysis of obtained results.

1 Introduction

There are many technical, economic, social, and other scientific applications which need optimal (i.e. the best in a certain sense) partition of some set into disjoint subsets. As a case in point there exist problems of territorial planning of services sectors; determining of election districts' boundaries; resources conservation problems, and so on. The variety of initial data involving some information about set properties, constraints of a certain problem parameters or performance criteria causes plenty of optimization partitioning problems. One marks out two groups: discreet [1] and continuous SPP [2, 3]. By virtue of difference of its mathematical formulating the mathematical apparatus used to solving these problems is essentially different.
Until now, problems from both mentioned groups were studied in static statement. Those problems are well-structured and systematized. In this paper we propose to extend the class of continuous SPP. We base ourselves upon the fact that a certain problem parameters or characteristics vary with time according to the fixed rule and, in this connection, optimal set partition will have a dynamic nature, and boundaries between subsets will change in the course of time.

2 Proposed Problem Formulation

Let a set of functions $u_i(x,t) \in L_2(\Omega \times [0,T]), i = \overline{1,N}$, be given, where $\Omega \subset R^2$ is bounded Lebesgue measurable set; $P_N(\Omega)$ is the class of every possible partitions of Ω into N subsets:

$$P_N(\Omega) = \{\bar{\omega} = (\Omega_1, ..., \Omega_N) : \bigcup_{i=1}^{N} \Omega_i = \Omega,\ \Omega_i \cap \Omega_j = \emptyset,\ i \neq j,\ i,j = \overline{1,N}\}.$$

We consider the case, when boundaries between subsets can change with time, and let's denote by $\bar{\omega}(t) = (\Omega_1(t), ..., \Omega_N(t))$ the partition of set Ω which corresponds to a time moment t, $t \in [0,T]$. This partition defines change of a certain system condition $\rho(x,t)$ in such a way that $\forall x \in \Omega_i(t)$:

$$\dot{\rho}(x,t) = \rho(x,t) + u_i(x,t) + f(x,t), \quad \rho(x,0) = \rho_0(x).$$

One needs to determine such partition $\bar{\omega}^*(t) = (\Omega_1^*(t), ..., \Omega_N^*(t))$, $t \in [0, T]$, of set Ω into N disjoint subsets for which the following functional reaches its minimal value:

$$J(\bar{\omega}(\cdot)) = \beta_0 \int_0^T \sum_{i=1}^N \int_{\Omega_i(t)} (c(x, \tau_i, t) + a_i)\rho(x, t)\, dxdt +$$

$$+\beta_1 \int_0^T \sum_{i=1}^N \int_{\Omega_i(t)} u_i^2(x, t)\, dxdt \to \min_{\bar{\omega} \in P_N(\Omega)},$$

where $\beta_0, \beta_1 \geq 0$ is constants which define a priority of respective item; $c(x, \tau_i, t)$ is a given function, $\tau_i \in \Omega$, $i = \overline{1, N}$, are given points called "centers" of subsets; $T > 0$, a_i, $i = \overline{1, N}$, are fixed values; $f(x, t)$, $\rho_0(x)$ are known functions of its arguments.

Let's call such problems *Dynamical Continuous Set Partitioning Problems with Moving Boundaries between Subsets.* Although in some cases (under a certain initial data) the optimal boundaries could be static by its nature. Stated problem differs from earlier studied ones by two facts: the function $\rho(x, t)$ which describes the state of a certain system varies with time under an appointed differential rule, and the rate of change of function $\rho(x, t)$ in each point $x \in \Omega$ depends on belonging of this point to current subset $\Omega_i(t)$. Accordingly, partitions of set Ω can be various in different time moments. The features of the problem mentioned above make impossible the use of pre-existing algorithms of solving SPP [2]. Note that a stated SPP could be classified as a problem of optimal control of system with lumped parameters. For that we introduce into consideration the class of function

$$U(t) = \{u(x, t; \bar{\omega}(t)) : u(x, t; \bar{\omega}(t)) = u_i(x, t)$$

$$for\ almost\ all\ x \in \Omega_i(t),\ i = \overline{1, N};$$

$$\bar{\omega}(t) = (\Omega_1(t), ..., \Omega_N(t)) \in P_N(\Omega)\}, \quad \forall t \in [0, T].$$

Let a function $\rho(x, t)$, which describe the system state, over all $x \in \Omega$ be a solution of Cauchy problem

$$\dot{\rho}(x, t) = \rho(x, t) + u(x, t) + f(x, t), \quad u \in U(t), \quad \rho(x, 0) = \rho_0(x). \quad (1)$$

Every admissible partition $\bar{\omega}(t) \in P_N(\Omega)$ defines $\forall x \in \Omega$ the function $u(x, t; \bar{\omega}(t)) \in U(t)$ such that its corresponding Cauchy problem (1) has the unique solution $\rho(x, t) \in L_2(\Omega \times [0, T])$.

Let's agree on $\rho(x, t; \bar{\omega}(t))$ for denoting the solution of problem (1) corresponding to function $u(x, t; \bar{\omega}(t))$.

In that case the problem consists of finding such a control function $u^*(x,t;\bar{\omega}(t)) \in U(t)$ for all $x \in \Omega$ and respective phase path $\rho^*(x,t;\bar{\omega}(t))$, $t \in [0,T]$ for which the functional

$$J(u(\cdot,\cdot),\rho(\cdot,\cdot)) = \beta_0 \int_0^T \sum_{i=1}^N \int_{\Omega_i(t)} (c(x,\tau_i,t)+a_i)\rho(x,t;\bar{\omega}(t))\,dxdt + \\ +\beta_1 \int_0^T \int_{\Omega} u^2(x,t;\bar{\omega}(t))\,dxdt \to \min_{u\in U}, \qquad (2)$$

would reach its minimal value.

3 Problem-Solving Procedure

Dependence of right-hand member of differential equation (1) on set partition complicates use of optimal control methods of systems describing by ordinary differential equations for solving problem (1)-(2). We propose the method of solving formulated SPP with moving boundaries between subsets which synthesizes both basic concepts of continuous set partitioning theory developed by Kiseleva [2] and optimal control theory.

Let's consider the problem (1)-(2). Introducing characteristic functions $\lambda_i(\cdot,t)$ of subsets $\Omega_i(t)$, $i=\overline{1,N}$, allows writing the control function in the form of

$$u(x,t;\bar{\omega}(t)) = \sum_{i=1}^N u_i(x,t)\lambda_i(x,t). \qquad (3)$$

Instead of problem (2) an equivalent problem is considered: one needs to find the vector-function $\lambda(\cdot,\cdot) = (\lambda_1(\cdot,\cdot),...,\lambda_N(\cdot,\cdot))$ such that

$$I(\lambda(\cdot,\cdot)) = \sum_{i=1}^N \int_0^T \int_{\Omega} [\beta_0(c(x,\tau_i,t)+a_i)\rho(x,t) + \\ +\beta_1 u_i^2(x,t)]\lambda_i(x,t)\,dxdt \to \min_{\lambda\in\Lambda}, \qquad (4)$$

where

$$\Lambda(t) = \{\lambda(x,t) = (\lambda_1(x,t),...,\lambda_N(x,t)) : \quad \lambda_i(x,t) = 0 \vee 1,\ i=\overline{1,N}, \\ \sum_{i=1}^N \lambda_i(x,t) = 1,\ \mathit{for\ almost\ all}\ x \in \Omega\}, \quad t \in [0,T],$$

we denoted by $\rho(x,t)$ a solution of problem (1) which corresponds to function (3).

Further, according to the theory of sets partitioning, a problem (4) amounts to the following problem

$$I(\lambda(\cdot,\cdot)) \to \min_{\lambda \in \Lambda_1}, \tag{5}$$

$$\Lambda_1(t) = \{\lambda(x,t) = (\lambda_1(x,t), ..., \lambda_N(x,t)) : \quad 0 \le \lambda_i(x,t) \le 1,\ i = \overline{1,N},$$

$$\sum_{i=1}^{N} \lambda_i(x,t) = 1,\ for\ almost\ all\ x \in \Omega\}, \quad t \in [0,T],$$

Problem (5) is an optimal control one, in which vector-function $\lambda(\cdot,\cdot)$ stands as control function. For its solving one needs write out the necessary conditions of optimality in the form of Pontrjagin's maximum principle:

i) - stationarity by $\rho(\cdot,\cdot)$: $\quad \forall x \in \Omega$

$$\frac{\partial \Psi}{\partial t} = -\frac{\partial H}{\partial t}, \tag{6}$$

where Hamiltonian-Pontrjagin function is given by

$$H(\lambda(\cdot,\cdot), \rho(\cdot,\cdot), \Psi(\cdot,\cdot)) =$$

$$= \int_{\Omega} (\rho(x,t) + \sum_{i=1}^{N} u_i(x,t)\lambda_i(x,t) + f(x,t))\Psi(x,t)\,dx -$$

$$- \int_{\Omega} \sum_{i=1}^{N} (\beta_0(c(x,\tau_i,t) + a_i)\rho(x,t) + \beta_1 u_i^2(x,t))\lambda_i(x,t)\,dx;$$

ii) - transversality by $\rho(\cdot,\cdot)$: $\quad \forall x \in \Omega$

$$\Psi(x,T) = 0; \tag{7}$$

iii) - optimality by $\lambda(\cdot,\cdot)$:

$$\lambda^*(x,t) : \quad \max_{\lambda \in \Lambda_1} H(\lambda(\cdot,\cdot), \rho(\cdot,\cdot), \Psi(\cdot,\cdot)).$$

Taking into account the structure of Λ_1 we can write: $\forall x \in \Omega$, $t \in [0,T]$

$$\lambda_i^*(x,t) = \begin{cases} 1, & q_i(x,\tau_i,t) = max_{k=\overline{1,N}}\, q_k(x,\tau_k,t), \\ 0, & otherwise; \end{cases} \tag{8}$$

where

$$q_i(x,\tau_i,t) = u_i(x,t)\Psi(x,t) - \beta_0(c(x,\tau_i,t) + a_i)\rho(x,t) - \beta_1 u_i^2(x,t).$$

Because of linearity and boundedness of objective functional (2) with respect to the function $\lambda(x,t)$, as well as linearity of right-hand member of differential equation (1) with respect to functions $\rho(x,t)$ and $u(x,t)$, the principle of the maximum stands as not only necessary condition, but also sufficient condition of optimality.
For the numerical solving obtained boundary problem Newton's method is applied.

4 Conclusion

In this paper we have presented a new continuous set partitioning problem. It is characterized by the presence of family of differential constraints and the dependence of right-hand members of differential equations on a partition of some set. We also have developed and implemented a method of solving the SPP with moving boundaries between subsets. The main idea is the following: original problem is amounted to optimal control problem using body of the theory of continuous set partitioning problems. One should apply the necessary conditions of optimality in the form of Pontrjagin's maximum principle to obtained problem. The results of multiple computational experiments confirm the possibility of applying proposed approach to solving the stated problem.

Acknowledgement. The authors would like to thank Alexandr Firsov for his valuable discussions, constructive comments, and helpful suggestions.

References

1. Balas E, Padberg MW (1977) Set partitioning: a survey. Comb. Optimiz. Lect. summer Sch. Comb. Optimmiz. urbino. PP. 151–210
2. Kiseleva E, Stepanchuk T (2003) On the Efficiency of a Global Non-differentiable Optimization Algorithm Based on the Method of Optimal Set Partitioning. 25:209–235
3. Corley HW, Roberts SD (1972) Duality relationships a partitioning problem. SIAM 23:490–494

Symmetry in the Duality Theory for Vector Optimization Problems

Martina Wittmann-Hohlbein

Institute of Mathematics, Martin-Luther-University Halle-Wittenberg, 06099 Halle, Germany,
martina.wittmann@student.uni-halle.de

Summary. This article deals with duality theory for the linear vector optimization problem and its geometric dual problem.

1 Introduction

The objective of this work is to obtain symmetry in the duality theory for linear vector optimization problems as known from the scalar duality theory. We derive results that are related to the concept of geometric duality as introduced in [1] and extend these results to a larger class of optimization problems. We emphasize that the dual problem for the linear vector optimization problem is naturally set-valued as it can easily be derived with results of the Lemma of Farkas.

2 Duality Theory in Linear Multiobjective Programming

In the following let $M \in \mathbb{R}^{q\times n}$, $A \in \mathbb{R}^{m\times n}$ and $b \in \mathbb{R}^m$ where q, m and n denote positive integers. Let K be any non trivial closed convex and pointed ordering cone in $\mathbb{R}^q$ then we denote by

$$K^* := \{y \in \mathbb{R}^q \,|\forall k \in K : y^T k \geq 0\}$$

the positive dual cone to K. For a subset $A \subseteq \mathbb{R}^q$ the set of weakly efficient elements w.r.t. the ordering cone K is given by

$$\text{w-Min}_K\, A := \{y \in A | (y - \operatorname{ri} K) \cap (A + K) = \emptyset\}$$

where $\operatorname{ri} K$ denotes the relative interior of K as defined in [2], p. 44. The set of weakly maximal elements of A is defined by

B. Fleischmann et al.(Eds.), *Operations Research Proceedings 2008*,
DOI: 10.1007/978-3-642-00142-0_87, © Springer-Verlag Berlin Heidelberg 2009

$$\text{w-Max}_K\, A := -\,\text{w-Min}_K(-A).$$

We consider the linear vector optimization problem

$$\text{(LVOP)} \quad \begin{array}{l} P := \text{w-Min}_{\mathbb{R}^q_+} \bigcup_{x\in\mathcal{X}}(\{Mx\} + \mathbb{R}^q_+) \\ \mathcal{X} := \{x \in \mathbb{R}^n \,|\, Ax \geq b\}. \end{array}$$

Duality assertions for (LVOP) can be found in [3, 4] and many other papers. To be in line with [4] we assign

$$\text{(SD)} \quad \begin{array}{l} D := \text{w-Max}_{\mathbb{R}^q_+}(\bigcup_{(u,c)\in\mathcal{U}}\{y \in \mathbb{R}^q \,|\, c^T y = u^T b\} - \mathbb{R}^q_+) \\ \mathcal{U} := \{(u,c) \in \mathbb{R}^m_+ \times \mathbb{R}^q_+ \setminus \{0\} \,|\, 1\!\!1^T c = 1, A^T u = M^T c\} \end{array}$$

with

$$1\!\!1 := (1, ..., 1)^T \in \mathbb{R}^q$$

to be the corresponding set-valued dual problem for (LVOP). Theorem 15 in [4] ensures strong duality, i.e. $P = D$.
We are interested in the dual problem for (SD) and apply a duality theorem for set-valued optimization problems as stated in Theorem 8, [4]. The objective map of (SD) is not convex w.r.t $\mathbb{R}^q_+$. However, convexity is a necessary condition to obtain strong duality assertions. As dualization of (SD) proves to be futile a second optimization problem is derived from (SD). We consider the vector-valued optimization problem

$$\text{(VD)} \quad \bar{D} := \text{w-Min}_{(-\bar{K})}(\bigcup_{(u,c)\in\mathcal{U}}\{(c_1, ..., c_{q-1}, u^T b)^T\} - \bar{K})$$

where

$$\bar{K} = \{y \in \mathbb{R}^q \,|\, y_1 = 0, ..., y_{q-1} = 0, y_q \geq 0\}$$

denotes the new ordering cone. The optimization problem (VD) is referred to as geometric dual problem to (LVOP).
An element $(u^0, c^0) \in \mathcal{U}$ is called a solution of (SD) if $\{y \in \mathbb{R}^q \,|\, c^{0T} y = u^{0T} b\} \cap D \neq \emptyset$ and it is called a solution of (VD) if $(c^0_1, ..., c^0_{q-1}, u^{0T} b)^T \in \bar{D}$. Immediately, we conclude that if $(u^0, c^0) \in \mathcal{U}$ solves (SD) it also solves (VD) and vice versa. For $v \in \mathbb{R}^q$ we set

$$H(v) := \{y \in \mathbb{R}^q \,|\, (v_1, ..., v_{q-1}, 1 - \sum_{i=1}^{q-1} v_i) y = v_q\}$$

and obtain the following relationship between (VD) and (LVOP). If v^0 is an element of $\bar{D}$, then is $H(v^0)$ is a supporting hyperplane to $\bigcup_{x\in\mathcal{X}}\{Mx\} + \mathbb{R}^q_+$ and the intersection $H(v^0) \cap (\bigcup_{x\in\mathcal{X}}\{Mx\} + \mathbb{R}^q_+)$ is a proper exposed weakly efficient face of the primal image set. Moreover, weakly maximal exposed faces of the image set of (VD) are mapped to weakly efficient exposed faces of the primal image set.

Lemma 1. *Let V be a be a proper weakly efficient exposed face of the image set of (VD), then*

$$\bigcap_{v\in V} H(v) \cap \left(\bigcup_{x\in\mathcal{X}} \{Mx\} + \mathbb{R}^q_+\right)$$

is a proper weakly efficient exposed face of the image set of (LVOP).

A proof of Lemma 1 is stated in [1].

3 Duality Assertions for the Geometric Dual Problem

Let

$$D^* = \begin{pmatrix} 0 & , E_{q-1} & , 0 \\ b^T & , 0 & , 0 \end{pmatrix} \in \mathbb{R}^{q\times q+m}$$

where E_{q-1} denotes the identity matrix in $\mathbb{R}^{q-1\times q-1}$. We rewrite (VD) in a way that more clearly resembles the form of a linear vector optimization problem, namely

$$\text{(VD)}\ \ \bar{D} = \text{w-Min}_{(-\bar{K})}(\textstyle\bigcup_{u^*:=(u,c)\in\mathcal{U}}\{D^*u^*\} - \bar{K}).$$

Note, that for $\bar{K}$ the concept of weakly minimal elements of a subset of $\mathbb{R}^q$ agrees with the stronger solution concept of minimal elements as defined in [3], p. 103. Further, for $y \in \mathbb{R}^q$ we define the vector

$$c(y) := (y_1, ..., y_{q-1}, 1 - y_1 - ... - y_{q-1})^T \in \mathbb{R}^q.$$

The definition of weakly efficient elements yields $y^0 \in \text{w-Min}_{(-\bar{K})}(D^*[\mathcal{U}] - \bar{K})$ if (A1) and (A2) are both satisfied where

$$\begin{aligned}
&\text{(A1)}\ \ y^0 \in D^*[\mathcal{U}] - \bar{K} \iff \\
&\qquad c(y^0) \geq 0 \text{ and } \exists u^0 \in \mathbb{R}^m_+ : A^T u^0 = M^T c(y^0), y^0_q \leq u^{0T} b, \\
&\text{(A2)}\ \ \forall\, y \in D^*[\mathcal{U}] - \bar{K} : y^0 \notin y - \bar{K}\backslash\{0\} \iff \\
&\qquad (c(y^0) \geq 0, u \in \mathbb{R}^m_+, A^T u = M^T c(y^0) \Longrightarrow u^T b \leq y^0_q)
\end{aligned}$$

We present a second pair of assertions (B1) and (B2) given by

$$\text{(B1)}\ \ \exists\, (x^0, l^0) \in \mathbb{R}^n \times \mathbb{R}^q_+ : Ax^0 \geq b,\ c(y^0)^T M x^0 + c(y^0)^T l^0 \leq y^0_q$$

$$\text{(B2)}\ \ \forall\, (x, l) \in \mathbb{R}^n \times \mathbb{R}^q_+, Ax \geq b :\ c(y^0)^T M x + c(y^0)^T l \geq y^0_q$$

The Lemma of Farkas and its direct conclusion lead to the following results.

Lemma 2.
Statements (A1) and (A2) imply (B1).

Proof. By Theorem II.1.2., [5], statement (B1) is equivalent to

$$(\overline{\text{B1}})\ \ v^1 \in \mathbb{R}^m_+, v^2 \geq 0, -A^T v^1 + M^T c(y^0)v^2 = 0, c(y^0)v^2 \geq 0 \\ \Longrightarrow -b^T v^1 + y^0_q v^2 \geq 0.$$

Assume that $(\overline{\text{B1}})$ does not hold true. Then there is a pair (v^1, v^2) with the above properties such that $-b^T v^1 + y^0_q v^2 < 0$. Let $v^2 > 0$ and set $w := \frac{1}{v^2} v^1$. We obtain $w \in \mathbb{R}^m_+, c(y^0) \geq 0, A^T w = M^T c(y^0)$ and $b^T w > y^0_q$, a contradiction to (A2). Let $v^2 = 0$, i.e. there is a $v^1 \in \mathbb{R}^m_+$ with $A^T v^1 = 0$ and $b^T v^1 > 0$. Taking into account (A1) there is a $u^0 \in \mathbb{R}^m_+$ with $A^T u^0 = M^T c(y^0)$ and $b^T u^0 \geq y^0_q$. Hence for $w := u^0 + v^1 \in \mathbb{R}^m_+$ we have $A^T w = M^T c(y^0)$ and $b^T w > y^0_q$, a contradiction to (A2). □

Lemma 3. *Statement (A1) implies (B2).*

Proof. By Theorem II.1.2., [5], statement (A1) is equivalent to

$$(\overline{\text{A1}})\ \ c(y^0) \geq 0 \text{ and} \\ (x^1 \in \mathbb{R}^n, x^2 \geq 0, Ax^1 - bx^2 \geq 0 \Longrightarrow c(y^0)^T M x^1 - y^0_q x^2 \geq 0).$$

Let $x \in \mathbb{R}^n, l \in \mathbb{R}^q_+$ and $Ax \geq b$. Then from $(\overline{\text{A1}})$ for $x^1 := x$ and $x^2 := 1$ it follows that $c(y^0)^T M x \geq y^0_q$ and $c(y^0)^T l \geq 0$. Together this implies (B2). □

Defining the sets

$$\begin{aligned} \bar{\mathcal{X}} &:= \{(x,l) \in \mathbb{R}^n \times \mathbb{R}^q_+ \,|Ax \geq b\} \\ H^*(z) &:= \{y \in \mathbb{R}^q |(z_1 - z_q, ..., z_{q-1} - z_q, -1)y = -z_q\} \end{aligned}$$

for $z \in \mathbb{R}^q$ then (B1) can be written as

$$(\text{B1})\ \ y^0 \in \bigcup_{(x,l)\in\bar{\mathcal{X}}} H^*(Mx + l) - (-\bar{K})$$

and one easily verifies that (B2) is equivalent to

$$(\overline{\text{B2}})\ \ \forall (x,l) \in \bar{\mathcal{X}} : y^0 \notin H^*(Mx + l) - (-\bar{K}\backslash\{0\}).$$

(B1) and $(\overline{\text{B2}})$ imply that $y^0 \in \text{w-Max}_{(-\bar{K})} \bigcup_{(x,l)\in\bar{\mathcal{X}}} (H^*(Mx + l) + \bar{K})$. A similar argumentation is used to prove the opposite inclusion, i.e. that (B1) implies (A2) and that (B1) together with (B2) imply (A1). We have shown that a weakly efficient element from the set

$$\bigcup_{u^* \in \mathcal{U}} \{D^* u^*\} - \bar{K}$$

is also a weakly maximal element of the set $\bigcup_{(x,l)\in\bar{\mathcal{X}}}(H^*(Mx+l)+\bar{K})$ and vice versa. Hence the set-valued dual problem to (VD) is denoted by

$$\text{(SDD)} \quad \text{w-Max}_{(-\bar{K})}(\textstyle\bigcup_{(x,l)\in\bar{\mathcal{X}}}(H^*(Mx+l)+\bar{K}).$$

Again, from the set-valued optimization problem (SDD) a vector-valued optimization problem (VDD) with respect to the ordering cone $\bar{K}$ is derived. We set

$$\begin{aligned}\text{(VDD)} \quad \text{w-Max}_{\bar{K}}(\textstyle\bigcup_{(x,l)\in\bar{\mathcal{X}}}(\{((Mx+l)_1-(Mx+l)_q, \ldots \\ \ldots, (Mx+l)_{q-1}-(Mx+l)_q, -(Mx+l)_q)^T\} - \bar{K})\end{aligned}$$

and refer to (VDD) as the geometric dual problem to (VD). We define a solution of (SDD) and of (VDD) in the same way as for (SD) and (VD) in the previous section. We obtain that $(x^0,l^0) \in \bar{X}$ solves (SDD) iff it solves (VDD). It is easy to show that

$$H^*(z) \subseteq H^*(\bar{z}) + \bar{K} \Longleftrightarrow \exists \lambda \geq 0 : z = \bar{z} + \lambda \, 1\!\!1$$

holds true which is used to prove the following lemma.

Lemma 4. *An element $(x^0,l^0) \in \bar{\mathcal{X}}$ solves (VDD) iff it solves $(\overline{LVOP})$ where*

$$(\overline{LVOP}) \quad \text{w-Min}_{\mathbb{R}^q_+}(\textstyle\bigcup_{(x,l)\in\bar{\mathcal{X}}}(\{Mx+l\}).$$

Moreover, if $(x^0,l^0) \in \bar{\mathcal{X}}$ solves $(\overline{\text{LVOP}})$ then $H^*(Mx^0+l^0)$ is a supporting hyperplane to $\bigcup_{u^*\in\mathcal{U}}\{D^*u^*\} - \bar{K}$ and $H^*(Mx^0+l^0) \cap (\bigcup_{u^*\in\mathcal{U}}\{D^*u^*\} - \bar{K})$ is a weakly efficient face of the image set of (VD). An analogous statement as in Lemma 1 applies for $(\overline{\text{LVOP}})$ and (VD).

Lemma 5. *Let V^* be a be a proper weakly efficient exposed face of the image set of $(\overline{LVOP})$, then*

$$\bigcap_{v^*\in V^*} H^*(v^*) \cap (\bigcup_{u^*\in\mathcal{U}} \{D^*u^*\} - \bar{K})$$

is a proper weakly efficient exposed face of the image set of (VD) .

Since $(\overline{\text{LVOP}})$ is equivalent to (LVOP) we regard (LVOP) itself as the geometric dual problem to (VD) and symmetry for the linear vector optimization problem is obtained in the sense of a geometric duality between polyhedral sets.

4 Geometric Duality for a Larger Class of Vector Optimization Problems

The Lemma of Farkas is a powerful tool which also enables us to derive the set-valued dual problem (SD) for (LVOP) directly. It can be used to obtain a corresponding dual problem to the linear vector optimization problems with respect to any non trivial polyhedral convex and pointed ordering cone in $\mathbb{R}^q$. The feasible set $\mathcal{U}$ of the set-valued and the geometric dual problem for this larger class of optimization problems reads as

$$\mathcal{U} = \{(u, c) \in \mathbb{R}^m \times K^* | u \geq 0, k^T c = 1, A^T u = M^T c\}$$

with arbitrary $k \in \operatorname{ri} K$. The choice of k affects the form of the hyperplanes H and H^*. The dual restriction $k^T c = 1$ is essential in order to deduce the geometric dual problem. Note, that the geometric dual problem derived from the set-valued optimization problem is always to determine weakly maximal elements w.r.t. $\bar{K}$, independent from the ordering cone of the underlying primal problem.
In the special case of the linear scalar optimization problem the set-valued and geometric dual problem agree and the generalized geometric duality theory reduces to the known scalar duality theory.

5 Conclusion

If we focus on symmetry in the duality theory for vector optimization problems this will always lead to the concept of geometric duality. Nevertheless we have shown its close relationship to set-valued duality theory. Our approach seems to be new in multiobjective programming and is also being investigated for general convex vector optimization problems.

References

1. F. Heyde; A. Löhne. Geometric Duality in Multiple Objective Linear Programming. to appear in SIAM Optimization.
2. R. T. Rockafellar. Convex Analysis. Princeton University Press, 1972.
3. J. Jahn. Vector Optimization. Springer Verlag Berlin Heidelberg, 2004.
4. M. Wittmann. Eine Dualitätstheorie für das lineare Vektoroptimierungsproblem. Martin-Luther-Universität Halle-Wittenberg, Diplomarbeit, 2006.
5. W. Vogel. Lineares Optimieren. Akad. Verlagsgesellschaft Geest & Portig, Leipzig, 1970.

A Simple Proof for a Characterization of Sign-Central Matrices Using Linear Duality

Rico Zenklusen

ETH Zurich, Institute for Operations Research, 8092 Zurich, Switzerland

Summary. We consider a problem described in 1992 by Davidov and Davidova, where for a given matrix $A \in \mathbb{R}^{m \times n}$ we want to know whether every matrix $B \in \mathbb{R}^{m \times n}$ with the same sign pattern as A has a nonnegative, nonzero element in its null-space. Such a matrix A is called sign-central. Davidov and Davidova gave a characterization of sign-central matrices that was proven by a rather long argument. In this paper we present a simple proof showing that the aforementioned characterization of sign-central matrices can be seen as a consequence of the weak duality theorem of linear programming.

1 Introduction

An interesting question in the context of linear systems is how much can be said about the structure of the solutions if the input is not precisely known but only some structure is imposed on the input. An interesting class of such problems is obtained by imposing a sign pattern on the input (c.f. [2] for more information on such problems). One problem of this type is to decide whether every matrix $A \in \mathbb{R}^{m \times n}$ that corresponds to some predefined sign pattern has a nonnegative, nonzero element in its null-space. This problem was first considered in 1992 [3] where a characterization of the sign patterns for which there is always a nonnegative, nonzero element in the null-space of all corresponding linear systems was given and a rather complicated prove for this result was presented. Later, in 1994 another proof was presented using the separation theorem for convex sets [1], thus leading to the presumption that the characterization is essentially a consequence of the classical theorems of alternatives of linear systems. In this work we present a short proof based on duality theory of linear programming showing that the characterization is actually a consequence of the weak duality theorem of linear programming.

B. Fleischmann et al.(Eds.), *Operations Research Proceedings 2008*,
DOI: 10.1007/978-3-642-00142-0_88, © Springer-Verlag Berlin Heidelberg 2009

For every real number $r \in \mathbb{R}$ we define its *sign* by

$$sign(r) = \begin{cases} 0 & \text{if } r = 0 \\ 1 & \text{if } r > 0 \\ -1 & \text{if } r < 0 \ . \end{cases}$$

Similarly, to every real matrix $A \in \mathbb{R}^{m \times n}$ we associate a *sign pattern* $sign(A) \in \{0, 1, -1\}^{m \times n}$ defined by $sign(A)_{ij} = sign(A_{ij})$ for $i \in \{1, \ldots, m\}$ and $j \in \{1, \ldots, n\}$. For a matrix $A \in \mathbb{R}^{m \times n}$, its qualitative class $\mathcal{Q}(A)$ is the set of all matrices with the same sign pattern, i.e.,

$$\mathcal{Q}(A) = \{B \in \mathbb{R}^{m \times n} \mid sign(B) = sign(A)\} \ .$$

A matrix $A \in \mathbb{R}^{m \times n}$ is called *central* if it contains a nonnegative, nonzero element in its null-space and it is called *sign-central* if every matrix in its qualitative class is central.

In the following, the two signs $\oplus, \ominus$, which we call *non-strict* signs, are used to represent nonnegative respectively nonpositive values. In particular, we say that a given vector $v \in \mathbb{R}^m$ corresponds to a given non-strict sign vector $w \in \{\oplus, \ominus\}^m$ if $v_i \geq 0$ when $w_i = \oplus$ and $v_i \leq 0$ when $w_i = \ominus$. A matrix $A \in \mathbb{R}^{m \times n}$ is called *closed* if for any non-strict sign vector $v \in \{\oplus, \ominus\}^m$, there is a column of A that corresponds to v. Notice that if A is a closed matrix, then all matrices in $\mathcal{Q}(A)$ are closed since only the sign pattern of A is used to determine whether A is closed.

In [3] the following characterization of sign-central matrices was given.

Theorem 1. *A real matrix is sign-central if and only if it is closed.*

In the next section we give an alternative proof of this theorem using duality theory of linear programming.

As observed in [1], the problem of deciding whether a given matrix is not sign-central can easily be shown to be NP-complete by a reduction from the satisfiability problem.

2 Proof of Theorem 1 Based on Linear Duality

Our proof of Theorem 1 is based on the weak duality theorem between the following linear program $LP(A)$, which is defined for any real matrix $A \in \mathbb{R}^{m \times n}$, and its dual.

$$\begin{aligned} \max\ & e^T x \\ s.t.\ & Ax = 0 \\ & x \geq 0 \end{aligned} \qquad LP(A)$$

Where $e = (1, 1, \ldots, 1)^T \in \mathbb{R}^n$. It is easy to check that we have the following property.

Property 1. A matrix $A \in \mathbb{R}^{m \times n}$ is central if and only if $LP(A)$ has a solution with strictly positive value.

The dual problem of $LP(A)$ which is stated below allows to decide whether or not $LP(A)$ has a solution with strictly positive value.

$$\begin{aligned} \min\ & 0 \\ s.t.\ & A^T y \geq e \\ & y \text{ free} \end{aligned} \qquad DP(A)$$

By weak duality of linear programming we have the following equivalence.

Property 2. For $A \in \mathbb{R}^{m \times n}$ we have that $LP(A)$ has a solution with strictly positive objective value if and only if $DP(A)$ is infeasible.

In the following we prove separately both implications given by the equivalence of Theorem 1.

2.1 A is Sign-Central $\Rightarrow$ A is Closed

This implication is proved by contradiction. Let $A \in \mathbb{R}^{m \times n}$ be a central matrix and we suppose that A is not closed. Thus, there is a non-strict sign vector $v \in \{\oplus, \ominus\}^n$ such that no column of A corresponds to v. We will raise a contradiction by constructing a matrix $B \in \mathcal{Q}(A)$ that is not central. By the Properties 1 and 2 we have to define B and a solution $y \in \mathbb{R}^m$ that is feasible for $DP(B)$. We start with defining the dual solution $y \in \mathbb{R}^m$ by

$$y_i = \begin{cases} 1 & \text{if } v_i = \ominus \\ -1 & \text{if } v_i = \oplus \end{cases}.$$

The matrix B will be obtained from $sign(A)$ by scaling at most one element in every column of $sign(A)$ as follows. We fix an index $j \in \{1, 2, \ldots, n\}$ and consider the term $(sign(A)^T y)_j$. Since there is no column in $sign(A)$ that corresponds to v, we have

$$\exists \tilde{i} \in \{1, 2, \ldots, m\} \text{ with } sign(A)_{\tilde{i}j} \cdot y_{\tilde{i}} = 1.$$

Therefore, by setting initially $B = sign(A)$ and then multiplying the element $B_{\tilde{i}j}$ by a sufficiently large positive value we get $(B^T y)_j \geq 1$. This procedure can be applied to all rows $j \in \{1, \ldots, n\}$ of B to obtain $B^T y \geq e$. Thus, we have as desired that y is a feasible solution for $DP(B)$.

2.2 A is Sign-Central $\Leftarrow$ A is Closed

This implication will be proven by contradiction, too. Let $A \in \mathbb{R}^{m \times n}$ be a closed matrix and we suppose that A is not sign-central. Thus, there exists a non-central matrix $B \in \mathcal{Q}(A)$. Furthermore, since A is closed, B must be closed, too. By Property 1 we have that the linear problem $LP(B)$ has no solution with strictly positive value. Property 2 then implies that $DP(B)$ is feasible, i.e. $\exists y \in \mathbb{R}^m$ with $B^T y \geq e$. Since B is closed there exists a column $j \in \{1, \ldots, n\}$ of B satisfying

$$B_{ij} y_i \leq 0 \quad \forall i \in \{1, \ldots, m\}.$$

This implies $(B^T y)_j \leq 0$ and is thus in contradiction with $B^T y \geq e$.

3 Conclusion

A proof for a characterization of sign-central matrices was presented which is considerable simpler than the original proof. The presented proof highlights that the characterization is essentially a theorem of alternatives that can be deduced by using duality of linear programming.

References

1. T. Ando and R. A. Brualdi. Sign-central matrices. Linear Algebra and its Applications, 208-209, 1994.
2. R. A. Brualdi and B. L. Shader. Matrices of Sign-Solvable Linear Systems. Cambridge University Press, 1995.
3. G. Davidov and I. Davidova. Tautologies and positive solvability of linear homogeneous systems. Annals of Pure and Applied Logic, 1992.

Part XII

Network Optimization

Vickrey Auctions for Railway Tracks*

Ralf Borndörfer[1], Annette Mura[2], and Thomas Schlechte[1]

[1] Konrad-Zuse-Zentrum für Informationstechnik Berlin, Takustr.7, 14195 Berlin, Germany,
{borndoerfer,schlechte}@zib.de

[2] ProCom GmbH, Luisenstr. 41, 52070 Aachen, Germany,
mura@procom.de

Summary. We consider a single-shot second price auction for railway slots, the Vickrey Track Auction (VTA), in which the winner determination problem is a complex combinatorial optimization problem. We show that the VTA is incentive compatible, i.e., rational bidders are always motivated to bid their true valuation, and that it produces efficient allocations, even in the presence of constraints on allocations. The results carry over to "generalized" Vickrey auctions with combinatorial constraints.

1 Introduction

We consider in this paper the design of an auction-based allocation mechanism for railway slots in order to establish a fair and non-discriminatory access to a railway network, see [1] and [7] for more details and background information. In this setting, *train operating companies* (TOCs) compete for the use of a shared railway infrastructure by placing bids for trains that they intend to run. The trains consume infrastructure capacity, such as track segments and stations, over certain time intervals, and they can exclude each other due to safety and other operational constraints, even if they would not meet physically (actually, to make that sure). An *infrastructure manager* chooses from the bids a feasible subset, namely, a timetable, that maximizes the auction proceeds. Such a mechanism is desirable from an economic point of view because it can be argued that it leads to the most efficient use of a limited resource.

* This work was partly funded by the German Federal Ministry of Economics and Technology (BMWi), project *Trassenbörse*, grant 19M4031A.

B. Fleischmann et al.(Eds.), *Operations Research Proceedings 2008*,
DOI: 10.1007/978-3-642-00142-0_89, © Springer-Verlag Berlin Heidelberg 2009

Vickrey argued in his seminal paper [8] for the importance of incentive compatibility in auction design, and he showed that a second price auction has this property. He, and independently Clarke [4] and Groves [6], also proposed a sealed-bid auction that generalizes the simple Vickrey auction for a single item to the multi-item case, the so-called Vickrey-Clarke-Groves (VCG) mechanism, which is also incentive compatible. This classical result pertains to a combinatorial auction, in which bids are placed for bundles of items, and two bundles can be allocated iff they do not contain the same item. This is, however, not sufficient for a railway track auction, in which more general constraints on the compatibility of slots arise, e.g., from minimum headway constraints. Whatever these constraints may be, a second price auction can of course also be conducted in such a setting. However, it is a priori not clear if such an auction is incentive compatible. Our main result is that this is indeed the case.

2 Railway Track Allocation

The *optimal track allocation problem*, also called *train timetabling problem* (TTP), can be informally described as follows: given an infrastructure and a set of bids for slots to run specific trains, construct a timetable of maximum value. What makes the problem difficult are the many and complex technical and operational requirements for the feasibility of a timetable, see [2] and [3] for a detailed description. At a high level, the TTP can be stated as the following integer program:

$$\begin{aligned} &(\text{TTP(M)}) \quad \text{(i)} \max \quad \alpha(M) := && \sum_{i \in M} \sum_{p \in P} b_p^i \\ &\text{(ii)} && \sum_{i \in M} \sum_{p \in P_q} x_p^i \leq \kappa_q \qquad \forall q \in \mathcal{C} \\ &\text{(iii)} && x_p^i \in \{0,1\}. \end{aligned}$$

Here, M is a set of TOCs that place bids on a set P of slots (paths) to run trains through some railway network. More precisely, TOC i places bid b_p^i on slot p (we set $b_p^i = -1$ if TOC i does not bid for p). x_p^i denotes a binary decision variable, which takes value 1 if slot path p is allocated to TOC i and 0 otherwise. The objective function (i) maximizes the value of the assigned slots; let us denote the optimum by $\alpha(M)$, and the problem by TTP(M), depending on the set M of bidders. Constraints (ii) guarantee the feasibility of a timetable using a set of cliques $\mathcal{C} \subseteq 2^P$ of bids and associated capacities κ, namely, by stating that at most κ_q of the slots P_q of clique q can be allocated simultaneously. This generalizes the classical combinatorial auction, in which all conflicts arise from competition for items. Finally, (iii) are the integrality constraints. Special purpose methods have been designed that can solve TTPs of medium size, see again [2] and [3].

The TTP can be used in a railway slot auction as the winner determination problem to compute an optimal allocation of slots to bidders. If all bidders would submit their true willingness to pay (or valuation) v_p^i as bids, i.e., $b_p^i = v_p^i$, TTP would assign the resources to the users with the highest utility. Such an allocation (i.e., the one that results from $b_p^i = v_p^i$) is called *efficient*. Bidders do, however, in general not easily reveal their true valuations. Hence, the problem arises to design an auction mechanism that produces efficient allocations without knowing the bidders' willingness to pay. One way to approach this problem is to charge from a bidder i a price $p(i)$ smaller than $\sum_{p \in P} b_p^i x_p^i$, the sum of the assigned bids, in such a way that it becomes attractive to *bid truthfully*, i.e., $b_p^i = v_p^i$. More formally, let the *utility* $u(i)$ of bidder i be defined as $u(i) := \sum_{p \in P} v_p^i x_p^i - p(i)$, i.e., willingness to pay minus price. Then a bidding strategy is *dominant* if it maximizes $u(i)$ no matter what any other bidder $j \in M \backslash \{i\}$ submits. An auction mechanism in which truthful bidding is a dominant strategy is called *incentive compatible*. For a standard combinatorial auction, the Vickrey-Clarke-Groves mechanism is incentive compatible, and it will turn out in the next section that an appropriate generalization, the VTA, is an incentive compatible railway slot auction.

3 A Generalized VCG Auction

Definition 1. *Consider the railway track allocation setting of Section 2. A* Vickrey track auction *(VTA) is a single shot combinatorial auction of railway slots in which the winner determination problem is solved using model TTP(M) and, given an optimal allocation $\hat{x}$, the price that bidder i is charged is defined in compliance with the Vickrey-Clarke-Groves mechanism as*

$$p_{vta}(i) := \alpha(M \backslash \{i\}) - \Big(\alpha(M) - \sum_{p \in P} b_p^i \hat{x}_p^i\Big).$$

Theorem 1. *Truthful bidding is a dominant strategy in a VTA.*

Proof. The proof is an extension of the standard one, see e.g., [5], to constrained winner determination. Denote by $X(M)$ the set of feasible allocations, i.e., the set of vectors x that satisfy TTP(M) (ii)–(iii). Focus on some bidder i and let the other bidders $j \in M \backslash \{i\}$ choose arbitrary bidding strategies $b_p^j \in \mathbb{R}_+$, $\forall p \in P$. Suppose bidder i bids truthfully, i.e., $b_p^i = v_p^i, \forall p \in P$, and denote by $\hat{x}$ the optimal allocation, by $\hat{p}_{vta}(i)$ the resulting price, and by $\hat{u}_i$ the utility. For any alternative bidding strategy $\overline{b}_p^i$, $p \in P$, there exists at least one $p \in P$ with $\overline{b}_p^i < v_p^i$. Suppose i makes such a bid, and let $\overline{x}$ be the optimal solution of the associated winner determination problem, $\overline{\alpha}(M)$ its value, $\overline{p}_{vta}(i)$ the associated price, and $\overline{u}_i$ the utility of bidder i in that alternative case. Then it holds:

$$\begin{aligned}
u(i) :&= \sum_{p\in P} v_p^i \hat{x}_p^i - \hat{p}_{vta}(i) \\
&= \sum_{p\in P} v_p^i \hat{x}_p^i - \alpha(M\backslash\{i\}) + \alpha(M) - \sum_{p\in P} b_p^i \hat{x}_p^i \\
&= \sum_{p\in P} v_p^i \hat{x}_p^i + \sum_{m\in M\backslash\{i\}} \sum_{p\in P} b_p^m \hat{x}_p^m - \alpha(M\backslash\{i\}) \\
&= \max_{x\in X(M)} \{\sum_{p\in P} v_p^i x_p^i + \sum_{m\in M\backslash\{i\}} \sum_{p\in P} b_p^m x_p^m\} - \alpha(M\backslash\{i\}) \\
&\geq \sum_{p\in P} v_p^i \overline{x}_p^i + \sum_{m\in M\backslash\{i\}} \sum_{p\in P} b_p^m \overline{x}_p^m - \alpha(M\backslash\{i\}) \\
&= \sum_{p\in P} v_p^i \overline{x}_p^i + \sum_{m\in M\backslash\{i\}} \sum_{p\in P} \overline{b}_p^m \overline{x}_p^m - \overline{\alpha}(M\backslash\{i\}) \\
&= \sum_{p\in P} v_p^i \overline{x}_p^i - \overline{\alpha}(M\backslash\{i\}) + \overline{\alpha}(M) - \sum_{p\in P} \overline{b}_p^i \overline{x}_p^i \\
&= \sum_{p\in P} v_p^i \overline{x}_p^i - \overline{p}_{vta}(i) \\
&= \overline{u}(i).
\end{aligned}$$

□

Note that this proof does not depend on the concrete structure of TTP, i.e., it generalizes to combinatorial Vickrey auctions with arbitrary combinatorial winner determination problems.

For example, it follows that a VTA with additional constraints on the number of slots that can be allotted to a bidder is also incentive compatible, because this rule can be dealt with by adding to TPP additional constraints of the form

$$\sum_{p\in P} x_p^m \leq \lambda \qquad \forall m \in M.$$

After these positive results on "winner determination constraints" or "allocation constraints" we now investigate two types of "bidding constraints" that are of interest in a railway auction.

3.1 Minimum Bids

Due to maintenance requirements, a railway network operator would be interested in stipulating lower bounds $\mu_p, p \in P$, on bids for slots in order to generate a minimum cash flow. Consider an according redefinition of auction prices as follows:

$$p^{lb}_{vta}(i) := \max\{\sum_{p \in P} \mu_p x^i_p,\ p_{vta}(i)\}.$$

Unfortunately, the following example shows that truthful bidding, i.e., $b^i_j = v^i_j$, if $v^i_j \geq \mu(j)$, is not a dominant strategy for the resulting auction.

Example 1. Consider the auction in Figure 1 i) for two conflict-free paths A and B, and two bidders r and s. Figure 1 ii) and iii) show that the pricing mechanism p^{lb}_{vta} is not incentive compatible, because truthful bidding is not a dominant strategy for bidder r.

i)	r	s
A	2	1
B	10	6
μ	3	3

ii)	r	s
A	-1	-1
B	10	9
p^{lb}_{vta}	9	0
u	1	0

iii)	r	s
A	3	-1
B	10	9
p^{lb}_{vta}	9	0
u	3	0

Fig. 1. i) willingness to pay, ii) truthful bidding, iii) best strategy.

3.2 Limited Number of Bids

Another reasonable auctioning constraint would be to limit number of bids on individual slots per participant; this is because of handling costs for bids for the auctioneer and because of the complexity to come up with these bids for the bidders. Again, we can give an example that truthful bidding is not a dominant strategy in such an auction.

Example 2. Consider the auction in Figure 2 i) for two conflict-free paths A and B, and two bidders r and s. Imagine a limit on the number of submitted bids of at most 1. Figure 2 ii) and iii) show that such an auction is not incentive compatible, because truthful bidding, i.e., in this case, bidding for the most valuable slot, is not a dominating strategy for bidder s. The reason is clearly that any bidder that wants to bid for more valuable slots than the upper limit allows, cannot guess which of the subsets that he can bid on produces the maximal utility (w.r.t. the other bids).

i)	r	s
A	1	8
B	10	9

ii)	r	s
A	-1	-1
B	10	9
p_{vta}	9	0
u	1	0

iii)	r	s
A	-1	8
B	10	-1
p_{vta}	0	0
u	10	8

Fig. 2. i) willingness to pay, ii) truthful bidding, iii) best strategy.

References

1. R. Borndörfer, M. Grötschel, S. Lukac, K. Mitusch, T. Schlechte, S. Schultz, and A. Tanner, An auctioning approach to railway slot allocation, Competition and Regulation in Network Industries, 1 (2006), pp. 163-196.
2. R. Borndörfer and T. Schlechte, Models for railway track allocation, in ATMOS 2007 - 7th Workshop on Algorithmic Approaches for Transportation Modeling, Optimization, and Systems, C. Liebchen, R. K. Ahuja, and J. A. Mesa, eds., Dagstuhl, Germany, 2007, Internationales Begegnungs- und Forschungszentrum für Informatik (IBFI), Schloss Dagstuhl, Germany.
3. V. Cacchiani, Models and Algorithms for Combinatorial Optimization Problems arising in Railway Applications, PhD thesis, DEIS, Bologna, 2007.
4. E. H. Clarke, Multipart pricing of public goods, Public Choice, 2 (1971), pp. 19-33.
5. P. Cramton, Y. Shoham, and R. Steinberg, Combinatorial Auctions, The MIT Press, 2006.
6. T. Groves, Incentives in Teams, Econometrica, 41 (1973), pp. 617-631.
7. A. Mura, Trassenauktionen im Schienenverkehr, Master's thesis, Technische Universität Berlin, 2006.
8. W. Vickrey, Counterspeculation, auctions, and competitive sealed tenders, The Journal of Finance, 16 (1961), pp. 8-37.

The Line Connectivity Problem

Ralf Borndörfer, Marika Neumann, and Marc E. Pfetsch

Zuse Institute Berlin, Takustr. 7, 14195 Berlin
{borndoerfer,marika.neumann,pfetsch}@zib.de

Summary. This paper introduces the *line connectivity problem*, a generalization of the Steiner tree problem and a special case of the line planning problem. We study its complexity and give an IP formulation in terms of an exponential number of constraints associated with "line cut constraints". These inequalities can be separated in polynomial time. We also generalize the Steiner partition inequalities.

1 Introduction

The *line connectivity problem* (LCP) can be described as follows. We are given an undirected graph $G = (V, E)$, a set of *terminal nodes* $T \subseteq V$, and a set of *lines* L (simple paths) defined on the graph G, see the left of Figure 1 for an example. The lines have nonnegative costs $C \in \mathbb{R}_+^L$ and cover all edges, i.e., for every $e \in E$ there is an $\ell \in L$ such that $e \in \ell$. The problem is to find a set of lines $L' \subseteq L$ of minimal cost such that for each pair of distinct terminal nodes $t_1, t_2 \in T$ there exists a path from t_1 to t_2, which is completely covered by lines of L'.
LCP is a generalization of the Steiner tree problem (STP) since we get an STP if all lines have length one. In contrast to the STP with nonnegative costs, see [4, 5] for an overview, the optimal solution of the line connectivity problem does not have to be a tree. There can be two lines that form a cycle, but both are necessary to connect two terminal nodes, see the right of Figure 1. However, an optimal solution of LCP is minimally connected, i.e., if we remove a line from the solution, there exist at least two terminals which are not connected.
LCP is a special case of the *line planning problem* in which passenger routes are not fixed a priori, see [2] and the references therein for a detailed definition. Line planning deals with finding a set of lines

B. Fleischmann et al.(Eds.), *Operations Research Proceedings 2008,*
DOI: 10.1007/978-3-642-00142-0_90, © Springer-Verlag Berlin Heidelberg 2009

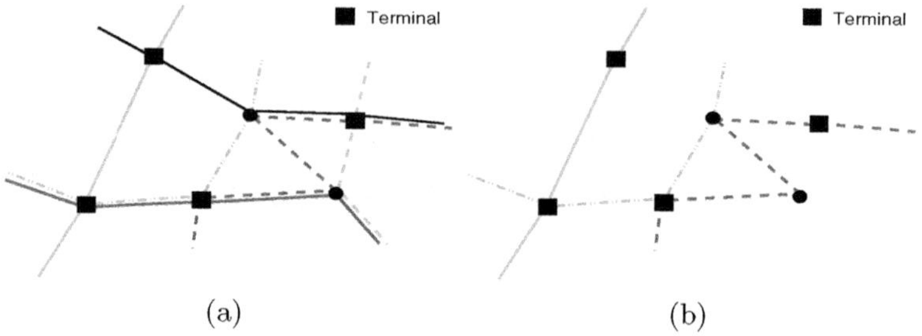

Fig. 1. Example of a line connectivity problem.

and corresponding frequencies such that a given demand can be transported. Usually, the objective is to minimize cost and/or travel times. If we neglect travel time, capacity, and frequency constraints, the line planning problem reduces to LCP, namely, all stations that are departures or destinations of a passenger trip have to be connected by lines. Since line planning problems can not be solved to proven optimality for medium-sized and large instances, it is of interest to analyze LCP. This article is structured as follows. In Section 2 we investigate the complexity of the LCP. An IP formulation and a polynomial time separation algorithm for a class of line cut inequalities associated with this formulation is proposed in Section 3. A polyhedral analysis is sketched in Section 4.

2 Complexity of LCP

Since the line connectivity problem is a generalization of the Steiner tree problem [5], it is strongly NP-hard in general. The complexity of two important special cases, for which the STP can be solved efficiently, is as follows:

Proposition 1. *1. LCP is polynomially solvable for* $|T| = 2$.
2. LCP is NP-hard for $T = V$.

Sketch of proof. 1. We can construct a directed graph D' similar to the one in Section 3 below. A shortest path in D' between two terminal nodes corresponds to a minimal cost connected line set in G.
2. We reduce the set covering problem to the line connectivity problem. In a set covering problem we are given a finite set S, a set $\mathcal{M} \subseteq 2^S$, and a positive integer k. The problem is to find a subset $\mathcal{M}' \subseteq \mathcal{M}$, $|\mathcal{M}'| \leq k$, such that for all $s \in S$ there exists an $M \in \mathcal{M}'$ with $s \in M$.

Given a set covering instance, we define a line connectivity problem in a graph $G = (V, E)$ as follows: The nodes are $V = S \cup \{v\}$ with v being one extra node. We first assume a complete graph and remove all edges that are not covered by a line after the construction of the lines. Let $V = \{v := s_0, s_1, s_2, \ldots\}$. For each set $M \in \mathcal{M}$ order the elements in M and construct a line beginning in node v and passing all nodes of M in the given order. The cost of this line is 1.
It can be easily seen that a cover $\mathcal{M}'$ with less than k elements exists if and only if we find a line set connecting all nodes with cost smaller or equal to k. □

3 An Integer Programming Formulation

An integer program for LCP can be formulated as

$$\begin{aligned} (\mathrm{LCP}_{cut}) \min & \sum_{\ell \in L} C_\ell \, x_\ell \\ \text{s.t} & \sum_{\ell \in L_{\delta(W)}} x_\ell \geq 1 \qquad \emptyset \subsetneq W \cap T \subsetneq T \\ & x_\ell \in \{0, 1\}. \end{aligned}$$

Here, $L_{\delta(W)} := \{\ell \in L \,|\, \exists e \in \delta(W) \cap \ell\}$ is the set of all lines that cross a cut $\delta(W)$ at least one time. If $\delta(W)$ with $\emptyset \subsetneq W \cap T \subsetneq T$ is an (s,t)-cut we call $L_{\delta(W)}$ an *(s,t)-line cut* or shortly *line cut*. We call L' a *minimal (s,t)-line cut* with respect to x if

$$\sum_{\ell \in L'} x_\ell = \min\{\sum_{\ell \in \tilde{L}} x_\ell \,|\, \tilde{L} \text{ is an } (s,t)\text{-line cut}\}.$$

We call the inequalities in (LCP_{cut}) *line cut constraints*. Their number can be exponential in the size of the input. We therefore propose an efficient separation algorithm that decides whether a given point x^* is valid for the LP-Relaxation of (LCP_{cut}) or finds a violated line cut constraint. It will turn out that this problem can be formulated as a max flow/min cut problem in a suitable auxiliary digraph. The construction is as follows: We are given a graph $G = (V, E)$, a set of lines L, and two distinct nodes $s, t \in T \subseteq V$. Each line $\ell \in L$ has a value $x_\ell \geq 0$. We construct a directed graph $D' = (V', A')$ with node set

$$V' = \{s\} \cup \{t\} \cup \{v_\ell, w_\ell \,|\, \ell \in L\}$$

and the following arcs $a \in A'$ and capacities c_a

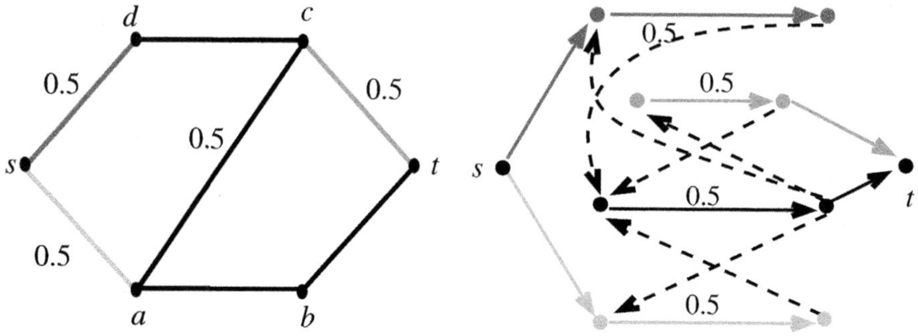

Fig. 2. *Left:* Graph G with four lines ($\ell_1 = \{s,d\}$, $\ell_2 = \{s,a\}$, $\ell_3 = \{d,c,a,b,t\}$, $\ell_4 = \{c,t\}$) with value 0.5 and two terminal nodes s and t. *Right:* Corresponding directed graph D'. Here, each arc has capacity 0.5. The dashed arcs are of the form $(w_{\ell'}, v_\ell)$. The minimal (s,t)-cut has value 0.5.

$$\begin{array}{lll}
(s, v_\ell) & c_{sv_\ell} = x_\ell & \text{if } s \in \ell,\ \forall\, \ell \in L \\
(v_\ell, w_\ell) & c_{v_\ell w_\ell} = x_\ell & \forall\, \ell \in L \\
(w_{\ell'}, v_\ell) & c_{w_{\ell'} v_\ell} = \min\{x_\ell, x_{\ell'}\} & \forall\, \ell, \ell' \in L,\ \ell \neq \ell',\ \ \ell \text{ and } \ell' \text{ have} \\
 & & \text{a node } v \in V \backslash \{s,t\} \text{ in common} \\
(w_{\ell'}, t) & c_{w_{\ell'} t} = x_{\ell'} & \text{if } t \in \ell',\ \forall\, \ell' \in L.
\end{array}$$

Figure 2 illustrates this construction.

Lemma 1.

1. *Each simple path from s to t has the form $(s, v_{\ell_1}, w_{\ell_1}, \ldots, v_{\ell_k}, w_{\ell_k}, t)$, $k \geq 1$.*
2. *The only arc with target node w_ℓ is (v_ℓ, w_ℓ), $\forall\, \ell \in L$.*
3. *The only arc with source node v_ℓ is (v_ℓ, w_ℓ), $\forall\, \ell \in L$.*
4. *There is a directed (s,t)-cut with minimal capacity in D' such that all arcs over this cut are of the form (v_ℓ, w_ℓ), $\ell \in L$.* □

Proof. The first three parts can easily be seen. Consider part 4. Assume (s, v_ℓ) is in a minimal cut. Then we can replace this arc by (v_ℓ, w_ℓ) with the same value because this is the only arc with source node v_ℓ (Part 3). With a similar argument we can replace $(w_{\ell'}, t)$ by $(v_{\ell'}, w_{\ell'})$. Assume $(w_{\ell'}, v_\ell)$, $\ell \neq \ell'$, is in the cut and $x_\ell \leq x_{\ell'}$. Then we can replace this arc by (v_ℓ, w_ℓ) with same capacity because of Part 3 and $c_{w_{\ell'}, v_\ell} = \min\{x_\ell, x_{\ell'}\}$. If $x_{\ell'} \leq x_\ell$, we can replace it by $(v_{\ell'}, w_{\ell'})$ with same capacity because of Part 2 and the definition of the capacities. □

Proposition 2. *There is a one-to-one correspondence between minimal directed (s,t)-cuts in D' and minimal (s,t)-line cuts in G of the same capacity.*

Proof. We only show the forward direction. Let $\delta(W')$ be a minimal (s,t)-cut in D'. After applying part 4 of Lemma 1, let $L' = \{\ell \in L \,|\, (v_\ell, w_\ell) \in A', v_\ell \in W', w_\ell \in V' \backslash W'\}$. Assume L' is not an (s,t)-line cut. Then there exists a path from s to t in G that is covered by lines in $L \backslash L'$. Let $\ell_1, \ldots, \ell_r$ be the lines that are used in this order when traversing the path. Then $(s, v_{\ell_1}, w_{\ell_1}, \ldots, v_{\ell_r}, w_{\ell_r}, t)$ is a path from s to t in D'. This is a contradiction to the assumption that $\delta(W')$ is a cut in D'.
It can be easily seen that L' and $\delta(W')$ have the same capacity. □

Theorem 1. *The separation problem for line cut constraints can be solved in polynomial time.*

Computing for every two terminals $s, t \in T$ the minimum (s,t)-cut in D' can be done in polynomial time. If and only if the value of this cut is smaller than 1, we can construct a violated line cut constraint. □

4 Polyhedral Analysis

Let $P_{\mathrm{LCP}} := \mathrm{conv}\{x \in \{0,1\}^L \,|\, x \text{ satisfies the line cut constraints}\}$ be the *line connectivity polytope.* We assume that the line connectivity polytope is non-empty, i.e., the graph G is connected.
Using the results for the set covering polytope of Balas and Ng [1], we get the following information about P_{LCP}.

Corollary 1.

1. *The LCP-polytope P_{LCP} is full dimensional if and only if there exists no valid cut $\delta(W)$ with $|L_{\delta(W)}| = 1$.*
   In the following we assume P_{LCP} to be full dimensional.
2. *The inequality $x_\ell \geq 0$ defines a facet of P_{LCP} if and only if $|L_{\delta(W)}| \geq 3$ for all W with $\emptyset \subsetneq W \cap T \subsetneq T$.*
3. *All inequalities $x_\ell \leq 1$ define facets of P_{LCP}.*
4. *All facet defining inequalities $\alpha x \geq \alpha_0$ for P_{LCP} have $\alpha \geq 0$, $\alpha_0 > 0$.*
5. *A line cut inequality is facet defining if and only if the following two properties are satisfied:*
   a) *There exists no W', $\emptyset \subsetneq W' \cap T \subsetneq T$, such that $L_{\delta(W')} \subsetneq L_{\delta(W)}$.*
   b) *For each two W_1, W_2, $\emptyset \subsetneq W_i \cap T \subsetneq T$, with $|L_{\delta(W_i)} \backslash L_{\delta(W)}| = 1$, $i = 1, 2$ and $L_{\delta(W_1)} \backslash L_{\delta(W)} = L_{\delta(W_2)} \backslash L_{\delta(W)}$, we have*
   $$|L_{\delta(W_1)} \cap L_{\delta(W_2)} \cap L_{\delta(W)}| \geq 1.$$
6. *The only facet defining inequalities for P_{LCP} with integer coefficients and righthand side equal to 1 are the line cut inequalities.*

Similar to the Steiner tree problem we can define *partition inequalities*. Let $P = (V_1, \ldots, V_k)$ be a partition of the node set V where $V_i \cap T \neq \emptyset$ for $i = 1, \ldots, k$ and $k \geq 3$, i.e., P is a Steiner partition. Let G_P be the graph that arises by contracting each node set V_i to a single node.

Lemma 2. *The* line partition inequality

$$\sum_{\ell \in L} a_\ell \cdot x_\ell \geq k - 1, \quad a_\ell := (\textit{number of nodes in } G_P \textit{ visited by } \ell) - 1$$

is valid for the line connectivity problem. □

Note that if $k = 2$ we get a line cut constraint.
Analogous to the properties which are necessary for a Steiner partition inequality to be facet defining, c. f. Grötschel and Monma [3], we can formulate the following Proposition.

Proposition 3. *Let* $\tilde{L} := \{\ell \in L \mid a_\ell = 0\}$. *The line partition inequality is facet defining if the following properties are satisfied.*

1. $G(V_i)$ *is connected by* $\tilde{L}$, $i = 1, \ldots, k$.
2. $G(V_i)$ *contains no line cut* $L' \subseteq \tilde{L}$ *with* $|L'| = 1$, $i = 1, \ldots, k$.
3. *Each line visits at most two nodes in* G_P, *i.e.,* $a_\ell \in \{0, 1\} \, \forall \ell \in L$.
4. *The shrunk graph* G_P *is 2-line-connected, i.e., if we remove any node with all adjacent lines, the resulting graph is connected.* □

Examples can be constructed in which a line partition inequality is facet defining, but does not satisfy all of the first three properties of Lemma 3. Indeed, only Property 4 is necessary.

Proposition 4. *If the shrunk graph* G_P *is not 2-line-connected, the partition inequality is not facet defining for* P_{LCP}. □

References

1. Egon Balas and Shu Ming Ng. On the set covering polytope: I. All the facets with coefficients in 0,1,2. Mathematical Programming, 43:57-69, 1989.
2. Ralf Borndörfer, Martin Grötschel, and Marc E. Pfetsch. A column-generation approach to line planning in public transport. Transportation Science, 41(1):123-132, 2007.
3. Martin Grötschel and Clyde L. Monma. Integer polyhedra arising form certain network design problems with connectivity constraints. SIAM Journal on Discrete Mathematics, 3(4):502-523, 1990.
4. Tobias Polzin. Algorithms for the Steiner Problems in Networks. PhD thesis, University of Saarland, 2003.
5. Hans Jürgen Prömel and Angelika Steger. The Steiner Tree Problem. Vieweg, Braunschweig/Wiesbaden, 2002.

Heuristics for Budget Facility Location–Network Design Problems with Minisum Objective

Cara Cocking[1] and Gerhard Reinelt[1]

Department of Computer Science, University of Heidelberg
cara.cocking@gmail.com,
gerhard.reinelt@informatik.uni-heidelberg.de

Summary. In this paper we present the first heuristics for discrete budget facility location–network design with minisum objective. Simple greedy heuristics, a custom heuristic that separates the problems of facility location and network design, a basic local search, and simulated annealing heuristics using two different kinds of neighborhoods are developed. The results of each heuristic are compared with known optimal solutions to a series of test problems.

1 Introduction

As is suggested by the name, facility location–network design (FLND) combines facility location and network design. Facility location deals with optimally locating facilities. There are two main parties involved in any facility location problem: the facilities themselves and the clients of the facilities. Typically we want the facilities to be close to the clients, which can be defined in several ways, such as minimizing total travel cost.

Because we deal only with discrete facility location, the problems are represented using graphs. The nodes of the graph are the union of the clients and the possible facility locations, and edges represent the ability to travel from one node to another and are the means by which clients reach facilities. A node may represent both a client and a potential facility location.

In network design, the basic problem is to optimally construct a network that enables some kind of flow, and possibly that satisfies additional constraints. The nodes are given and the network is constructed from a set of potential edges, or links. In our case, the flow involved is that between clients and facilities.

B. Fleischmann et al.(Eds.), *Operations Research Proceedings 2008*,
DOI: 10.1007/978-3-642-00142-0_91, © Springer-Verlag Berlin Heidelberg 2009

The objective in any of these problems, or how optimality is determined, can vary. A common objective in facility location is to minimize the total travel costs (i.e., minisum), and this is the objective that we consider in this paper. In facility location–network design, the objective may be met using the means of both facility location and network design: by building both facilities and links. In a budget problem, a budget, or limit that may not be exceeded, is given for the construction costs. The problem we study here is discrete budget FLND with minisum objective, an $\mathcal{NP}$-hard problem.

2 Previous Work

Facility location–network design problems of this type were first considered by Melkote and Daskin [8]. They developed an IP formulation for the problem that is based on that of the fixed charge network design problem [1]. In a fixed charge problem, the sum of the construction costs and travel costs is to be minimized and there is no budget constraint. The fixed charge FLND formulation can be modified slightly to produce a budget FLND formulation. This is a tight formulation whose LP relaxation produces good lower bounds.

In another paper [10], Melkote and Daskin develop polynomial time algorithms for solving two special cases on graphs where there are no existing links and the set of candidate links forms a tree. The first case involves locating exactly two facilities with no fixed costs. The second case considers fixed charge FLND–facilities have fixed costs–and an unknown number are to be located.

The same authors have done work on related problems including FLND with a maximum covering objective [11] and capacitated fixed charge FLND [9]. Other work on related problems includes approximation algorithms for capacitated cable facility location [12, 3], variations on network design that involve facilities [6, 4, 5], and variations on facility location that involve network construction [7].

To our knowledge, there are no heuristics in the literature for budget facility location–network design with minisum objective on a general graph.

3 Heuristics

An FLND problem instance is represented with the following components:

$G = (V, E)$	graph
$K \subseteq V$	set of clients
$J_e \subseteq V$	set of existing facility sites
$J_p \subseteq V$	set of potential facility sites
E	set of existing edges (or links)
L_p	set of potential edges
$a_k, k \in K$	demands at each client
$f_j, j \in J_p$	construction costs for each potential facility
$c_{ij}, ij \in L_p$	construction costs for each potential edge
$d_{ij}, ij \in E \cup L_p$	travel costs on each edge and potential edge
B	budget

A solution is a subset of potential facilities $J \subseteq J_p$ and a subset of potential edges $L \subseteq L_p$ whose total construction cost is within budget, i.e.,

$$\sum_{j \in J} f_j + \sum_{ij \in L} c_{ij} \leq B. \tag{1}$$

The quantity we want to minimize is the sum of the shortest paths from each client $k \in K$ to its nearest facility $j \in J_e \cup J$ using the graph with edges $E \cup L$. We assume that the original graph, G, is connected.

3.1 Greedy Heuristics

We developed two simple greedy heuristics. In GREEDY-ADD, elements (links or facilities) are selected one at a time to add to the solution, until the budget is reached. Each time, the element that produces the greatest improvement in the objective (reducing total travel cost) per unit construction cost is selected.

In GREEDY-SUB, a subtractive greedy heuristic, we start with a solution that contains all the potential elements (which is most likely infeasible because of its construction cost) and remove one-by-one those elements whose removal causes the least harm to the objective per unit construction cost, until the total construction cost of those elements in the solution falls at or below the budget. However, a small twist is required: If all nodes have a facility in the solution, then the objective value is 0 and removing any or all of the links will not hurt the objective. Thus, there is a first stage where we consider the construction cost of facilities only, and remove facilities from the solution one-by-one until the facility cost falls at or below the budget. In the second stage we consider the construction cost of all the elements, and remove elements one-by-one, be they links or facilities, until the solution is within budget.

3.2 Custom Heuristic

Imagine a spectrum of feasible solutions, laid out such that at one end are solutions where the entire budget is spent on links and no facilities are built, and at the other end are solutions where the entire budget is spent on facilities and no links are built. In between are solutions that build some facilities and some links. The idea in CUSTOM is to start at the first end, with no facilities, and methodically proceed to the other end, selecting the best solution found along the way. (If there are no existing facilities in the problem, then start with one facility.) The steps along the way are determined by the number of facilities in the solution: start with 0, then one facility, then two, up to the number of facilities that may be built if we spent the entire budget on facilities alone.

In this heuristic the subproblems of facility location and network design are solved independently. At each step, a p-median facility location problem is solved first, locating the specified number of facilities, and then with these selected facilities, a network design problem is solved to choose links with the remaining money. The p-median heuristic used is from [14], using the fast implementation described in [13]. The network design heuristic is based on [2], but with some modifications that improve the results.

3.3 Neighbor Operators

We have developed two different types of neighborhoods: Hamming neighborhoods and step neighborhoods. In Hamming neighborhoods we associate a bit with each potential element (link or facility), and a solution is simply a bit string indicating which elements are in the solution. Then neighborhoods are defined using Hamming distances. The Hamming distance between two bit strings is the number of bits in which they differ. Any Hamming distance could be used, but we use Hamming distance 2. Thus the neighbors of a given solution are those with 2 additional elements, 2 fewer elements, or one element swapped for another, with the caveat that the total construction cost of the elements does not exceed the budget.

In step neighborhoods the neighborhood of a given solution includes those solutions that differ in *step* money's worth of elements. So the neighbors of a solution are those that differ in a certain value's worth of elements rather than a certain number of elements. As *step* we use the maximum construction cost of an element (assuming this is less than the budget), thus allowing all elements to enter and leave the solution.

3.4 Local Search

In LOCAL-SEARCH, we start with a random solution and explore the entire neighborhood of that solution, moving to the best solution found if it is better than the current, and repeating the process until no more improvement is possible. Hamming neighborhoods are used because it is easy to enumerate all the neighbors of a given solution in order to find the best, which is not the case with step neighborhoods.

3.5 Simulated Annealing

Simulated annealing is a standard metaheuristic that generates random neighbors in a given neighborhood, moving to the neighbor if it is better, and possibly moving to the neighbor even if it is not better. The probability of moving to a not-better neighbor depends on a decreasing temperature, allowing the solution to jump around a lot early and slowly settle down.
In applying simulated annealing to budget FLND, we use both step and Hamming neighborhoods since random (and not best) neighbors are to be generated. These heuristics are SIMANN-HAMMING and SIMANN-STEP.

4 Results

Table 1 shows the results of the heuristics on a test suite of 54 problem instances with varying characteristics. For all problems, $|V| = |K| = 40$, $|J_e|$ ranges from 0 to 8, $|J_p|$ from 32 to 40, $|E|$ from 39 to 80, and $|L_p|$ from 80 to 182. For heuristics with random selections involved, the best result of 10 runs for each problem instance was used.
The performance of the simulated annealing heuristics is dependent on the parameters used, in particular the temperature reduction function. The results we present use slow temperature reduction ($f(t) = t * 0.9999$) and these heuristics take much longer to run than any of the others. The results produced, however, are better for it. Interestingly, SIMANN-STEP did much better than SIMANN-HAMMING, obtaining solutions within 1.1% of optimal on average. The step neighborhoods as we have defined them may allow more "reach" than the Hamming neighborhoods, resulting in more of the solution space being explored.

Table 1. Heuristics results on 54 problem instances, giving percent over optimal of solutions produced, on average and in the worst case

Heuristic	Average	Worst Case
Greedy-Add	9.2%	25.9%
Greedy-Sub	8.4%	24.9%
Custom	4.7%	17.5%
Local-Search	2.5%	10.0%
SimAnn-Hamming	6.7%	30.2%
SimAnn-Step	1.1%	4.2%

References

1. Balakrishnan A, Magnanti TL, Wong RT (1989) A dual-ascent procedure for large-scale uncapacitated network design. Operations Research 37:716–740
2. Berman O, Ingco DI, Odoni AR (1992) Improving the location of minisum facilities through network modification. Annals of Operations Research 40:1–16
3. Chen X, Chen B (2007) Approximation algorithms for soft-capacitated facility location in capacitated network design. Algorithmica Online First
4. Current JR (1988) The design of a hierarchical transportation network with transshipment facilities. Transportation Science 22:270–277
5. Current JR, Pirkul H (1991) The hierarchical network design problem with transshipment facilities. European Journal of Operational Research 51:338–347
6. Drezner Z, Wesolowsky GO (2003) Network design: Selection and design of links and facility location. Transportation Research Part A 37:241–256
7. Klincewicz JG (1998) Hub location in backbone/tributary network design: A review. Location Science 6:307–335
8. Melkote S, Daskin MS (2001) An integrated model of facility location and transportation network design. Transportation Research 35:515–538
9. Melkote S, Daskin MS (2001) Capacitated facility location/network design problems. European Journal of Operational Research 129:481–495
10. Melkote S, Daskin MS (2001) Polynomially solvable cases of combined facility location-network design problems. Technical Report
11. Melkote S, Daskin MS (1998) The maximum covering facility location-network design problem. Technical Report
12. Ravi R, Sinha A (2006) Approximation algorithms for problems combining facility location and network design. Operations Research 54:73–81
13. Resende MGC, Werneck RF (2007) A fast swap-based local search procedure for location problems. Annals of Operations Research 150:205–230
14. Teitz MB, Bart P (1968) Heuristic methods for estimating the generalized vertex median of a weighted graph. Operations Research 16:955–961

Combinatorial Aspects of Move-Up Crews

Holger Flier[1], Abhishek Gaurav[2], and Marc Nunkesser[1]

[1] Institute of Theoretical Computer Science, ETH Zürich, Switzerland
{holger.flier,marc.nunkesser}@inf.ethz.ch
[2] Department of Computer Science and Engineering, IIT Kanpur, India
abhiga@iitk.ac.in

Summary. In railway planning, a frequent problem is that a delayed train necessarily delays its crew. To prevent delay propagation along the crew's succeeding trips, a move-up crew may take over the rest of the crew's duty on time. We address two problems in this context: First, during the planning stage, one has to make a tradeoff between robustness gained and additional costs incurred by introducing move-up crews. We suggest different heuristics to solve this problem, depending on the size of the instance. Second, during operations, dispatchers have to make optimal crew swap decisions, i.e., use available move-up crews to minimize overall passenger delay. For the local case, we give an efficient algorithm. However, optimizing crew swaps over the whole railway network is NP-hard.

1 Introduction

Delay propagation is a serious problem in railway operations. One reason for delay propagation can be a crew which is delayed from its preceding trip. A practical recovery operation is to swap the remaining duties of the delayed crew and a move-up crew, as introduced by Shebalov and Klabjan [7] for the airline context. A crew arrives late at a hub, such that it cannot reach its follow-up flight. Instead, it is assigned a later flight and a second crew, called the *move-up crew*, covers the flight of the delayed crew. Such an exchange is of course only possible if the two crews have the same crew base and other regulatory rules are not violated (for example the maximum labor time is not exceeded for any crew after the exchange). While Shebalov and Klabjan propose a large scale integer program for the construction of crew schedules in the airline case, we focus on the underlying algorithmic questions that arise in the railway case. In section 2 we develop algorithms to maximize the number of possible swaps in a crew schedule during the planning phase, and in section 3 we state results regarding the operational phase, i.e. how to optimally swap crews.

B. Fleischmann et al.(Eds.), *Operations Research Proceedings 2008*,
DOI: 10.1007/978-3-642-00142-0_92, © Springer-Verlag Berlin Heidelberg 2009

2 Maximizing Crew Swapping Possibilities

Many crew scheduling systems [1] work in two[3] phases: In a first phase for a given set of crew bases B and trips $T = \{t_1, \ldots, t_\tau\}$, a very large set of potential duties $D = \{d^1, \ldots, d^n\}$ with $d^j \subset T$, $1 \leq j \leq n$, is created. These duties are compatible with the regulatory rules for a single duty (maximum work time, pauses etc.). In a second phase a constrained set cover solution on (T, D) is created, i.e., one wants to cover all trips by a minimum cost set of duties. The constraints are related to labor rules across the duties (e.g., the average shift time at Depot A must not deviate too much from the average shift time at Depot B). Our goal is to incorporate into the second phase a maximization of crew swaps: For each pair of duties (d_i, d_j) and a given common station that both crews cover, we can determine in advance whether a crew swap is possible or not (e.g., both crews have the same home depot, the exchange does not violate the maximum work time, etc.). Therefore, for each set of duties that covers the complete set of trips, one can count the attained number of crew swapping possibilities. The combination of the NP-complete set cover problem [6] and crew swapping leads to several natural problems, one being the set cover with pairwise profits trade-off cost problem:

Definition 1 (SCPP-tc). *Given a ground set T of tasks, a set system D with set costs $c_i (\in C)$ for each duty $d_i \in D$, and for each pair of sets (d_i, d_j) a profit $p_{ij} (\in P)$ for choosing this pair of sets. Find a set cover $S \subset D$ for T that minimizes the trade-off cost*

$$\sum_{d_i \in S} c_i - \sum_{d_i \in S, d_j \in S} p_{ij} \ . \tag{1}$$

For a detailed discussion of several SCPP variants and their relation to the dense k-subgraph problem, refer to [5]. All these variants are NP-complete and seem to be hard to approximate. However, a restricted version of SCPP-tc is solvable in polynomial time: For a given SCPP instance (T, D, C, P) of tasks, duties, set costs, and pairwise profits let us define the graph G_{SC} on the vertex set D and edge set P. If we ignore the set-cover aspect, i.e., assume that any subgraph of G_{SC} covers T and assume unit set costs, SCPP-tc translates into a problem where one can buy vertices at a uniform cost of α and obtain unit profit from edges in the chosen induced subgraph. We call the resulting profit maximization problem *Maximum Profit Subgraph* (MPS) problem.

[3] The crew scheduling is typically followed by a crew rostering phase [4], in which *rosters*, i.e., a sequence of duties to be performed by one crew, are created out of the duties. This phase is out of the scope of this paper.

Definition 2 (MPS). *The* maximum profit subgraph *problem asks for the subgraph* $G_{opt}^{MPS} = (V', E')$ *for which* $|E'| - \alpha|V'|$ *is maximum.*

Theorem 1. *The MPS problem on a graph* $G = (V, E)$ *with* n *vertices and* m *edges for a vertex cost of* α *can be solved in time* $O(MBF(n + m, m))$*, where* $MBF(x, y)$ *denotes the time to solve a maximum bipartite flow problem on a directed acyclic graph of* x *vertices and* y *edges.*

Proof. First, we transform G into a bipartite graph B, see Figure 1 (a) and (b). The first partition P_1 of B consists of the vertices of G each of weight α, the second partition P_2 has one unit weight vertex for each edge of G. The edges in B indicate the incidences in G. Now, the maximum profit subgraph S_{opt} of G corresponds to a maximum weight independent set I in B as follows: $I \cap P_2$ corresponds to all the edges in S_{opt}, $I \cap P_1$ corresponds to all the vertices in G that are *not* in S_{opt}. The above claim is correct because the sketched transformation is a bijection between induced subgraphs in G and inclusion maximal independent sets in B; it maps induced subgraphs of n' vertices and m' edges to independent sets of weight $\alpha(n - n') + m'$.
Maximum weight independent sets can be found in polynomial time in bipartite graphs. More precisely, the complement of a maximum weight independent set is a minimum weight vertex cover. It is folklore that a weighted vertex cover in a bipartite graph B can be found by computing a minimum capacity s-t-cut in an acyclic graph, as shown in Figure 1 (b) and (c). □

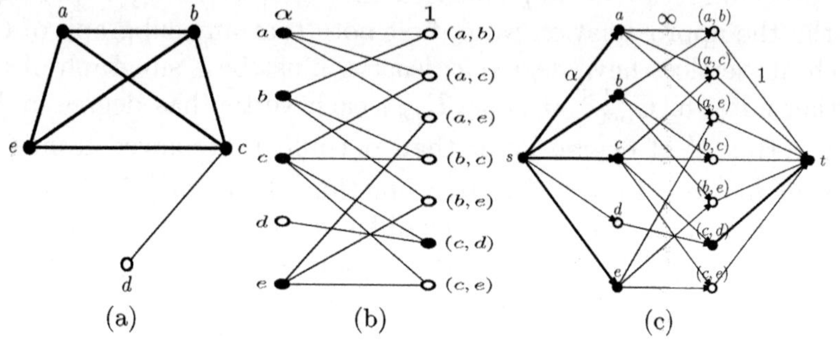

Fig. 1. Transformation from MPS to minimum capacity s-t-cut. Example with $\alpha = 1.3$. (a) Graph G with MPS indicated by thick edges and black vertices. (b) Corresponding graph B with maximum weight independent set I indicated by white vertices. Respectively, the minimum weight vertex cover is indicated by black vertices. (c) The corresponding minimum capacity s-t-cut, indicated by thick edges.

2.1 Computing Fast Approximate Solutions for MPS

For large real-world crew scheduling instances, which typically have n in the order of one million, the running time to solve MPS exactly can be prohibitive.[4] Therefore, we present an approximation algorithm for the following reformulated (but trivially equivalent) version of MPS:

Definition 3 (MPS2). *Given a graph $G = (V, E)$ and a parameter α, find an induced subgraph $G_{opt}^{MPS} = (V', E')$ of G, for which $|E'| + \alpha|V - V'|$ is maximum.*

Denote by $G^{\geq\alpha} = (V^{\geq\alpha}, E^{\geq\alpha})$ the maximum induced subgraph of G in which each vertex has degree greater or equal to α, potentially $G^{\geq\alpha} = \emptyset$. In the following theorem we show that $G^{\geq\alpha}$ is a 2-approximation for MPS2 and can be constructed much faster than an optimal solution.

Theorem 2. *For a graph $G = (V, E)$ an induced subgraph $G^{\geq\alpha}$ in which the degree of each vertex is at least α can be constructed in time $O(|E| + |V| \log |V|)$. This subgraph is a 2-approximation to an optimal MPS2 solution on G.*

Proof. In order to construct $G^{\geq\alpha}$ we keep deleting vertices from G that have degree less than α in the remaining graph until no such vertex exists. On termination the remaining graph has trivially the property that no vertex has degree less than α. By an inductive argument, none of the deleted vertices could have been part of $G^{\geq\alpha}$, therefore the found graph is of maximum size. Using a Fibonacci heap, this procedure can be implemented in $O(|E| + |V| \log |V|)$. Regarding the the approximation ratio, first note that any subgraph of G, in which all vertices have degree at least α must be a subgraph of $G^{\geq\alpha}$. Furthermore, in $G_{\text{opt}}^{\text{MPS}} = (V_{\text{opt}}, E_{\text{opt}})$ each vertex has degree at least α: A vertex v' of degree lower than α could be removed from $G_{\text{opt}}^{\text{MPS}}$ with an increase in the objective function. Thus, $G_{\text{opt}}^{\text{MPS}} \subseteq G^{\geq\alpha}$ and $|E_{\text{opt}}| \geq \alpha|V_{\text{opt}}|/2$. To measure the differences between the two graphs, we contract $G_{\text{opt}}^{\text{MPS}}$ to a single vertex inside $G^{\geq\alpha}$, yielding a new multigraph $\tilde{G}^{\geq\alpha} = (\tilde{V}^{\geq\alpha}, \tilde{E}^{\geq\alpha})$, with $|E^{\geq\alpha}| = |E_{\text{opt}}| + |\tilde{E}^{\geq\alpha}|$. In $\tilde{G}^{\geq\alpha}$ all but the newly generated vertex have degree greater than or equal to α, so $|\tilde{E}^{\geq\alpha}| \geq \frac{\alpha(|\tilde{V}^{\geq\alpha}|-1)}{2} = \frac{\alpha(|V^{\geq\alpha}|-|V_{\text{opt}}|)}{2}$. The approximation ratio becomes

[4] If we plug in the bipartite FIFO preflow-push algorithm of Ahuja, Orlin, Stein and Tarjan [2], we get a running time of $O(nm + n^3)$, with n and m being the number of vertices and edges in the original MPS instance.

$$
\begin{aligned}
r &= \frac{|E_{\text{opt}}| + \alpha(|V| - |V_{\text{opt}}|)}{|E^{\geq\alpha}| + \alpha(|V| - |V^{\geq\alpha}|)} \\
&= \frac{|E_{\text{opt}}| + \alpha(|V| - |V_{\text{opt}}|)}{|E_{\text{opt}}| + |\tilde{E}^{\geq\alpha}| + \alpha(|V| - |V^{\geq\alpha}|)} \\
&\leq \frac{|E_{\text{opt}}| + \alpha(|V| - |V_{\text{opt}}|)}{|E_{\text{opt}}| + \frac{\alpha(|V^{\geq\alpha}| - |V_{\text{opt}}|)}{2} + \alpha(|V| - |V^{\geq\alpha}|)} \\
&\leq \frac{\frac{\alpha|V_{\text{opt}}|}{2} + \alpha(|V| - |V_{\text{opt}}|)}{\frac{\alpha|V_{\text{opt}}|}{2} + \frac{\alpha(|V^{\geq\alpha}| - |V_{\text{opt}}|)}{2} + \alpha(|V| - |V^{\geq\alpha}|)} \\
&= \frac{|V| - \frac{|V_{\text{opt}}|}{2}}{|V| - \frac{|V^{\geq\alpha}|}{2}} \leq 2
\end{aligned}
$$

□

2.2 Heuristics

In this section, we shortly present heuristics for SCPP-tc that build on the algorithmic discussion above. First, medium sized SCPP-tc instances might be solved via Lagrangian relaxation by exploiting the polynomial time solvability of MPS, for details refer to [5]. Second, for large instances, we suggest the following heuristic, which enriches a set cover solution by choosing additional sets that yield many crew swaps.

---

1. Compute an approximate set cover solution to (T, D, C), using a crew scheduling algorithm, see [3, 1] for references.
2. Construct an SCPP-tc instance, where all chosen sets in the set cover get cost 0, the remaining c_d and p_e values are set to reflect additional costs and profits from move-ups.
3. Execute an MPS2 approximation algorithm (or use the exact MPS algorithm if possible).
4. Do local improvements.

---

3 Choosing Optimal Crew Swaps

Regarding operations, we are interested in finding optimal crew swap decisions given a crew schedule that could have been developed using the methods from above. We call this the *minimum delay propagating crew swapping* (MDCS) problem. The objective is to minimize the weighted total delay depending on these decisions. The weights correspond to the importance of the trains, which could be determined by the expected number of passengers. In the following, we only state the main results. For proofs and further details, refer to [5]. If only a single station is considered, optimal crew swaps can be computed efficiently. In the weighted case, it suffices to compute a perfect weighted matching in a bipartite graph. In the unweighted case, we can do even better:

Lemma 1. *An optimal crew swap at a single station with unit weights can be computed in $O(n \log n)$ by matching trains first-in-first-out (FIFO), according to their actual arrival times and planned departure times.*

In general, however, delays might propagate through a network of stations, incurring dependencies among decisions at different stations. Solving the MDCS for a network of stations is NP-hard.

Theorem 3. *MDCS is NP-hard even in the unweighted case of only two consecutive stations with an arbitrary number of trains.*

Theorem 4. *MDCS is NP-hard even in the unweighted case of only two trains driving along an arbitrary number of consecutive stations.*

Together, these two results indicate that the complexity of the network MDCS lies both in the number of trains as well as in the dependencies among local crew swap decisions, even in the most simplistic network topologies. It would be interesting to find approximation algorithms for the network MDCS or to study the problem in an online setting.

References

1. E. Abbink, M. Fischetti, L. Kroon, G. Timmer, and M. Vromans. Reinventing crew scheduling at Netherlands Railways. Interfaces, 35(5):393-401, 2005.
2. R. K. Ahuja, J. B. Orlin, C. Stein, and R. E. Tarjan. Improved algorithms for bipartite network flow. SIAM Journal on Computing, 23(5):906-933, 1994.
3. A. Caprara, P. Toth, and M. Fischetti. Algorithms for the set covering problem. Annals of Operations Research, 98:353-371, 2000.
4. A. Caprara, P. Toth, D. Vigo, and M. Fischetti. Modeling and solving the crew rostering problem. Operations Research, 46(6):820-830, Nov. - Dec. 1998.
5. H. Flier, A. Gaurav, and M. Nunkesser. Combinatorial aspects of move-up crews. Technical report, ARRIVAL Project, 2007.
6. R. Karp. Reducibility among combinatorial problems. In R. Miller and J. Thatcher, editors, Complexity of Computer Computations, pages 85-103. Plenum Press, New York, 1972.
7. S. Shebalov and D. Klabjan. Roubst airline crew pairing: Move-up crews. Transportation Science, 40(3):300-312, 2006.

Computational Complexity of Impact Size Estimation for Spreading Processes on Networks

Marco Laumanns and Rico Zenklusen

Institute for Operations Research, ETH Zurich, 8092 Zurich, Switzerland

Summary. Spreading processes on networks can often be mapped onto network reliability problems. The computational complexity of computing the probability that the spreading process reaches some given subset K of the nodes is well studied as it reduces to the classical K-terminal reliability problem. Often one is not interested in a particular set K, but more global properties of the spreading process, such as the expected spreading size or the probability of a large spreading. We show that the direct Monte Carlo approach is an FPRAS for the expected spreading size, but unless $NP \subseteq BPP$, there is no randomized constant-factor approximation for the probability of large spreadings. When nodes are weighted to represent their importance, we show that estimating the expected spreading impact is of the same computational complexity as estimating s-t reliability.

1 Introduction

Especially for the study of disease spreading, various models were introduced to describe the dynamics of spreading processes. Agent-based models can reflect various mixing properties of the population and different infection rates between every pair of individuals by an underlying contact network. As pointed out by Grassberger [5], these models can often be mapped onto bond percolation problems allowing to eliminate the time dependency of the spreading process. The spreading process can then be described as follows. Let $G = (V, E)$ be a directed network where for every arc $e = (v, w) \in E$ a *spreading probability* $p(e)$ is given corresponding to the probability that the disease will spread from node v to node w if v is infected. Let $S \subseteq V$ a set of initially infected nodes. A possible outcome of this spreading process can be simulated by flipping a biased coin for every arc $(v, w) \in E$ to determine whether it is an *active arc* meaning that the disease will spread from node v to node w if v

B. Fleischmann et al.(Eds.), *Operations Research Proceedings 2008*,
DOI: 10.1007/978-3-642-00142-0_93, © Springer-Verlag Berlin Heidelberg 2009

gets infected. Let E' be the set of active arcs. A particular node $v \in V$ gets infected if there is a path in (V, E') from at least one node of S to v. We assume w.l.o.g. that the set of initially infected nodes S contains exactly one node s. We call the triple (V, E, p) a *reliability network* and the quadruple (V, E, p, s) a *spreading network*. For a spreading network $G = (V, E, p, s)$ we denote by I_G the random element containing all the nodes reached by the disease (I_G is thus a random subset of V). As a natural extension we can associate with every node $v \in V$ an integer weight $w(v) \in \{0, 1, 2, \dots\}$ representing the impact of the spreading when reaching node v and denote the weight of any subset of nodes $V' \subseteq V$ by $w(V') = \sum_{v \in V'} w(v)$.

Computing the probability that a specific node, all nodes, or, more generally, some given set of nodes $K \subseteq V$ will be covered by the spreading process are well studied problems known as the *two-terminal*, *all-terminal* and *K-terminal reliability* problems. All of them are known to be #P-complete even on very restricted classes of networks [1, 10, 11, 12]. In the context of spreading processes, however, one is often interested in more global properties instead of the probability that some given set of nodes will be reached by the process. We thus consider the following two questions:

Expected spreading size: What is the expected sum of weights of the nodes covered by the spreading process?

Probability of large spreadings: For a given $\alpha \in (0, 1)$, what is the probability that the sum of weights of the nodes covered by the spreading is at least $\alpha \cdot w(V)$?

The most frequently used method for obtaining estimated solutions to the above problems is a direct Monte Carlo approach, where the spreading process is simulated multiple times and the fraction of outcomes where the event of interest occurs is returned. This approach is known to be efficient only if the quantity to estimate is not too small. Thus, the difficult cases are the estimation of small expected spreading sizes and small probabilities of large spreadings, i.e., the estimation of rare events. When looking for estimation algorithms we are generally interested in ϵ–δ approximations, which are algorithms returning a value accurate up to a relative error of ϵ with probability at least $1 - \delta$. A fully polynomial randomized approximation scheme (FPRAS) is an ϵ–δ approximation with a running time bounded by a polynomial in the input size and $1/\epsilon$.

We show that the problem of computing the expected spreading size exactly is a difficult problem even when the underlying network is acyclic and unweighted. However, whereas the direct Monte Carlo approach is an FPRAS for the unweighted version, the estimation of the expected spreading size in the weighted version is computationally of the same difficulty as the *s*-*t* reliability problem, for which no FPRAS is known to date. Finally, we show that, unless $NP \subseteq BPP$, there is no randomized constant-factor approximation for the probability of large spreading sizes even in the unweighted case.

2 Estimating the Expected Spreading Size

Theorem 1. *Computing the expected spreading size in the unweighted case is $\#P$-complete, even when restricted to acyclic networks and a uniform spreading probability $\overline{p} \in (0,1)$, $p(e) = \overline{p}\ \forall e \in E$.*

Proof. To show that the problem is $\#P$-hard, we build a reduction from the *s*-*t* reliability problem with uniform failure probabilities, which is known to be $\#P$-complete [12]. Let $G = (V, E)$ be an acyclic network, $s \in V$ be the starting node of the spreading process, $\overline{p} \in (0, 1)$ some fixed uniform spreading probability and $t \in V \backslash s$. Furthermore, let $G' = (V', E')$ be the acyclic graph obtained from G by adding a node w and an arc from t to w (with spreading probability $\overline{p}$). By construction of G' we have $\mathrm{E}[|I_{G'}|] = \mathrm{E}[|I_G|] + \mathrm{P}[w \in I_{G'}] = \mathrm{E}[|I_G|] + \overline{p} \cdot \mathrm{P}[t \in I_G]$. Thus that the *s*-*t* reliability $\mathrm{P}[t \in I_G]$ in G can be determined as a function of $\mathrm{E}[|I_G|]$, $\mathrm{E}[|I_{G'}|]$ and $\overline{p}$, implying that computing the expected spreading size in an acyclic graph with uniform spreading probability is $\#P$-hard. Furthermore, the problem lies in $\#P$ as it can be reduced to the *s*-*t* reliability problem by observing that $\mathrm{E}[|I_G|]$ can be expressed in terms of *s*-*t* reliabilities as $\mathrm{E}[|I_G|] = \sum_{v \in V} \mathrm{P}[v \in I_{G'}]$. □

Despite being $\#P$-complete, it is easy to obtain an FPRAS in the unweighted case just by applying a direct Monte Carlo approach since the expected spreading size is at least $1/|V|$ as the node s is always infected. This can be seen by applying the Generalized Zero-One Estimator Theorem [4], which shows that the direct Monte Carlo approach for estimating $\mathrm{E}[|I_G|]$ is an ϵ–δ approximation if the number of iterations N satisfies $N \geq 4(\mathrm{e} - 2)\ln(2/\delta) \cdot (|V|/\epsilon^2)$. The weighted case, however, is of the same difficulty as the *s*-*t* reliability problem, for which the existence of an FPRAS is still unresolved.

Theorem 2. *Computing or estimating the expected spreading size in a weighted spreading network is of the same computational complexity as computing or estimating s-t reliabilities on the same underlying graph.*

Proof. The s-t reliability of a reliability network $G = (V, E, p)$ with $s, t \in V$ is the expected spreading size on the spreading network $G' = (V, E, p, s, w)$ where the node t has a weight equal to one and all other nodes have zero weight. On the other hand, the expected spreading size on a spreading network $G = (V, E, p, s, w)$ can be expressed as the weighted sum of $|V|$ s-t reliabilities as in the proof of Theorem 1 as $\mathrm{E}[|I_G|] = \sum_{v \in V} w(v)\mathrm{P}[v \in I_G]$. Thus, an ϵ–δ approximation for $\mathrm{E}[|I_G|]$ can be obtained by getting for every $v \in V$ an ϵ–δ' approximation X_v of $\mathrm{P}[v \in I_G]$ with $\delta' = \delta/|V|$ and estimating $\mathrm{E}[|I_G|]$ by $Y = \sum_{v \in V} w(v)X_v$. Hence, Y is an ϵ-approximation of $\mathrm{E}[|I_G|]$ if X_v is an ϵ-approximation of $\mathrm{P}[v \in I_G]$ for all $v \in V$, which happens with probability at least $1 - |V'|\delta' = 1 - \delta$ □

We propose a simple hybrid algorithm for estimating the expected spreading size, which is based on the fact that an ϵ–δ approximation of $\mathrm{E}[|I_G|]$ can be obtained by ϵ–δ' approximations of the s-v reliabilities for all nodes $v \in V$ as shown in the proof of Theorem 2. The idea is that we do not have to estimate the s-t reliability separately for all nodes in V. As usual, the s-v reliabilities that are not very small can easily be estimated by a direct Monte Carlo algorithm. After a given number of iterations we determine those nodes whose Monte Carlo estimate is not yet an ϵ–δ' approximation and improve these values by appying one of the known s-t reliability estimation algorithms [2, 3, 6, 7, 8, 9].

3 Estimating the Probability of Large Spreadings

Let $\alpha \in (0, 1)$ be the threshold at which an outcome of a spreading process is considered as large, we call this an α-spreading. The main idea in this section is to reduce the K-terminal reliability problem to the problem of estimating the probability of large spreading sizes. For this reduction we need the following simple observation.

Lemma 1. *For every weighted spreading network* $G = (V, E, p, s, w)$ *with positive integer weights* w *an unweighted spreading network* $G' = (V', E', p', s')$ *can be constructed in* $\mathcal{O}(\mathrm{size}(G'))$ *time such that* $w(I_G)$ *and* $|I_{G'}|$ *have the same distribution and* $\mathrm{size}(G') = \mathcal{O}(\mathrm{size}(G) + w(V))$.

Proof. G' can be constructed on the base of G by adding for every node $v \in V$ a set of $w(v) - 1$ nodes and arcs from v to the added nodes that are active with probability one. □

It is natural to expect that finding the probability of large spreadings should not be easier than finding the expected spreading size because by solving (respectively approximating) the probability of large spreadings for different values of α, we could determine (respectively estimate) the whole distribution of $|I_G|$ and not only its expected value. The following theorem shows that it is hard to approximate α-spreading probabilities and implies that, unless $NP \subseteq BPP$, there is no randomized constant-factor approximation for this problem.

Theorem 3. *Unless $P = NP$, there is no constant-factor approximation for estimating the probability of α-spreadings for any fixed $\alpha \in (0, 1)$, even when the underlying network is unweighted.*

Proof. Let $G = (V, E, p)$ be a reliability network, $s \in V$ and $K \subseteq V \setminus \{s\}$. We begin by reducing the K-terminal reliability problem on G, which asks to determine the probability that all nodes in K can be reached from s after the edge failures, to the problem of determining the probability of an α-spreading in a weighted spreading network G' with positive integer weights. The network G' is obtained from G by adding an additional isolated vertex with a weight of $\lfloor 2\frac{1-\alpha}{\alpha}|K||V|\rfloor$. Furthermore, we assign a weight of $\lceil \frac{\alpha}{1-\alpha}|V|\rceil$ to the vertex s, a weight of $2|V|$ to each node in K, while other vertices have unit weight. It is easy to check that a spreading in G' is an α-spreading if and only if the spreading reaches all nodes in K. Thus, the probability of an α-spreading in G' is exactly the K-terminal reliability in G. Furthermore, the input size of G' is still bounded by a polynomial of the input size of G. Lemma 1 implies that the role of G' can be replaced by an unweighted spreading network with input size being polynomial in the size of G. Finally, since there is no constant-factor approximation for K-terminal reliability unless $P = NP$ [1], the theorem follows. □

4 Conclusions

The complexity of estimating the expected final size of a spreading process and the probability of large spreadings in graphs with unweighted and weighted nodes was analyzed. All of the considered problems are difficult when an exact solution has to be found. However, when we are interested in efficient approximations, the problems can be divided into three groups: problems for which an FPRAS is known (expected spreading size in the unweighted case), problems for which no FPRAS is known but we neither have an argument that it is hard to find one (expected spreading size in a weighted network) and problems for which no FPRAS is known, and we have an argument showing that it is hard to find one (probability of large spreadings in the weighted and unweighted case). It would be of particular interest to close the gap for the problem of estimating the expected spreading size in a weighted network, that is, either to find an FPRAS or to prove that this problem is hard. As we have seen in Section 3, this is equivalent to answer the same question for the *s*-*t* reliability problem. Another interesting direction would be to develop practically useful algorithms for the different problems we studied for instances that cannot be solved efficiently by direct Monte Carlo simulation.

References

1. M. O. Ball. Complexity of network reliability computations. *Networks*, 10(2):153–165, 1980.
2. W. W. Bein, J. Kamburowski, and M. F. M. Stallmann. Optimal reductions of two-terminal directed acyclic graphs. *SIAM Journal on Computing*, 21(6):1112–1129, 1992.
3. M. Chari, T. Feo, and J. Provan. The delta-wye approximation procedure for two-terminal reliability. *Operations Research*, 14:745–755, 1996.
4. P. Dagum, R. Karp, M. Luby, and S. Ross. An optimal algorithm for Monte Carlo estimation. *SIAM Journal on Computing*, 29(5):1484–1496, 2000.
5. P. Grassberger. Critical behavior of the general epidemic process and dynamical percolation. *Mathematical Biosciences*, 63(2):157–172, 1983.
6. D. D. Harms and C. J. Colbourn. Renormalization of two-terminal network reliability. *Networks*, 23:289–298, 1993.
7. R. Karp and M. Luby. Monte Carlo algorithms for the planar multiterminal network reliability problem. *Journal of Complexity*, 1:45–64, 1985.
8. M. Laumanns and R. Zenklusen. Monte-Carlo estimation for s-t reliability in acyclic networks. In *Proceedings of the European Conference on Complex Systems (ECCS), Dresden, Germany*, 2007.
9. C. Lucet and J. Manouvrier. Exact methods to compute network reliability. In *Proceedings of 1st International Conference on Mathematical Methods in Reliability, Bucharest*, 1997.
10. J. S. Provan. The complexity of reliability computations in planar and acyclic graphs. *SIAM Journal on Computing*, 15(3):694–702, 1986.
11. J. S. Provan and M. O. Ball. The complexity of counting cuts and of computing the probability that a graph is connected. *SIAM Journal on Computing*, 12:777–788, 1983.
12. L. G. Valiant. The complexity of enumeration and reliability problems. *SIAM Journal on Computing*, 8(3):410–421, 1979.

GPSR Compliance

The European Union's (EU) General Product Safety Regulation (GPSR) is a set of rules that requires consumer products to be safe and our obligations to ensure this.

If you have any concerns about our products, you can contact us on ProductSafety@springernature.com

In case Publisher is established outside the EU, the EU authorized representative is:

Springer Nature Customer Service Center GmbH
Europaplatz 3
69115 Heidelberg, Germany

Batch number: 10755047

Printed by Printforce, the Netherlands